Global food security: ethical and legal challenges

Global food security: ethical and legal challenges

EurSafe 2010
Bilbao, Spain
16-18 September 2010

edited by:
Carlos M. Romeo Casabona
Leire Escajedo San Epifanio
Aitziber Emaldi Cirión

Wageningen Academic
Publishers

ISBN: 978-90-8686-154-5
e-ISBN: 978-90-8686-710-3
DOI: 10.3921/978-90-8686-710-3

Photo cover:
Nauris Paulins, Jelgava, Latvia

First published, 2010

EurSafe 2010 Committees

Organising Committee

Inter-University Chair of the BBVA Foundation – Provincial Government of Biscay in Law and the Human Genome (University of Deusto and University of the Basque Country) – Spain

Prof. Dr. Carlos M. Romeo Casabona (Director)
Prof. Dr. Leire Escajedo San Epifanio (Coordinator – University of the Basque Country)
Prof. Dr. Aitziber Emaldi Cirión (Coordinator – University of Deusto)
Dr. Pilar Nicolás Jiménez
Dr. Iñigo de Miguel Beriaín
Dr. María Jorqui Azofra
Dr. Ainhoa Lasa Lopez
Emilio Armaza Armaza
Aroa Galán Rojas
Elena Atienza Macías
Leire Larrazabal Laraudogoitia
Claudia G. Salas Pro

Scientific committee

Coordinators

Carlos Mª Romeo Casabona, Director of the Congress' Organising Committee. Professor of Criminal Law, Faculty of Law, University of the Basque Country and Chair Director of the Inter-University Chair of the BBVA Foundation – Provincial Government of Biscay in Law and the Human Genome (University of Deusto and University of the Basque Country) – Spain

Matthias Kaiser, President of EurSafe. Director of the National Committee for Research Ethics in Science and Technology (NENT), Professor at the Centre for the Study of Science and the Humanities at the University of Bergen – Norway

Members (alphabetically ordered):

Brena Sesma, Ingrid (Coordinator of the Centre for Legal Studies concerning Health, Institute for Legal Research of the National Autonomous University of Mexico)

Castro Jover, Adoración (Professor of Ecclesiastical and Roman Law, Faculty of Law, University of the Basque Country – Spain)

Cortina Orts, Adela (Professor of Ethics and Political Philosophy, Faculty of Philosophy, University of Valencia; Member of the Bioethics Committee of Spain)

De Renobales Scheifler, Mertxe (Professor of Biochemistry and Molecular Biology, Faculty of Pharmacy, University of the Basque Country – Spain)

De Tavernier, Johan (Centre for Science, Technology and Ethics, Catholic University of Leuven – Belgium)

Emaldi Cirión, Aitziber (Senior Researcher of the Interuniversity Chair in Law and Human Genome; Lecturer of Law and Biotechnology, Faculty of Law, University of Deusto – Spain)

Escajedo San Epifanio, Leire (Senior Lecturer of Constitutional Law, University of the Basque Country – Spain)

Freire De Sá, Fatima (Faculty of Law Sciences, Federal University of Minais Generais, Belo Horizonte – Brazil)

Gjerris, Mickey (LIFE – Faculty of Life Sciences, University of Copenhagen – Denmark)

Gómez Córdoba, Ana Isabel (Director of the Medicine Program, Faculty of Medicine, El Rosario University, Bogotá – Colombia)
Gonzalez De Cancino, Emilssen (Director of the Centre for Genetics and Legal Research, Faculty of Law, University of Externado de Colombia, Bogotá – Colombia)
Hagen, Kristin (Europäische Akademie zur Erforschung von Folgen wissenschaft-lich-technischer Entwicklungen; Member of the EurSafe's Executive Committee – Germany)
Herbosa, Inmaculada (Senior Lecturer of Private Law, Faculty of Law, University of Deusto – Spain)
Meijboom, Franck L. B. (Secretary of the EurSafe Executive Committee; Lecturer and Research Coordinator, Ethics Institute, Utrecht University – Netherlands)
Millar, Kate (Director of the Centre for Applied Bioethics, University of Nottingham; Vice-president of the EurSafe's Executive Committee – United Kingdom)
Nicol, Dianne (Faculty of Law, University of Tasmania – Australia)
Olsson, I. Anna S. (Executive Board Member of EurSafe; Researcher, Laboratory Animal Science, Institute for Molecular and Cell Biology, Porto – Portugal)
Pérez De Las Heras, Beatriz (Senior Lecturer of European Union Law, Faculty of Law, University of Deusto – Spain)
Rebato Ochoa, Esther (Professor of Biological Anthropology, Faculty of Science and Technology, University of the Basque Country – Spain)
Willemsen, Ariane (Executive Secretary, Swiss Ethics Committee on Non-human Gene Technology, Federal Office for the Environment – Switzerland)

Table of contents

EurSafe 2010 Committees 7

Preface 19

Keynote papers

Human development and food: a global vision 25
R. Clotet Ballús, Y. Colomer Xena and F. Mayor Zaragoza

Designing ethical strategies for global food security 31
M. Kaiser

Why patenting life is a controversial economical activity: European legal implementation and penal related issues 37
C.M. Romeo Casabona

Ethical dimensions in the social perception of agrofood biotechnology: a case for an interethical approach 46
E. Muñoz

Part 1. Climate change policies and food security

Taming the unruly side of ethics: overcoming challenges of a bottom-up approach to ethics in the areas of climate change and food security 55
R. Anthony

Precautionary principle and global food safety 65
A. Urruela Mora and A. Libano Beristain

Food system and climate change: the false premises of antitrust policy 73
V. Sodano

Climate change and food insecurity: an ethical problem that cannot be postponed 79
M. Nari Lloveras

Part 2. Sustainable agriculture and aquaculture for global food security

Ethical reflection on a practical proposal of sustainable poultry farming 87
F.L.B. Meijboom

Behavioural change as a key to sustainable agriculture 91
C. de Lauwere, E. ten Pierick, V. Beekman and M. van Asseldonk

Balanced sustainable welfare friendly animal breeding 97
A.-M. Neeteson-van Nieuwenhoven and J.W. Hardiman

The use of certain transgenic seeds in a *more sustainable* organic agriculture 104
M. de Renobales and L. Escajedo

Nanotechnology and trust in aquaculture 110
B.K. Myskja

BENEFISH: benefits and costs of welfare interventions in aquaculture 116
C.M. Mejdell, S. Kadri and B. Damsgård

Food sovereignty, a right to life: the case of the Peruvian Andes 121
L.A. Hallazi Méndez

Part 3. The role of local breeds and traditions in sustainable agriculture and aquaculture

Traditional part-time grazing: a more sustainable sheep milk and cheese production in the Basque country 129
J.R. Mauleón, L.J.R. Barrón, M. Virto, N. Mandaluniz, G. Amores, F.J. Pérez Elortondo, J. Arranz, E. Abilleira, I. Beltrán de Heredia, J.C. Ruiz de Gordoa, A.I. Nájera, R. Ruiz, M. Albisu and M. de Renobales

Sambucana sheep: ethical and socio-economical successful outcomes of a 20-years *in situ* preservation program 132
P. Cornale, M. Renna, R. Fortina and L.M. Battaglini

The Latxa dairy sheep in the Basque Country: importance, challenges and opportunities for a traditional livestock activity 138
R. Ruiz, B. Díez-Unquera, I. Beltrán de Heredia, N. Mandaluniz, J. Arranz and E. Ugarte

A bio-integrated model of food production and knowledge: a study case in Ciudad del Carmen, México 141
L. Domínguez-Trejo, M.R. Morales-Garza and W. Cano

Autochthonous breeds, sustainable local exploitation, and quality of animal products: the case study of Piemontese cattle 144
M. Renna, P. Cornale, C. Lussiana, A. Mimosi and L.M. Battaglini

Part 4. The role of biotechnology in food security and ethical and legal aspects of bioscience patenting

Synthetic biology, ethical challenges and its impact on food security 149
M. Engelhard

Climate-ready GM crops, intellectual property and global justice 153
C.A. Timmermann, H. van den Belt and M.J.J.A. Korthals

Transgenic livestock: a contribution to global food security? 159
R. Huber, C.B.A. Whitelaw and I.A.S. Olsson

Communication and biotechnology: a decade of misunderstandings 165
E.F. Guiral

Part 5. Eating, nutrition and health: ethical and legal considerations

Conflicting concerns in fighting obesity 171
K.K. Jensen

Morbid obesity in children and the law 176
S. Romeo and P. Nicolás

Parental views on public-policy options regarding healthy eating 184
D. Wickins-Drazilova, S. Nicholls, G. Williams, C. Börnhorst, L. Grafström, S. De Henauw, S. Marild, D. Molnar, L.A. Moreno Aznar, I. Pigeot, A. Siani, M. Tornaritis, T. Veidebaum and W. Ahrens, on behalf of the IDEFICS consortium

Human rights, development and hunger in the world: ethical and legal proposals 189
F.J. Blázquez Ruiz

Dietary disorders and mental health: a contemporaneous debate 194
M.F. Freire de Sá and A.C. Brochado Teixeira

Obesity in ethnic minorities and low-income populations: social and public policy considerations from a comparative study 198
A. Poveda, A. Jelenkovic, M.E. Ibañez and E. Rebato

Eating and drinking: challenges for global bioethics 202
T. Ausín

Food policy and institutional conflict between food and health: a sociological approach to obesity in Spain 207
C. Díaz-Méndez

Part 6. Food labelling and consumer protection

Certification and ethical standards in aquaculture 215
A.S. Haugen, M. Kaiser and V. Almaas

Buyer be-aware: the ethics of food labelling reform and 'mobilising the consumer' 221
X. Frohlich

Nutrient profiling: are dairy products really unhealthy? 228
L. Escajedo, A.M. Rocandio and M. de Renobales

Framing omega-3: the politics of essential fatty acids in functional foods 234
M. Döring and S. Stirn

Seafood safety, authenticity and analytical methodologies to detect fraud 239
C.G. Sotelo, R.I. Pérez-Martín, M. Aursand, Y. Yamashita, M. Yamashita and I. Martinez

GMO, consumption and consumer vulnerability in Brazilian Consumer Law: the right to be duly informed 245
A.P. Myszczuk and F. Glitz

Part 7. The role of ethical engagement and public participation in EU food governance

Food citizenship: is there a duty for responsible consumption? 251
J. De Tavernier

Willingness to pay for farm animal welfare: consumers' motives and expectations analysed at the example of dairy cows in Germany 257
M. Henseleit

The role of NGOs in food security 263
E. Essien

Governance addressing societal values 269
V. Almaas and M. Kaiser

Reducing meat consumption: ethical arguments in the context of today's consumer society 275
E. de Bakker and H. Dagevos

Making sense of demands for the integration of ethics into the regulation of GMOs 281
F. Lord and L. Létourneau

Sociable swine: engaging society's values into a sustainable pig breeding research programme that aims for balanced improvement in productivity and welfare 287
M. Benard and T. de Cock Buning

Part 8. Ethics of bioenergy production

Bioenergy and agricultural land use: examining the framing of the ethical debate 295
C. Gamborg, K. Millar and P. Sandøe

Searching words for sustainable biofuels 301
L. Landeweerd, M. van der Veen and P. Osseweijer

Bioenergies: unveiling the ethos of the agrarian research 307
M. de Lattre-Gasquet, D. Vermersch, M. Bursztyn and P.-H. Duée

Biomass production for energy usage in Bavaria: an ethical assessment using a case study comparison of rapeseed and sorghum 313
C. Dürnberger, B. Formowitz, H. Grimm, A. Uhl and M. Zichy

Part 9. Animal welfare

Designing a computer game for pigs, to explore and redefine the interface between animal welfare, science and ethics 323
C.P.G. Driessen, M.B.M. Bracke and M. Copier

Specifying animal welfare standards: ethical issues 329
P.B. Thompson

Animal welfare: between governmental and individual responsibility 337
L. Heerwagen

The ethical basis of using animals in agricultural production: consumer understandings 342
S. Kupsala, P. Jokinen and M. Vinnari

Assessment of farm animals and intersubjectivity 348
F. Karlsson

Animal welfare regulations in the EU: a matter of ethics or a matter of human economic interest? 354
M. Landera-Luri

Effects of feed supplements on meat quality: an ethical analysis that includes the welfare and health of production animals and human consumers 360
F.L.B. Meijboom, M.B.M. Bracke and F.J. van der Staay

Part 10. Animal biotechnology

Animal ethics committees, democracy and decision-making 366
H. Röcklinsberg

Is there more to life than welfare? How the concept of animal integrity can contribute to discussions of animal ethics 372
M. Gjerris and C. Gamborg

Some legal and ethical aspects of xenotransplantation in light of the proposed revision of the Directive on the protection of animals used for scientific purposes 378
M. Jorqui Azofra

Transgenic animals: Spanish legislation 384
A. Emaldi Cirión

Part 11. Animal research and animal disease control

Moral status of the African catfish 390
B. Bovenkerk and F. Meijboom

Why don't they just pull themselves together and eat less? Public perceptions of animal use in obesity research 395
T.B. Lund, J. Lassen and P. Sandøe

Rite of passage to brute science with bias at the bench? Widening the ethical discussion of animal-based research 400
I.A.S. Olsson, O.E. Varga and P. Sandøe

Assessing the benefit of animal experimentation 406
O.E. Varga, P. Sandøe and I.A.S. Olsson

Attempt to typologize relations between different worlds (human, animal, living) based on legal and ethical questions surrounding the status of GM fish 412
L. Coutellec and I. Doussan

Genomics and animal health: use of animals in biomedical research 418
B.M. Jugo, A. Larruskain, K. Garcia-Etxebarria, L.I. Mazón and Miren Gorrotxategi Azurmendi

Part 12. General questions on animal ethics

Do animals count for less? 423
T. Visak

Doing better: John Dewey's meliorism as a method for bringing animal ethics into practice 429
H. Grimm

The argument from marginal cases: a response 435
I. de Miguel Beriain

Part 13. Teaching ethics, animal welfare and law for the biosciences and agri-food sector

How *ethics* is taught by European veterinary faculties: a review of published literature and web resources 441
M. Magalhães-Sant'Ana, I.A.S. Olsson, P. Sandøe and K. Millar

Food ethics dilemma: an open learning resource for teaching ethics 447
A. Algers, M. Kaiser, C. Kallum, H. Loor, K. Wahlgreen and S. Welin

New version of animal ethics dilemma allows students to make use of science, drama and images in ethics learning 453
A.J. Hanlon, A. Algers, T. Dich, T. Hansen, S.B. Christiansen, H. Loor and P. Sandøe

Shifting minds? The effect of training in laboratory animal science on researchers' attitudes to animal use in biomedical research 456
N.H. Franco, J.A. Nunes and I.A.S. Olsson

Policy, ethics and animals use; the need for training: a Dutch example of training civil servants in ethics 459
F.R. Stafleu, H.M.J. van Rij and F.L.B. Meijboom

Transgenic foods: science, safety, law, and ethics 464
M. de Renobales, L. Escajedo and M. Martínez de Pancorbo

European higher education area and teaching law and ethics in the biosciences 467
M. Gorrotxategi Azurmendi

Part 14. Ethics, food policies and the case of GMO's

Widening the perimeter: economics, philosophy, ethics and the quest for vital questions 472
A. Tsioumanis

Redefining the EU agricultural policy under market constitutionalism: from protectionism to competitiveness 478
A. Lasa López

The dignity of living beings with regard to plants: a brief commentary 483
I. de Miguel Beriain

Shifting schemes of naturalness 488
P.F. van Haperen, B. Gremmen and J. Jacobs

Ignored bioethics? On the role ethical concerns in GM food policies 494
J. Lassen

Part 15. Ethical, legal and social issues concerning new systems, components and foodstuffs

The farm-to-fork approach of food security concerning chemicals (with special reference to REACH, a 'new' European Community Regulation) 501
E. Atienza-Macías

In vitro meat: promises and responses: cooperation between science, social research and ethics 507
C. van der Weele

Legal and ethical challenges regarding edible *in vitro* meat production 513
E.J. Armaza-Armaza and J. Armaza-Galdos

Safe and socially robust development of nanofood through ISO standards? 521
E.M. Forsberg

Authors index 527

Keyword index 529

Preface

What will happen when the limited resources of our globe are not sufficient to feed the growing population of the world? Will the scarceness of essential food not necessarily lead to conflicts, even war between people? This is the question Condorcet posed in 1795 at the end of his otherwise extremely optimistic little book *Sketch for a Historical Picture of the Progress of the Human Mind.* Condorcet's belief in the progress of humankind remained unaffected of this scenario. He believed there are two good answers to this challenge. The first one is the technology-fix argument: new solutions will most probably be offered by future technology, and this may alleviate the problem. His second argument is the ethics-argument: one can not assume that humankind will have made that much scientific and technological progress towards living conditions that make this growth possible, without at the same time assuming that the ethics of humankind have progressed significantly as well. And the ethics of it is clear for him: 'Our moral duty is not to make sure that unborn life is born, but that those that are born are secured a life in reasonable welfare, dignity and happiness'.

A bit more than 200 years after Condorcet we can state that the scenario Condorcet describes is more and more a reality: already now we seem unable to feed 1 billion of people living today, and population growth and other factors, like e.g. climate change, indicate that the problem will become worse in our immediate future. Global food security poses an immense challenge to the ability of humankind to manage life on earth. And what are our answers to the problem? Unfortunately it does not seem that the ethics argument has won much ground. Our ethics has not yet progressed alongside the technological progress, in any case not sufficiently to avoid the problem. We are left with the technology-fix argument. But we have heard that one before, and we hear it currently a lot in connection to climate change. Many of us do not believe that innovation and technology alone can provide all the answers we need to solve our current problems.

The *European Society for Agricultural and Food Ethics* (EurSafe; www.eursafe.org) is designed on the concept that ethics is indeed an essential part of our food culture(s) and the way we produce food for people. Modern technology confronts us with new dilemmas, many of them based on conflicts of values and ethical norms. Many times the ethical dimensions in relation to food seem to drown in mere technicalities or they are poorly alluded to in political rhetoric. The mission of EurSafe is to bring the ethical dimensions to the surface and make them subject to scholarly debate and democratic deliberation. The papers collected in this volume and presented at EurSafe's 9th Congress in Bilbao are a good example of the variety of the challenges and viewpoints.

Preceded by other EurSafe conferences in Nottingham (2009), Vienna (2007), Oslo (2006), Leuven (2004), Toulouse (2003), Florence (2001), Copenhagen (2000) and Wageningen (1999), EurSafe 2010 was organized by the Inter-University Chair of the BBVA Foundation – Provincial Government of Biscay in Law and the Human Genome (Universities of Deusto and of The Basque Country), whose Director is Professor Carlos M. Romeo Casabona. Founded in 1993, the Chair in Law and the Human Genome aimed to be an Institution that fosters a rigorous and cross-disciplinary analysis of legal implications of the advances in molecular biology, while at the same time making available the benefits of its work for the University Community, professionals of different fields, and society at large. Over the last 17 years, its research field has been continuously developed and adapted to new issues and challenges, such as the Congress we are referring to now, where the 'Human Right to Food' gains special prominence. Each and everyone of the Chair's members feels extremely fortunate because the experience of organizing this 9th EurSafe Congress has opened new thematic horizons. Moreover and importantly, it enables us and the EurSafe community to maintain close links with researchers from more than 20 countries around the world.

2010 provides us with a privileged observatory for the Food policies of the 3rd millennium. Several factors led to the creation of strategic fronts for feeding the world at the end of the last century. It was necessary to face up to challenges e.g. as the paradox of hunger versus the increase in the pathologies caused by overeating. Recent experiences and scholarly studies have made the immensity of issues related to global food security clear to most of us, including decision makers in high ranking positions. The increase in prizes for rice in 2008 has led to social unrest and fear of political instability in several parts of the world. At the same time we learn that hunger is the number one global health problem affecting roughly 1 in 6 people on earth. While some progress towards reduction of chronic hunger was made in the 1980's and the first part of the 1990's, hunger has been slowly but steadily rising during the past decade again. This is in stark contrast to the Millennium Development Goals which aim at halving the problem worldwide. Furthermore, recently we have learned that our food consumption and production causes close to one third of our total greenhouse gas emissions. Emissions from animal production alone are ca 18% of the total. And looking at climate change scenarios, we learn that climate change will have adverse effects on food production through more severe weather events, rise of sea level, salination of freshwater and soil erosion.

These problems seem mainly to affect the developing world, but it is clear that the industrialised countries can hardly ignore the issue either. Food is a major factor in public health and one has learned more about how our diet must take part of the blame for several common health problems like coronary diseases and even cancer. The increase in obesity just adds to the list of known problems. But aside from the internal problems in the industrialised countries, we have to acknowledge that globalisation leaves no part of the world unaffected of the troubles of others. World trade and the global financial system react highly sensitive to disturbances and social unrest. Regional wars and social crises create migration and instability in the balance of power. Industrialised countries like the EU being the main importers of food products, and (still) being the main sources of scientific and technological capabilities, are also in the possession of policy tools that can either alleviate or increase global food security.

Legal instruments are crucial for food safety already. World trade without a surrounding web of internationally agreed legal measures is unthinkable. The world of economy is only partially the result of pure market mechanisms; a lot is the result of regulation through legal mechanisms. These are also main tools to tackle our environmental problems and move humanity towards a more sustainable development. In some areas, the law is designed to set minimal standards for people's ethical intuitions as e.g. in animal welfare. What legal measures do we have in store to improve global food security, and is our legal framework really adjusted to the challenges we now face?

As it turns out, many of the proposed solutions and answers hinge crucially on implicit ethical stances. New technology that is developed and marketed as benefits and 'solutions' in regard to food security sometimes raises more problems than it solves, or ignores basic societal values. Debates about biotechnology or designer food based on nanotechnology are good illustrations of this. Industrialised food production promises more effectiveness in the use of natural resources, but it also meets sustained scepticism in large parts of the population. Is the turn to traditional food, regional production, slow food, eco- or organic foods an answer that can ensure food security in the long run? Food security provides substantial problematic content to ethical principles like human equity, justice between current and future generations, respect for human dignity, or attributing moral status to animals and the environment. While we strive to maintain our basic ethical convictions and engage in societal debates about other important values, we may have to change our ways of life and learn new priorities in the face of global responsibility. Science and technology needs to ask questions about ethics and provide us and decision makers with relevant information and new options.

In Europe one has come to realize that matters of food often also are matters of ethics. EurSafe is an important forum to further this discussion. The contributions found in this volume bring together presentations given by a diverse group of authors from academia, public sector professionals, through to representatives from non-governmental organisations (NGOs). This diversity exemplifies the philosophy of the EurSafe Society that aims to bring together different groups to discuss the ethics of agriculture and food production.

The contributions in this book have been peer-reviewed by the scientific committee before being accepted for presentation at the Congress. The editors are extremely grateful to the many colleagues who provided their expertise during the reviewing process and editing. We are convinced that the present volume of scholarly papers will advance our understanding of the complexity of global food security and highlight its legal and ethical challenges.

Prof. Dr. phil. Matthias Kaiser
President of EurSafe; University of Bergen

Prof. Dr. iur. Dr. med. Dr. hc. mult. Carlos María Romeo-Casabona
Director of the Inter-University Chair in Law and the Human Genome

Keynote papers

Human development and food: a global vision

R. Clotet Ballús, Y. Colomer Xena and F. Mayor Zaragoza
Triptolemos Foundation, Barcelona, Spain

Abstract

The food fact forms part of life and thus is complex. Efforts have been made to give an overall vision by grouping the themes into four very wide conceptual axes, which in turn contain many subsystems. Thus, each axis has its own articulations and any progress in each of these is positive, but the overall solution to human feeding (technical and social) corresponds to a solution that is simultaneously tran-sectorial and trans-disciplinary. Evidently, each region will have its own statistically priority axes and in some cases the actions to be carried out vary between environments for the same objective. It is clear that one model, with various systems of equations with multiple variables, has no single response, either today or tomorrow. There is a wide range of possible actions, and imaginative solutions must be sought to obtain results on the four axes. However, the chosen ones must be guided by the principles of the ethics of the culture of peace that people and nations need.

Keywords: human feeding, ethics of the culture of peace, human development

Introduction

Human development is closely linked to resolving the basic problem of the food essential for sustenance. Thus, the more that this is ensured and the less effort and time dedicated to this producing it, more time is available for all those other activities that generate civilisations and cultures.

Without mentioning our ancestors' change from vegetarians to omnivores, for millennia, humans were hunter-gatherers as the only way of obtaining food resources. More than 2 million years of hunting and gathering. Over 2 million years during which the main daily concern for humans was the search for food. This system was an evolutionary dead-end.

Approximately 10,000 years ago, the first indications of the great revolution appeared. Food production was carried out by a fraction of society while the rest could do many other activities given that their food supply as a basic biological necessity was covered. This led to the birth of trade, money and the economy, techniques and science, and also to the political concept of empire; in short, civilisation. At that time, the first farming and animal husbandry appeared in various places, together with the earliest pottery recipients. The success of these first attempts led to the definitive implantation of agricultural communities. This first great food revolution was fully consolidated by 7,000 years ago, the date of the beginning of the Neolithic age.

Thus, the Neolithic represents the first step that enabled humankind to develop all its capacities. Freed from the daily search for food for subsistence, a significant part of the population could then spend their time of other more creative tasks that supplied solutions to challenges. Limiting ourselves to the world of food, the Neolithic saw the birth of the concept of food production (basic ingredients) by a few, but also saw the development of the concept of cooking in the sense of food processed ready to consume (Triptolemos Foundation, 2004).

However, the Neolithic unfortunately introduced a problem into the incipient society: hunger. *The Neolithic revolution meant a social pact*: while some produced food, the rest had to produce something relevant for society; everyone should contribute an added value to the system: bartering appeared followed by money, the market and the economy. The greater availability of food favoured the growth of the population, which then required greater food resources, and led to the concept of conquest to obtain new productive lands. From then on, there was a fragile social balance between necessities, availability and acquisitive capacity given its sensitivity to climatic conditions, population, war, economy, etc. Yet, society has made continuous scientific and technological efforts to increase productivity and quality in the primary sector as well as legal measures to ensure the right to food for humans. Even so, the balance has not been reached and there are still around 1,000 million people (out of 6,000 million) who still live under the shadow of hunger or malnutrition.

The current situation

If we focus this global conceptual approach on the situation in the first decade of the 21st century we find a situation that can be summed up as follows:

According to data from the UN (2009), in 2008 the encouraging tendency towards the eradication of hunger in the developing countries changed and began to increase again rising from 16% (2006) to 17% (2008). Regionally, in sub-Saharan Africa there were than 100 million more extremely poor persons than in 1990. Globally, the UN (2008) estimated that in 2015 there will be around one thousand million people living in extreme poverty. One of the factors that contribute to this problem is the growth in the urban population in the large cities (Figure 1). The UN (2009) highlighted that the improvement of living conditions in the slums has not followed the rapid growth in the cities. Finally, obesity reappears as a global factor that also now increasingly affects the developing countries (Figure 2).

Hunger and/or malnutrition are not exclusively found in the developing countries, although they have a greater and more dramatic impact there. In the developed countries, in a setting of frequent local farm surpluses, they are cushioned by social assistance (in the USA, 36.5 million people (11% of the population), received 'food stamps' in 2009), but this does not hide the lack of freedom to the basic right to have and organise one's own diet. The proportion and solution is similar in Europe. However, these rich countries also have serious problems in malnutrition whose most paradigmatic example is that of obesity. In the USA, at the end of the 20th century, a third of adults were obese, in Canada, one in every four, a similar rate to that found in most rich European countries.

From this brief and conceptual presentation of the world situation it can be concluded that hunger and malnutrition are global, in various forms, but all of them significant.

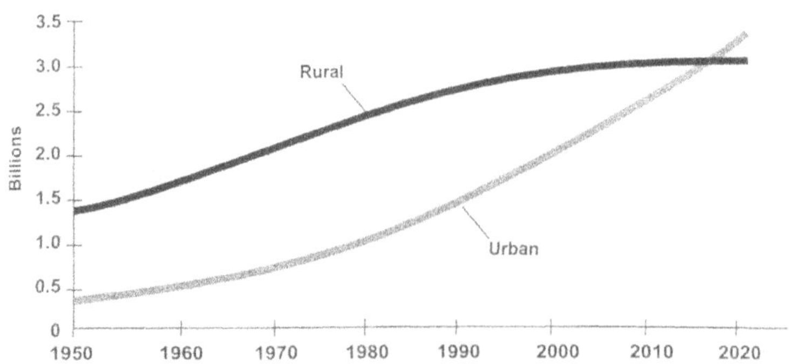

Figure 1. Urban and rural population levels in developing countries, 1950-2020 (United Nations, 1996).

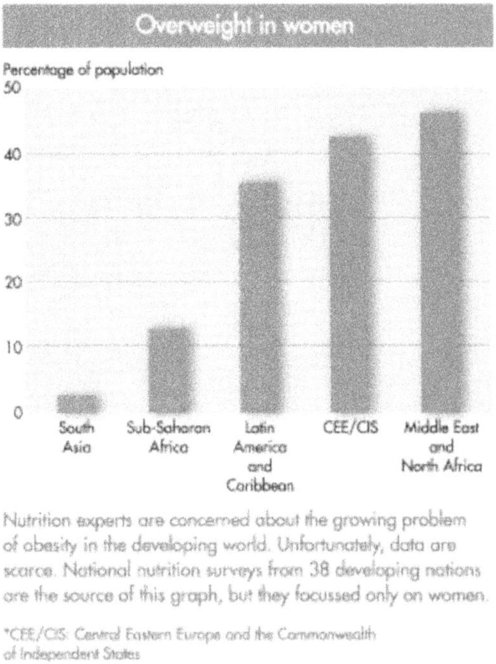

Figure 2. Overweight in women (after R. Martarell, International Food Policy Research Institute, 2001).

Global vision and prospective

The solution to each case of hunger and malnutrition could seem to be specific, but the reality is that each of these situations is conditioned by four global themes that come together in each problem. Only a balance between these can contribute to a stable solution. The four axes for action are presented in Table 1.

What fundamental questions are covered by each of the systems? *Availability* deals with ensuring that there is technically enough food for all humans. It must not be forgotten that all food comes from primary production based on photosynthesis. Important gains have been made in the increase in production, but the basic indicator is the *per capita increase* and the outlook in this aspect is not encouraging (Figure 3), and even less so given the tendency towards a diet with a high consumption of meat from stabled animals as this means a higher consumption of primary food energy. Fisheries are in danger, the cultivable hectares have reached the limit and water is scarce, and then there is the

Table 1. Axis of the food balance (Triptolemos Foundation, 2010).

Axes	Area of influence
Availability	Production and distribution of enough food
Economic	Adequate agricultural/food prices received from the small producer to the consumer
Political and social	Purchasing power for the population to acquire a basic nutritional food supply
Food culture	Knowledge (education / information) + confidence in the food system

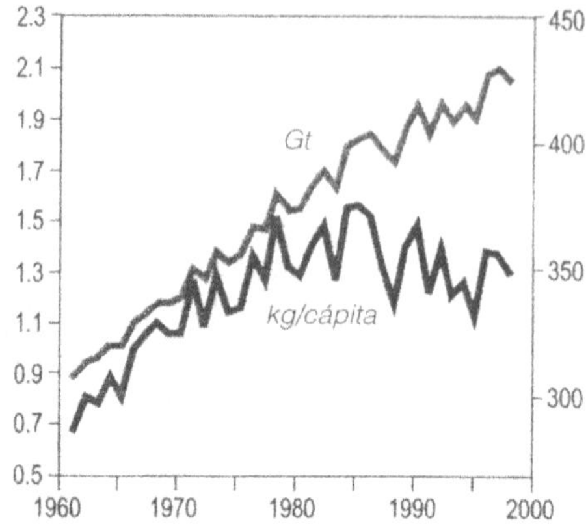

Figure 3. World grain production, 1961-1998 (obtained and calculated from FAOSTAT data – FAO, 1999). Grain production (Gt) Grain production (kg/head)

determining factor, population. This system has two means of regulation: science and technology converted into innovation, and demographic policy (Lamo de Espinosa, 2007). And finally, a worrying indeterminacy: the evolution of the biofuel markets and climate change can affect the reliability of the forecasts in the medium and long term (FAO, 2009).

Another aspect of the *availability* system is that generally availability is counted simply on the basis of the primary production. However, nowadays, with the growing urbanisation of the planet, food reaches the consumer in the form of products processed to a greater or lesser degree and this step is fundamental for increasing availability and ubiquity.

In a rural structure (until the 19th century), progress led to a tendency for population growth that was balanced by military conquests of new lands whose repopulation absorbed the population surplus. Food was prepared in rural pantries and kitchens. The Industrial Revolution totally changed this model. The expansion of the rural population changed to urban concentration, from the pantry and kitchen to the shortage of food from the country (problems of conservation and transport) and a lack of adequate housing. This setting led to hunger and influenced Malthus (1798) in his 'An Essay on the Principles of Population'.

A set of visionaries appeared, supported by science and technology and impelled by an ethical sentiment that did not annul their economic sense, and they founded a set of innovative groundbreaking companies, which helped to palliate the problem (conservation and externalisation of cooking) and allow the harmonious development of the population in an urban environment. During the 18th century, science renewed agronomy: fertilisers, irrigation, mechanisation, genetic improvement, is the need for raw material. The food industry acquired its technological dimension around 1850 (Liebig) 52 years after Malthus's manifesto was published and contributed to the failure of his postulates and allowed food stability, until the present, in the urban world (approx. 80% of the population of the developed countries). Its philosophy continues to be permanent research-innovation, which leads us to believe that its contribution to this axis will be fundamental.

In an ever more urban society, in both the globally rich and the developing countries, innovations based on science and technology for both production and service (quality + nutrition) and applied by groundbreaking companies will again acquire the same or greater importance than they had in the

19th century. In the developing countries, rural production is important for those who still live in the country, but help is increasingly needed for the creation of local agrifood companies, with technology, to cover the growing needs for processed products for the cities. With all respects, we cannot struggle to return to a 17th-century agrifood technology, as we know what size of population that was able to support. This is even truer when this return to the past is enthusiastically backed by those who are also enthusiastic about applying biotechnology and other technologies to medical progress. But is food not the first 'global medicine' for life?

On the *economic axis*, fair prices from the primary production, passing through distribution and the companies (products and services) are the leading factor. Bearing in mind the small producers and the influence of climate on agrarian production, a certain control of the system by governments, or even better, United Nations organisms, is fundamental. Financial speculation on agrifood resources must be avoided (remember the price increases in 2008-2009 throughout the foodstuff market) or protectionism and quotas that distort the market. An economic prospective must be developed based on the sustainability of the resources that sets tendencies so that they can be taken into account in the business world and that specifically consider the influence of the economic axis on the global food situation and its relation with society. There are encouraging signs in this sense. For example, the direct influence of food on economic history and the health costs provoked by malnutrition for Society have been studied by R.W. Fogel (2004).

The *political and social axis* in the noble sense of the concept is mainly responsible for the coordination and simultaneous resolution of the whole. The UN (2009) highlights its importance among the millennium targets: to strengthen social policies. Food security and hunger have been used politically throughout history either through blockades of the civilian population or in massive and publicised assistance in emergency situations. Prolonged food crises have everywhere been the popular spark for revolutions including those with the greatest ideological content (the French and Russian Revolutions).

The political system should establish rules that allow a harmonious development of everything together. Basically:

- security (availability);
- safety (harmlessness);
- sustainability of the global system;
- basis economic regulation (employment and others);
- policies for development of science, innovation, education/information.

The axis of *the culture on food* and its profound resolution with the rest is no less important. Man has always linked food to beliefs in a subconscious demonstration of its importance. A lack of preoccupation for food is also a factor in health. Truthful, global, non-partial education that allows the information to be understood. One example, how many people interpret the labels on the products in relation with their diet as a whole? Nor is gastronomic culture (an important but partial aspect) mistaken with food culture. On the axis of communication of knowledge, it is essential to avoid the global development of a double food track in matters of health. Distrust through poor education can destroy a scientific programme, a company or an end market final with greater possibilities of availability. And also coherence: obesity, apart from other factors, is the result of high-calorie feeding, which nowadays is cheap. Thus what are needed are not only dietary information but also modification of the economic parameters.

Final abstraction

The food fact forms part of life and thus is complex. Efforts have been made to give an overall vision by grouping the themes into four very wide conceptual axes, which in turn contain many subsystems. Thus, each axis has its own articulations and any progress in each of these is positive, but the overall solution to human feeding (technical and social) corresponds to a solution that is simultaneously trans-sectoral and trans-disciplinary. Evidently, each region will have its own statistically priority axes and in some cases the actions to be carried out vary between environments for the same objective.

It is clear that one model, with various systems of equations with multiple variables, has no single response, either today or tomorrow. There is a wide range of possible actions (Smil, 2000), and imaginative solutions must be sought to obtain results on the four axes. However, the chosen ones must be guided by the principles of the ethics of the culture of peace that people and nations need.

References

FAO (2009). Expert Meeting on 'How to feed the world in 2050' – (Rome, 24/26 June).
Fogel, R.W. (2004). 'The Escape from Hunger and premature death, 1700-2100' Europe, America and The Third World. Cambridge University Press.
Lamo de Espinosa, J. (2007). 'Como vivíamos: Alimentos y alimentación en la España del Siglo XX'. ed. Fundación Triptolemos. Introducción. Lunwerg Editores – Barcelona.
Smil, V. (2000). 'Feeding the World: A Challenge for the Twenty-First Century'. Massachusetts Institute of Technology Press.
Triptolemos Foundation (2004). 'Alimentos: La Conquista Humana. Lunwerg ed. – Barcelona
UN (2009). Objetivos de desarrollo del milenio. Report 2009.

Designing ethical strategies for global food security

M. Kaiser
University of Bergen, Centre for the Study of the Sciences and the Humanities, Allegt. 34, 5020 Bergen, Norway; matthias.kaiser@etikkom.no

Abstract

Global food security is facing five major problems: (1) the growth of the world population which is expected to reach roughly 9-10 billion people by 2050, and the resulting food needs; (2) the impact of food production on climate change and the impact of climate change on food production; (3) the physical limits of arable land, together with the limits of natural resources, where we already today are exploiting resources near to or above the renewable limits; (4) health issues related to malnutrition (in poor countries) on the one hand, and obesity (in rich countries) on the other, together with nutritional relations to heart and coronary diseases, etc.; and (5) the imperfections of the global food markets, together with its entanglement with stock markets and global financial transactions, which (a) accounts for sudden jumps in prices for basic food products, (b) transforms traditional food production areas for other uses, and (c) stimulates non-robust and partially inconsistent consumer attitudes such as scepticism to new industrial food products coupled with demands for low food prices. We face complexity so that no optimal leverage point for global action can be identified. Firstly one needs to acknowledge differences in scope between the global, regional, national and the local. Secondly one needs to do justice to: agricultural production (crops and domesticated animals), aquatic productions (capture fisheries and aquaculture) and feed production – all of them subject to different stresses, constraints and sustainability issues. There arises also ethical complexity: there is good reason to assume that values differ among people and cultures, and all global strategies may be vulnerable to defection by some parties. But we all need also to make trade-offs for the global benefit. There might be reason to fear that the best may stand in the way of the good. I thus argue for a strategy of satisfycing rather than of optimizing. We need simple heuristics which define strategies so that various actors can act and cooperate.

Keywords: bounded rationality; heuristics; decision models; global food security

The question

In this paper I pose the following question:

> Recognizing the inherent complexity of current and future problems of global food security, are there any general characteristics of the type of policies and strategies that we need to employ now in order to tackle these problems and that at the same time acknowledge the practical, real-world constraints under which our decision-making takes place?

One should perhaps note at the outset that I am mainly concerned with general characteristics of these policies and strategies, rather than their concrete content. The latter would by far exceed my competencies. Thus, I focus on the type of rationality that we should strive for, and try to say something about what follows from this.

In order to anticipate the conclusion of the paper, let me state it in general terms:

> A plea is made to take on board the concept of bounded rationality, seeking satisfycing rather than optimizing strategies and decisions, and acknowledging the important role of decision heuristics. This implies among others decision-making on vertical and horizontal levels which deliberately sidesteps demands for certainty in favour of heuristics that bring about some positive change of practice.

The major problems

I see five major problem-areas that affect our global food security, given our present state of knowledge:

The first problem is global population growth. According to Wikipedia, accessed 19 March 2010, 'the human population of the world is estimated by the United States Census Bureau to be 6,809,300,000'. World population has been growing steadily but with different growth rates since around 1400. Parts of the population growth during the last hundred years are due to improved medical care, hygiene, and the Green Revolution which increased agricultural productivity. It is expected that the world population at 2050 will be somewhere between 9 and 10 billion people. This is roughly 50% more people to provide food for during the next 40 years than currently.

The second problem is climate change. Several studies have concluded that our present global food production is in fact a major contributor to climate change through the emission of climate gasses (Steinfeld *et al.*, 2006; EIPRO, 2006; Sonesson *et al.*, 2009). On a global basis greenhouse gas (GHG) emissions from animal production alone are estimated to be ca 18% of total GHG emissions, while in rich countries like the EU total consumption of food items may contribute up to 29-30% of GHG emissions. Methane and nitrous oxide from livestock are more potent GHG than CO_2. Aquatic products, especially from capture fisheries, show high GHG emissions of fossil CO_2 (through energy use). The other side of the problem is that climate change is expected to impact negatively on our food production systems through various mechanisms (FAO, 2007). Increase in global temperature and precipitation, increase in frequency and intensity of extreme weather events, the rise of sea level, etc. are all relevant effects of climate change with severe consequences for food production. Fresh water availability and soil quality are key parameters.

The third problem is that we are already facing the limits of exploitable natural resources for food production. This is very clear in fisheries where we globally are already at the brink of sustainable utilisation of natural fish stocks, and for some fish stocks already exceeding these limits. There is simply no more fish in the oceans to take out in significant numbers (FAO, 2008). But similar constraints apply also to agriculture on a global scale. There is by and large no more arable land available that we could use for increased food productions. Rather the opposite: many areas face already the serious problem of soil erosion due to bad management of agriculture and climate change will assumedly increase this problem. Thus the main option to provide more food for a growing world population is through increased and improved effectiveness in current productions and new technologies.

The fourth problem is due to our growing awareness about the relation between dietary habits and human health. The food sector adds to the expected health problems which are directly due to climate change (such as increase in epidemic diseases, etc.). One side of the problem is the growing issue of malnutrition (FAO, 2009). Estimates show that close to one billion people are chronically or acutely hungry, and the main contributing causes to this are thought to be changes in food and energy prices. This calls for major changes in food and development aid policies (Sanchez, 2009). The other side of the coin is the growing problem of obesity (Wikipedia, 2010b). Even though its more specific causes are still debated and largely uncertain, it seems clear that dietary habits, lack of physical activity and genetic factors all play a role. Obesity contributes to other related health problems such as increase in type 2 diabetes. Food habits are also more and more drawn in as explanatory factors in relation to other health issues such as coronary diseases where the type of fat intake is seen as a major problem in many rich countries, which again is partly due to an excess of meat in the diet.

The fifth problem is seen to be the imperfections of the global food market. In spite of the optimism of believers in free-market and efforts through WTO to liberalise food trade, one has to conclude that market mechanisms alone seem not to be efficient tools to tackle any of the above problems above, rather to amplify them. Recent episodes in connection to the price of rice and other basic food stuff just underline the need to find more satisfactory regulative frameworks for trade. The rise in market value for bio-fuels contributed to weaken important food productions. Generally speaking, the global trade of food and the attractions of high export value have not always been of benefit for poor producer countries that have to satisfy the food needs of a large and poor population. On the other hand, consumers in rich countries show little trust is industrialised food productions and remain sceptical in regard to new food technologies, such as biotechnology. Thus the typical response of the food industry to meet the increased future food demand through new technology, intensification of production and global markets is not a totally credible response and meets assumedly resistance among consumers.

The problem then arises how to tackle the above problems in a holistic and promising way. I do not think that I exaggerate when I claim that nobody currently has a fully credible answer to all of these problems at hand. I certainly do not have one or know of such an answer. But given that the problems are imminent and that the Precautionary Principle would urge us to act now in order to prevent that things get even worse, what general strategies should we look for to get on top of these problems?

The problem is also ethical

Many people realise that the problems here described are not purely or merely scientific or technical problems, but that they indeed are based on some major ethical problems. They involve among others the well-known problems of human dignity and equity among current people, of justice between current and future generations, of attributing ethical value to animals and nature, and of acting with imperfect information. In this respect, food ethics would be largely parallel to climate ethics.

So far, parts of this discussion have been in terms of economics and its relation to ethics. The rationale has been the widespread view that sensible policies should be based on estimates of costs and benefits, or more specifically, on a detailed risk-cost-benefit analysis. However, one of the problems of this position is that recommendations can differ significantly. John Broome (2008) analyzes two such important attempts in relation to climate change ethics: the Stern review and the opposing Nordhaus' report (for precise references see: Broome, 2008). He attributes the difference between the two reports to them applying a different discount rate for future harms and benefits. Thus they end up with opposing recommendations.

Discount rates for the future have been discussed in the literature widely and it is known that small differences easily lead to large consequences in the evaluation of future effects. Broome contrasts here two positions: utilitarianism and prioritarianism. In his argument, utilitarianism operates on the assumption that future people will – due to economic growth – on average posses more goods than current generations, and thus it is reasonable to discount such goods. They have diminishing marginal value. Prioritarianism on the other hand argues that a benefit for a rich person should be assigned less social value than the same benefit for a poor person. In other words, the less well off are given ethical priority. Given prioritarianism, Broome argues, one ends up with a relatively higher discount rate than utilitarianism. He himself favours lower discount rates for various reasons, but underlines that the choice between what is seen as 'correct' discount rate should be a democratic choice.

Broome is obviously right in a number of his claims, among them the important claim that any choice of a discount rate ultimately rests on an ethical stance. Yet, I believe he is wrong in his portray of the basic role of ethics in our decision making for a climate and food future. I think his focus on discount rates is buying in to a paradigm of decision making that is ill conceived in this connection. I shall explain why I think so.

Complexity and uncertainty

System theory has for a long time been studying what they term complex systems. Such systems are not only characterized by the large numbers of parameters that enter the system, but also by the fact that the whole system may show properties not obvious from the study of its individual parts, so-called emergent properties. Ant colonies or human economies are good examples of this. Mathematical models of such systems may involve a number of non-linear equations that e.g. show rapid shifts between multiple possible equilibria points. Typically, when such shifts occur the system cannot easily be transformed back into a previous equilibrium by simply working backwards through the most recent changes of parameters. Furthermore, the precise change of the state of the system is impossible to predict on the basis of individual parameters. Thus, complex systems involve major system uncertainties. It is important to realize that these uncertainties are normally not a simple lack of knowledge (in the sense of not yet knowing enough), but are supposed to be inherent features of the system.

Complex systems are usually open systems in interaction with other open systems, such as e.g. ecological systems in interaction with human behaviour. Many areas of our life have been subjected to analyses based on complex systems, such as the growth of big cities, and it is generally assumed that standard paradigms of scientific thinking, such as e.g. those stemming from Newtonian physics, are poor guides for managing these systems. The basic flaw lies in the assumption of predictability based on more or less complete information. Decision strategies that estimate future states either with certainty or probability fail both to acknowledge the systems' interactions with other systems and the systems' potential for rapid non-linear change of basic properties.

My claim is that global food security as portrayed in the five major problems mentioned above exhibits all characteristics of such a complex system. There are major inherent system uncertainties and interplays with other complex systems on various levels, from individuals to states, from specific production sites to global climate, from micro- to macro-economy, and across the various sectors of food production and consumption. We do not even have models of global food production similar to the models of global climate change, but one can clearly see interactions between the various factors, each of which subject to inherent complexity. In other words, we know pretty well that these factors interact and in part depend on each other, we have some rough ideas about possible causal pathways, and we can reasonably ascertain some possible future scenarios, but we are still faced with major uncertainties that make predictions, even probabilistic predictions for long term scenarios in practice impossible. If I am correct in assuming that our global food production by and large should be portrayed as a number of interconnected complex systems, then this has immediate consequences for the kind of decision-making that we should strive for.

The first such consequence is that risk-cost-benefit analyses as major decision tool is ill-adapted to this situation. The rationale of this kind of analysis is to find the strategy which optimizes the possible outcomes, given the available information. This works as long as we have good probabilistic estimates of possible long term developments. In complex systems we do not have such information. Furthermore, in our case we do not even know what to count as costs and what as benefits since most effects may be a cost for some and a benefit for others. Assumptions of economic growth, in themselves perhaps a rather dubious starting point, tend to hide the nature of the ethical issues at stake. I see it also a major problem that risk-cost-benefit analyses seeks to find the overall global optimal solution. If it is applied only locally or individually one tends to open up for the problem of free-riders, i.e. decisions that are based on a large benefit for some given that the problems of others are ignored. Thus risk-cost-benefit analysis could only work if it were applied globally and included all aspects of the problem. The result of this is that individual agents and groups may feel powerless and admit a moral responsibility only to the degree that it is a part of a larger global strategy. Thus one ends up with potential agents waiting for each other and doing nothing.

Bounded rationality

In his book *Models of Man* of 1957 Herbert Simon suggested an alternative to the concept of rationality as optimization. He suggested that one should take account of our cognitive limits, of the limits of information typically available to us, of the emotional and irrational drives that are with us, and of the finite amount of time we have for our decision making. In many contexts, these factors will make optimization impossible. Simon proposed the term *bounded rationality* for this type of decision making. Part of his analysis was based on the model of decision making in court cases. Often one has to come to a decision on the basis of very limited and incomplete information, the costs of further information gathering is high and will often exceed the time frame for decision making, and often the weighing of the pros and cons needs to be accompanied by principles as *in dubio pro reo,* etc. Courts make their decisions not as optimizing but as *satisfycing* (a term also coined by Simon), i.e. judging when, all factors considered, one has enough information to act on. Companies act on a similar basis. They use simple *heuristics* for deciding on business strategies in time rather than perfect information that usually comes at too high a cost. Bounded rationality thus indirectly implies a critique of the rational agent model dominant in economic literature.

A central term in bounded rationality is decision heuristics. Amos Tversky, Daniel Kahnemann and Paul Slovic (Kahnemann *et al.*, 1982) have all been scholars who have contributed largely to the empirical study of such heuristics. Heuristics in general are simple but efficient rules of decision making, often anchored in our cognitive framework through evolutionary processes, and often though not always leading to good results. Sometimes they lead to cognitive bias or systematic errors.

Some prominent heuristics are e.g. the availability heuristic (judging the frequency of an event by the ease with which it comes to mind), anchoring heuristic (adjusting the judging of probability depending on the initial fixation), attribute substitution (judging a complex problem, e.g. a moral one, by relating to a known but different problem), framing (a tendency to react to a problem based on how it is presented), recognition heuristic (selecting an object from a set of alternatives based on one or few recognised values of the object), etc. It is perhaps noteworthy that Kahnemann and Tversky by and large were more concerned about cognitive fallacies while social psychologists such as Gerd Gigerenzer (Gigerenzer, 2006; Gigerenzer and Todd, 1999) emphasize the positive use and the power of such heuristics. They have among others shown that Take-the-best heuristics (i.e. decisions based on the first cue that makes a difference between the available alternatives) outperforms more sophisticated and demanding decisions tools such as multiple regression analysis or Bayesian networks (Gigerenzer *et al.*, 2002).

Outlook and conclusion

Decision making based on bounded rationality is a more attractive framework for our global food production than traditional economic frameworks due to the complexity of the issue at hand. It has as one of its attractions the ability to strengthen the immediate responsibility of various actors. Thus it seems also ethically more attractive. Simple heuristics can point to actions which seem right given some few alternatives even though not all consequences can be evaluated at the time of decision. A prominent example could be the decision of the city council of Gent in Belgium to introduce a vegetarian weekday. Realizing that the 240,000 citizens thus could reduce GHG emissions equal to the annual emission of 18,000 cars seems a good enough reason to endorse such a policy. Simple heuristics may also help consumers to make better dietary decisions, e.g. by encouraging them to introduce weekly plans for lunches and dinners. It is easier for individuals to judge the climate and health qualities of an overall plan and then act according to the plan, rather than evaluate each individual meal. It also allows for less radical changes in dietary habits and thus increases the chances of real commitment. The various food sectors and the producers may be encouraged to put to life some voluntary policies without waiting for governmental action. Best management practices and ethical frameworks as soft law may be an important step in the right direction. Simple heuristics may tell the aquaculture sector that reducing the fish meal and fish oil in feed is a good thing regardless of how much use the agricultural sector may make of such resources.

There is no overall master plan for global food security and due to the nature of the problem there will not be one. In spite of this we need to act now and we need to make hard decisions. This we need to do horizontally, i.e. involving all levels from individuals to governments and international organizations, and vertically, i.e. involving all sectors of the food sector. The model for this kind of necessary decisions does not come from traditional economy, but rather from bounded rationality and simple heuristics. On this basis there is no excuse not to act now.

References

Broome, J. (2008). 'The Ethics of Climate Change', Scientific American June 2008, p. 97-102; accesible at http://ele.arizona.edu/files/broome.pdf (accessed 25. March 2010).

EIPRO (2006). Environmental Impact of Products (EIPRO), European Commission Joint Research Centre, available at http://ec.europa.eu/environment/ipp/pdf/eipro_report.pdf (accessed 25. March 2010).

FAO (2007). 'Climate Change and Food Security: A Framework Document' Rome 2007; accessible at: ftp://ftp.fao.org/docrep/fao/010/k2595e/k2595e00.pdf (accessed 25. March 2010).

FAO (2008). 'The State of World Fisheries and Aquaculture', accessible at: http://www.fao.org/docrep/011/i0250e/i0250e00.htm (accessed 25. March 2010).

FAO (2009). 'FAO at work 2008-2009 – Hunger and crisis', accessible at ftp://ftp.fao.org/docrep/fao/012/i1023e/i1023e00.pdf (accessed 25. March 2010).

Gigerenzer, G. (2006). 'Bounded and Rational' in R.J. Stainton (ed.), Contemporary Debates in Cognitive Science, Blackwell, p. 129.

Gigerenzer, G., Czerlinski, J. and Martignon, L. (2002). 'How good are fast and frugal heuristics?', in: T. Gilovich, D. Griffin & D. Kahnemann (eds.), Heuristics and Biases – The Psychology of Intuitive Judgement, Cambridge.

Gigerenzer, G. and Todd, P.M. (1999). Simple Heuristics That Make Us Smart, Oxford.

Kahneman, D., Tversky, A. and Slovic, P. (eds.) (1982). Judgment under Uncertainty: Heuristics & Biases. Cambridge, UK, Cambridge University Press.

Sanchez, P.A. (2009), 'A smarter way to combat hunger', Nature, vol. 458, p. 148.

Simon, H. (1957), Models of Man, Social and Rational: Mathematical Essays on Rational Human Behavior in a Social Setting. New York: Wiley

Steinfeld, H., Gerber, P., Wassenaar, T., Castel, V., Rosales, M. and de Haan, C. (2006). Livestock's Long Shadow. Environmental issues and options. FAO, Rome (accessible at: http://www.fao.org/docrep/010/a0701e/a0701e00.htm) (accessed 25. March 2010).

Sonesson, U., Davis, J. and Ziegler, F. (2009). ' Food Production and Emissions of Greenhouse Gases – An overview of the climate impact of different product groups', Report SIK The Swedish Institute of Food and Biotechnology, at: http://www.se2009.eu/polopoly_fs/1.23297!menu/standard/file/foodproduction.pdf (accessed 25. March 2010).

Wikipedia, (2010a). 'Human population of the world estimated by US Census Bureau' http://en.wikipedia.org/wiki/'World_population (accessed 19 March 2010).

Wikipedia, (2010b). 'Obesity' accessible at: http://en.wikipedia.org/wiki/Obesity (accessed 25. March 2010).

Why patenting life is a controversial economical activity: European legal implementation and penal related issues

C.M. Romeo Casabona
Inter-University Chair in Law and the Human Genome, University of Deusto and University of the Basque Country, Bilbao, Spain; cromeo@genomelaw.deusto.es

Abstract

The European patents system has responded to the demand of commercial developers and researchers concerning the need for adequate legal protection for biotechnological innovations related primarily to plants, animals and parts of the human body. However, the legal confusion regarding the scope of the regulation in some cases (e.g. embryonic stem cells) and its extensive application to other products has been highlighted. The protection against the infringement of industrial property rights through criminal law must continue to be subjected to the principles of *ultima ratio* and subsidiarity.

Keywords: property rights, biotechnology, European law, European patent office, criminal law, plant varieties, animal welfare, embryonic stem cells

The concept and functioning of the invention patent (industrial property)

The invention patent is the legal instrument which aims to stimulate scientific research and the technological process. This objective is achieved through the recognition of certain specific legal powers over inventions, and giving these inventions special protection through the patent mechanism. As will be seen below, biotechnological inventions, despite referring to living organisms, are patentable. The classic requirements for patentability are the following:

- to concern inventions that are new (a previously non-existent object, procedure or technique: originality);
- they have to truly involve an invention i.e. an invention that is an advance compared to the previous situation (they must not be obvious or self-evident);
- they must have useful application or industrial application (practical use, as opposed to being simply of aesthetic or intellectual merit).

Patentability also requires that the object of the invention may be reproduced or repeated by a third party, an average expert, on the basis of the documentation presented by the person claiming the patent. This condition, and the fact that the issuing of the patent is public, makes it a valuable instrument for guaranteeing food security, by virtue of its status in law, as the process of the production of new foods can be controlled through the information that is made available to the public. In addition, the fact that plant varieties and animal breeds are excluded from patentability is an indirect factor in guaranteeing biodiversity, as these objects cannot be appropriated and exploited exclusively.

Patents can concern products or procedures, or may be derivative patents. The patent gives[1] the inventor of a patentable invention an exclusive right of monopoly for a limited period of time (twenty years). This right is transferable to the inventor's heirs until the time period referred to expires. Research activity is thereby encouraged since the monopoly on exploiting inventions can provide researchers with a profit

[1] In relation to the aspects that follow see Amelia Martín Uranga, 'Biopatentes biotecnológicas' (Biotechnological Biopatents), Romeo Casabona, C. M. (ed.), *Enciclopedia de Bioderecho y Bioética (The Encyclopaedia of Biolaw and Bioethics)*, BBVA Foundation Inter-University Chair in Law and the Human Genome & ed. Comares, Granada (forthcoming).

from the investment made in research through the development of their results into saleable products or processes, enabling them to obtain a financial return on the funds used. In turn these can be used on new research projects. The concession of the monopoly is granted to the inventor or his/her heirs in return for the research carried out but when the period for which the exclusive rights ends the content of the biotechnological inventions, like any others, has to be made public, and its exploitation becomes unrestricted. It is felt that in this way society's technological wealth is enriched.

The legal protection of inventions is strengthened through the instrument of the patent, since this confers to its holder an exclusive right to industrial property, in such a way that the holder will be able not only to exploit the patented invention but also prevent any third party who does not have his/her consent from manufacturing, offering, introducing onto the market or using a product that is the subject of the patent or importing or possessing the product for any of the purposes mentioned.

Some infringements of industrial property rights can even result in an offence being committed, with the corresponding criminal liability. In these cases patent legislation becomes, to a certain extent, the regulation that completes the description of the criminal offence, the patent legislation thereby becoming a 'ley penal en blanco' (in Spanish law, a criminal law that defines a penalty for a criminal offence that is in turn defined in another law) since the description of the criminal offence implicitly refers to patent legislation.[2] In this way criminal law protects legal property of which the patent is a part, that is to say, industrial property. This covers an important part of business economic activity aimed at the commercial exploitation of patented products and procedures. The observance of the principles of *ultima ratio* and subsidiarity in criminal law must be stressed in these matters.

In accordance with the 1973 European Patent Convention, 'European patents will be awarded for any invention that is susceptible to industrial application, that is new and that involves an inventive advance' (art. 52.1). As can be seen, this precept contains the characteristics required for a patent to be approved. The doubt that remained regarding this regulation was its compatibility with the patentability of live material of plant, animal or human origin.

The important Directive 98/44/EC of the European Parliament and Council, BOE (Official State Bulletin) of 6th July 1998 clarified this matter supporting such compatibility, under the conditions which are described below. Just as has occurred with other European regulations with priority over the internal law of the member states of the European Union, this directive underwent transposition into national law with the aim of adapting the respective legislations regarding patents to it, although in this case it was not a smooth process. Similarly, the 1973 European Patent Convention was also revised in 1999 to adapt it to the new legal regime established by the Directive.

Spanish legislation was based fundamentally both on the European Patent Convention, in its general framework, and on the European Directive which refers specifically to biotechnological patents.[3]

Patents as a form of legal protection of biotechnological inventions: the European regulatory framework

In response to the extensive conjecture that this matter has raised, current European regulation provides for biotechnologies to be protected by law, principally or exclusively through patents. In the end, any

[2] Cf., e.g. art. 273.1 and 273.2 of the Spanish Penal Code.

[3] Cf. Law 10/2002, of 29th April, which modifies Law 11/1986 of 20th March on patents, for the incorporation into Spanish law of Directive 98/44/CE of the European Parliament and Council of 6th July, relating to the legal protection of biotechnological inventions.

sort of legal solution other than patents has been ruled out, as patents establish the relevant aspects that differentiate a discovery from an invention. This, as shall be seen, was one of the most controversial points regarding the arguments for or against patents.

In addition, in the statement of reasons it is stated that no attempt is being made to replace the general system established in each country regarding patents, but rather that the starting point continues to be the existing basic, general regulation through patents (recital no. 8). However, many have been keen to point out that it must be demonstrated in every case that the alleged biotechnological advance or innovation fulfils each and every one of the traditionally recognised requirements of a patent. And, specifically, it is expressly required, both in the recitals in the Directive's statement of reasons (recital no. 22), and in its articles (arts. 3.1 and 5.3), that whoever presents the request for the patent must reliably demonstrate the industrial application of the biotechnological product. That is to say, its usefulness must be fully defined and clearly demonstrated although it is true that the Directive makes the understanding of the two other requirements of the patent (new inventions and not being obvious) more flexible.

The authors of the Directive, and therefore the European Council and Parliament, have also been concerned to clarify in the greatest possible detail which biotechnological material may or may not be patentable, above all in relation to human biological elements or when the human body is involved in some way (art. 5); and also when animals or plants are affected (arts. 2 and 4).

This repeated concern for conceptual clarity explains why, unusually, the statement of reasons consists of fifty six recitals, while the section on provisions consists of only eighteen articles. That is to say, it is more concerned about clarifying what it is trying to regulate and its meaning or scope than clarifying the regulation itself.

What, then, is the European regulatory framework that establishes the Directive for the protection of biotechnological inventions? The framework explicitly refers to the member states' national laws on patents, without prejudice to the necessary adaptations to that which is set out in the Directive (art. 1). Consequently, in accordance with this requirement regarding the law on patents, national law must lie within the legal framework for the protection of biotechnological inventions, that is, of inventions that involve the manipulation of live material, whether plant, animal or human.

The framework declares, as a general principle, that products that are composed of or contain biological material or the procedures through which they are produced are patentable, provided that they conform to the traditional requirements of patents: that is, that they are new inventions that involve inventive activity and are susceptible of industrial application (art. 3.1). Consequently, biological material isolated from its natural environment or produced by means of a technical process is patentable even if it already exists in its natural state (art. 3.2).

This precept is the basis of the stance taken by the Directive in relation to the patentability of animals and plants, and to human genes or parts of them, since it admits the possibility under the conditions just described. This does not prevent there being a desire for it to be clarified yet more specifically, as will be seen below.

Provisions regarding non-human biological material: animals and plants

The object of the Directive is not limited to biological components of human origin, and so inventions that concern plants or animals are patentable, if the technical feasibility of the invention is not confined to a particular plant or animal variety (art. 4.2), which would be excluded. That is, the patenting of plant and animal varieties and essentially biological processes for the production of plants or animals (art. 4.1), as stated above, continue to be governed by existing laws concerning these matters.

According to the Directive, inventions which concern a microbiogical or other technical process or a product obtained by means of such a process are patentable (art. 4.3). This solution is not new either, because before being approved the Directive had already accepted their patentability. However, this recognition does not constitute, at least formally, an exception to the exclusion from patenting of essentially biological processes for the production of plants or animals (art. 4.1, b). The explanation for this system can be found in that once a microorganism has been obtained it is capable of rapid and abundant self-replication or self-reproduction.

As an effect of this more flexible regulatory system, the European Patents Office (EPO) has accepted the patenting of transgenic animals (like the Harvard rat, which was the first case of an animal patent in 1988 by the United States Patents and Trade Marks Office (USPTO)), despite the opposition to this by opponents based on the argument that this patent would be contrary to art. 53. a of the European Patents Convention (which excludes the patenting of products contrary to morality or public order)[4]. The patenting of plants, specifically certain varieties of maize, soya and medicinal plants, has also been recognised.

Another aspect that gave rise to concern, this time by farmers, was the danger that they would become dependent on the companies that produced grain, seed and cattle through genetic procedures, in such a way that for each new planting season or reproductive period they would have to buy the corresponding patented products at onerously high prices (for example seeds or animals for rearing), or pay a levy or some other sort of remuneration to these companies, with a consequent increase in the cost of their farming. There has been a desire to resolve this fear with the so called 'farmers' privilege', according to which the farmer will be able to hold back some of the seeds from his harvest for sowing in the following season, and similarly if other products, such as cattle or animal reproduction materials are concerned (art. 11).

The specific regulation on elements of the human body

Regarding human biological material, the Directive expressly indicates that 'an element isolated from the human body or otherwise produced by means of a technical process, including the sequence or partial sequence of a gene... even if the structure of that element is identical to that of a natural element may constitute a patentable invention' (art. 5.2). In all cases, the Directive insists that the industrial application of a sequence or a partial sequence of a gene must be disclosed in the patent application (art. 5.3).

By contrast, the human body, at the various stages of its formation and development, and the simple discovery of one of its elements, including the sequence or partial sequence of a gene, cannot constitute patentable inventions (art. 5.1).

The European Patents Office accepted the patenting of some human protein (e.g. the hormone relaxin) even before Directive 98/44/CE was approved.[5]

[4] Decision of the EPO of 6th July 2004, which accepted the patenting of the Harvard rat in a modified version. Bear in mind that the claim for the patent had been presented to the EPO in 1985, nearly twenty years previously.

[5] EPO, resolution of 18th January 1995.

Regarding the stated conditions for patentability, there continues to be discussion concerning whether the sequencing of a gene or part of one isolated in the laboratory or obtained by another technical procedure really constitutes an inventive activity. In relation to this, the Directive indicates that a gene or a functional fraction of DNA can be likened to a chemical molecule, but an essential difference persists. This is based on what is really interesting about this molecule, namely the genetic information it carries and not the medium as a whole. If the structure of the information 'is identical to a natural element, it is a discovery and not an invention'. Consequently, according to contrary opinion, and in contrast to the position of the Directive, the technical reproduction or the isolation of that information would not constitute an inventive activity, which is an essential prerequisite for the recognition of the patent, without prejudice to the fact that the technical procedure of reproduction or isolation of the human body element (the DNA sequence) can, in itself, constitute an inventive activity.

This discussion is not yet closed. In fact, in 1998 the USPTO accepted the patenting of two human genes responsible for breast cancer in women (BRCA1 and BRCA2) the early detection of which is highly effective in preventing this cancer. However, recently a North American District Judge[6] annulled the patent of both genes, because it is considered that the mere purification of these genes does not on its own modify an essential characteristic of the DNA; namely, its nucleotide sequence, or more precisely the capacity to detect mutations in these genes. The susceptibility of these patients to develop a cancer depends on whether the essential characteristics of the genes remain unaltered. Consequently, the purified genes are not something different from that which exists in nature.

The public order and morality limit

Independently of whether or not they agree with the requirements of patents, the European system excludes inventions whose commercial exploitation is contrary to the public order or morality from the European legal framework (art. 6.1). This is an excluding clause that was already contained in the previous conventional European regulations (the 1973 European Patent Convention) and has been renewed in accordance with the regulation introduced by the Directive. However, in reality it has never been applied, neither by the EPO nor by any other national court within the European Union, except as described below. The failure to apply it is both for pragmatic reasons resulting from increasing flexibility regarding the awarding of patents and because of the very difficulty of applying it, as it concerns undefined legal concepts, which are difficult to unravel.

Specific cases of exclusion from patenting

As a development of this general exclusion clause or to make it more precise some circumstances for exclusion are specifically mentioned. Thus, in relation to human biological material, the processes for cloning human beings, processes for modifying the germ line genetic identity of human beings and the uses of human embryos are excluded (art. 6.2a, b and c).

In regard to animals, processes for modifying the genetic identity of animals which are likely to cause them suffering or physical harm without any substantial medical benefit to man or animal, and also animals resulting from such processes are not considered to be patentable (art. 6.2d).

These cases mentioned explicitly in the Directive do not constitute a closed list of the limits of patenting for reasons of public order or morality. Instead, only a few possible situations which are clear cases of the contravention of such limiting principles are mentioned, by way of example.

[6] District Court ruling of 29th March 2010 (Association for Molecular Pathology *v.* UPSTO).

Some problematic cases

The case that has probably generated most controversy about the exclusion or otherwise of patenting through the application of some of the cases of exclusion described is the patenting of embryonic human stem cells. In fact, added to the Directive's lack of precision in relation to this, is an ill thought out decision of the EPO, which it had to subsequently rectify, and the existence of powerful research groups aimed at obtaining new therapies with these cells (the transfer of these cells once they have been differentiated in the laboratory). If this research were to be successful it would provide incalculable financial benefits thanks to the legal instrument of the patent.

The question has been posed as to whether the exclusion from patenting related to the methods in which human embryos are used includes the use of embryonic stem cells, since although the origin of these human embryo cells is beyond doubt, the subsequent processing of the extracted cells in the laboratory does not involve the use of embryos as such, but rather only cells separated from them. However, according to another of the clause's interpretative criteria, which is relevant in this case as it originates from the EPO[7], the fact that in general the process of obtaining stem cells involves the destruction of the corresponding embryo and therefore the process involves the use of embryos. This in itself would be a reason for excluding stem cells from the patenting mechanism. Finally, regarding whether the mere use of embryo cells falls within the exclusion clause, in the light of their origin it is argued that only the use of totipotent stem cells would fall under the exclusion regulation, given their capacity to lead to a complete human being, and therefore also a human embryo. This would not occur with multipotent and pluripotent cells, which do not have this capacity.

The trend which is increasing is that of considering that this case is not included in the reason for exclusion described, and therefore the patenting of stems cells derived from and modified in the laboratory should be recognised. This does not have any solid legal base, but the same can be said regarding the other interpretations described.

In addition, within the framework of the member states of the European Union there are various new trends. Thus, the German Federal Patent Tribunal in 2006 declared that no product obtained from human tissue is patentable, a restriction that seems to go even beyond European regulation.

Finally, the 'Edinburgh patent' is well known, where a patent was awarded for the genetic modification of breast stem cells with the aim of giving them a survival advantage compared to other differentiated, unwanted cells (1999). The patent did not specify that it related to non-human breasts, as in English the word breasts is used in a generic sense, including human breasts if not otherwise specified. Consequently it could be interpreted that the patent could also be extended to humans. This was contrary to the clause on public order and led to the patent being challenged by the EPO. After a lengthy process the EPO maintained the patent with amendments, with the aim of restricting its scope.

The penal discussion on this point revolves around whether the creation of human embryos for research purposes must be prohibited and, in turn, if this is an offence together with the question of whether it is permissible to use surplus or supernumerary stem cells obtained from assisted reproduction techniques for research purposes. The European Convention on Human Rights and Biomedicine prohibits the former (art. 18.2) and leaves the latter to the discretion of the member states, provided it is formally authorised by means of a law (art. 18.1). The possibility of cloning to create clones from which to obtain stem cells for research has also been considered. There are very different positions in the domestic laws of member states of the EU. Thus, in Spain the Penal Code considers the fertilisation of human ovules

[7] EPO, Decision of the Enlarged Board of Appeal dated 25th November 2008 G 2/06.

for any other purpose than human procreation to be an offence (art. 161.2); non-penal legislation does allow the technique of activating ovocites through nuclear transfer for research purposes, provided that certain conditions are fulfilled.[8]

Effects of the clauses excluding patenting

It must not be forgotten that the awarding of a patent in reality only represents the ability to prevent third parties from exploiting the invention for commercial purposes without the consent of, or authorisation from, the patent holder. There is probably a relationship between these issues and the clause on public order and the provision indicated in the Directive itself which states that the Commission must report to the European Parliament and Council every five years any problems encountered with regard to the relationship between this Directive and international agreements on the protection of human rights to which the Member States have acceded (art. 16a).

Consequently, the awarding of the patent must not be confused with whether the activity in it, or the use of the product or procedure obtained, are permitted or prohibited in another section of the legal system, such as, for example, in penal law. That is to say, by way of an example, patenting techniques for cloning human beings is not recognised, as it is considered to be contrary to public order or morality, as has been established. However there could be a state law that permitted it, or to the contrary prohibited it. For example, reproductive human cloning constitutes an offence in some European legal systems[9], which is an indication that its commercial exploitation may be contrary to the public order and a reason for refusing the patent. Conversely, the exclusion of the patent by specific legislation does not mean that the activity (e.g. its commercial exploitation) must be considered to be prohibited, only that it will not enjoy the protection provided by patenting.

Consequently, it must be stressed that whether or not the activity or the use of the product is permitted is one thing, but whether or not the patent on the product or procedure on which this activity is based is recognised is quite another.

National legislations and European biopatent regulation

It is true to say that currently, all member states of the European Union have transposed Directive 98/44/CE into their respective national legal systems, although at first they were not particularly diligent in doing so. In addition, the member states were not unanimous in accepting the Directive, once it was approved by the relevant EU bodies. I shall mention the most notable cases.

In France, the National Consultative Ethics Committee for Health and Life Sciences openly ruled against the incorporation of the Directive into French legislation (the law of 29th July 1994 in force at the time) as it was thought at that time that the Directive was ambiguous and would not help to clarify the position of the researchers[10]. The Committee asked for a public debate before moving away from the principles that inspire French legislation, namely, knowledge of the sequence of a gene cannot under any circumstances be assimilated into an invented product, and therefore is not patentable. Its use, like all knowledge, which is the common property of humanity, cannot be limited by the existence of patents, which would seek in the name of the law to protect the exclusivity of such knowledge. That

[8] Law 14/2007 of 3rd July on Biomedical Research (art. 33.2).

[9] Cf., e.g. art. 160.3 of the Spanish penal code.

[10] National Consultative Ethics Committee for Health and Life Sciences, Avis no. 64 (2000) on Preliminary draft law in the intellectual property code of Directive 98/44/CE of the European Parliament and Council, 6th July 1998, regarding the legal protection of biotechnological inventions.

is to say, the French advocated in their *Avis* freedom of access to knowledge concerning genes or parts of genes. However, this legal solution is not unproblematic. A private company in the consortium that made the final results of the human genome map project public renounced the patent on the parts of the map it obtained from the consortiums' research. It did not do so, however, to sell this information as commercial information to those who could make a financial profit on the basis of the research that they had carried out. It is true that the rulings of this committee are not binding, and that it seems that the French government had already started the transposition process (which was finally completed in the 2004 Law). But it is no less true that their rulings reflect a state of opinion that is very influential in France and generally highly respected.

Another position of rejection of the European legal system regarding biopatents can be seen in the appeal lodged by two EU member states before the European Court of Justice. It concerns an appeal for annulment presented in October 1998 by Holland against Directive 44/98/CE. Soon after, (December 1998) the Italian government presented a similar appeal for annulment, as did Norway, although the latter fell victim to procedural errors, which resulted in their appeal not progressing. The arguments of the first appeal were based on various issues. It was argued that the Directive contravened fundamental rights: specifically the right to human dignity, by permitting the patenting of material of human origin. It was further argued that no provision was established in relation to the recipient of material treated biotechnologically, which went against the principle of free will by patients.

The European Court of Justice ruled on this motion, which it rejected in each and every one of its terms[11]. To cite some of the most relevant arguments of this judgement: it is not a question of creating a new law, since it concerns national patents to which national procedure and law is applicable; the Directive was the only means of guaranteeing the correct functioning of the internal market, in view of the legislative divergences of the member states regarding this matter, therefore the principle of subsidiarity is not violated; the directive does not represent a violation of human dignity, since it limits the patent sufficiently so that the human body remains effectively unavailable and inalienable, for example, on excluding the human body and its parts in its different states of its constitution and development from the patent, and on expressly considering certain procedures to be against the public order and morality, and therefore not patentable.

Some criticisms opposed to the Directive

Despite it appearing that the process of uniform regulation of the protection of biological inventions in the European Community has been completed since the Directive was approved by the corresponding community institutions, this is not the situation, as has been shown above. Instead, some important criticisms have arisen of the system for the regulation of patents that has been established. Some are listed below:

- There has not been sufficient thought about deciding what can be patented and what cannot. This is reflected in the confusion there is regarding the patenting of modified embryonic stem cells or more broadly the patents conceded regarding plants with medicinal properties.
- The exclusion from the Directive of the series of articles on the requirement for consent from the people who provide biological samples from which a patentable invention is obtained has also been criticised. Without entering into whether this requirement is sensible (although I would like to at least point out that it is a question that should be taken into account), it certainly does not seem an appropriate concern for the patenting process and that, consequently, it should be resolved by industrial property law.

[11] European Court of Justice, judgement of 9th October 2001.

- Let us remember that it has also been argued that the genes and their parts (the DNA sequences), even although they may have been isolated in the laboratory, contain information, and this is what is really interesting as an object for possible financial exploitation. This information already exists in nature, and so from this point of view it would be a discovery and not an invention. It has been suggested that a decision in favour of the patenting of mere information would be harmful to research. As seen above, discussion about this has been revived. For this reason patenting must either be accepted or rejected.
- Some patented inventions concerning health are already a reflection of the inequalities of access to them, since they lead to financial rights that are branded as exorbitant by some sectors (e.g. the case of products for the diagnosis of a predisposition to breast cancer in the United Kingdom).

Acknowledgements

This work is a part of a Spanish Science and Innovation Ministry funded research project (DER 2008-00954/JURI).

References

Byk, C., (2009). La patente sobre células madre humanas (The Patenting of Human Stem Cells), Romeo Casabona, C. M. (ed.), Retos de la investigación y comercialización de nuevos fármacos (The Challenges of Researching and Commercialising New Pharmaceuticals), Inter-University Chair of the BBVA Foundation – Provincial Government of Biscay in Law and the Human Genome & ed. Comares, Bilbao – Granada, 117-131.

Kesselheim, A.S. and Mello, M.M. (2010). Gene Patenting – Is the Pendulum Swinging Back?, The New England Journal of Medicine, 10.1056, 1-4.

Lobato García-Luján, M. (1995). European Patents Office: Decision of 18th January 1995 (The Relaxin Issue), Law and the Human Genome Review, 3.

Martín Uranga, A. (2003). La protección jurídica de las innovaciones biotecnológicas. Especial consideración de su protección penal (The Legal Protection of Biotechnological Innovations. Special Consideration of their Legal Protection), Inter-University Chair of the BBVA Foundation – Provincial Government of Biscay in Law and the Human Genome & ed. Comares, Bilbao – Granada.

McManis, C. (ed.) (2007). Biodiversity and the Law. Intellectual Property, Biotechnology and Traditional Knowledge, Earthscan, London.

Paredes Castañón, J.M. (2001). La protección penal de las patentes e innovaciones tecnológicas (The Legal Protection of Technological Patents and Innovations), Mc Graw Hill, Madrid.

Romeo Casabona, C.M. (2001). La protección jurídica del genoma humano y de las innovaciones biotecnológicas: la cuestión de su patentabilidad (The Legal Protection of Technological Patents and Innovations: The Question of their Patentability), Comunicaciones IDEI. Ten lectures on new technologies and industrial property, special issue.

Ethical dimensions in the social perception of agrofood biotechnology: a case for an interethical approach

E. Muñoz
Institute of Philosophy, CCHS, CSIC, Spain and EVI Chair, Escuela de Minas, UPM, Brazil

Keywords: biotechnology, public perception, social controversy

Introduction

Agriculture and livestock, what we usually call the 'agricultural industry,' are some of the most relevant activities in human development, but, at the same time, the most negative for conservation of the natural environment. They are one of the major innovative milestones of the Neolithic culture, which brought important transformations in society and in the mental structure of prehistoric populations. Nevertheless, its evolution, in conflict with the availability of resources and with the demographic explosion, has brought up relevant issues for reflection and ethical debate from those posed by English economist and demographer Thomas R. Malthus regarding the difficulties in food production to keep up with the needs of population growth, assumed in the approaches of the Club de Roma ('The limits to growth') up to current times, which focus on the concern for 'food safety,' an ambivalent term if seen from the perspective of the developing world-ensure food for everyone-or the developed world-guarantee the health of the well fed.

In this complex and contradictory context, biotechnological agriculture emerged almost two decades ago and puts its first product on the market fifteen years ago based on an important scientific achievement-the possibility of applying genetic engineering in plants (and animals)-in order to improve productivity and reduce environmental costs, but at the centre of a profound social controversy, with particular impact in Europe (Muñoz, 2009).

Public perception and biotechnology in Europe: 2002 and 2005.

The relevance of this conflict and the strategic importance of biotechnology in science and technology policies have determined that surveys will be carried out regarding citizens' attitudes and perceptions toward biological sciences and biotechnology. The surveys, through the formula called Eurobarometers, represent a group of demoscopic work carried out since 1991.

In 2002, the fifth Eurobarometer was particularly intense in the public debates regarding biotechnology, and with particular focus on agro-biotechnology. At that time, increased optimism was observed among European citizens regarding biotechnology, after a decade (1991-1999) of marked decline. Nevertheless, genetically modified foods are a black spot for some European citizens. These biotech products are considered of little use and dangerous, they are not socially acceptable. Paradoxically, the situation regarding genetically modified crops (what is usually referred to as GMO, genetically modified organisms) is much less negative. It would seem that European citizens do not associate what is rationally so, that GM foods come essentially from GM plants.

Another interesting fact is that Europe is split in regard to the attitude toward transgenic plants. Their use is socially supported in Spain, Portugal, Belgium, Ireland, United Kingdom, Finland, Germany and Holland; but these plants are rejected by countries that had opposed the sale of the same (seeds), supporting the extension of the moratorium on cultivating them, such as France, Italy, Sweden, Denmark, Austria and Luxembourg.

This fact allows for an initial orientation to the ethical dimensions of the problem: it points to the importance of modulation of social attitudes of political actions or decisions with the consequent media impact.

The survey corresponding to 2005 is a product of a special Eurobarometer. This initiative, in the context of the Lisbon strategy of March 2000, directed toward the creation of the European Research Area (ERA), and articulated around the 'Science and Society' action plan, was completed with another promoted under the same 'Science and Society' plan, and carried out during the same period of field work. The objective of this second consultation under the banner 'Social values, Science and Technology,' was to 'assess the view of Europeans about social values and ethics, the perceptions of the citizens in regard to the actors involved in (the development of) science and technology, as well as decision making procedures. The final objective was to evaluate the perceived influence of ethics on science and technology in the future (TNS Opinion & Social, 2005).'

In the first of these two consultations, the following argument was formulated: 'foods from genetically modified organisms are dangerous.' A majority (54%) of those surveyed agree with the statement, while 23% were ambivalent, neither agree nor disagree, which suggests that it reflects a situation of uncertainty. Again, the results are very different among the different countries (Cyprus and Greece lead the belief in the unhealthiness of GM foods, followed by Croatia and Austria, with percentages between 88% and 70%). However, the citizens of Holland and the United Kingdom show the lowest percentages in agreement with this statement and the highest in disagreement. Nevertheless, it is interesting to remember that neither Holland nor the United Kingdom were countries that, in prior surveys, revealed the greatest support for GM foods.

This fact points to a new ethical dimension in the analysis of social perceptions: the potential risk introduced by the questionnaires and questions (content and form) formulated therein.

In the second survey, oriented toward exploring the relationships between social values and science and technology, the impact of biotechnology and genetic engineering is analyzed. Even with the limitations of the referenced questionnaires, the data show that two thirds of the surveyed European population believe that biotechnology and genetic engineering will have positive effects for the future of our lives in the next 20 years. The countries with the highest rates of positive response are: Iceland (86%), Norway (81%) among EFTA countries and in the EU the order is: Hungary (74%), Spain and Denmark each with 72%, the Czech Republic and Estonia with 71%, followed by Sweden and Italy with 70%, with Italy's figure as an absolute surprise considering previous surveys, specifically oriented toward biotechnology. The countries that show the lowest percentages in regards to the benefits that could come from life technologies are: Austria (43%), Greece (53%)-data coherent with the results seen up to now-and Lithuania (54%).

When the specific field of the connection between 'agriculture and high technology' is investigated, a concept that extends beyond biotechnology, even though it includes it, at the general level of the European Union, a result is obtained similar to that previously reported for biotechnology: 66% of EU citizens believe that these technologies applied to agriculture will bring benefits to our quality of life in the next 20 years. This favorable position is the majority in all the countries surveyed with the sole exception of France.

It is important to point out that among those surveyed, the prevailing idea is that decisions about the application of new emerging technologies will be made based on Consequentialist Ethics (risk and benefit analysis), as opposed to ethics based on morals or principles.

In the last section of the study, dedicated to the future applications of science, which informs of a notable slant toward Biology, there is a section dedicated to genetically modified organisms as a source of meat production and agricultural crops.

The question about the use of GMO to obtain meat from cell cultures was opposed to the option of animal slaughter but in spite of, more than 50% of European citizens expressed rejection to the first option. Nevertheless, once again, the differences between countries are very broad. In the EU, the greatest opposition came from Cyprus with 88% and the least opposition was shown by Spain (39%) and Bulgaria (23%).

The question regarding the approval of genetically modified agricultural crops, modulated to offer the increased variety of agricultural products on the regional level as a positive objective, also did not receive significant approval, although with a much more favorable situation for this option of agricultural production than for the previous referring to livestock production. Of all the Europeans surveyed, 37% stated that they would 'never' approve the option, while 31% would be in agreement with it, if the production were regulated and controlled. Only 6 of the 32 countries surveyed rejected by majority the agro-biotechnological alternative under the conditions of the question (Croatia, 60%; Switzerland, 58%; Cyprus, 56%; Greece, 54%; Slovenia, 53%; and France, 52%).

A new indication on the level of ethics regarding the conditioning factors introduced by the formulation and content of the questions.

Other interesting data have been obtained in the course of the work done in our group about consumer attitudes and the governance of food safety in Spain (Todd *et al.*, 2009).This work reports the analysis of a recent study of public perception of food safety governance in Spain, using genetically modified (GM) foods as an indicator. The data make clear that Spanish food consumers are aware of their rights and role in the marketplace. They are critical of current regulatory decision making, which they perceive to be unduly influenced by certain social actors, such as industry. In contrast, consumers demand decisions to be based primarily on scientific opinion, as well as consumer preferences. They want authorities to facilitate informed purchasing decisions, and favor labeling of GM foods mostly on the grounds of their right to know. However, consumers' actual level of knowledge with respect to food technology and food safety remains low. There are several ambivalences as to the real impact of these attitudes on actual consumer behavior (specifically when it comes to organizing themselves or searching out background information).

Social controversy about agro-biotechnology in Europe (a personal analytical viewpoint)

From the cursory analysis offered, it seems clear to conclude observing the complexity that surrounds the social controversy about genetically modified foods in Europe.

I present a personal viewpoint already developed in previous works (Muñoz, 1998, 2001; Ponce *et al.*, 2005), which I summarize here. In those previous works I have tried to apply a socio-historical, political and philosophical analysis under the holistic perspective, colored by some drops of scientific-technical reflection.

In the first place, the boomerang effect of the awareness of scientists in the 1970s of the technological potential of genetic engineering (Asilomar Conference, 1975) has to be verified. This awareness among scientists has brought highs and lows to the development of biotechnology based on transgenesis. On the one hand, it armed the critics of new technologies to point out that if the scientists themselves are reflecting on the potential risks, they might be huge. On the other hand, it has served to demonstrate, although it's hard to get the message across, that the precautionary attitude of scientists has given rise to one of the technologies with the least number of accidents through human history, in other words, it is a safe technology.

In any case, it seems obvious that there are multiple factors involved in the controversy. Among them, I will mention the following:

- The (unfortunate) coincidence of the sale of genetically modified foods with food crises (epitomized by the case of 'mad cow' crisis).
- A growing perception – mobilized by certain facts and sub debates that have taken place in the heart of the scientific community – regarding the scientists' lack of rigor to manage the consequences of the new technologies.
- Consumer rejection of taking risks when there is no perceived direct benefit.
- The lack of trust in the responsible food regulatory agencies in Europe and other parts of the world.
- European governments' protectionist interests that are reflected in the barriers to the trade of genetically modified crops (and foods).
- The obstinate attitude of the United States, resisting (European) consumer demands regarding labeling of such foods, and the 'right to know demanded by consumers.'
- Anti-American sentiments very present in the European socio-political dynamic, which have capitalized the movements of the 'anti-globalisationists.'
- Biased and sensationalist treatment of these topics by the media.

Main conflicts with ethical repercussions

The conflicts with ethical repercussions, within the social controversy about agri-biotechnology, allow the landscape to be outlined with the diversity of interests, beliefs and values that are at stake.

a. Risk and benefit comparison.
   - From the methodological point of view, there is bias in the surveys that seek to identify social attitudes and the measurement of public perception in regard to the different applications of biotechnology. In the majority of the general surveys done in Europe – Eurobarometer type – the questions about the applications in the agrofood sector tend to be formulated about the possible risks, while there are no questions about the potential benefits. The situation is reversed when questions regarding applications for biotechnology in the health sector are analyzed.
   - From the cognitive point of view, there is an insistence in the idea that optimistic and pessimistic positions to accepting risk in biotechnology applications are based on different levels of knowledge regarding the sources of risk and their possible impact.

   The assumption that understanding the science and technology requires one to have at least, basic knowledge of the facts and adequate information about the scientific method and the way it approaches the reality of the facts is correct. The problem is in the distance that separates the experts from lay people in the different ways and models of approaching the evaluation of technologies. This distance grows if we take into account the asymmetry that exists in regard to the sources of information: in summary, we find ourselves faced with a problem of 'trust and credibility.'
b. Conflicts of interests.
   - Conflicting interests among the different types of agriculture (organic, conventional or industrial and biotechnological).
   - Conflict between the agrofood sectors: seed producers, farmers, processors and distributors.
   - Geostrategic conflicts with agriculture as a hostage (United States versus Europe, Europe versus Asia, developing countries versus the Western world and versus each of the large blocks).
   - Conflicts in the heart of the European Union: predominantly consumer countries versus producing countries, the problem of the Common Agricultural Policy.

- Conflicts in the heart of large firms: evolution of the mergers of large agricultural chemical companies and pharmaceutical companies that occurred a few years ago, to the separation in a period of a few years. 'Retracing the steps'.
- Possible conflicts derived from the new uses of agriculture to produce substances with high added value (the case of 'pharming', obtaining drugs from plants).

c. Opposing rationalities.
   - Different ways of approaching the implications of new biotechnology in the agriculture and livestock sector (see Muñoz, 1998).

   This study evaluated the appropriateness of framing the debate in a well established conceptual framework (definitions). This framework developed in line with the view of 'society of risk: and the resulting conflict between citizens and experts in view of their different approaches to the matter of trust (opposing rationalities). This exercise in assessment converges in a social critique regarding the technical scientific arguments.

   As a personal proposal I have proposed the coexistence of rational debate with the integration of modern ways and formulas in the practice of social dialogue as they are trying to apply in Spain ('governance').

d. Values.

   The values that are under debate are several: moral, religious, socio-economic and political, mostly conceived and proposed as entitative (Menéndez Viso, 2005), which enormously complicates rational debate.

   The relevance of the values surfaces as soon as social attitudes toward the new technologies are analyzed in depth.

   In order to move toward identifying the possessive values of the citizenry, focused surveys have been proposed – carried out by our group on agri-biotechnology in Spain – to identify those values, in pursuit of the following objectives: to characterize usage habits, the degree of data, and the influence of the data.

   As a result it has been proposed that a distinction be made between consumers and users (Ponce *et al.*, 2005).

Conclusions of the analysis of social controversy

1. A first and important conclusion, derived from the analysis of public perception to genetically modified foods and crops is the importance of the context. In this new society, society is increasingly participating in scientific and technological development. In the global society arranged around the market, consumers are decisive judges, and therefore, they achieve a fundamental value for the applications of technology (perception and social evaluation).
2. Surveys are the methodological instrument on which the analyses of public perception of technologies are based. Despite its value, this instrument has notable limitations.

   Among these problems the following must be mentioned: the samples should be representative but not excessive; the framework in which the questionnaires are created – errors in the approach or design can have regrettable consequences; the type of survey – personal interviews according to itineraries (door to door), telephone, virtual; the depth of the analysis going from descriptive approaches to those supported with robust statistical treatment.
3. In principle, the framework for demoscopic study had its roots in the theory of the 'cognitive deficit' which quickly showed its little explanatory efficacy. Although changes were introduced, the study focused on surveys directed to citizens as such and as users of biotechnology, not as consumers.

   The results reveal a high percentage of those surveyed who positioned themselves in the 'Don't know/no answer' section. This percentage is in the limits of 30% and it has not changed over time. There is notable ambivalence in the responses according to sectors, with the applications in health receiving medium/high acceptance, agricultural uses deserving of medium acceptance, and the use in food earning the lowest approval (low/medium).

This confusing situation is made even more complex when compared by country and application. Among the 15 traditional EU countries, Finland, Spain and Portugal showed the most positive attitudes; Holland, United Kingdom, Sweden, Denmark, Ireland and Belgium were in the middle; France showed great variability between very critical attitudes and lukewarm attitudes; Germany, Austria, Luxembourg and fundamentally Greece are countries where the citizens show the most negative attitudes toward the use of biotechnology.

The 12 new Member States add diversity to the panorama, making it practically impossible to formulate a general rule about European attitudes toward Biotechnology. There is no common position, rather almost as many facets as there were member countries of the EU in 2008.

Regarding the scale of approval of organisms to be subjected to modification through genetic engineering, there is a consensus on the following scale from greater to lesser approval: microoganisms, plants, animals, human beings. And in regard to the type of existing applications, likewise, a very generalized agreement again on the scale of more to less in: therapy, diagnosis, agriculture, food.

Final reflection

In the same way as for the effective dynamic of Ethics in the general management of the knowledge produced by the activity of scientific and technical research, I have proposed the need to apply ethics in plural, by virtue of the plurality of subjects and, consequently the use of the term 'Interethics' (Muñoz, 2008), I suggest this same attitude in the analysis and use of ethics in the field of 'agrofood biotechnology.'

References

Menéndez Viso, A. (2005). Las consecuencias y el origen de los valores, Madrid: Siglo XXI, 2005.

Muñoz, E. (1998), 'Nueva biotecnología y sector agropecuario: el reto de las racionalidades contrapuestas. In: Durán, Alicia and Riechmann, Jorge, coords., Genes en el laboratorio y en la fábrica. Madrid: Editorial Trotta, Fundación 1° de Mayo, 1998, p.119-140'.

Muñoz, E. (2001), Biotecnología y sociedad. Encuentros y desencuentros. Madrid: Cambridge University Press, OEI, 2001,176 p.

Muñoz, E. (2008), 'Dinámica y dimensiones de la ética en la investigación científica', Arbor, no. 730, p. 197-206, marzo-abril 2008.

Muñoz, E. (2009), 'Visión actual de los transgénicos en Europa: una controversia social siempre abierta'. In: da Costa, M.A.F. and Barrozo da Costa, M. de F. (eds.), Biosegurança de OGM.Uma visão integrada. Rio de Janeiro: Publit soluções editoriais, 2009, p. 33-53.

Ponce, G., E. Muñoz Ruiz, D. Santos Benito, O. Todt and M. Plaza (2005), 'La opinión de los consumidores españoles sobre los alimentos transgénicos y su seguridad'. Revista Internacional de Sociología (RIS), Córdoba (España), n. 41, p.93-108, mayo-agosto 2005.

TNS Opinion & Social (2005), Europeans, Science and Technology, Special Eurobarometer 224, Wave 63.1, Brussels: Directorate General Research, Directorate General Press and Communication, 2005, 125 p.

TNS Opinion & Social (2005), Social values, Science and Technology, Special Eurobarometer 225, Wave 63.1, Brussels: Directorate General Research, Directorate General Press and Communication, 2005, 97 p.

Todt, O., E. Muñoz, M. Gonzáles, G. Ponce and B. Estévez (2009), 'Consumer attitudes and the governance of food safety' Public Understanding of Science 18, p. 103-114.

Part 1. Climate change policies and food security

Taming the unruly side of ethics: overcoming challenges of a bottom-up approach to ethics in the areas of climate change and food security

R. Anthony
Department of Philosophy, University of Alaska, Anchorage, USA

Abstract

For my presentation, I will investigate the challenges involved in addressing ethical questions related to climate change and food insecurity in a public engagement atmosphere where scientists, policy-makers and laypersons interact. Ethics in the service of public policy is plagued by 'unruliness', disruptive mechanism that can impede an open discussion on ethical issues involving climate change and food security. This includes the recent factual controversy surrounding the impact of industrialized food production on anthropogenic climate change. In considering how to mitigate unruliness in empirical ethics, through a bottom-up approach, I explore the dynamic between the laypublic and experts in the areas of climate ethics and food security and how concerns are posed, framed, discussed, displaced and resolved. Next, I elaborate on the possibilities and limits of dealing with ethics in a participatory setting that prioritizes citizen input and deliberative discourse. I conclude by offering some modest suggestions on what should be taken into count when confronting issues of science and ethics vis-à-vis policy-making.

Keywords: empirical ethics, food security, climate ethics, public engagement

Introduction

Empirical ethics in the service of public policy is no stranger to 'unruliness', understood as mechanisms that impede, displace or effectively shut down public engagement and involvement in ethical discourse. A certain amount of unruliness is to be expected when dealing with controversial issues like anthropogenic climate change and food security. The ethical dimensions of these two areas of empirical ethics are complex and include difficulties associated with apprehending, posing, framing and debating specific concerns; courtesy of scientific uncertainty, the process of eliciting public meaning and support for action, and aspects of politicking.

In considering how to mitigate unruliness in empirical ethics, and after a brief digression that considers the merits of the bottom-up paradigm in empirical ethics, I will consider the challenges involved in addressing ethical questions related to climate ethics and food security in a public engagement atmosphere where scientists, policy-makers, ethicists and laypersons interact. For empirical ethicists and others working in these areas, questions related to risk and scientific uncertainty are par for the course. Adopting a bottom-up approach to ethics, which has been the model en vogue to tackle these questions lately (and which has proven to be more successful than principalist approaches) is not without its pitfalls, however. My investigation will follow in this vein in order to address some of the pitfalls and to stimulate further ways in which ethical aspects of climate ethics can be unearthed and addressed more effectively. Within a bottom-up paradigm there are a number of challenges or complexities that can be identified which impede an open discussion on ethical issues involving climate change and food security. This includes the recent controversy surrounding the research methodology, factual claims, and policy recommendations concluded from them on the impact of industrialized animal food production on anthropogenic climate change. In considering how to mitigate unruliness in empirical ethics, and after a brief digression that considers the merits of the bottom-up paradigm in empirical ethics, I begin by exploring how ethical issues in climate ethics and food security are posed, framed, discussed, displaced and resolved. Next, I elaborate on the possibilities and limits of dealing with ethics in a participatory setting that prioritizes citizen input and deliberative discourse. I conclude by offering central ideas on what should be taken into count when confronting issues of science and ethics vis-à-vis policy-making, especially as they intersect with the abovementioned areas of empirical ethics.

The relevance of a bottom up approach to policy-making: why public engagement is key

The pivotal roles of empirical ethics like climate ethics and food ethics (with an eye to food security) for policy making hinge primarily on ethics' functions and influence in society, its reach and power to motivate appropriate reflection on central social issues, and its ability to inspire change in attitude and action. In liberal democracies around the globe, it is believed that conscientious citizens are able to aid in the policy-making endeavor by evaluating the relative merits of scientific claims and value of arguments about morality and consider the scale and impact of their consumption choices in order to overcome moral blind spots in both their personal and collective lives. Given this orientation, an ethics discourse that does not engage and involve the public in the evaluation and discursive exercise is not only ineffective. It is also irrelevant. As biomedical ethicists and environmental ethicists have encountered in the last few decades, approaching ethical dilemmas from a principalist or top-down approach is less attractive as a model for public engagement (Hedgecoe, 2004). From the point of view of ethics in the service of public policy-making, not only is the principalist model unattractive for theoretical reasons (such as not being able to handle the full range of concerns at stake, or framing the issues narrowly, or advocating a one size fits all solution, and potentially discounting important facts and the situatedness of the stakeholders), it has lost favor with ethicists because it does not pay enough attention to the procedural side of ethics discourse. As Burgess (2004: 10) laments, 'The institutionalization of ethics [per principalism] as a source of policy analysis encourages narrow definitions of issues and terms of reference in response to short time frames.'

From a procedural point of view, the principalist approach seemingly advocates for a paradigm of estrangement, i.e. it does not encourage meaningful engagement between scientists, ethicists, policy makers and the public at the outset or 'upstream'. Furthermore, by its design, the principalist framework is not meant to include public voices in earlier phases of ethics discourse such as during the conception phase of the intersection between ethics and policy-making. As a consequence, the public may not be involved in the 'deeper questions about values, visions, and vested interests' (Wilsdon and Willis, 2004: 18), where scientific uncertainty and controversy persist. The fix is in, as it were, amidst the absence of a more cooperative and collaborative ethics cum policy-making encounter. Thus, certain interests/concerns may never see the light of day. Instead, the public is at once considered to be passive and thoroughly in need of education. Public consultation as it were, is relegated to later moments of the policy innovation process. The principalist paradigm *talks at* the public and *not with* them as is needed in policy matters that require public input, support and buy-in.

In advocating for procedural discourse at earlier stages of the ethics encounter, Burgess (2004) contends that an integration of ethical reflection and public consultation is central in order that a more fine-grained understanding of the societal issues at stake can come through. This in turn will likely deepen the ethical analysis process to include a wider spectrum of the values involved and can provide a wider range of morally viable policy recommendations or alternatives. In environmental/agricultural ethics, similar line of thought is implied in Thompson and Nardone (1999) on the interplay between descriptive and normative views of sustainability (see also Norton, 2007).

Where uncertain and controversial facts are involved, Burgess and others argue that a bottom up approach to ethics is more reliable at getting at the public's own brand of wisdom. 'Lay expertise', which has its own legitimate value and perspective on empirical and normative matters (Epstein, 1996) is integral for promoting public trust in science, and the scientific and policy-making processes (Irwin, 2006).

Robust public engagement is integral for 'buy-in' of public policies in areas where scientific uncertainty and a plethora of public moral sentiments remain in contention in public meaning formation regarding controversial issues, and where citizen or consumer sovereignty are valued at a premium. When citizens are actively more involved in the process of public policy-making through engagement mechanisms that afford transparency and visibility they are more confident that they have some semblance of control over how complex and controversial scientific issues are discussed, framed and resolved.

As a paradigm for ethical discourse, a bottom-up approach helps to keep other relevant stakeholders and so called experts in check, while promoting a wider notion of risk that includes aspects of welfare above and beyond the conventional expected value interpretation often adopted by members of the scientific community and policy makers (Thompson and Dean, 1996). Furthermore, if members of the public are invited to engage actively in the 'deeper questions about values, visions and vested interests' earlier on in the process and in relation to the consequences of adopted policies and not just on questions of risk assessment of the application or implementation of institutional commitments that have already been ratified, they will likely have the opportunity to challenge the validity of epistemological and methodological aspects of the scientific and policy-making enterprises (Borner and Menz, 2005). They will also have the opportunity to expose vulnerabilities and raise criticisms of the processes involved in considering the applicability of scientific knowledge to environmental and food policy that will affect them and future generations intimately. This has the effect of keeping all parties more honest, in theory (at least). A bottom-up approach that actively galvanizes or invites public engagement at earlier phases helps the public to see more lucidly how uncertainty in science and policy-making are dealt with and how scientific research is integrated into environmental and food policy (Thompson, 1986; Sarewitz, 2004). This in turn can create more opportunity for debate or dialogue concerning more basic questions related to fundamental values or virtues such as 'What kind of agriculture do we want to shape in response to climate change?', and avert the feeling that the 'fix is in'.

So, how then should ethicists *et al.* create effective and responsible sites of public engagement, ones which ensure that public involvement will yield good policy advice that is ethically legitimate and addresses the important questions of fundamental values and visions?

Actual public engagement instead of leaving the ethical dimensions of climate change and food policies to expert ethics committees is necessary as a way to maintain transparency in the policy-making process and to anticipate complacency by expert bodies (Borner and Menz, 2005). However, public involvement can be messy and unruly. Where does the unruliness lie and what should and can ethics do about minimizing said unruliness?

Portraits of unruliness in the sciences of climate and food security

Unruliness in empirical ethics can manifest itself in many forms and thus hold up publicly inspired policy-making. One example, in the case of climate change, is captured vividly in novelist Ian McEwan's latest literary effort *Solar* (2010). Inspired by a weeklong trip he took to the Arctic with a group of artists and scientists and sponsored by a British-based project called Cape Farewell, McEwan's farcical portrayal of a delinquent scientist in *Solar* caricatures how hard it is for 'spoiled, indolent, self-centered' and limited by their respective private and professional silos, human beings to take on the challenges required to come to terms with and reverse the effects of anthropogenic climate change. McEwan reflects on and laments the fact that the best of intentions often takes a back seat and can go awry when human beings do not cooperate well in the midst of a total breakdown in our social contract with others. McEwan alludes to the complex nature of public discourse between experts and the laypublic when it comes to ferreting through and vetting the science and ethics of climate change. It is a process ensconced in disarray and disagreement over the present narrative and what must be done about it. According to McEwan, many of us continue to struggle to understand climate science, its politics and the human institutions marshaled to respond effectively and justly to it. Reliance on the notion of 'sound science' to lead the way and on experts has just not 'cut it'. *Solar* echoes that ethics is an important resource for science and public policy-making and that more reflection on and better understanding of the mechanisms and complexities associated with the apprehending, framing, and displacing of ethics discourse amidst scientific uncertainty in a highly charged political arena, is necessary.

Public interest forums such as blogs (as a window to the public's concerns), can highlight the failures of the recent Copenhagen Summit (December 2009) and dissatisfaction with the Copenhagen Accord and the unruliness that persists in this area of empirical ethics. In one such example, George Dvorsky (2010) (see also Rajamani, 2009) contends that the failure of Copenhagen is symptomatic of 'a deeper malaise that is currently paralyzing the countries of our warming planet' and that 'this crisis is revealing, that our social and political institutions are ill equipped to deal with a pending catastrophe such as this.' Apart from highlighting the disturbing trend of nation states to only deal with climate threats that are considered local, Dvorsky points to the structural difficulties of effectively engaging the democratically governed 'people[s] of the world' to be 'universally educated about the intricacies of the issues (including scientific, economic, cultural and political considerations) and disarmed of their petty selfishness and local biases'. How should the public be consulted (especially when there is weak consensus on the reasons for rising temperatures and disappointment with the effort leading up to the Summit), on what they feel should be done about climate change when there is a 'tremendous unwillingness to deal with a problem that has yet to show any real tangible negative effects' for the West, and when anthropogenic climate change denalists are seemingly increasing in number and are resistant to human intervention in 'fixing' the climate?

In the intersection of climate change and food security, the recent calls from leading figures in the climate change establishment, such as Intergovernmental Panel on Climate Change (IPCC) chairman Rajendra Pachauri (2009) for a more vegetarian based diet and by various government officials lately for a 'less meat = less heat' campaign, have come under attack from members of both the laypublic and scientific communities. Based on Livestock's Long Shadow (FAO, 2006), 'the livestock sector emerges as one of the top two or three most significant contributors to the most serious environmental problems, at every scale from local to global. [How we farm animals] should be a major policy focus when dealing with problems of land degradation, climate change and air pollution, water shortage and water pollution and loss of biodiversity... The impact is so significant that it needs to be addressed with urgency. Major reductions in impact could be achieved at reasonable cost'. The FAO report argues that people should cut back on their meat consumption since meat production is responsible for 18% of greenhouse gas emissions. The report arrived at this figure by tallying all greenhouse-gas emissions associated with meat production from farm to fork, such as on site farm vehicle use, fertiliser production, land clearance, and methane emissions from the animals' digestion. These findings not only raise policy considerations. They also challenge certain folkways about food (see also FAO, 2006).

However, according to Pitesky *et al.* (2009) United Nations specialists should reconsider their conclusion about the contribution of meat production to climate change. Contrary to the FAO report, Pitesky *et al.* argue that authors of the report presented a 'lopsided analysis' of the issue when they failed to be consistent in how they calculated transport emissions (which only includes fossil fuel burning), relative to how the committee arrived at the meat emissions quotient. Pitesky *et al.* argue for smarter animal farming practices and not less farming in order to curb global warming. Curtailing meat production and consumption would not really be as beneficial for the climate than has been published. They conclude that producing less meat and milk will only mean more hunger in poor countries. For other descriptions of contested science see Borenstein (2010) and Cederberg *et al.* (2009).

When thinking of how we feed people, the environmental cost of feeding 9 billion people does concern many. A recent FAO report (2009) indicated that 1 billion people go hungry worldwide, amidst escalating food prices and the global financial crisis. How should we consider Olivier De Schutter's (UN Special Rapporteur on the Right to Food), claim that our present high meat consumption is 'unethical' since it comes at an exorbitant environmental cost? In a recent EURSAFE newsletter (2009), Lennert questions meat's role in the global hunger problem and climate change. 'While roughly one billion people in the world are malnourished, we convert about 40% of the world's cereals into meat at a highly inefficient rate. The high demand for meat makes grain less affordable for the poorest, and requires inefficient use of valuable agricultural land.' Other discussion further compound the unruliness regarding how to think about the effects of animal agriculture on climate and vice versa. See for example, Röcklinsberg (2009) and Ilea (2009).

As briefly touched on above, there is quite a bit of messiness involved in ferreting through the science as well as the moral sentiments associated with the areas of empirical ethics. The issues of climate change and food security are riddled with scientific uncertainty and politics. The unruliness here reflects how the specific concerns regarding these areas are 'broached, framed, understood, and discussed' with the public by 'experts'.

Sources of unruliness

Scholars (for example Felt *et al.*, 2009; Haimes, 2002; Borner and Menz, 2005; Burgess, 2004) who investigate unruliness in empirical ethics where lay expertise meets scientific expertise, have identified a number of factors, which for the purposes of this presentation, I will describe only briefly.

The status of facts

A major source of unruliness is connected to the status of facts in a climate of scientific uncertainty. Facts come into play in three major ways:

1. Sound or 'the right' facts as trump

Facts can appear in either scientific or social form and can be the site of political struggle (Irwin, 2006). Scientists as experts, may contend that laypersons' arguments are more often than not built on erroneous facts or facts which are not obtained through rigorous means. There may be truth to this, but there is also concern about professional hubris leveled by the laypublic that is justified (see for example, Thompson and Dean, 1996; Wohlforth, 2004). In instances where laypersons feel subservient to the authority of so-called experts, according to Felt *et al.* (2009: 361), they may devalue their own empirical assessments or chalk it up to armchair science or mere anecdotal evidence, and thus, view their position and experiences as less significant. Laypersons often defer to the experts as being the ones who possess the 'right' facts and that they have little to offer. Scientists may wield their authority and employ 'their facts' in ways that can influence not only the path of discourse broached, but also as a mechanism to shut down discussion altogether. They may contend that ethics problems evaporate when the 'right 'facts' are employed and that only they are in possession of the 'right facts'. It should be noted that other forms of unruliness can emerge from within the scientific community. In controversial areas like climate change, there has been and continues to be contention that has led to unruliness between different groups of scientists over who has the 'right facts'.

In the case of Pitesky *et al.* above, this debate about uncertain science on the impact of animal food production on climate change highlights the varying techniques different organizations (i.e. IPCC, 2007 and FAO, 2006) employ for apportioning emissions between different sectors of the economy. As a member of the laypublic who is concerned about whether or not to support the animal food system, unruliness can come in the form of whose science is 'the right science or fact?' Since meat is a necessary source of protein in some societies with small food resources, especially where crop plants are unable to thrive, a meat-based diet is the only option for being food secure. The differences in the way in which the IPCC and FAO's methodology has meant confusion and perhaps frustration with the scientific and policy making communities.

2. The fact-value divide

There seems to be a deeply rooted cultural belief that scientific facts are stronger than any other type of value-based lay expertise. Felt *et al.* (2009) observed that the divide between facts and values remains sharp. There seems to be a natural proclivity to rank scientific knowledge (facts) above values (in general) and higher on the totem pole than laypublic's own knowledge. They argue that epistemic asymmetry leads to knowledge hierarchies and both laypeople and scientists seemed to have internalized the model that facts are a better argumentative resource than mere values (Felt *et al.*, 2009). Here, it is assumed that facts are like 'givens' and speak for themselves. As has been pointed out by others, however, social values influence how scientific facts are shaped (Pinch and Bijker, 1984), and if Latour (1991, 1992) is correct, conceptions of social orders are inscribed in these facts that may benefit the status quo at jeopardy to the wellbeing of the vulnerable. Scientific facts are associated with values and preferences and judgments that influence the ways they may be conceptualized (Limoges, 1993; Heymann *et al.*, 2009). The laypublic may be suspicious of experts who have a limited view of how the public is placed at risk by research, new technologies, and policies that are concluded from uncertain science (Thompson and Dean, 1996).

3. Facts as frame-makers

A related way in which facts are trump in the process of public engagement, i.e. as impacting the identification of what should be the ethical issues raised and how they ought to be addressed, is the way facts are employed to reframe the issues at hand to the point of erasing ethically motivated doubts (Felt *et al.*, 2009)

Facts can also be deployed in a manner akin to 'throwing down the gavel' to shut down an ethical discussion. This is done when normative discussions are reformulated into scientific ones. The question above whether meat consumption is unethical can be reframed into a scientific one. What Olivier De Schutter seems to suggest is that a line of argumentation that rests purely on scientific measures and assessment, i.e. an assumed public ideal to live ethically in collusion with environmental facts – renders alternatives to food security hardly arguable. Pitesky *et al.* (2009) on the other hand appeal to a different methodology and thus reframe the issue again in a new light.

The advantage to the scientists who reframes an ethical problem in terms of a scientific problem, is that they are able to manage the discussion by redefining the framework in which, for example, climate change is seen (or not) to be a problem, and by closing discussion in order to avoid challenge from lay people based on their lay expertise. For example, suspending negotiation on animal farming because of hard facts that suggest inimical and irreversible implication to the planet may no longer be seen as an ethical dilemma but as a political or practical one with primarily economic costs.

Displacing ethical questions

Unruliness can also take the form of displacement. This occurs when ethical concerns and questions are '[displaced] to the margins or outside the realm of what should and can be discussed here and now' (Felt *et al.*, 2009: 362). Displacement can be instantiated as follows:

1. Displacement to the ethical expertise of others

Given the seeming pervasiveness of knowledge hierarchies around scientific uncertainty, deferment and reliance on others in this way can be seen as a spoiler to engagement that seeks greater public involvement. Here, laypersons who are fatigued by, apathetic or disenchanted by public discourse may assume a learned helplessness stance and defer their ethical jurisdiction and competency to so-called experts or those changed with making up the rules or policies. While acknowledging that there is a certain level of self-responsibility when it comes to climate and food matters, they may refer to the necessity for more institutionalized forms of ethical reflection and responsibility. This has been pointed out in conjunction with the failure of Copenhagen (Rajamani, 2009). 'Experts', i.e. people and institutions seen as holding epistemic authority on ethics and science, and policy, are the ones meant to deal with it as part of their pre-defined role responsibility. The upshot of this is that concentrated and honest policies (and their sponsors) may not get credit for trying to bring action in the face of failure, as in Copenhagen.

Furthermore, deflection of this sort of ethical responsibility to the regulatory space may mask the necessity and usefulness of ethical inquiry into specific issues. If the ethical is identified with the legal, then once regulation is in place and policies comply with them, further ethical debate is not necessary (Felt *et al.*, 2009).

The side-lining of ethical questions by the scientists or laypersons this way raises questions about the ethical legitimacy of sources giving advice to policy makers and whether layinterests are being marginalized as a consequence of a lack of representation and omission of a broad range of the public's concern on issues concerning uncertain science and its impact on society (Haimes, 2002).

2. 'Downstreaming' ethical questions

The question of the right moment to ask ethical questions should also occasion pause for those involved in the ethics of policy-making from a bottom up approach. Should climate ethics and food security policies be discussed more 'downstream' in the policy-making process? It may be too late or irreversible if geotechnologies are employed to cool the planet or as a solution to food deserts without adequate public input on the fundamental ethical questions, for example. Shouldn't the debate on the impact of policy occur earlier, then? Ethics is surely not just about consequence. According to Felt *et al.* (2009) if the end of public involvement in the policy-making chain is meant to heighten legitimacy and confidence in the outcome, then the public should be consulted more 'upstream'. An advantage of 'upstreaming' ethical discussions is that the public is freer to ask basic questions about the value of food security research including possible trajectories of policy and technological development and its societal consequences, for example.

Public consultation of the ethical consequences or application of policy should not be relegated to later stages of the policy innovation process but perhaps also at formation's start. Addressing the issue of when laypersons should enter the conversation may allow societal aims to be explicitly addressed by the public themselves and not by proxy of (self) interested experts. Scientists and policy-makers, who, even though may be motivated by societal goods and goals, may fall short of what the lay-expertise may advise.

Taming unruliness

As described above, participatory settings where scientific uncertainty meets public engagement can be unruly. The unruly interaction between scientists, policy makers and laypersons on a given controversial issue such as the ones discussed above, is punctuated by power inequilibrium and asymmetrical knowledge hierarchies. Given the potential pitfalls described above regarding public engagement and what is at stake (i.e. can we continue to spew greenhouse gas emissions into the atmosphere with reckless abandon when we are faced with questions about the morality of raising and killing 60 billion terrestrial animals a year, of farming practices and policies that promote converting plant sources of food for food animals (and perhaps) at cost to a billion people who go hungry, and possibly resulting in irreparable harm to our planet?), it is urgent that empirical ethics find a way to provide guidance or tame this sort of unruliness. How should our policy-making go about raising questions about what and how much we should put on consumers' plates, (especially if it means unearthing parts of our culture, curtailing jobs, livelihoods and ways of life)? How should ethics overcome the mechanisms that result in the closure or displacement of public engagement?

Briefly, those of us who engage in empirical ethics should consider how to improve the 'mechanisms of voicing' associated with public consultation in different stages of the policy-making chain. We may start by more effectively:

1. Appreciating the extent to which 'facts' and 'values' confound the engagement process. Considering this epistemic hierarchy is central to understanding the interaction processes between laypeople and scientists in ethical settings. By encouraging laypersons, policy-makers and scientists to equally engage in a process of mutual learning and to reflect on the view that science or fact is more authoritative over values, experts (especially, and other with more political clout), are challenged to find ways to reduce hierarchical inequilibrium inherent in these relations. Reduction strategies should provide greater opportunity for experts and laypersons alike to consider the 'other' as co-citizen and not as 'interested expert' or as mere consumer. As long as scientific knowledge is automatically accepted as superior to any value position and as long as simple regulations are longed for as a central aim, public input into policy innovation will be frustrated.
2. Creating and strengthening the bonds of mutual trust earlier in the engagement stream so that potentially controversial perspectives and issues may be fairly considered and brought to the table for a measured response. Ethicists can help here by helping to make assumptions that entail unruliness explicit to the participants, in order to facilitate a reflexive discussion space. In this light, ethicists can help to structure engagement opportunities that are strongly-process oriented and where the structure of the engagement design is not preordained by a select few but rather developed by the participants in an ongoing process of collaboration and cooperation (see Felt *et al.*, 2009: 358 for an example).
3. Situating discussions of values and visions more 'upstream' (Wilsdon and Willis, 2004) so that policies can have greater ethical legitimacy and support beyond also being derived as a function of a collaborative endeavor. These discussions will likely reveal a better view of the range of possibilities and limits open in a participatory setting, which, in turn, can inform policy innovation. It may also help to anticipate episodes of moral psychology of denial, i.e. instances in which the public is overwhelmed by the crises at hand, despite being educated thoroughly educated on the empirical premises, and subsequently, opts out altogether form the political engagement arena.

Conclusion

The renewed interest in ethics as an integral arbiter in the debate over anthropogenic climate change (Gardner, 2006, 2010; Posas, 2007) is due, in part, to the perceived need for a different kind of relationship between science, politics and society. The failure of the 2009 Copenhagen Summit on

Climate Change highlights the need for more effective public engagement that is not only transparent but involves the public at earlier stages of policy-making. Science alone cannot produce adequate public policy advice and buy-in when questions of values and the ethical legitimacy of policy trajectories are decided upon unilaterally by interested experts.

In overcoming unruliness in public engagement setting regarding climate ethics and food security, the mechanisms that produce knowledge hierarchies and displacement must be confronted in order that a bottom-up approach to ethics can deliver greater bang for its buck in the service of public policy. Ethicists have an integral role to play in bridging the dialogue between layperson, scientists and policy-makers and in helping to make the process of public engagement more transparent, visible and accountable.

References

Borenstein, S. (2010). Scientists seek better way to do climate report. Washington Times 10.02.2010. http://www.washingtontimes.com/news/2010/feb/10/scientists-seek-better-way-do-climate-report/. Retrieved on 18 Feburary 2010.

Cederberg, C., Sonesson, U., Henriksson, M., Davis, J. and Sund, V. (2009). Greenhouse gas emissions from production of meat, milk and eggs in Sweden 1990 and 2005. SIK-Rapport 793. Göteborg: SIK – Institutet för Livsmedel och Bioteknik.

Borner, A. and W. Menz. (2005). Bioethical controversies and public advice: The Production of ethical expertise and its role in the substantiation of political decision-making, in S. Maasen and P. Weingart (eds.) Democratization of Expertise: Exploring Novel Forms of Scientific Advice in Political Decision-Making. Springer: Dordrecht, 21-40.

Burgess, M.M. (2004). Public consultation in ethics: An experiment in representative ethics, Journal of Bioethical Inquiry 1(1): 4-13.

Dvorsky, G. (2010), January 8. Five reasons the Copenhagen climate conference failed. Institute for Ethics and Emerging Technologies, Retrieved from http://ieet.org/index/php/more/dvorsky20100110/.

Epstein, S. (1996). Impure science: AIDS, activism and the politics of knowledge. University of California Press: Berkeley.

Felt, U., M. Fochler, A. Muller, and M. Strassnig. (2009). Unruly ethics: On the difficulties of a bottom-up approach to ethics in the field of genomics, Public Understanding of Science18(3): 354-371.

Food and Agriculture Organization of the United Nations (FAO). (2009). Climate change and food security. Retrieved on 18 February 2010 from http://www.fao.org/climatechange.

Food and Agriculture Organization of the United Nations (FAO). (2009). 1.02 billion people hungry: One sixth of humanity undernourished – more than ever before. Retrieved on March 30 2010 from http://www.fao.org/news/story/en/item/20568/icode/.

Food and Agriculture Organization of the United Nations (FAO). (2006). Livestock's Long Shadow: Environmental Issues and Options. Retrieved on March 25, 2010 at http://www.fao.org/docrep/010/a0701e/a0701e00.htm.

Food and Agriculture Organization of the United Nations (FAO). (2006). 'Livestock a major threat to environment.' http://www.fao.org/newsroom/eb/news/2006/1000448/index.html. Retrieved on 23 February 2010.

Gardner, S. (2006). A Perfect Moral Storm: Climate Change, Intergenerational Ethics and the Problem of Moral Corruption. Environmental Values 15:3, 397-413.

Gardner, S. (2010). Ethics and climate change: an introduction. Wiley Interdisciplinary Reviews: Climate Change 1(1), 54-66.

Haimes, E. (2002). What can the social sciences contribute to the study of ethics? Theoretical, empirical and substantive considerations. Bioethics 16(2): 89-113.

Hedgecoe, A.M. (2004). Critical bioethics: Beyond the social science critique of applied ethics. Bioethics 18(2): 120-43.

Heymann, M., Andersen, H. and Sandøe. P. (2009): What is climate science all about Philosophical perspectives. In: M. Gjerris, C. Gamborg, J.E. Olesen, & J. Wolf, J. (eds.), Earth on fire. Climate change from a philosophical and ethical perspective, University of Copenhagen, Copenhagen, 69-88. Available at www-earthonfire.foi.dk.

Ilea, R.C. (2009). Intensive livestock farming: Global trends, increased environmental concerns, and ethical solutions. Journal of Agricultural and Environmental Ethics 22 (2), 153-167.

Irwin, A. (2006). The politics of talk: Coming to terms with the 'new scientific governance'. Social Studies of Science 36(2): 299-320.
Latour, B. (1992). Technology is society made durable, in J. Law (ed.) A Sociology of Monsters: Essays on Power, Technology and Domination. Routledge: London, 103-131.
Latour, B. (1991). Where are the missing masses? Sociology of a few mundane artifacts. In: W.E. Bijker and J. Law (eds.) Shaping Technology, Building Society: Studies in Sociotechnical Change. MIT Press: Cambridge, MA, 225-258.
Law, J. (2004). After Method: Mess in Social Science Research. Routledge: NY.
Lennert, T. (2009). The ethical merits of vegetarianism promotion. European Society for Agricultural and Environmental Ethics (EURSAFE) Newsletter 11(3), 1-4.
Limoges, C. (1993). Expert knowledge and decision-making in controversy contexts. Public Understanding of Science 2(4): 417-426.
McEwan, I. (2010). Solar. Random House: UK.
Norton, B. (2007). Sustainability: A Philosophy of Adaptive Ecosystem Management. University of Chicago Press: IL.
Pinch, T.J. and W.E. Bijker (1984). The social construction of facts and artifacts: Or how the sociology of science and the sociology of technology might benefit each other, in WE Bijker, TP Hughes and PT Trevor (eds.) The Social Construction of Technological Systems. MIT Press: Cambridge, MA. 17-50.
Pitesky, M., K. Stackhouse and F.M. Mitloehner (2009). Clearing the Air: Livestock's Contribution to Climate Change. In: D. Sparks (ed.) Advances in Agronomy, Vol. 103, Burlington: Academic Press, 1-40.
Posas, P. (2007). Poles of religion and ethics in addressing climate change. Ethics in Science and Environmental Politics Vol. 2007, 31-49.
Rajamani, L. (2009). Addressing the 'post-Kyoto' stress disorder: Reflections on the emerging legal architecture of the climate regime. International and Comparative Law Quarterly 58: 803-834.
Röcklinsberg. H. (2009). Animal ethics in a time of climate changes'. In: Millar, K., Hobson West, P. Nerlich, B. (eds.), Ethical futures: bioscience and food horizons. Wageningen: Academic Publishers: Wageningen, the Netherlands, pp. 92-96.
Sarewitz, D. (2004). How science makes environmental controversies worse. Environmental Science & Policy, 7, 385-403.
Thompson, P.B. and A. Nardone (1999). Sustainable livestock production: methodological and ethical challenges. Livestock Production Science, 61:111-119.
Thompson, P.B. and Dean, W. (1996). Competing conceptions of risk. Risk: Health, Safety and Environment 7, 361-384.
Thompson, P.B. (1995). Risk and responsibility in modern agriculture, In B. Mepham, G.A. Tucker, and J. Wiseman (eds.) Issues in Agricultural Bioethics. Nottingham University Press, Nottingham, 216-240.
Thompson, P.B. (1986). Uncertainty arguments in environmental issues, Environmental. Ethics, 8(1): 59-75.
United Nations Intergovernmental Panel on Climate Change, IPCC. (2007. Climate Change 2007, The physical science basis. Cambridge University Press, Cambridge.
Wilsdon, J. and R. Willis. (2004). See-through science: Why public engagement needs to move more upstream. London: Demos.
Wohlforth, C. (2004). The Whale and the Supercomputer: On the Northern Front of Climate Change. Farrar, Straus and Giroux: NY.

Precautionary principle and global food safety

A. Urruela Mora[1,2] *and A. Libano Beristain*[3]
[1]*University of Zaragoza, Department of Criminal Law, Philosophy of Law and History of Law, Calle Pedro Cerbuna, 12, 50.009 Zaragoza, Spain;* [2]*Inter-University Chair in Law and the Human Genome (Universities of the Basque Country and Deusto), Avenida de las Universidades 24, 48024 Bilbao, Spain;* [3]*Universidad Autónoma de Barcelona, Department of Private Law, Campus UAB, 08193 Bellaterra, Spain; asier@unizar.es*

Abstract

The precautionary principle was born in Germany at the end of the sixties under the name of *Vorsorgeprinzip*. During the seventies it became one of the essential principles which defined the environmental policy. The aforementioned principle, according to Olivier Godard, was initially based on a broad vision of prevention against the menaces for the environment, combining preventive action criteria, early detection of risks and activation of the environmental protection without needing to have scientific certainties. The principle was later extended to the rest of Europe, being incorporated to the international environmental legal texts. Furthermore the precautionary principle has been incorporated to the European Communitarian Law and has been included in the EC Treaty (Treaty establishing the European Community): the Treaty of Maastricht included it as one of the basic principles of the European policy at environmental level (art. 130-R-2 of the mentioned treaty) and the version according to the Treaty of Amsterdam expressly confirmed it in art. 174 (nowadays the Treaty on the Functioning of the European Union includes it in art. 191). At Communitarian level one of its most important applications in the last decades has been associated with the embargo decisions taken by the European Commission in March 1996 related to meat and products derived from cattle coming from the United Kingdom as a result of the well known crisis of the Bovine Spongiform Encephalopaty (BSE). This paper pretends to analyse the relevance of the precautionary principle with relation to the global food safety, on the basis of the legal regulations and case-law existing up to now at European Union level.

Keywords: food security, risk management, EU legislation

The precautionary principle: general framework

The increasing possibilities of intervention in the environment and in the human health in general, together with the developments in the fields of Genetics and Biotechnologies basically in the last two decades, make necessary to consider a new scientific scenario: the one associated with the existence of activities related to which an important degree of uncertainty about their consequences does exist. The mentioned situation gave raise to a new (legal) principle called the precautionary principle.

Historically, as underlined by Olivier Godard (Godard, 2000, 2001, 2002), the precautionary principle was born in Germany at the end of the sixties under the name of *Vorsorgeprinzip*. During the seventies it became in this country one of the basic principles which defined the environmental policy. With this principle a broad vision of prevention in face of the risks for the environment was established, combining preventive action criteria, early detection of risks, as well as activation of the environmental protection without the need of having scientific certainties.

The principle was lately extended to the rest of Europe, as well as incorporated to international texts in the field of environmental law. Its legal recognition at an international level was produced by the Ministerial Declaration of the Second International Conference on the Protection of the North Sea (London, 24th-25th November 1987). Progressively it appeared in the Rio Declaration on Environment and Development (Rio de Janeiro, 3rd-14th June 1992 (UN)), as well as in other international texts: Declaration of the Bergen Conference on Sustainable Development (7th May 1990), the Convention on Biological Diversity (Rio de Janeiro, 5th June 1992 (UN)), the United Nations Framework Convention on Climate Change (New York, 9th May 1992), the Convention on the Protection of the Marine Environment of the Baltic Sea Area (Helsinki, 9th April 1992), the Convention for the Protection of the Marine Environment of the North-East Atlantic (Paris, 22nd September 1992), Protocol of the Convention on the Prevention of Marine Pollution by Dumping of Wastes and Other Matter (7th November 1996) and more recently the Cartagena Protocol on Biosafety to the Convention on Biological Diversity (Montreal, 29th January 2000 (UN)).

According to the Wingspread (Wisconsin, USA) Statement on the Precautionary Principle (January 1998) '...it is necessary to implement the Precautionary Principle: where an activity raises threats of harm to the environment or human health, precautionary measures should be taken even if some cause and effect relationships are not fully established scientifically. In this context the proponent of an activity, rather than the public bears the burden of proof. The process of applying the Precautionary Principle must be open, informed and democratic, and must include potentially affected parties. It must also involve an examination of the full range of alternatives, including no action' (on this topic see also the Lowell Statement on Science and the Precautionary Principle).

Furthermore the precautionary principle has been incorporated to the European Communitarian (Union) Law and has been positivised first in the Treaty establishing the European Community: the Treaty of Maastricht (1992) includes it as one of the basic principles of the European policy at environmental level (art. 130-R-2) and the version according to the Treaty of Amsterdam expressly confirms it in its art. 174.2. Nowadays the Treaty on the Functioning of the European Union includes it in art. 191.

On this aspect, at Communitarian level one its most relevant applications in the last decades has been connected with the embargo decisions adopted by the European Commission in March 1996 related to meat and bovine products coming from Great Britain as a consequence of the well-known Bovine Spongiform Encephalopathy (BSE).

The main two topics arisen by the precautionary principle (Ramos Torre, 2000a) – once its existence established and its general recognition obtained – are the ones related to its content and its legal status. No general consensus has been reached in the scientific community on the extension of the precautionary principle. According to Ramos Torre (Ramos Torre, 2000a) three main lines of interpretation of the principle can be drawn:

1. First of all, a more radical position aware of the risks that menace the environment and the human health. These authors apply for a zero risk policy, or even ask for an abstention or for a moratorium for all those activities which do not guaranty their inocuity. This position is absolutely utopical in a society like ours, characterized by the assumption of certain level of risks as totally necessary for the maintenance of our level of life and in order to allow the technologic development.
2. Secondly, other authors adhere to the doctrine that Ramos Torre (Ramos Torre, 2000a) calls conservative or minimalist. That is characterized, according to this author, by the restriction of the application of the precautionary principle to the risks that are probable and imply serious or even irreversible consequences, rejecting the inversion of the burden of the proof, and admitting the assumption of risks at certain levels.
3. In the scientific community a third group of authors maintain a leading position which is intermediate between the two previous ones. They consider that the precautionary principle should only be applied on the basis of a scientific hypothesis of certain relevance, based on the criteria of a significant sector of the scientific community, or at least on the ground of a solid argumentation at

> a methodological level. The inversion of the burden of the proof in a radical way is rejected and the burden of the proof tends to be distributed, according to opportunity criteria, between the one who proposes the innovation and the one who rejects it. In no case the certification of the inocuity of the activity (zero risk) is demanded, having to consider the intrinsic vulnerability of the environment and the need of an *ex ante* protection. Finally, a combination of abstention measures (moratorium) and positive ones (increase of research activities, more vigilance, etc.) is proposed.

Secondly, from a legal point of view the nature of the precautionary principle is discussed. Does it constitute a real principle with normative efficacy which can be claimed at a jurisdictional level or only directive criteria of the legislative decisions in the environmental, health, genetic and biotechnological fields basically?

Initially, there was no consensus in the scientific community related to this particular and the positions were confronted. But it is evident that the progressive recognition of the normative character of the principle in the positive legislation of a number of states, as well as at international level, has increased its importance and the normative rank of the principle. This is the main cause of the broaden of the field of action of the precautionary principle, and having being initially recognised to secure the protection of the environment and the conservation of the natural resources, the principle has been extended to the fields of public health and (linked to the previous one) food safety by the politic instances of the European Union (Godard, 2001).

In conclusion, the application of the precautionary principle will be adequate in cases where uncertainty about the level of risk combined with a possibility of serious and irreversible damages does exist. On the basis of these fundamental premises the traditional criteria for the adoption of decisions are insufficient considering the dimension of the consequences.

Related to the conditions for the application of the precautionary principle, two basic texts establish the criteria that inform the principle: on the one side, the report of Philippe Kourilsky and Geneviève Viney to the French First Minister (2000), on the other side, the Communication of the European Commission on the precautionary principle (Brussels, 2.2.2000. COM(2000) 1 final).

Therefore, the conditions of application of the precautionary principle are, on the one side, the uncertainty about the risk, and, on the other side, the possibility that the risk will lead to a serious and irreversible damage. In this case, the necessary scientific evaluation has to be made so as to determine the precautory measures to adopt.

Once the existence of a risk derived from a scientific activity is determined, what has to be done is to apply concrete measures so as to avoid that the mentioned risk leads to a serious and irreversible damage. In principle, and even when the risk has to be uncertain (because that constitutes the foundation of the application of the precautionary principle) the entity of the damages hypothetically associated with the development of an activity makes necessary to adopt measures to avoid its concretion.

Nevertheless it has to be pointed out that the adopted measures do not have to consist necessarily in a penalty or a moratorium of the activity. It is true that this possibility cannot be absolutely excluded, as it can be necessary when the rest of measures do not show their efficacy in order to avoid the concretion of the damage. Nevertheless, the application of the precautionary principle does not make it necessary to penalize every activity with risk in which the level of risk is uncertain, but imply the adoption of the adequate measures to avoid the creation of a serious and irreversible damage. Only in this last case those ones will consist in the paralysis of the activity.

Global food safety: the European Union framework

The acceptance of the relevance of the precautionary principle in the field of food safety

The precautionary principle has been incorporated to the European Communitarian (Union) Law and has been included in the EC Treaty: the Treaty of Maastricht mentioned it as one of the basic principles of the European policy at environmental level (art. 130-R-2 of the mentioned treaty) and the Treaty of Amsterdam expressly maintained its relevance in art. 174. Nowadays the Treaty on the Functioning of the European Union includes it in art. 191. Nevertheless, and even when its origin is associated to environmental law, nowadays it has become a central principle in the fields of human health protection and, connected to this one, of food safety (Manteca Valdelande, 2007).

At communitarian (EU) level one of its most important applications in the last decades has been associated with the embargo decisions taken by the European Commission in March 1996 related to meat and products derived from cattle coming from the United Kingdom as a result of the well known crisis of the Bovine Spongiform Encephalopathy (BSE). The European Commission Decision of 27th March 1996 on emergency measures to protect against bovine spongiform encephalopathy (96/239/EC) expressly established on the basis of the precautionary principle (even when it was not mentioned with this name) that the United Kingdom shall not export from its territory to the other Member States or third countries:

- live bovine animals, their semen and embryos;
- meat of bovine animals slaughtered in the United Kingdom;
- products obtained from bovine animals slaughtered in the United Kingdom which are liable to enter the animal feed or human food chain, and materials destined for use in chain, and materials destined for use in medicinal products, cosmetics or pharmaceutical products;
- mammalian derived meat and bone-meal.

The arguments to maintain the inspiration of the decision in the precautionary principle can be easily understood when analysing the fifth, six and seventh recitals of the decision.

The Court of Justice of the European Communities (nowadays, Court of Justice of the European Union) had the possibility of expressing his opinion related to this decision (and consequently, to the application of the precautionary principle in the mentioned framework) in two judgments: 'United Kingdom of Great Britain and Northern Ireland vs. Commission of the European Communities' and 'National Farmers' Union'. In both judgments the Court of Justice expressly accepting the application of the precautionary principle in the 'mad cow disease' case (food safety field) states: 'At the time when the contested decision was adopted, there was great uncertainty as to the risks posed by live animals, bovine meat and derived products. Where there is uncertainty as to the existence or extent of risks to human health, the institutions may take protective measures without having to wait until the reality and seriousness of those risks become fully apparent. That approach is borne out by Article 130r(1) of the EC Treaty, according to which Community policy on the environment is to pursue the objective *inter alia* of protecting human health. Article 130r(2) provides that that policy is to aim at a high level of protection and is to be based in particular on thee principles that preventive action should be taken and that environmental protection requirements must be integrated into the definition and implementation of other Community policies' (grounds 99 and 100 of the case 'United Kingdom of Great Britain and Northern Ireland vs. Commission of the European Communities' and grounds 63 and 64 of the case 'National Farmers' Union').

On the basis of the case law established by the Court of Justice of the European Union, even when the precautionary principle is only mentioned in the Treaty on the Functioning of the European Union in the framework of the environment protection (art. 191.2), this disposition has to be interpretated systematically in relation with articles 6.a Treaty on the Functioning of the European Union ('the Union shall have competence to carry out actions to support, coordinate or supplement the actions of the Member States. The areas of such action shall, at European level, be: (a) protection and improvement of human health...'), art. 12 Treaty on the Functioning of the European Union ('Consumer protection requirements shall be taken into account in defining and implementing other Union policies and activities'), art. 168.1 Treaty on the Functioning of the European Union ('a high level of human health protection shall be ensured in the definition and implementation of all Union policies and activities') and art. 169.1 Treaty on the Functioning of the European Union ('in order to promote the interests of consumers and to ensure a high level of consumer protection, the Union shall contribute to protecting the health, safety and economic interests of consumers, as well as to promoting their right to information, education and to organize themselves in order to safeguard their interests') (González Vaqué, 2006). Consequently the precautionary principle can be applicated in the health field and therefore in the context of food safety and has to be considered a general principle of the Communitarian (European Union) law (González Vaqué, 2006). Other decisions of the Court of First Instance have confirmed the mentioned criteria (see Judgment of the Court of First Instance – Third Chamber – of 11th September 2002 – Pfizer Animal Health SA vs. Council of the European Union, Case T-13/99 – and Judgment of the Court of First Instance of 26th November 2002 – Artegodan GmbH and Others vs. Commission of the European Communities –).

The Conclusions of the European Council of Nice (7-10 December 2000) included in their point 35 (Section V. Citizens' Europe, A. Consumer health and safety) the following reference to the precautionary principle: 'the European Council affirms the need to implement rapidly and fully the principles introduced by the Treaty of Amsterdam, which provides for a high level of human health protection in the definition and implementation of all Community policies and activities. In this context, the European Council notes the Council Resolution on the precautionary principle'. The mentioned resolution was approved and included as Annex III of the referred document. Particular relevance in the field of food safety has to be afforded to point B and point C of the Resolution. The Conclusions of the Presidency of the European Council of Barcelona (2002) following the conclusions of the European Council in Nice (December 2000) on the precautionary principle, recalled the need to clarify arrangements for its application. On top of that the Resolution on food safety in the European Economic Area (EEA) adopted at the 12th meeting of the Joint Parliamentary Committee in Keflavik, on 16 March 1999 (1999/C 304/05) emphasises in its point 5 the importance of application of the precautionary principle and reaffirms in its point 13 the overriding need for a precautionary approach within the EEA to the assessment and evaluation of applications for the marketing of Genetically Modified Organisms (GMOs) intended to enter the food chain and calls for the development of clear identification of genetically modified foods or ingredients in foods by consistent, comprehensive and informative labelling where they have been approved for use (Oanta, 2007).

Manteca Valdelande (2007) underlines that the application of the precautionary principle (referred basically to the food safety field) can be done both in an active or positive way or in a passive or negative one. The active application consists in the adoption of preventive measures in relation to a potential risk in absence of a scientific or technical certainty. The passive or negative application would consist in restricting the authorization of certain substances or foodstuffs until the proof of its inocuity is obtained (for example, in the case of inclusion of food additives in the so called positive lists). Nevertheless it is discussed if the last referred example would be a proper application of the precautionary principle.

Legal references

Specifically referred to food safety we have to mention the European Commission Green Paper on the General Principles of Food Law in the European Union (COM (97) 176. 30th April 1997) which expressly states that 'the Treaty requires the Community to contribute to the maintenance of a high level of protection of public health, the environment and consumers. In order to ensure a high level of protection and coherence, protective measures should be based on risk assessment, taking into account all relevant risk factors, including technological aspects, the best available scientific evidence and the availability of inspection sampling and testing methods. Where a full risk assessment is not possible, measures should be based on the precautionary principle'. On its basis three years later the European Commission issued the White Paper on Food Safety of 12th January 2000 whose objective is to outline a comprehensive range of actions needed to complement and modernise existing EU food legislation, to make it more coherent, understandable and flexible, to promote better enforcement of that legislation, and to provide greater transparency to consumers, guaranteeing, in addition, a high level of food safety. Its origin is related with a series of crises concerning human food and animal feed (BSE, dioxin, etc.) which exposed weaknesses in the design and application of food legislation within the EU. This led the Commission to include the promotion of a high level of food safety among its policy priorities. As was stressed at the Helsinki European Council in December 1999, particular attention must be focused on improving quality standards and reinforcing systems of checks throughout the food chain, from farm to table. Among the general principles set out by the Commission on which European food safety policy should be based we have to underline the express reference to the application of the precautionary principle in risk management. Even when the White Paper has not binding character as such for the Member States, it provides basic criteria in order to ensure food safety, expressly recognising the role of the precautionary principle in this field.

The application of the precautionary principle in texts with legal character in the last decade has been permanent. The basic European Union instrument in the field of food safety as it establishes a set of essential definitions and regulations for the legal systems of the Member States (García Rivas, 2004), Regulation no. 178/2002 of the European Parliament and of the Council of 28th January 2002 laying down the general principles and requirements of food law, establishing the European Food Safety Authority and laying down procedures in matters of food safety, establishes in its article 7 (precautionary principle) that 'in specific circumstances, where, following an assessment of available information, the possibility of harmful effects on health is identified but scientific uncertainty persists, provisional risk management measures necessary to ensure the high level of health protection chosen in the Community may be adopted, pending further scientific information for a more comprehensive risk assessment. Measures adopted on the basis of paragraph 1 shall be proportionate and no more restrictive of trade than is required to achieve the high level of health protection chosen in the Community, regard being had to technical and economic feasibility and other factors regarded as legitimate in the matter under consideration. The measures shall be reviewed within a reasonable period of time, depending on the nature of the risk to life or health identified and the type of scientific information needed to clarify the scientific uncertainty and to conduct a more comprehensive risk assessment'. On the basis of this article some Commission Decisions have been adopted. Among others we can mention Commission Decision of 3rd April 2008 on emergency measures regarding the unauthorised genetically modified organism 'Bt 63' in rice products and Commission Decision of 5th September 2006 on emergency measures regarding the non-authorised genetically modified organism 'LL RICE 601' in rice products. Following the mentioned criteria it is evident that the precautionary principle has been incorporated to the risk management in the field of food safety at European Union level. Particular importance has to be afforded to article 53 Regulation no. 178/2002, which establishes an emergency measures' clause; the majority of experts consider that the precautionary principle can be invoked by via art. 53.1 Regulation no. 178/2002 (Rodríguez Font, 2007).

As signaled by González Vaqué (2006), other references to the precautionary principle in the European Union Law in the context of food safety are included in Commission Directive 1999/50/EC amending Directive 91/321/EEC on *infant formulae* and follow-on formulae whose recital 4 states that 'whereas, taking into account the Community's international obligations, in cases where the relevant scientific evidence is insufficient, the precautionary principle allows the Community to provisionally adopt measures on the basis of available pertinent information, pending an additional assessment of risk and a review of the measure within a reasonable period of time'. After the approbation of Regulation 178/2002 some norms of the European Union refer directly to its art. 7: concretely Commission Directive 2004/1/EC of 6th January 2004 amending Directive 2002/72/EC as regard the suspension of the use of azodicarbonamide as blowing agent and Regulation no. 852/2004 of the European Parliament and of the Council of 29th April 2004 on the hygiene of foodstuffs. On top of that, Regulation no. 1831/2003 of the European Parliament and of the Council of 22nd September 2003 on additives for use in animal nutrition states in its recital 6 that action by the Community relating to human health, animal health and the environment should be based on the precautionary principle and Regulation no. 854/2004 of the European Parliament and of the Council of 29th April 2004 laying down specific rules for the organization of official controls on products of animal origin intended for human consumption mentions specifically the precautionary principle as part of the veterinarians' necessary backgrounds and qualifications.

Finally we have to point out that some of the existing regulations in the field of Genetically Modified Organisms (GMOs) – Directive 2001/18/EC of the European Parliament and of the Council of 12th March 2001 on the deliberate release in the environment of genetically modified organisms (recital 8, articles 1 and 4), Regulation no. 1830/2003 of the European Parliament and of the Council of 22nd September 2003, concerning the traceability and labeling of genetically modified organisms and the traceability of food and feed products produced from genetically modified organisms and amending Directive 2001/18/EC (recital 3), and Regulation no. 1946/2003 of the European Parliament and of the Council of 15th July 2003 on transboundary movements of genetically modified organisms (recital 2 and 22, and article 1) – do include specific references to the precautionary principle (González Vaqué, 2006).

To sum up we can conclude that the precautionary principle has nowadays a full normative (legal) character even when this aspect was initially controversial. Its application at a European Union level is not only restricted to the environmental field (even when the Treaty on the Functioning of the European Union only mentions it in art. 191 related to this particular sphere) but also to other fields as human health (including food safety). In conclusion, the relevance of the precautionary principle with relation to the global food safety, on the basis of the legal regulations and case law existing up to now at a European Union level cannot be discussed.

References

García Rivas, N. (2004). Influencia del principio de precaución sobre los delitos contra la seguridad alimentaria. Revista General de Derecho Penal 1. Available at Internet http://www.iustel.com/v2/revistas/detalle_revista.asp?id=8&numero=1. Accessed 1 May 2010.

Godard, O. (2000). Le principe de précaution, une nouvelle logique de l'action entre science et démocratie. Philosophie politique 11: 17-56.

Godard, O. (2001). Le principe de précaution entre débats et gestion de crises. Régards sur l'actualité 274: 33-48.

Godard, O. (2002). Le principe de précaution face au dilemme de la traduction juridique des demandes sociales – Leçons de méthode tirées de l'affaire de la vache folle. In: C. Leben and J. Verhoeven (eds.) Le principe de précaution: aspects de droit international et de droit communautaire. Paris, Panthéon-Assas, pp. 29-63.

González Vaqué, L. (2001). La aplicación del principio de precaución en la legislación alimentaria: ¿una nueva frontera de la protección del consumidor? Available at internet http://www.porticolegal.com/articulos/pa_112.php. Accessed 1 May 2010.
González Vaqué, L. (2006). El derecho alimentario de la Unión Europea: la aplicación del principio de precaución. Revista de Derecho del Consumo y de la Alimentación: 5. Available online http://sites.google.com/site/aibadaredeco/numero5_2006. Accessed 1 May 2010.
Kourilsky, Ph. and Viney, G. (2000). Le principe de précaution: rapport au Premier ministre présenté par Philippe Kourilsky, Geneviève Viney. Odile Jacob. La Documentation française, Paris.
Manteca Baldelande, V. (2007). El principio de precaución alimentaria. Distribución y consumo. Available online http://www.mapa.es/ministerio/pags/biblioteca/revistas/pdf_DYC/DYC_2007_96_94_99.pdf. Accessed 1 May 2010.
Oanta, G.A. (2007), La política de seguridad alimentaria en la Unión Europea, Tirant lo Blanch, Valencia, 2007.
Ramos Torre, R. (2000a). El retorno de Casandra. Modernización ecológica, prevención e incertidumbre. In: García Blanco, J.M. and Navarro Susaeta, P. (eds.) ¿Más allá de la modernidad? Las dimensiones de la información, la comunicación y las nuevas tecnologías, Centro de Investigaciones Sociológicas, Madrid, pp. 403-455.
Ramos Torre, R. (2000b). Prometeo y las flores del mal: el problema del riesgo en la sociología contemporánea. Perspectivas sociológicas y principio de precaución. Available online at: http://www.proteccioncivil.org/ceise/ceisevirtual/ceisecv2000_p02-01.htm. Accessed 1 May 2010. Accessed 1 May 2010.
Rodríguez Font, M. (2007). Régimen jurídico de la seguridad alimentaria, Marcial Pons, Madrid.
Romeo Casabona, C.M. (ed.) (2004). Principio de precaución, Biotecnología y Derecho, Comares, Bilbao-Granada.

Food system and climate change: the false premises of antitrust policy

V. Sodano
University of Naples Federico II, Department of Agricultural Economics, via Università 96, 80055 Portici, Italy; vsodano@unina.it

Abstract

The global food system is currently accountable for at least 30% of the global greenhouse gas (GHG) emissions that cause climate change. Influencing the behaviour of food Trans-National Corporations (TNCs) in such a way as to shift towards a more sustainable food model may greatly contribute to tackling global warming. Nevertheless, because reducing GHG emissions is a public good, firms do not provide it unless with direct state intervention. With regards to state intervention the paper analyzes the role of competition policy, showing how it can help in fighting global warming, provided that it overcomes the over thirty year lasting dominance of the 'Chicago paradigm'. The general claim of the paper is that it is necessary to return to the original spirit of antitrust legislation which endorses a view of corporation as an artificial powerful legal entity created by the state in order to serve the public interest. According to this view, large firms, and especially TNCs, in the food sector as well in other sectors, may be imposed strict environmental regulations and human rights duties.

Keywords: global warming, corporation, competition policy, food

Introduction

According to recent estimates (IAASTD, 2008), the global food system is currently accountable for at least 30% of the global GHG emissions that cause climate change. Considering also emissions by indirect activities associated with food production and distribution, such as home storage and refrigerators, waste disposal, transportation by final consumers and so on, this estimate may rise dramatically to as high as nearly 50% of total emissions (Grain, 2009). Agribusiness corporations, backing a model of food production and distribution that functions by converting oil into food, are largely responsible for these huge emissions. Influencing the behaviour of food TNCs in such a way as to shift towards a more sustainable food model may greatly contribute to tackling global warming. Actions to induce food corporations to assume a more sustainable form of conduct come from both the private and the public sector. On the private sector side initiatives come from consumers (individuals and consumer associations), environmental associations and non governmental organizations. On the public sector side, there are at least three kinds of intervention: (1) direct regulation, based on a command-and-control approach; (2) 'soft regulation', including self-regulation, use of incentives, awards and accreditation systems, market-based initiatives, disclosure obligations and educational campaigns; (3) definitions of duties of corporations, through corporate law and competition policy. The paper stresses that, given that reducing GHG emissions is comparable to a public good, only state intervention may be expected to be effective. Moreover, given that corporations cannot be granted the same moral status as natural persons, even soft regulation, which requires some form of corporate social responsibility and therefore of corporate morality, cannot be effective. With regards to state intervention the paper analyzes the role of competition policy, showing how it can help in fighting global warming, provided that it overcomes the over thirty year lasting dominance of the 'Chicago paradigm'.

Global warming mitigation: the role of public and private sector

It is a matter of fact that induced climate change is representative of a tragedy of the commons, a typical collective action problem. Maintaining a stable climate has the structure of a public good exhibiting both the property of non excludability and non rivalry. The free riding problem, i.e. the fact that non contributors can benefit from others' GHG reductions without taking on costs themselves, prevents private rational actors from engaging in mitigation efforts. Beyond being a public good, the protection of a stable climate that fits human biological and economic needs, can be considered to be a human right. In particular, it is of the kind of second generation human rights, i.e. economic and social rights, grounded in the notion that government has affirmative obligations to protect individuals from deprivation of the basic material necessities of life.

In the case of public goods, economic and social theories based on rational choice models hold that the market (i.e. the private sector) fails to supply them. Therefore, the only effective provider is the state, as the latter has the precise political mandate to accommodate for general public welfare against scattered private interests. With regards to human rights the general view is that the state has the ultimate duty to uphold them. The state can intervene either directly or indirectly. Direct interventions include: public investments in global warming mitigation; setting compulsory standards in defence of low emission production and consumption activities; imposing human rights duties on corporations for climate change and environmental harm; implementing tort liability laws that make private actors pay for damage to climate and environment. Indirect interventions include: market based incentives aimed at promoting private climate friendly behaviour; embracing a voluntary corporate social responsibility (CSR) approach that shifts the burden of public interest onto corporations, which are deemed to possess other-regarding preferences and moral values. In this paper it is claimed that only direct intervention can be effective because, in the case of market-based instruments, it may apply the same sources of market failure that the intervention seeks to correct. The voluntary CSR approach is not viable because it hinges on the false premise that corporations have the same moral status as natural persons. The moral status of corporations endorsed by scholars like French (French, 1984) is to be rejected when the three necessary conditions for moral agency are examined: the ability to intend an action; the ability to perform an action; the ability to autonomously choose an intentional action. In the case of conglomerate collectives, such as corporations, these conditions are not fulfilled (Ronnegard, 2006: 82) and therefore they do not qualify as moral agents conceived as distinct from their members. Consequently, corporate moral responsibility attributions to collectives as distinct from their corporate members are illegitimate.

Competition policy and climate change: the perspective of the Chicago school

Given that only direct intervention by the state can assure adequate levels of global warming mitigation, the issue to be addressed is what role competition policy, among other forms of public intervention, can have in promoting corporate climate friendly behaviour.

Competition policy originated in the US in 1890 with the Sherman Act. In the European Union the first antitrust regulation was set by the treaty of Rome in 1957. There are commonly described three historical phases of US antitrust law implementation, the first dating from 1890 to 1940, the second from 1945 to 1975 and the third from 1970 to the present (Viscusi *et al.*, 2005). These three phases have been characterized by different economic and political theories incorporating two different ideologies of the market and the state: the evolutionary vision and the intentional vision (Page, 2008). The evolutionary vision views the market, framed solely by laws on property and contracts, as a mechanism for facilitating free exchanges among countless individuals in the pursuit of their best interests. In this vision, the market without state intervention naturally tends to a perfect competition ideal form destroying monopoly. On the contrary, the intentional vision views the market as a mechanism within which powerful interests can coerce consumers, labour and small businesses. In this vision markets tend toward monopoly unless government intervenes. The political economic theories corresponding to these two visions are the laissez-faire and the welfare state theories. The more the intentional vision is preferred to the evolutionary vision, the greater is the scope and the enforcement of antitrust law, and vice versa.

The Sherman Act and the first period of antitrust law implementation embodied a compromise between the two visions. Notwithstanding the faith in the market, coherent with a strong liberal theory of the state, it was recognized as a matter of fact that monopolies and extreme economic power concentrations actually occur in the real world, producing social inequalities and injustice. At that time, state intervention was intended as a way to promote social justice and mediate among class conflicts in society. In the second period, the intentional view prevailed. Stemming from the disillusionment with markets during the Great Depression, the New Deal initiated the era of the welfare state based on the idea, supported by the growing economic literature on market failure, that economic state intervention should be legitimated by efficiency more than by equity concerns. The years between 1950 and 1970 are the golden era of antitrust legislation. The view of the markets taken up by the Court was that of recognition that coercion is the reality of market relationships. That is to say that in contrast with the previous ideological faith in the freedom of contracts, it was acknowledged that in a market transaction each party may be forced by the bargaining power of the other to accept unfair payments and obligations. Along with these views, the then prevailing theory of industrial organization, the structure-conduct-performance paradigm, facilitated a strong enforcement of antitrust legislation, holding that the mere measure of market share was sufficient to witness the presence of market power and monopoly inefficiencies. By the mid-1970s the evolutionary view completely dismissed the intentional view with the uprising of the so called Chicago school of antitrust. Chicago scholars applying neoclassical economics maintained that unfettered markets always lead to the best social outcomes. They pointed out that many of the practices that the courts had been viewing as harmful to competition and economic welfare, such as vertical restraints, may instead improve economic efficiency. Moreover they contested the structuralist view by claiming that a firm's large market share may signal superior efficiency and that, consistently with the contestability theory (Baumol *et al.*, 1982), freedom of entry is the only parameter to be scrutinized by antitrust laws. The general wisdom of the Chicago school was that state intervention and regulation is always harmful to the general interest.

The Chicago 'revolution' has made competition policy a useless instrument for reaching goals of general interest such as providing public goods and promoting social justice. In order to make competition policy a useful instrument against global warming, it is necessary to reject some assumptions of the Chicago antitrust school and revive instead the conventional wisdom of the previous approaches in the wake of the intentional view. Among the assumptions to be scrutinized are those related to the three following issues: the theory of the firm; the nature of corporation; the goals of antitrust policy.

The Chicago approach endorses a neoclassical theory of the firm where the firm is defined by a technical production function. The neoclassical theory of the firm, even in its modern neo-institutional version that accounts for transaction costs, explains a firm's behaviour exclusively through the efficiency argument (exploitation of scale and scope economies). According to Chicago scholars, large size and above-normal returns must be due to efficiency differentials between firms. In their world made of equilibria and complete contracts, power-seeking behaviours are not conceivable (Raghuram and Zingales, 1998). Organizational, institutional and cognitive problems addressed by alternative theories (such as managerial, evolutionary, property rights, and behavioural theories) are dismissed as trivial.

With regards to the legal debate on the nature of corporations (the latter defined as economic organizations whose members are granted limited liability by incorporation statutes), the Chicago view is consistent with the Nexus-of-Contracts theory, which contrasts the two alternative theories, namely the Legal Fiction and the Real Entity theories (Ronnegard, 2006). The Nexus-of-Contracts theory depicts the corporation as a web of contracts among all the members, which implies that it should not be regarded as a separate legal entity from the shareholders and that rights and duties can be defined only with regards to its members. Because the corporation is the result of a free contract, it is not dependent upon state grants and the same act of incorporation (granted by the state) is only a shorthand way of obtaining a contractual situation equivalent to that which could materialize through the private contracting of individuals. This conception of corporation is based on a libertarian ideology that says that corporations ought to merely be a commercial instrument for furthering the ends of the incorporating parties. Because corporations are not autonomous entities, any moral status (and therefore social responsibility) is ruled out, and because they are not a 'creature' of the state but the result of free contracts, they cannot be given rules and duties by the state. Therefore, one cannot expect them to provide public goods, such as climate stabilization, either voluntarily or compulsorily.

Finally, as regards the goals of antitrust, the Chicago school states that antitrust policy ought to deal only with consumer losses due to high prices and/or output restrictions (Burns, 2006). Any equity concern about wealth distribution or unfair business practices is dismissed. For instance, in the Chicago view low final prices generally signal efficiency and practices like predatory pricing, reciprocal selling and cross-subsidization by conglomerates, unfair procurement contracts, and so on, are given little attention.

All these three sets of assumptions entail that corporations pertain to the private more than to the public sphere and that antitrust pertains to the economic more than to the political sphere. In consequence, corporations should not be required to seek public goals (like providing public goods such as climate stabilization) and antitrust should not be required to seek goals like equity and justice (among which climate justice) but should only pursue economic efficiency in terms of low consumer prices.

Competition policy and climate change: reversing the false premises of the Chicago school

Stemming from the intentional vision, and in opposition to the evolutionary vision of the Chicago school, the previous assumptions can be reversed in such a way as to justify a wider scope of antitrust policy able to encompass the goal of climate stabilization.

The first hypotheses to be reversed are those concerning the theory of the firm and the nature of the corporation. Firms cannot be described purely as technical production functions but as institutions (as economic theory had to acknowledge after the seminal work of Coase of 1937) that in some way substitute the market with power as means of resource allocation. Like states, firms exercise power in various forms, either inside their organizational boundaries or outside, over their competitors, their suppliers, their customers and the same state, through lobbying and bribing. Modern corporations are firms which, through the limited liability and other rights granted by the state (such as unlimited life span, unlimited asset acquisition, complete flexibility and mobility in business conduct, constitutional rights equal to those of natural persons), possess even superpowers (Nace, 2003; Korten, 2001), i.e. powers that cannot be enjoyed by a single individual and even less (because of territorial limits) by a single state. Because corporations are legal persons, with specific rights granted by the state, their nature cannot be described through the Nexus-of-Contract theory endorsed by the Chicago School. Their nature is better described by the Legal Fiction theory. The Legal Fiction theory essentially says that the corporation is merely an abstract creation of law which is granted to an association of individuals. The corporation is an artificial legal entity with an existence distinct from the incorporating members and exists entirely at the discretion of the state. The Legal Fiction theory differs from the Nexus-of-Contract theory which does not recognize the corporation distinct from its members and does affirm that it is independent from the power of the state because it is the result of free contracts by individuals. The

Legal Fiction theory also differs from the Real Entity theory that considers corporations to be real, social organisms that possess a will and life of their own, with characteristics that are distinct from their individual members. Similar to the Nexus-of-Contracts theory, the Real Entity theory rejects the notion that corporation is a creation or grant from the state. However, differently from the Nexus-of-Contracts theory, the Real Entity theory claims that corporations ought to be granted legal rights as natural persons, rights which are owed to the corporation itself as a separate organism and are not derived from the rights of the individual members. The Legal Fiction theory is the only theory on the nature of corporations that is consistent with the advocacy of an antitrust regulation aimed at directly controlling and limiting the scope of activity of corporations. Because corporations are legal persons they can be given rights and duties. Nevertheless, because they are not natural persons, as instead envisaged by the Real Entity theory, they do not automatically enjoy basic rights (like the rights to free speech and due process of law) and do not possess moral responsibility. Because they are creatures of the state, they do not have their own life and in the divide between the private and public sphere they can be put somehow on the public side.

Shifting from the idea of corporations as private efficiency-seeking organizations to the idea of corporations as social bodies enjoying large powers by virtue of state grants allows us to recognize that corporations may have an important role in addressing general social problems like global warming. Two arguments must be considered. First, because the power of corporations, including the power to affect global warming, depends on state grants, state regulations and obligations imposed on corporations in order to contribute to climate stabilization cannot be considered as illegitimate limitation to private freedom (as envisioned by Chicago scholars and neo-liberalists). Such regulations and obligations should instead be considered a due act of governance involving subjects (state regulators and corporations) that both pertain to the public sphere. Secondly, obligations imposed on corporations may be of the kind of human rights duties in case of environmental harm (Mabaquiao, 2002). It is worth noticing that rights are, after all, a response to the problem of power; in particular human rights are asserted in order to protect individuals from abuse of power by states. When one recognizes that many TNCs are really as powerful as or more powerful than many states, it does make sense to treat them as duty-holders, with the same obligations as the states to uphold human rights (Sinden, 2007). It is also important to notice that, because according to the Legal Fiction theory corporations do not possess moral responsibility, we cannot rely on CSR or voluntary codes of conducts as ways to protect the public from environmental harm and any power abuse made by corporations.

The second set of hypotheses to be reversed is that concerning the definition of the scope of antitrust policy. It is general wisdom that antitrust policy should prevent excesses in exercise of power by large firms. The difference between the Chicago School and alternative approaches based on the intentional view is with the kind of power at stake. The Chicago school only considers market power in the form of high consumer prices. Alternative approaches instead look at different kinds of power: the bargaining power towards suppliers and employees; the power to choose technologies and products with different environmental impacts; the power to influence the political arena; the power to 'capture' regulators; the power to influence cultural and social values; and even more. If antitrust policy has to deal with all these kinds of power then it must widen its scope, adding to the economic goal efficiency, social and political goals, such as business fairness, distributive equity, environment protection, enforcement of human rights and so on. In this perspective, antitrust policy should provide incentives (either positive or negative) for business firms to pursue public goals, such as global warming mitigation.

Conclusion

The global food system is populated by many large TNCs (Etc.Group, 2008). These corporations have de facto become a key part of the fabric of global environmental governance. In their role as investors, polluters, experts, manufacturers, lobbyists and employers, corporations are central players in environmental issues. While necessary, voluntary action on the part of corporations and consumers is not alone sufficient to mitigate the worst effects of global warming. However, in the food sector, voluntary actions have been weak and sparse so far (Cogan, 2006). For instance in the Ceres report (CERES, 2008), which rates firms by their achievements in climate-related corporate governance, there are no companies from the food sector among the top ten firms. Among the bottom twelve there are instead three food giants: ConAgra, Bunge, and PepsiCo. Climate stabilization, as in general environmental protection, is a public good and as such is not provided by the private sector but needs public intervention. Among the many kinds of public intervention, the paper has focused on antitrust legislation. At its origin, antitrust legislation was conceived as a means to mitigate power wielded by large corporations in society. With the spread of neo-liberalism from the mid-1970s, the Chicago School radically changed the meaning and the scope of antitrust laws, with drastic changes in its enforcement (Mueller, 2009). The general claim of this paper is that it is necessary to go back to the original spirit of antitrust legislation which endorses an idea of corporation as an artificial powerful legal entity created by the state in order to serve the public interest. Only in this way can large firms, in particular TNCs in the food sector, be expected to comply with environmental regulations and guarantee human rights.

References

Baunol, W., Willing, R. and Panzar C.J. (1982). Contestable Markets and the Theory of Industry Structure. Harcourt Brace Jovanovich, San Diego, 510 pp.

Burns, J.W. (2006). Challenging the Chicago School on vertical restraints. Utah law Review 3:913-919.

CERES (2008). Corporate Governance and Climate Change: Consumer and Technology Companies. Available at: http://www.ceres.org.

Cogan, D. (2006). Corporate governance and climate change: making the connection. Boston, CERES. Available at: http://www.ceres.org/netc.ommunity/Document.Doc?id=90.

Etc.Group (2008). Who owns nature? Corporate power and the final frontier in the commodification of life. Available at: www.etc.group.org.

French P.A. (1984). Collective and corporate responsibility. Columbia Univ. Press, 215 pp.

Grain (2009). The International food system and the climate crisis, Seedling October 2009. Available at: http://www.grain.org.

IAASTD (2008). Global report. Available at: http://www.agassessment.org/.

Korten, D.C. (2001). When Corporations rule the world. Berret-Koehler Publ., 384pp.

Mabaquiao, N.M. (2002). Corporations and the cause of environmental protection. Eubios Journal of Asian and International Bioethics, 12 11-15.

Mueller, W.F. (2009). Fighting antitrust policy, Xlibris Corporation, 285 pp.

Nace, T. (2003). Gangs of America. Berret-Koehler Publishers INC. San Francisco, 281pp.

Page, W.H. (2008). The ideological origins and evolution of U.S Antitrust Law. In: Issues in competition Law and policy. ABA section of antitrust law.

Raghuram, G.R. and L. Zingales (1998). Power in a theory of the firm. The Quarterly Journal of Economics, 113 2:387-432.

Ronnegard, D. (2006). Corporate Moral Agency and the role of the corporation in society. The London School of Economics and Political Science, 263 pp.

Sinden, A. (2007). The power of rights: imposing human rights duties on transnational corporations for environmental harm. In: McBarnet, D. *et al.* (eds.) The new corporate accountability: corporate social responsibility and the law. Cambridge University Press.

Viscusi, W., Harrington, J.E. and Vernon, J.M. (2005). Economics of regulation and antitrust. The MIT Press, 927 pp.

Climate change and food insecurity: an ethical problem that cannot be postponed

M. Nari Lloveras
Universidad de la República, Motivos de Proteo 1365, 11400 Montevideo, Uruguay; mnari@adinet.com.uy

Abstract

One of the big challenges that we face during this century is to guarantee the world's food security, considering the impact of climate change. More than 1000 million people around the world suffer from hunger. Most of them live in developing countries, which are expected to be more affected by climate change, particularly those countries located in regions which suffer from severe droughts such as northeast and central Africa, southeast Asia, and even Latin America, where regions in Peru, Brazil and Chile, and other countries, are harmed by the impact of droughts and in other cases, by floods that are caused by intense rainfalls. During the World Summit on Food Security of the Food and Agriculture Organization of the United Nations, which took place in Rome on November 2009, UN Secretary-General Ban Ki Moon stated: 'There can be no food security without climate security'. International organizations acknowledge climate change and its impact on the world's food insecurity situation as an ethical problem. It is being recognized as damaged caused by human beings. Damage that we are responsible for, and to which we cannot remain indifferent. This paper presents the magnitude of the problem and makes an analysis of the impact of climate change in the food insecurity situation from all its dimensions: food availability, access, use and nutritional state. The ethical perspective will contribute to understand the problem in a comprehensive way and will help to find responsible answers that are based on the human rights defense, a human duty that cannot be postponed.

Keywords: climate, food system, hunger, ethics

Introduction

One of the most urgent challenges we face during this century is to respond to the critical food insecurity situation in which millions of human beings live; which combined with the phenomenon of climate change will increase the hunger in the world significantly in the upcoming years, even in relative terms, in other words, in proportion to the projected population. It is expected that in 40 years the number of people that suffer from hunger will increase from 10 to 20% due to climate change (WFP, 2009). In addition, most of the people suffering from hunger in the world live in underdeveloped countries, which are the countries that are expected to be most affected by climate change.

In 2007, 197 million people, most of them living in developing countries were affected by extreme weather conditions in particular by floods caused by heavy rainfalls, and in other cases by the opposite conditions. An average of 500 weather-related disasters occur every year compared to the 120 that occurred in the 1980s (FAO, 2007).

It's a reality that extreme weather events and the disasters that are caused by them are happening with increasing frequency and can have a great impact on food supply, as we recently witnessed in the earthquakes in Haiti (January 2010) and Chile (February 2010) and the dramatic situation experienced by these countries' populations.

Despite these alarming figures, only in 2008 (FAO, June 2008), it was recognized internationally the close links between climate change and the global food insecurity. It was the beginning of a series of international efforts to link climate change, bioenergy, and food security, with the aim of strengthening networks and active partnerships for development and contribute to protect the environment and eliminate world hunger.

Subsequently, two major world summits were held, one on Food Security (Rome, November 2009) and the other on Climate Change (Copenhagen, December 2009). However, beyond the international acknowledgement of the problem, the scandal of hunger, combined with climate change poses a serious threat to achieving the Millennium Development Goals (MDGs) declared by the United Nations in 2000, where one of the goals is to halve between 1990 and 2015 the proportion of people who suffer from hunger.

According to the United Nations Food and Agriculture Organization (FAO/WFP, 2009), out of the 1020 millions of people currently suffering from hunger, 642 million are located in Asia and the Pacific, 265 million in northeast and central Africa, 53 million in Latin America and the Caribbean, 42 million in the Middle East and 15 million in industrialized countries.

The World Summit Declaration on Food Security (FAO, November 2009), especially in sections 2 and 5, states: 'We are alarmed that the number of people suffering from hunger and poverty now exceeds 1 billion. This is an unacceptable blight on the lives, livelihoods and dignity of one-sixth of the world's population ... while strides have been made, the overall efforts so far have fallen short of achieving the Millennium Development Goals and the commitments of the World Food Summits (Section 2). 'Climate change poses additional severe risks to food security and the agriculture sector. It's expected impact is particularly fraught with danger for smallholder farmers in developing countries ... and for already vulnerable populations' (Section 5).

The persistence of hunger and poverty in a world characterized by an enormous accumulation of wealth and the existence of food in sufficient quantities to feed all humanity, aggravated by climate change, shows that we are facing a problem that mankind has not yet resolved and that despite the international declarations and commitments made by governments, shows a tendency to get worse and demonstrates that the current solutions have been insufficient.

Taking into consideration the reasons stated above, we can affirm that climate change and its impact on global food insecurity, is one of the most serious moral issues of our century, since it threatens the survival of our species and increases the injustice and inequality among the people and populations of our planet, given the fact that the ones who suffer more the impact of climate change are the ones that are less responsible for causing the problem.

Interaction climate change: food system and its impact on food insecurity

Climate change not only impacts on food production, but it also influences the functioning of the local, national and global food system. The effects of climate change, which have a direct impact on the system are clearly mentioned on the Fourth Assessment Report of the Intergovernmental Panel on United Nations Climate Change (IPCC, 2007), which forecasts the consequences of climate change. We have extracted those that are most significant to help understand our problem:

- Frequency of heat waves will likely increase, reducing crop yields in warm zones and increasing forest fires. Air quality will also decline in many cities.
- Catastrophic rainfalls will likely increase. The change in the processes of evaporation-condensation of water would cause changes in wind patterns, in the frequency and intensity of hurricanes, tornadoes or typhoons, and also changes in the hydrological cycle and disturbing phenomena like the Southern Oscillation phenomena such as El Nino or La Nina.

- Floods, especially in Southeast Asia, from Pakistan to Vietnam. Risk of disappearance of islands in the Caribbean, the Indian Ocean and the Pacific.
- Increased water shortages, especially in southern Europe and Africa. In southern and eastern Australia and parts of New Zealand, water shortages will be especially significant after 2030. Drinking water will also be scarce in India and the Middle East.
- Negative consequences for human health due to the following causes: heat waves, increased malnutrition and diseases associated with it, storms and floods, with adverse effects on freshwater due to pollution, and increased diarrhea; spread of endemic diseases like malaria to areas where it is not spread yet; increase of asthma, increased concentration of ozone levels, which will have an impact on cardio-respiratory diseases.
- Changes in agricultural and fishery productivity. Many areas that can be cultivated and also cropping periods will be reduced, especially in Africa where it is estimated that crops in some areas could be reduced by 50% by 2020.

In addition, the global food system is not only being unprepared to deal with climate change but is also one of the main engines. The model of industrial agriculture that supplies the global food system operates primarily through the conversion of oil in food, producing huge amounts of greenhouse gases in the process.

Agricultural production releases greenhouse gases emissions into the atmosphere and produces most of the methane emissions (through cattle and wetlands, especially rice fields) and nitrous oxide (because of fertilizers usage). Changes in land use such as deforestation and land degradation – devastating effects of unsustainable agricultural practices – emit large amounts of carbon into the atmosphere and contribute to climate change.

Agriculture is not only a victim of climate change but also a source of greenhouse gases. The emissions of these gases by the food industry and agriculture account for over 30% of the total annual emissions.

Taking into account what was stated before, the Group of Cordoba (Cordoba Declaration, 2009) calls for consistency and action on food security and climate change: 'Governments are currently dealing with food security and climate change with disconnected policy agendas. The Copenhagen negotiations only see input-intensive agriculture in relation to greenhouse gas emissions and as a potential source of carbon credits. Conversely, in Rome negotiations are focused on increasing agricultural production, without giving enough consideration to the impact of climate change or to the resilience of small producers to safeguard food security in a time of multiple crises and risks. Our key concern is that future food security is imperiled by climate change. States must ensure that food security and climate change policies are harmonized and the outcomes of the negotiations in both forums shall be mutually supportive'. This is supported by UN Secretary-General Ban Ki Moon in the World Summit on Food Security of the Food and Agriculture Organization of the United Nations in 2009 when he stated 'There can be no food security without climate security'.

In short, the negative effects of climate change have a dramatic impact on the human level, all related to food insecurity, hunger, diseases and poverty.

Anthropogenic dimension of the problem

The issue raised is not a product of nature or an event happening due to chance, there is full recognition of human intervention. We are responsible for risking the quality of life, welfare and the integrity of millions of people that are and will be affected by hunger and the consequences of climate change, especially the most vulnerable ones.

It is important to quote a reflection from Velayos (2008) who notes that children that are born these days, and especially future generations, will come into the world with a 'new original sin' in a metaphorical sense. They will be born wounded, stigmatized and fearful of the future and the development of their living conditions (Velayos, 2008: 57).

More specifically, Cortinas (2009) refers to the responsibility of humans to solve the problems of poverty and hunger and affirms that they are not a natural event, but a social product, for which we are, therefore, responsible. Two factors support this responsibility: today we have a social conscience, widespread, that is our responsibility to end poverty and hunger, and we also have the resources that are necessary to eradicate them. And yet we continue allowing their existence, therefore, we are responsible... Because the responsibility for poverty is not only of those who cause it directly, but also those who, for lack of interest, do nothing to prevent it, despite they know that there are resources available to stop it (Cortinas, 2009: 16).

We are facing a moral problem which we are responsible for

Taking into consideration what was stated above, we can affirm that climate change and its impact on food insecurity, is both an ethical and political problem as it is related to damages, interest, judgments, and fairness. The acknowledgement of climate change as a problem attributed directly or indirectly to human activity, makes it a caused damage, which we are responsible for.

This acknowledgement implies rising questions about accountability and justice. According to Velayos this damage is not produced by everyone in the same way, but affects other people now and in the future. 'If we do not solve this problem now, the consequences of climate change can be atrocious and irreversible' (Velayos, 2008: 36).

The Declaration on the Ethical Dimensions of Climate Change (Buenos Aires, 2004) clearly states that there is not enough reflection on the ethical dimensions of many aspects involved in the changes in climate. This situation is worrying because the policies regarding climate change generate ethical issues about which human beings, societies, communities, plants, animals and ecosystems will survive and what people and countries will have to deal with the burden of climate change.

The approach of climate change from an ethical perspective is essential because if the ethical dimensions are not considered, the international community can adopt responses that are ethically wrong or unfair. It also requires a perspective of fairness in climate change policies in order to break down the boundaries that currently impede progress in negotiations.

Moreover, a global consensus on climate that has an ethical foundation can prevent future disparities between the rich and the poor, as well as reducing the potential international tension that would arise from the scarcity of food and water or perceived injustices in the use of global atmospheric assets.

The reflection on the ethical dimensions of the impact of climate change on the world's food security situation and the search for answers to the problem makes it easier for both policy makers and professionals linked to the environment or the general public, to understand all the practical problems involved. It also encourages interdisciplinary dialogue, the pursuit of an ethical consensus, and integrates the ethical analysis in the research on climate change and food insecurity.

Therefore, it is urgent to consider the ethical dimension of the problem in order to contribute to a better and holistic understanding of the concepts, to critical thinking, and to responsible guidance of personal and social action, understanding that moral responsibility does not just repair the damages that are caused. It mainly consists of facing our opinions and our actions or lack of actions, and requires both individual and collective responses.

The search for responsible responses to climate change and its impact on food insecurity should include the participation of citizens, organizations, social movements, and governments worldwide, especially those countries that are most vulnerable to climate change. Solving it is already an urgent moral duty of humanity.

References

Cortina, A. and Pereira, G. (eds.) (2009). Pobreza y Libertad. Erradicar la pobreza desde el enfoque de Amartya Sen, Tecnos, Madrid, 290 pp.

EHAP. Chair of Studies on Hunger and Poverty. (2009). Córdoba Declaration on the Right to Food and governance of the global agricultura and food. October 4, 2009. Available at: www.fao.org/righttofood/news22-es.htm.

FAO. Food and Agriculture Organization (2007). Climate change and food security: a framework document. FAO, Roma. Available at: www.fao.org/docrep/010/i0145s/i0145s00.htm.

FAO. Food and Agriculture Organization (2008). World Food Security, challenges of climate change and bioenergy. FAO, Roma. Available at: www.fao.org/foodclimate/conference/en.

FAO. Food and Agriculture Organization (2009).The State of Food Insecurity in the world. FAO/PMA, Roma. 2009. Available at: www.fao.org/docrep/011/i0291e/i0291e00.htm.

FAO. Food and Agriculture Organization (2009). The World Summit on Food Security. Roma, 16-18 nov. 2009. Available at: www.fao.org/fileadmin/templatest/wsfs/Summit/WSFSO9_Declaration.pdf.

Gómez-Heras, J.M. and Velayos, C. (Coords.) (2004). Tomarse en serio la naturaleza, Biblioteca Nueva, Madrid, 326 pp.

IPCC. Intergovernmental Panel on United Nations climate change (2007). Fourth Assesment Report: Summary for Policymakers. Available at: www.ipcc.ch/pdf/assessment.report.

Velayos Castelo, C. (2008). Ética y cambio climático Desclée de Brouwer, Bilbao, 160 pp.

Part 2. Sustainable agriculture and aquaculture for global food security

Ethical reflection on a practical proposal of sustainable poultry farming

F.L.B. Meijboom
Ethics Institute, Utrecht University, Heidelberglaan 8, 3584 CS Utrecht, the Netherlands; f.l.b.meijboom@uu.nl

Abstract

The introduction of sustainable food production implies an assessment of different aspects at the same time. A mere focus on animal welfare or the use of natural resources is not good enough. Sustainability asks for production systems that integrate aspects of welfare, respect for nature, justice, aesthetics, etc. This paper addresses one specific proposal: a sustainable housing system for laying hens. During the last five years an interdisciplinary team of researcher and entrepreneurs worked on the development and introduction of a sustainable housing system for laying hens. They managed to integrate different aspects in one system, including human and animal health, product safety, economic efficiency, animal welfare and aesthetics. This, however, implied numerous moral assessments. In practice most of these assessments were taken implicitly. In the current stage of introduction, fine tuning and necessary alterations in order to build the housing system confront the team with explicit ethical questions. This paper aims to explicate and analyze some of these questions, and show the need for professionals and commercial organizations in practice to deal with such moral problems.

Keywords: sustainability, animal welfare, food safety, landscape

Laying hens: a changing practice

The practice of poultry farming is rapidly changing in the Netherlands. On the one hand, the sector is still in the aftermath of the outbreak of avian flue in 2003. Only recently, the total number of laying hens reached 45 million hens, which is almost similar to the situation before the outbreak. At the same time, the number of poultry farms is rapidly decreasing. The first decrease was, of course, during and directly after the outbreak of bird flue, but since 2005 there has been a continuous decline of the number of enterprises in this sector. As a consequence, the still existing companies start to grow. In the period 2000-2009 the number of laying hens per farm increased with 65% up until an average of 32,000 animals per farm (Statistics Netherlands/CBS, 2009). These rather big companies are confronted with the implementation of all kind of regulations and societal changes. For instance, the Council Directive 1999/74/EC formulates minimum standards for housing of laying hens and entailed that unenriched cage systems are due to be phased out by 2012 and are no longer allowed to be introduced since 2003 (EC, 1999). Related to this directive, animal welfare organisations made some further steps and demanded in a whole range of campaigns that eggs from unenriched cage systems have to be banned completely, not only in the direct selling of eggs, but also in food processing. As a consequence, most Dutch supermarkets have a variety of eggs from alternative housing systems, organic eggs or eggs with health claims. Also some big retailers banned the products from unenriched cages in the food production.

The need for new housing systems: focus on one example

The above context stresses the need for the development of alternative housing systems for laying hen husbandry. One of these initiatives in the Netherlands is the *Rondeel* housing system, which is the focus of this paper.

The choice for this specific housing system in this paper and the underling research project is threefold. First, I treat this system as an example of a variety of initiatives that aim to take policy and public requirements regarding laying hen husbandry seriously. Second, by focussing on a specific case, the ethical questions become more explicit than in a general discussion of initiatives that react on or participate in the above-mentioned public and policy process. Finally, the *Rondeel* initiative is from an ethical point of view an interesting case as it aims to integrate a variety of aspects in the housing system that are all related to basic moral values, such as welfare, health, and care for the environment.

The *Rondeel* housing system is the result of close cooperation between a commercial organization that has a strong track record in developing and building housing systems for the poultry husbandry and a consortium of academics from different disciplines representing a variety of Dutch Universities. In practice it resulted in a housing system that consists of four key components. First, there are night quarters, where the hens can eat, sleep, rest and lay eggs. Second, it contains a day quarter. This provides the hens the opportunity to scratching and take dust baths. The aim is 'to bring nature indoors'. Third, there is wooded fringe that provides an additional opportunity to scratch, but also to explore and look for shelter. The final element is the central core. This is a three-floor part of the building that contains the working space for the poultry farmer, a floor that is open to visitors and to sell the eggs, and a top floor that houses two heat exchangers which are used for climate control in the system and for pre-and re-drying the manure (Rondeel, 2010).

This innovative housing system is not only interesting from a technological point of view, but also in terms of ethics. It aims to combine corporate social responsibility with sustainable husbandry and food production.

Sustainability: beyond animal welfare

The *Rondeel* project has its origin in the acknowledgement that sustainability in laying hen husbandry should include more than attention to animal welfare. A report on system innovation of laying hen husbandry forms the start of this process. The report defines three challenges for a sustainable housing system: (a) it should be a socially desirable system with a truthful and positive image as well as happy chickens, (b) it should include explicit attention to the robustness and naturalness of the system and (c) is has to find room for free range areas, where hens can forage and explore (WUR, 2004). This implies that sustainability is mainly conceived in terms of the integration of several valuable aspects. It tries to prevent either/ or choices and search for combinations. The final aim or moral ideal was defined in terms of 'a happy hen life, proud farmers and a satisfied society' (WUR, 2004). The *Rondeel* system tries to translate this ideal in practice.

Apart from conceptual questions whether or not this is an adequate definition of sustainability, this view directly raises questions regarding the evaluation and weighing of situations of conflict. It fits a commonsense view that – if possible – we choose for housing systems that fit to the needs of the laying hen, are the best working environment for the poultry farmer, is completely environmental friendly and keep the public satisfied. In practice, however, these aspects are not only complementary, but also can be in competition with each other. For example, animal welfare may imply more work for a farmer, or better working conditions for the farmer may have a consequence on the environmental impact of the system. Referring to sustainability as such will of course not suffice in such situations. This is not a flaw of this approach as such, but it can become a serious vulnerability both in theoretical as in practical terms. As long as these moral assessments and decisions are taken mainly implicitly and would be communicated as the self-evident result of striving for sustainability in general, the theoretical underpinning is poor and one's position in the public debate becomes weak.

Therefore the main aim of the ethics projects was twofold: to explicate the (often implicit) moral beliefs and ideals and to formulate step stones for a framework that help the professionals in practice to deal with the ethical dilemmas at stake.

Explicating moral beliefs

In a number of in-depth interviews with key persons in the process of development and introduction of the *Rondeel* system, we tried to explicate the moral beliefs and ideals. The first questions focus on (a) why the project is valuable and (b) how this value of the system is related to moral values. The above-mentioned sustainable character of the housing system is often defined as what makes this concept valuable. The innovation is considered to be the integration of a genuine care for animals, environment, landscape, but also a clear eye for the business and marketing dimensions.

A follow-up question was on the hierarchy of the involved values. Quite often the interviewees indicate that all values in this concept are of equal value. However, in the focussing on dilemma and problems, it appears that there nonetheless is a hierarchy and that not all values are at the same level. Even though animal welfare, human health, care for the environment, etc. were all considered as intrinsically valuable, the hierarchy often tend to be: human-related values, animal-related values, environment-related values. The human-related values are valued most. This choice for the human perspective is often justified as a *sine qua non*: the sector of laying hen husbandry cannot exist if it does not care for public health, the health of the professional farmer, but also economy. According to a majority this equally holds for the attention and care for animals. The reason, however, differs. Some argue that we should care for the animal for its own sake, while others mainly refer to the current public opinion that does not accept problems with the welfare of animals.

Finally, the environmental values are mentioned. Mostly it is stressed that these are important, but if one had to choose these are evaluated less. This is also reflected in the answers on the question of why the attention to the environment is necessary. Only few respondents refer to an intrinsic value of nature. Mostly, it is mentioned in relation to possible health effects or climate change related to, for example, emission and manure management.

Thus, on a *prima facie* level all agree on the relevance and importance of the values related to human health and safety, animal health and wellbeing, nature and environment and even the esthetical values and its implications for the integration of the housing system in the landscape. However, in practice there are serious differences in the interpretation and ranking of these values. This especially becomes clear in discussions on the importance of free range facilities and the relevance of naturalness. At this point, the questions pops up for whose sake free range facilities are relevant. Some argue that this is in the interest of the general public rather than that it is essential for the well-being of the hen. Others consider this as a crucial step towards a more respectful, animal-centred way of keeping animals. In the development of a new housing system such moral differences complicates the process, because choosing for free range cannot without costs. Allowing hens to run free outside complicates a number of discussions on animal health and product safety, but also with respect to the workload of the poultry farmer, space requirements and the economic feasibility of the housing system. This implies that value-free choices in this discussion do not exist.

When confronted with this problem, the *Rondeel* concept decided to include a free range element that provides the hen the opportunity to explore and look for shelter. However, it is not a real outdoor free-range farming. Without explicating the underlying moral beliefs and ethical assessment, this choice might be conceived as a weak compromise. This may hold for both extremes in the discussion. Some will stress the additional health, safety and economical issues, while others will only emphasize the fact that the free-range element is relatively small and the hens are not really allowed to go outdoors. To make corporate social responsibility operational in this context, it implies that one is able to formulate the considerations and values that underlie the choice. In this specific case, one can argue that the choice is not just a weak compromise, but that it is in line with the aim to integrate the interests of all involved parties as much as possible. It fits in the moral ideal that underlies this concept. Making this explicit does not imply that the choice is beyond discussion, but it essential to discuss the moral choices with others in a fruitful and constructive way.

An ethical balance board

Given the moral ideal of the *Rondeel* concept to integrate the interests of the animal, the farmer, the public, and the consumer, it is crucial to have tools that help to make the ethical that are continuously needed. The discussion on free-range is only one element. Especially, in the current stage of introduction, fine tuning and necessary alterations in order to build the housing system are needed and raise new ethical questions. Suppose, it turns out that the energy system can further be improved from an environmental perspective, but this implies a reduction of the ability of the hens to use the free range element or has direct implications for the workload of the poultry farmer. Or suppose new financial investors ask for alterations to the concept that do – at least at face value – not directly fit to the ideal of integration. What would be the best reply? The answer to such questions is not self-evident given the ideal to integrate the interests of all involved parties. It asks for a careful balance and assessment that – once again – can be substantiated and communicated to others. Therefore, the final aim of this project is to define an ethical framework that helps professionals in housing systems and poultry farming to (a) clarify the ethical questions and to (b) help making the necessary assessments by relating the problem to the moral ideal of integration.

References

CBS (2009). 'Iets meer kippen, flink minder kippenhouders', 20 November 2009, http://www.cbs.nl.

European Union (1999). Council Directive 1999/74/EC laying down minimum standards for the protection of laying hens. Official Journal of the European Communities. L 203, 53-57.

Rondeel (2010). The public farmers and hens. http://www.rondeel.org.

WUR (2004). 'Laying Hen Husbandry, Towards a happy hen life, proud farmers and a satisfied society', Wageningen University and Research center, www.houdenvanhennen.nl.

Behavioural change as a key to sustainable agriculture

C. de Lauwere, E. ten Pierick, V. Beekman and M. van Asseldonk
LEI Wageningen UR, P.O. Box 35, 6700 AA Wageningen, the Netherlands; carolien.delauwere@wur.nl

Abstract

Shifting towards a more sustainable agriculture requires a behavioural change of many stakeholders. A structural improvement of sustainability, for example, depends on producers who have to be able and willing to take measures to improve sustainability on their farms, retailers who have to offer sustainable products, and consumers who have to buy these sustainable products. The theory of planned behaviour (TPB) is a widely applied model to explain behavioural change related to conscious choices and to give insight into interventions needed to stimulate desired behaviour. TPB might also be useful to understand stakeholders' choices with regard to sustainability and to explore which interventions are needed to stimulate more sustainable behaviour. In the current study behavioural change of farmers with respect to the transition towards a more sustainable group housing for their pregnant sows was explored (which is obligatory from 2013 in the Netherlands). A computer assisted personal interviewing approach was applied to elicit behavioural beliefs of 105 farmers. Data analysis shows that TPB might be useful indeed to understand the farmers' choices with regard to sustainability and that it gives indications for possible interventions to make farmers who still have to change to group housing feel more comfortable about this. Farmers who did not change to group housing yet appeared to be more concerned that group housing was not good for animal welfare and that it would lead to more tail biting than farmers who did change already. They also appeared to have more doubts about their skills with regard to group housing. So, for these farmers providing knowledge about group housed sows might be a useful intervention.

Keywords: theory of planned behaviour, behavioural change, sustainability

Introduction

Sustainability is an important issue in European agriculture of today. McGlone (2001) defines it as follows: 'If our systems of production are in harmony with the environment, the animals, the workers and the community and if they are efficient and economically competitive then the system may be said to be sustainable'. A structural improvement of sustainability requires a behavioural change of many stakeholders. The theory of planned behaviour (TPB) is designed to predict and explain human behaviour in specific contexts (i.e. specific behaviours rather than aggregate behaviour) (Ajzen, 1991). In the underlying study it is used in the context of farmers who have to change to group housing for their pregnant sows. TPB argues that behaviour is the product of an intention to perform the behaviour. This intention is a function of (a) the attitude towards the behaviour, (b) the subjective norm to engage or not to engage in the behaviour and (c) perceived behavioural control which refers to a person's perception of his/her ability to perform the behaviour. Attitude, subjective norm and perceived behavioural control are based on beliefs. The attitude of the respondents is determined by the strength of their behavioural beliefs concerning the outcomes of the behaviour and the evaluation of those outcomes (e.g. do farmers consider animal welfare as important and, if so, do they belief that group housing might improve animal welfare?). The subjective norm is determined by the normative beliefs concerning the expectations of important referents of the respondents (e.g. do the farmers belief that their veterinarian expects them to change to group housing?) and their motivation to comply with those referents. Perceived behavioural control is determined by control beliefs concerning the presence of factors that may facilitate or impede the behaviour and the perceived power of those factors (e.g. do farmers belief that they have enough knowledge to keep sows in group housing and how important is this knowledge?). Perceived behavioural control is argued to influence behaviour as an (additional) antecedent of intention (e.g. the perception that one is able to learn how to keep sows in group housing influences one's intention to do so) but also as a factor mediating the relationship between intention and behaviour (e.g. among two individuals holding equally strong intentions to change to group housing, the one with the confidence that he/she can master this activity is more likely to persevere than the one doubting his/her ability).[1] To the extent that perceptions regarding behavioural control are unrealistic, inclusion of an additional variable should be considered: actual behavioural control. Despite a person's intention to perform a given behaviour and his/her perception that he/she is able to do so, lack of actual control may frustrate the realisation of his/her intention. However, when a person's perceptions are realistic, perceived control may serve as a proxy of actual control (Ajzen and Fishbein, 2005).

An overview of the conceptual model of TPB is given in the results section (Figure 1). The model not only explains behavioural change; it also gives insight into interventions needed to stimulate desired behaviour (Ajzen, 2002).

Material and methods

A questionnaire based on TPB was constructed to test whether the theory was useful to understand farmers' choices with regard to sustainability and to find out which interventions are needed to stimulate more sustainable behaviour. The change to group housing for pregnant sows was taken as a case since pig farmers are obliged to change to group housing in 2013 according to Dutch legislation. The hypothesis was that farmers who already did change to group housing entirely or partly perceive the underlying components in TPB differently than pig farmers who did not change to group housing yet. This would indicate that TPB might be applicable in farmers' situations. Moreover, it would give insight into the obstacles experienced by farmers who did not change yet (or only partly) and also into possible interventions to stimulate desired behaviour. In total, 105 pig farmers were interviewed by means of CAPI (Computer assisted personal interviewing), of whom 42 already had changed to group housing entirely, 32 had changed partly and 31 did not change yet.

The questionnaire consisted of questions based on the following elements of TPB (Ajzen, 2002):

- Intention – does/did the farmer intend to change to group housing?
- Attitude – how did/does the farmer feel about changing to group housing? Did/does he for example think that changing to group housing is disadvantageous or advantageous or not realistic or very realistic?
- Behavioural beliefs – how does the farmer think of group housing? Did/does he for example think that group housing was good or bad for animal welfare, animal health, tail biting, control over the sows or work pleasure?
- Perceived behavioural control – does/did the farmer think that he is able to change to group housing and can he decide to change by himself?
- Control belief strength – does/did the farmer expect that he has/had sufficient financial resources, time, knowledge and/or space to change to group housing before 2013?
- Control belief power – are money, time, knowledge or space facilitating or impeding the change to group housing in 2013?
- Subjective norms – how do persons who are relevant for the farmer think about changing to group housing before 2013?
- Motivation to comply – to what extent does the farmer comply to the opinions of persons who are relevant for him with regard to changing to group housing?

[1] Derived from an example of Ajzen (1991) about learning to ski.

- Normative belief strength – to what extent do persons who are relevant for the farmer think that he has to change to group housing?

Each element was interpreted as a construct of TPB and consisted of several variables. A 7 point Likert scale was applied for all variables, with 1 being the most negative score (fully disagree, very unlikely, absolutely untrue) and 7 being the most positive score (fully agree, very likely, totally true). A reliability analysis was carried out to check the consistency of the constructs.

The relationships between different elements of TPB were analyzed with a (multiple) linear regression analysis and Spearman rank correlations. Besides this, data were analysed by means of linear regression analysis with the elements of TPB (both the constructs and the underlying variables of each construct) as dependent variables and the fact whether a farmer was entirely, partly or not changed to group housing as qualitative explanatory variable.

Results

The perceptions of farmers with regard to changing to group housing for pregnant sows

Farmers' intentions to change to group housing for pregnant sows were higher for farmers who entirely or partly changed than for farmers who did not change yet and higher for farmers who entirely changed than for farmers who partly changed (Table 1). According to TPB, this is related to the farmers attitude, subjective norms and perceived behavioural control. With regard to attitude, the results show that farmers who did not change to group housing yet were more negative about group housing than farmers who partly or entirely changed, and that farmers who partly changed were more negative than farmers who entirely changed. The attitude of people is affected by their behavioural beliefs and the evaluation of behavioural outcome[2]. Farmers who did not change to group housing yet had lower scores for behavioural belief strength than farmers who did entirely change (Table 1). A closer look at the separate items showed that this was caused by the fact that farmers who did not change to group housing yet had

Table 1. Behavioural beliefs of farmers with regard to the change to group housing for pregnant sows (mean scores on a 7 point Likert scale).

	Cronbachs alpha	Entirely	Partly	Not	Overall
Intention	0.88	6.3^{c}	5.3^{b}	4.2^{a}	5.4
Attitude	0.92	5.0^{c}	4.4^{b}	3.0^{a}	4.2
Behavioural belief strength	0.80	4.1^{b}	3.6^{ab}	3.1^{a}	3.6
Subjective norms	0.72	5.4^{b}	5.1^{ab}	$4,5^{a}$	5.0
Normative belief strength	0.93	3.8	3.5	3.8	3.7
Motivation to comply	0.90	3.9	4.0	3.9	3.9
Perceived behavioural control	0.81	6.5^{b}	5.5^{a}	$5,2^{a}$	5.8
Control belief strength	0.63	5.2^{b}	4.0^{a}	4.4^{a}	4.6
Control belief power	0.57	5.4	5.3	5.2	5.3

[a,b,c] Different characters in a row mean a significant difference.

[2] The evaluation of behavioural outcome is left aside in the analyses for practical reasons.

more doubts about animal welfare and their work pleasure and worried more about the occurrence of tail biting of pregnant sows in group housing[3].

The results also showed that farmers who did not change to group housing yet had lower scores for subjective norm than farmers who did entirely change (Table 1). Looking at separate items of this construct clarified that farmers who did not change to group housing think to a lesser degree that persons relevant to them expect them to change to group housing or that pig farmers comparable to them already changed themselves. The related constructs 'normative belief strength' and 'motivation to comply' appeared not to be very helpful in explaining why the pig farmers who did not change yet had lower mean scores. In these constructs, it was asked to the farmers interviewed what specified persons or organizations (e.g. veterinarian, bank, food supplier, chain partner, family) expected them to do with regard to changing to group housing (normative belief strength) and to what extent the farmers interviewed wanted to comply to these expectations (motivation to comply).

Perceived behavioural control of farmers who partly or not changed to group housing appeared to be lower than that of farmers who entirely changed (Table 1). A look at the separate items showed that farmers who changed partly or not thought that they were less able to change to group housing and that they had less control of the decision to change. It appeared that they had more doubts about their skills and knowledge with regard to group housing and the building process. The related construct 'control belief strength' showed that farmers who partly or not changed expect to be less able to spend enough money and time on the realisation of group housing than the farmers who entirely changed. They also expect to have less space for building a new stable.

Association between elements of TPB

Correlations between the different elements of TPB are presented in Figure 1, while the results of (multiple) linear regressions are given in Table 2. TPB predicts the intentions of farmers to change to group housing in a significant way. All three underlying elements were important. The regression coefficient of 0.444 for the construct attitude implies that having a more positive attitude, increases intention with 0.444 points, both on the 7-point scale range. For the constructs subjective norm and perceived behavioural control this was 0.252 and 0.509 respectively. On its turn, attitude was affected by a positive perceived behavioural belief and increased 0.7306 points on the 7-point scale ranging from no behavioural belief to a strong behavioural belief. Regression models for subjective norm and perceived behavioural control explained the observed variance to a lesser degree, but were still statistical significant.

Discussion and conclusions

Former research has shown that TPB is reasonably successful in predicting and explaining behaviour in various domains (Ajzen, 1991; Armitage and Conner, 2001). Among those domains is that of agriculture. For instance, Beedell and Rehman (2000) show the usefulness of TPB in understanding farmers' conservation behaviour, Bergevoet *et al.* (2004) explained farm size-related entrepreneurial behaviour of dairy farmers and Fielding *et al.* (2008) explained farmer's decisions to engage in riparian zone management. And also in the case under study, TPB seems to provide a useful framework to study the efforts of Dutch pig farmers' to change to group housing for their pregnant sows as an example of sustainable agriculture. The pig farmers who already changed entirely to group housing answered numerous questions of the TPB based questionnaire differently than their colleagues who changed partly or did not change at all. This is not so surprisingly as they already made the investment which the others are still (partly) facing. More interesting are the farmers who partly changed to group housing. They do

[3] Due to the restricted number of pages in this paper results of separate items of the constructs of TPB had to be left out.

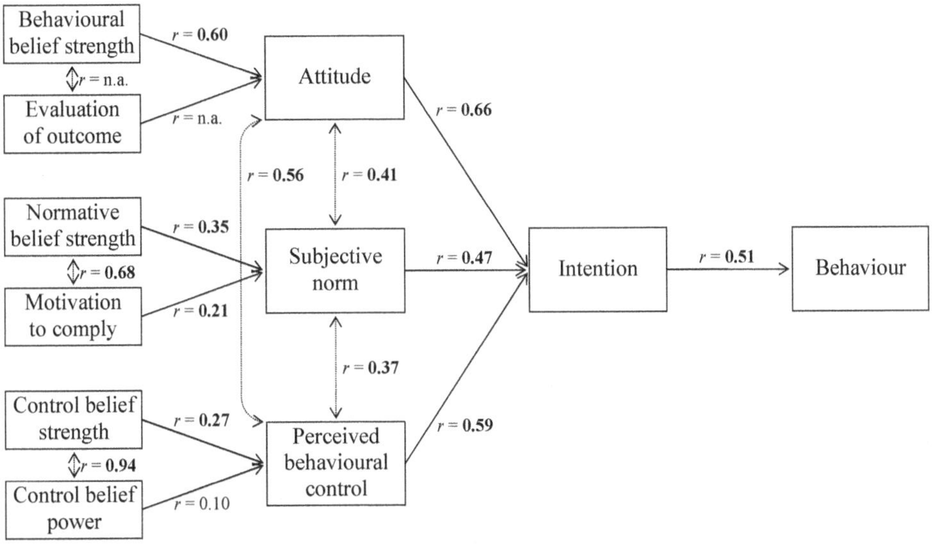

Figure 1. Overview of the theory of planned behaviour (TPB) with the correlations between the constructs of TPB; significant correlations are printed bold; n.a. = not analyzed because the evaluation of behavioural outcome has been left out the analyses for practical reasons.

Table 2. Linear regression models of the different elements of TPB with their predictor(s).

Elements of TPB	Linear regression models[2]	$R^2_{adj.}$
Intention	-0.725 + **0.444**×attitude + **0.252**×subjective norm + **0.509**×perceived behavioural control	48.8
Attitude[1]	**1.519** + **0.7306**×behavioural belief strength	34.8
Subjective norm	**3.672** + -0.006×motivation to comply + **0.380**×normative belief strength	14.9
Perceived behavioural control	**3.650** + 0.0831×control belief power + **0.3731**×control belief strength	14.8

[1] According to TPB, attitude depends on behavioural belief strength and the evaluation of behavioural outcome; the latter has not been elicited in the questionnaire.
[2] Significant regression coefficients are printed bold ($P<0.05$).

not seem to be convinced about group housing although they must be more or less experienced with it. Like the farmers who are not changed yet, they have doubts about their skills and knowledge of the building process (both items of perceived behavioural control). On the other hand, and contrary to the farmers who did not change yet, they seem to be less worried about animal welfare, the occurrence of tail biting and possible decrease of work pleasure (items of behavioural belief strength). Probably they have experienced that group housing turned out to be better than they expected with regard to these aspects, while they underestimated it with regard to the skills and knowledge needed for a successful implementation.

TPB not only helps to explain behaviour; it also gives insight into interventions which might be useful to stimulate sustainable behaviour. In the case of the pig farmers, emphasis should be on communication of knowledge with regard to animal welfare, the occurrence of tail biting, skills needed to keep pigs in group housing and providing knowledge about the building process. This knowledge can be provided by experts, but it also might be useful to organize discussions between pig farmers with more or less (or no) experience with group housing. The latter approach will be tested in a follow-up of the underlying study. About half of the farmers interviewed have indicated that they are interested to join such a follow-up meeting.

References

Ajzen, I. (1991). The Theory of Planned Behavior. Organizational Behavior and Human Decision Processes 50: 179-211.

Ajzen, I. (2002). constructing a TpB questionnaire: conceptual and methodological considerations. http://people.umass.edu/aizen/pdf/tpb.measurement.pdf).

Ajzen, I. and Fishbein, M. (2005). The influence of attitudes on behavior. In: Albarracín, D., Johnson, B.T. and Zanna, M.P. (eds.) The handbook of attitudes. Erlbaum, Mahwah, NJ, USA, pp. 173-221.

Armitage, C.J. and Conner, M. (2001). Efficacy of the Theory of Planned Behaviour: A meta-analytic review. British Journal of Social Psychology 40: 471-499.

Beedell, J.D.C. and Rehman, T. (2000). Using social-psychology models to understand farmers' conservation behaviour. Journal of Rural Studies 16: 117-127.

Bergevoet, R.H.M., Ondersteijn, C.J.M., Saatkamp, H.W., Van Woerkum, C.M.J. and Huirne, R.B.M. (2004). Entrepreneurial behaviour of Dutch dairy farmers under a milk quota system: Goals, objectives and attitudes. Agricultural Systems 80: 1-21.

Fielding, K.S., Terry, D.J., Masser, B.M. and Hogg, M.A. (2008). Integrating social identity theory and the theory of planned behaviour to explain decisions to engage in sustainable agricultural practices. British Journal of Social Psychology 47: 23-48.

McGlone, J.J. (2001). Farm animal welfare in the context of other society issues: toward sustainable systems. Livestock Production Science 72, 75-81.

Balanced sustainable welfare friendly animal breeding

A.-M. Neeteson-van Nieuwenhoven[1] *and J.W. Hardiman*[2]
[1]*European Forum of Farm Animal Breeders, Benedendorpsweg 98, 6862 WL Oosterbeek, the Netherlands;*
[2]*Cobb-Vantress, P.O. Box 1030, Siloam Springs, Arkansas, USA; anne-marie.neeteson@effab.info*

Abstract

The criteria in which animal breeding can make a difference define what is sustainable breeding: food safety and public health, product quality, genetic diversity, efficiency, health and welfare, environment. Not one aspect makes breeding sustainable, but the whole set – the balanced choices for the various markets and the weighing of food safety, welfare and health, etc. in the context of balanced, cost effective, sustainable, welfare friendly breeding. These are all interlinked. Similarly animal welfare is the result of the different aspects that define it. Animal selection has evolved over time from mainly yield related breeding goals to multi-trait selection, in poultry tens of traits. Critical conditions which are shown to have heritability have been included as selection traits. It has improved the possibilities to counteract the antagonism between various traits (e.g. yield and mastitis in cattle). There is feedback of results from production or consumption back into the selection process. The various ethical responsibilities in animal breeding and reproduction include: balanced sustainable breeding with different outcomes, depending on e.g. customer requirement, or region; not compromising real welfare for perceived welfare; not compromising democratic welfare for welfare of happy few animals; and the responsibility to be transparent about not only possible and imaginable undesirable effects, ànd about positive aspects of new developments.

Keywords: animal breeding, animal welfare, sustainability, balanced choices

Introduction

Animal breeding professionalises the attempts of mankind to choose the best animals for our food or for other purposes. Selection for most animals was started by domestication many centuries ago. For aquaculture the domestication of species is still ongoing. Selection in prehistoric times meant keeping the strongest animals through the winter time, choosing those feeling comfortable with men, reproducing in captivity. From these 'adapted' animals, those looking best fit for a certain goal were selected as parents and mated together. Thus farm animal breeds and lines were developed. It took until the 18th century before animal performance was systematically measured. In the 20th this resulted, combined with genetics and statistics, into the estimation of the internal make up of animals. Farm animal breeding and reproduction is in the midst of an era with evolving (bio)technologies and the opportunities to developing ever more sustainable answers to breeding that answer global and regional needs.

The aim of this paper is to discuss (1) what is balanced sustainable welfare-friendly animal breeding, (2) how can sustainable balances be realised, and (3) what ethical dilemmas and responsibilities do breeders meet?

What is balanced sustainable welfare-friendly animal breeding?

The Brundtland commission (Brundtland *et al.*, 1987) stated that sustainable development should fulfill the needs of the present generation, without decreasing the possibility for future generations to fulfill their needs. This definition is correct yet wide.

The sustainability concept in animal breeding has been worked out in the beginning of this century by cooperating groups of breeders, scientists, welfare organisation, ethicists, sociologists and economists in European Commission funded projects on Breeding and Society (BIO4-CT98-0055), and Sustainable Farm Animal Breeding and Reproduction (SEFABAR, QLG7-CT-2000-01368). Sustainable farm animal breeding and reproduction has been defined as the extent to which animal breeding and reproduction, as managed by professional organisations, contribute to maintenance and good care of animal genetic resources for future generations. The ethicists made clear the definition would require a real effort and honesty of the breeders – it was not just answering the society demands for improving (the perception of) animal welfare. They had to investigate where they can make a difference, and take responsibility to concentrate their efforts on these aspects, instead of focusing on the feel good issues. Not one aspect makes breeding sustainable, but the whole set – the balanced choices for the various markets/circumstances. This set consists of: food safety, welfare and health, efficiency, environment, product quality and genetic diversity (Neeteson *et al.*, 2006). They are all interlinked.

Sustainable breeding programmes 'maintain a balance in selection amongst traits with economic significance as well as those traits related to better fitness and well-being. It is illogical and unwise to expect successful long term breeding progress if ones breeding goals are focused on short term returns and biological needs of the organism are ignored. Compassionate, health oriented and science based decisions are key to sustainable successful breeding systems' (Katanbaf and Hardiman, 2009). This does include additional responsibilities, e.g. professionally skilled technicians, clear guidelines for customers on animal, farm management and, at a larger scale, responsibility to answer global food requirements.

How can sustainable balances be realised in animal breeding and reproduction?

The increased (computing and genomics) technologies assist breeding programmes to evolve from selection for 'single' criteria (e.g. growth, leanness, milk production, egg production) to multiple-trait selection balancing production, reproduction, product quality, and animal robustness characters to neutralise possible 'antagonistic effects' (Rauw *et al.*, 1998).

(Feedback) data and heritable traits

Measurable data and heritable traits are the first requirements for informed decision making. The importance of data for animal breeding can hardly be underestimated – it determines the *possible* improvements. This concerns the availability, reliability, accuracy, feasibility, repeatability and comparability of the data on which breeding estimations are based, ànd feedback mechanisms to capture what genetic progress is expressed at the production level. Breeders make extensive efforts to identify and develop science based heritable traits for individual measurements and selections. As an example the Cobb (broiler) breeding programme records 56 individual observations per pedigree candidate. More than ½ of these are refinements or improvements in trait recording added over the last decade, most are some measure of individuals health and fitness. Various traits in place and under development relate to e.g. bone structure, walking ability, cardiovascular fitness. Non-invasive surveys of bird joints by X-ray scanners are very effective in improving joint integrity in broiler age birds (see also Figure 1). To compensate for the low heritability of actual gait scores in pedigree broiler lines of chickens, poultry breeding organisations have devoted much research to the identification of component traits with higher heritability which affect leg health. These include e.g. routine pedigree individual bird measurements of Tibial Dyschondroplasia, varus, valgus, leg rotation and femoral head necrosis disorders each of which has some affect on bird standing or walking ability and also show at least low to moderate heritabilities as opposed to the very low to zero heritabilities produced by gait score analyses. More leg health related research in-house on pedigree farms and externally at multiple universities to identify more posture,

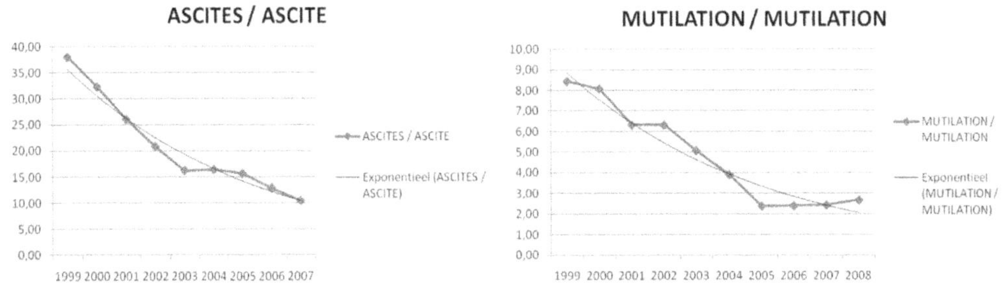

Figure 1. Trend in ascites and mutilation of broilers in Canada (Statistics Canada, 2009).

walking and stability related traits that can be measured objectively is expected to produce more traits with moderate heritabilities resulting in higher rates of genetic improvement for leg health and walking ability in the future. Furthermore, concomitant individualised cardiovascular measurements via blood oximeter machines and physical examination of birds for skin colour indication of compromised oxygen circulation of each individual pedigree candidates aid in identification of susceptible individuals and families to Ascities and Sudden Death Syndrome.

Next to in company data gathering the comparability of data and availability of feedback data is important. The organisations involved in Interbull (cattle) manage to agree international breeding values for bulls, thus enhancing international trade of semen. This process requires trust, accuracy, negotiation – continuous care. Also, since farmers use management programmes, and are linked to internet, breeding organisations can feed the farming data from e.g. pig finishing back into their breeding programme – if they manage to arrange access and to capture them in a feasible way. Poultry breeding companies use amongst others Agristats data, collected by most U.S. poultry companies, for the same type of feedback. The official data from Statistics Canada show health and welfare trends in e.g. chickens (Figure 1).

In Europe there is a need for comparable data across countries on health and performance at the end level (slaughterhouse) – these would enable society to follow health/welfare improvements of the animals in the field, companies to prove health/welfare developments, and scientists to improve health and welfare in research in a much more thorough way.

Broader breeding programmes

Animal breeding programmes have broadened over the last decades (Figure 2), including health and welfare related traits next to the 'traditional' production related traits. Pig breeding programmes have evolved from production related categories of traits in the 1980s to categories covering production, robustness, quality and longevity. Welfare related critical conditions known to be heritable have been included in poultry selection programmes for at least twenty years as selection traits. Dairy cattle breeding programmes include the antagonism between milk yield and the incidence of lameness, mastitis, reproductive disorders and metabolic disorders. There is a global recognition, since the end 1990s/2000, that dairy cattle breeding programmes should be more functional trait/health related, including outcrossing, and this has been put in the practice. In total also the cattle breeding programmes can include up to 56 traits. The balance of production : type : health/welfare is now around 40 : 30 : 30.

Molecular genetics is a new tool used or under development by animal breeders to better survey the gene pool and genome for selection to enhance well-being and performance. The pig and poultry genome are fully sequenced, the cattle genome is on its way, and other genomes will follow as genotyping is becoming

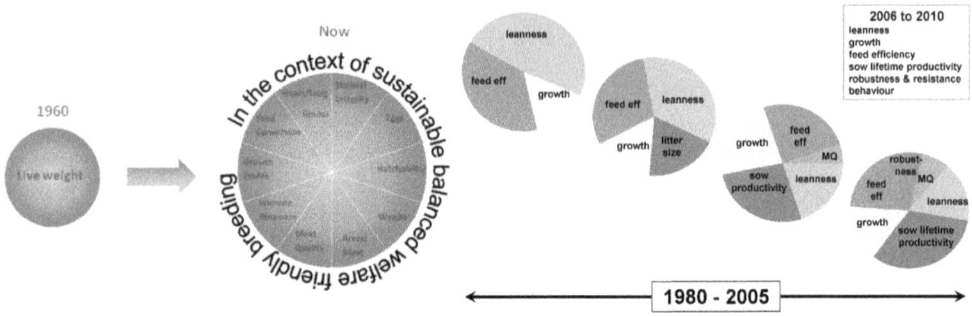

Figure 2. Trends in poultry (left; Laughlin, 2007) and pig (right; Knap, 2006) breeding.

increasingly cheaper and comparisons with related species can be used to 'upgrade' the information of a species. Breeding companies are researching and using the association between animal traits and DNA based Single Nucleotide Polymorphisms (SNP) to analyse the genomes of their pedigree pure lines. Thousands of informative SNP markers are being developed or identified. Associating these SNPs with desirable and undesirable trait provides a new genotypic rather than phenotype driven enhanced selection process to be used by geneticists in better development of pedigree pure lines. Industry SNP marker research includes SNP markers related to e.g. livability, ascites and egg production.

Balanced breeding programmes, welfare and diversity

The work on transparency and sustainability (Breeding and Society, SEFABAR) has resulted in a Code of Good Practice for farm animal breeding organisations, Code-EFABAR®, which is transparent and based on sustainability. In Code-EFABAR® the breeding company defines its way of sustainable breeding and the technologies it uses.

Animal welfare is the result of the many different aspects that influence it. Animal health and welfare was still mentioned separately in the sustainability definition of SEFABAR. Also, there is still a tendency to discuss welfare assessment separately from animal production assessments. However, developments in animal production and animal breeding make the integration of welfare in balanced animal production (assessments) and animal breeding more logical. In the future, welfare is to be considered in the context of balanced sustainable breeding, including also the environment and realistic minimum levels in balance with the business and societal interests and considerations. Animal welfare is an integral natural aspect of animal breeding – as the welfare of the animal is an intrinsic part of every aspect and the result of the balanced choices and care for the animal. Also, the spectrum of animal welfare cannot be properly described by one or a few specific welfare parameters. For example, there are at least 15 separate conditions in broiler chickens that can result in lameness. All traits need to be based on objective data – only relevant traits that have been shown to be heritable can be (and are being) included as selection traits.

Breeding companies maintain important genetic diversity within their gene pools and develop numerous basic lines to serve a wide range of markets globally. In poultry breeding this diversity is available within the company, in pig and cattle breeding the diversity is found within and between companies. Globally operating breeders produce multiple products for different segments of the market, with different requirements. For instance, some broiler integrators demand the optimum performance at light body weight for whole bird and KFC type cut-ups. Others require products that peak in performance at later ages for deboning segment. Some integrators require an enhanced chick number yield. These requirements are being met with different products by different primary breeders. Breeders are mating up complementary lines from their wide array of pure lines to produce a product with enhanced welfare complements. With organic and free-range markets being developed globally, the breeding companies are evaluating stocks that are more suited for environments with greater disease challenges and environmental stressors. New selection techniques/technologies are constantly being developed and evaluated for this specific purpose. Production of multiple products requires a larger arsenal of lines and gene pool which is costly to breeders but necessary for successful sustainable breeding programmes. The ultimate decision on the use of which of these products will be in large dictated by both our customers and consumers.

The basis of animal breeding, the capital breeders work from, is genetic diversity – diversity management is an important item within organised breeding programmes. The robustness of the organisation to answer changing demands is determining for its survival. Although economic evaluation is useful for the indication of the broad direction, the breeding programme needs to have a broad and long term focus. Decisions in breeding need to be made with perspectives of at least 10 years ahead, and taking all feedbacks into account continuously. This requires also that all possible future needs stay within reach. Many breeding organisations use the inbreeding management tools to optimise the genetic diversity in their breeding programme. Smaller programmes and populations can use specific adapted inbreeding programmes to limit inbreeding and retain greater genetic diversity over time. The importance of outcrossing is being recognised, and is being applied in the daily practice. This means that, for instance in broiler breeding, three companies may provide the world with breeding stock, but there are more lines and diversity available than 20 years ago with single product companies.

Stockmanship

Thorough examination of animals at the time of selection by experienced leads and by experienced selectors for various anatomical and physiological phenotypes allow breeding company staff to conduct the initial screening of healthy versus unfit individuals within populations. For instance, in poultry breeding, birds with leg or toe defects, skeletal abnormalities or poor feather cover are purged by specialized selectors skilled in individual bird physical evaluation. Animals chosen to be the parents of future generation need to fulfill minimum requirements – thresholds that every individual selected animal has to meet. Recording of various traits throughout animal life and from relatives allows families with elevated levels of undesired health status to be identified and removed from the programme.

Stockmanship of the multipliers and farmers is looked after via extensive outreach programmes and breeding management guides.

Global and environmental responsibilities

The *demand driven* livestock revolution (Delgado *et al.*, 1999) – the growing global demand for animal food in the coming 20 years – and the need to provide for animal food efficiently to minimise environmental impact will require that breeders continue their efforts to optimise the production animal, and that genetic progress is available globally. The exchange of genetic material between countries is important to ensure the availability of animal food at reasonable cost whilst minimising environmental cost. Animal agriculture has a carbon-nitrogen foot print with nitrous oxide dominating in global warming potential. The improvements in efficiency in animal breeding over the last decades have helped reduce carbon-nitrogen footprint (methane, nitrous oxide, ammonia, etc.). A recent report from a DEFRA funded short term project (Williams *et al.*, 2006) suggests that there has been a 23% reduction in broiler emissions over the last 20 years attributable to genetic improvements in the efficiency of growth. In all species, research and efforts are under way for further improvement.

What ethical dilemmas and responsibilities do breeders meet?

The domestication of animals in itself implies a dynamic process of the adaptation of species to the environment, and vice versa. Over time, the farmed animal and the pet animal have been differentiating from their wild ancestors: the pig from the wild boar, the dachshund from the wolf, and evoluated, in the case of pig, from *Sus scrofa scrofa* to *Sus scrofa domesticus*. Human demand leads to changes in breeding goals. The demand for lean meat, for instance, has led to a change in breeding goals in pigs (changing the 'lard-type' into the 'meat-type' pig) due to a change of human lifestyle. Thus, in consuming meat, milk or eggs, we take the responsibility for farming animals. We can do that as good as we can, balancing the various sustainability items described above. The choice for balanced sustainable breeding will lead to different outcomes, depending on e.g. customer requirement, or region.

Another ethical choice is choosing for real animal welfare and always being critical to oneself not to go for perceived welfare only. Sow group housing has urged companies to find a simple and heritable solution to reduce the undesired side-effect post-mixing aggression in pigs (Turner *et al.*, 2008). The same goes for the reduction of the increased parasitic, infectious and predatory challenges in production systems that look emotionally appealing such as outdoor sow systems. Furthermore, the ethics of democratic welfare improvements (welfare improvements for all the animals) need to be weighed against (maybe per animal larger) welfare improvements for happy few animals. Improving animal welfare of all production animals will not enlarge market shares and margins of stakeholders in the food chain – whilst the per animal larger welfare improvements of a few animals is a commercially attractive 'increased welfare label' concept.

Also, an important aim is to avoid the export of animal welfare issues to less stringently regulated areas, so that unsustainable products will be exported back unnoticed. Connected to this is the need for setting priorities. Any action that would lead to an improvement of animal welfare could be judged by the product [a×b×c] of (a) the number of individual animals it would affect; (b) the magnitude of the welfare improvement for each animal; and (c) the sentience level of the species (Kirkwood, 2006). Such a prioritisation system requires objective descriptors of welfare for a species, staying away from interpretations based on anthropomorphism.

We have to be transparent, balance and explain what we do, but also we need to look forward in an open and balanced way. This goes for new developments, and also for new technologies. It is ethical to have an open eye for the effects of new technologies, e.g. on food safety, animal health, animal welfare, and to realise when technologies will improve and overcome health and welfare issues. It is also ethical to be open about the possible benefits, as it would be unethical to keep this information away from the general public. We have the responsibility to be transparent about not only possible and imaginable undesirable effects, but also about the positive aspects of new developments.

References

Brundtland, G.H., Khalid, M., Agnelli, S., Al-Athel, S.A., Chidzero, B., Fadika, L.M., Hauff, V., Lang, I., Shijun, M., Marino de Botero, M., Singh, N., Nogeir-Neto, P., Okita, S., Rhamphal, S.S., Ruckelshaus, W.D., Sahnoun, M., Salim, E., Shaib, B., Sokolov, V., Stanovnnik, J., Strong, M. and MacNeil, J. (1987). Our common future. Report United Nations World Commission Environment and Development (WCED) 42nd session, Nairobi.

Delgado, C., Rosegrant, M., Steinfeld, H., Ehui, S. and Courbois, C. (1999). Livestock to 2020. The Next Food Revolution. IFPRI. FAO and ILRI. FAO discussion paper 28. 83 pp.

Katanbaf, M.N. and Hardiman, J.W. (2009). Primary broiler breeding – Striking a balance between economic and well-being traits. Poult. Sci. 2010 89: 822-824.

Knap P.W. and A.M. Neeteson (2006). Breeding programmes for animal genetic resources: the role of animal breeding organizations. EC-workshop on 'Legal and strategic aspects of animal genetic resources in Europe', Brussels.
Kirkwood, J.K. (2006) The distribution of the capacity for sentience in the animal kingdom. In: Turner, J. & D'Silva, J. (2006) Animals, ethics and trade: the challenge of animal sentience. Earthscan, London. pp. 12-26.
Laughlin, K. (2007). The evolution of genetics, breeding and production. Temperton Fellowship No. 15. 55pp.
Neeteson-van Nieuwenhoven, A.M., Merks, J.W.M., Bagnato, A. and Liinamo, A.E. (2006). Sustainable transparent farm animal breeding and reproduction. Livestock Sci. 103: 282-291.
Statistics Canada (2009). www.statcan.gc.ca. Available at http://cansim2.statcan.ca/cgi-win/cnsmcgi.exe?Lang=E&ResultTemplate=CST&CORCmd=Getc.Rel&CORId=PRIM55A&CORRel=4.
Turner, J. and D'Silva, J. (2006). Animals, ethics and trade: the challenge of animal sentience. Earthscan, London.
Williams, A.G., E. Audsley, and D.L. Sandars (2006). Determining the environmental burdens and resource use in the production of agricultural and horticultural commodities. Main Report. Defra Research Project IS0205. Bedford: Cranfield University and Defra. Available on www.defra.gov.uk.

The use of certain transgenic seeds in a *more sustainable* organic agriculture[1]

M. de Renobales[1] and L. Escajedo[2]
[1]Biochemistry and Molecular Biology, Faculty of Pharmacy, Universidad del País Vasco Vitoria-Gasteiz, Spain; [2]Constitutional Law, Faculty of Labour Sciences, and Inter-university Chair in Law and Human Genome, Euskal Herriko Unibertsitatea, Bilbao, Spain; mertxe.derenobales@ehu.es

Abstract

Taking into consideration Jonas' *Imperative of Responsibility*, sustainable food production implies that: (1) foods should be produced in sufficient quantity and nutritional quality, preferably where needed to be readily accessible to local populations and to contribute to their development; (2) it should be economically profitable for all involved, included the farmer, and of affordable price for all; (3) it should respect the environment and other non-human living beings. From the moment that genetically modified crops became commercially available in 1996, organic agriculture prohibited their use both in the United States of America and in the European Union, although no scientific reasons were given. A bitter discussion has ensued because organic and transgenic foods are presented to the public as mutually excluding opposites, intrinsically associated to different social and ethical options, in short, incompatible. Yet, genetically modified foods are as safe as conventional foods and the content of nutritional compounds in organic foods is equivalent to that of conventional foods. Also, the philosophy of organic agriculture of not using synthetic products (pesticides and soluble fertilisers) to limit environmental contamination and damage to other non-human organisms is very much in line with the reality of some genetically modified crops. In this paper we consider that acceptance by organic agriculture of these genetically modified crops will be beneficial because it will: (1) increase the productivity of organic agriculture (preferably *in situ*) thus reducing the amount of extra land (and the resulting destruction of habitats) needed to feed increasingly larger populations; (2) increase productivity and economic returns for subsistence and small farmers in developing countries (primarily Sub-Saharan Africa) without compromising exports of organic foods to Europe; (3) end the current sterile stalemate allowing people to actively look for viable solutions to food production using all available technologies because the challenge is great.

Keywords: transgenic seeds, organic agriculture, sustainable agriculture

Introduction

According to official data offered by different UN Agencies, more than eighty countries count with some sort of governing system for BT crops and there are about fifty binding and non-binding International tools to manage the issue with. It has been considered as a key technology that aims at the solution of problems related to food supplies, health provision and environmental protection in the Third Millennium (OECD). However, it is not a lesser truth that this scientific and technological progress arouses economical, social and ethical issues. The end-products resulting from modern biotechnology, particularly foodstuffs, are being subject to extrinsic objections which find no reference in the history of food ethics. It has been said that GM foods are one of the best examples of what has been described as *potentially problematic foods* (Zwart, 2000). Transgenic crops and foodstuffs are overtly criticised not in terms of how they are produced, but in regard to their potential socio-economic impact at every scale

[1] Report awarded the 2009 Prize of the International Society of Bioethics and House of Representatives of the Principality of Asturias. Report can be downloaded from: http://www.sibi.org/ingles/jpg/p2009.htm.

(global scale included) which comes associated to the way in which some transgenic food is produced, distributed and offered to consumers.

Convinced that other more constructive ways to debate about the contributions of modern agricultural biotechnology are possible, our proposal herein analyses the possibility that organic agriculture use certain transgenic seeds which may agree with its philosophy, trying to find points on which to arrive at a consensus. Perhaps this proposal may surprise, or even generate rejection, due to the tension public opinion is being subjected. Our objective is to participate in a current line of thought within the scientific community (Amman, 2008) which, in our opinion, underlines an interesting ethical and social dimension.

Sustainable food production

Perhaps the best known definition of sustainability is that of the Brundtland report (1987) which extended this concept to all activities that promote the development of a society: '*it consists in meeting the needs of the present without compromising the ability of future generations to meet their own needs*.' To meet their needs, humans must interact with Nature remembering that, as integral part of Nature, and because of their superior intellectual capacity, they are responsible for all living beings which populate Planet Earth (Jonas, 1985). When it comes to food production, and considering that about 9000 million people are expected to live on Earth in 2050, the *Keystone Alliance for Sustainable Agriculture* proposes: (1) to increase agricultural productivity decreasing, at the same time, its environmental impact; (2) to improve human health through abundant and nutritious foods; and (3) to improve social and economic conditions of rural agricultural communities (Keystone Center, 2009). The increase in agricultural productivity should occur, according to Dr. J. Diouf, FAO Director, through an increase in yields rather than in arable lands, and in those places where it is more necessary and by those who need it most (2009). An increase in productivity will reduce the need to increase the amount of land, with the concomitant destruction of habitats and ecosystems it implies. Thus, we understand that sustainable food production implies:

- foods should be produced in sufficient quantity and nutritional quality, preferably where needed to be readily accessible to local populations and to contribute to their development;
- it should be economically profitable for all involved, included the farmer, and of affordable price for all;
- it should respect the environment and other non-human living beings.

Organic foods

Organic agriculture is a way to produce food which is, in principle, less aggressive towards the environment than intensive conventional agriculture. It considerably reduces the amount of energy, inputs (such as fertilizers), and the level of field mechanization (Altieri, 1999). It specifically prohibits the use of chemically synthesized pesticides and soluble fertilizers. All this implies profound changes in agricultural practices. In the European Union, organic agriculture is regulated by Regulation 834/2007 which specifies what agronomic practices can, and cannot, be used for the produce to receive the organic label. It certifies that the food was produced according to the specifications established by this Regulation.

The control of weeds, pests and diseases, as well as providing the necessary amount of nitrogen to the growing crops are the main problems for organic agriculture to obtain. There are a number of agricultural practices such as crop rotations, the so-called 'green manures', animal manure, integrated pest management techniques, certain accepted pesticides such as the soil bacterium *Bacillus thuringiensis* and microbial products, etc. that allow for a certain level of weed and pest control. At the same time, many of these methods increase soil fertility and its water retention capacity.

The productivity of organic agriculture as compared to that of conventional agriculture is a highly controversial point which does not have an easy answer. Productivity depends on a number of interrelated factors such as the type of crop, soil, crop used in the rotation, amount of manure per hectare, green manure used, time the soil is fallow, etc. Thus, it is quite difficult to carry out comparative studies in commercial fields (not in experimental fields) for sufficiently long periods of time to average in the production of green manure and the time the field is fallow, so widely different productivities can be found. For example, Denison *et al.* (2004) obtained 66% (organic/conventional) for corn over 9 years, Mäder *et al.* (2002) got an average productivity of 80% over 21 years for several crops, and Goulding and Trewavas (2009) obtained and average value of 60-70% for wheat. In general, it is considered that the productivity of organic agriculture could be up to 20 or 30% lower than that of conventional agriculture, depending on the crop and the circumstances (Ronald and Adamchack, 2008). FAO Director recently stated that 'while organic agriculture contributes to hunger and poverty reduction and should be promoted, it cannot by itself feed the rapidly growing population' (cited ref.). Thus, if organic agriculture methods had to be used to produce most of the food needed by humanity, on average, 20 or 30% more land had to be dedicated to food production, and this would imply the destruction of natural habitats and forests with their ecosystems.

Transgenic foods

In 2009, several transgenic crops were grown in 25 countries on 134 million hectares (about 9% of all cultivated land). These crops can be grouped into two major types: (1) those with improvements in their agronomic characteristics; and (2) those with nutritional improvements. The majority of transgenic crops under cultivation today belong to the first group, and the major traits are insect resistance, herbicide tolerance, and both together (James, 2010). At present, high oleic acid canola and high lysine maize are the only transgenic crops with nutritional improvements authorised in the USA and Canada.

For a given transgenic crop to be authorised in any country in the world, it must pass a safety evaluation by the corresponding authority. In the European Union, safety and environmental evaluations are regulated by Directive 2001/18. No transgenic crop is authorised for food or feed use without a positive safety evaluation conducted by the European Food Safety Authority. The transgenic crop is compared, side by side, with its conventional isogenic variety. Until now, all transgenic crops bearing agronomic improvements have passed all safety tests, including allergenicity. From a nutritional point of view, all of them have been found to be 'substantially equivalent' to conventional foods. Insect resistant corn (Bt corn) has repeatedly been found to be safer than conventional corn because of its much lower fumonisin content. Indeed, in the 13 years of increasing presence of transgenic crops in the world, no health problem related to their consumption has been detected, and no negative long term health effects are expected.

The productivity of transgenic crops, particularly insect resistant and virus resistant crops, is 10% higher, on average, than conventional crops, and can be up to 50% or more in cases of severe damage by pests and parasites (Gonsalves, 2004; Huang *et al.*, 2005), due to reduction in losses. Insect resistant crops significantly and unquestionably reduce the amount of insecticides used on a crop, in variable amounts depending on the circumstances (PG Economics). Herbicide tolerant crops use herbicides which are significantly less toxic and less aggressive for the environment than products used on conventional crops. In addition, they facilitate crop management, substantially reducing the number of applications (fuel consumption and CO_2 emissions). As a result, farmers (both industrial and subsistence) make a net profit in spite of the higher price of the transgenic seed (Raney and Pingali, 2007). Herbicide tolerant crops facilitate the use of no-till, or low-till, agriculture, reducing soil erosion and preserving soil ecosystems.

The concept of *natural*

Organic agriculture rejected the use of transgenic seeds from the very beginning, and European Regulation 834/2007, in Whereas Clause 9, explicitly establishes that organic farmers cannot utilise genetically modified products, or products obtained thereof. The reason is that they are incompatible with the concept of organic production, without further explanation about this incompatibility. It is a commonly accepted idea, albeit false, that the products of organic farming are *natural* whereas transgenic crops have been extensively manipulated in the laboratory. Amman (2008) cites other authors who state that the concept of *natural* implies the integrity of the plant genome. However, recent studies of the genome of plants with sophisticated techniques (Amman, 2008) clearly demonstrate that conventional methods to improve crop plants introduce significantly larger modifications in the plant genome than transgenic techniques. We conclude that transgenic crops are as natural, or as unnatural, as conventional and, therefore, organic crops.

Quite often, organically grown foods are also described as *of high nutritive value*, as if other conventionally grown foods – not considered 'natural' – were of less nutritive value. Like in the comparative study of productivity, the concentration of nutrients depends on many different factors which are not always possible to standardise. According to the extensive study by the *Food Standards Agency* (2009), there are no statistically significant differences between organic and conventional foods in the content of vitamins, minerals, specific proteins, antioxidants, and fibre. Other, less extensive, reviews available in the scientific literature give contradictory results which, to our understanding, confirm that of the *Food Standards Agency*.

Towards a *more* sustainable organic agriculture through the use of *certain* transgenic seeds

Legal documents and public declarations from various authorities transmit the idea that the principles of organic agriculture are incompatible with the use of transgenic seeds. A bitter discussion among proponents of each one has ensued even in the scientific literature leading to a sterile stalemate – in our opinion – which severely hinders finding positive solutions to the challenge of feeding over 9,000 million people by 2050. Thus, we consider that there is no scientific justification for organic agriculture to prohibit the use of *certain* transgenic seeds because (1) organic foods are nutritionally equivalent to conventional foods, and so are transgenic foods; (2) all of them are equally natural, or equally artificial.

The philosophy of organic agriculture of avoiding the use of synthetic pesticides and highly soluble fertilizers to reduce its impact on the environment is, in our opinion, a very positive environmental contribution. Yet, at the same time, as explained above, this is the main reason for its lower productivity, which would lead to destruction of habitats if organic agriculture methods were to be used for generalised food production.

We realise that panaceas do not exist and that it will not be possible to find a food production system which is completely free of all kinds of problems. Both organic methods of food production and transgenic seeds are different tools to help humanity produce the food it needs, where it needs it, according to local characteristics. Therefore, we propose that transgenic crops other than herbicide tolerant ones (which require the use of a given herbicide) be accepted by organic agriculture to improve its productivity. In addition to current insect resistant and virus resistant crops, and those with higher nutrient concentrations, drought resistant and marginal land tolerant crops will improve productivity under difficult stress conditions in a few years. Considering that due to climate change, a substantial decrease in available water is anticipated in developing countries (particularly in Sub-Saharan Africa), these crops could be very important for their inhabitants.

Our responsibility to current and future generations

The generalised rejection of transgenic seeds in Europe has important negative consequences for several groups such as European farmers and meat producers, and farmers and peoples of various developing countries.

The authorization process for new transgenic crops in Europe lags behind those in other countries, resulting in many fewer transgenic crops allowed in Europe either for feed and food use or for cultivation. Shipments or raw materials which contain traces of non-authorized transgenic crops are rejected with increasingly higher costs for the food industry in general and feed and meat producers in particular (Aramyan *et al.*, 2009). Ultimately, meat prices will increase resulting in increased risk for the less affluent members of society to obtain a proper diet. European farmers are losing competitiveness because of the absurdity of not being able to grow the transgenic foods the commercialization of which is, nevertheless, authorised.

Beyond the borders of Europe, the negative attitude towards transgenic crops is seriously affecting food security in many developing countries. The Nuffield Council on Bioethics (2003) clearly states it is a moral imperative to make transgenic crops available to developing countries due to their potential to help alleviate hunger and malnutrition. Agriculture development in these countries has been seriously hindered in the last decade (Paarlberg, 2008). Many countries do not accept transgenic crops not to jeopardise their exports (many of them organically produced) to Europe. Even vitamin A-accumulating *Golden Rice* is still not commercialized in spite of the benefits it could provide for people whose staple food is rice.

Conclusions

Knowing that humanity faces a very important challenge to provide food for an ever increasing population on an ever decreasing good arable land, our proposal tries to combine low input agronomic methods to promote care for the environment, with the use of *certain* transgenic seeds to increase the productivity of these methods everywhere for the benefit of many. It is a proposal to encourage people to find the best methods to produce the necessary food without rejecting any of the tools we currently have, because humanity cannot afford to do so.

References

Altieri, M. (1999). Agroecología – bases científicas para una agricultura sustentable. Nordan-Comunidad, Montevideo.

Amman, K. (2008). Integrated farming: why organic farmers should use transgenic crops. New Biotechnology 25(2-3), 101-107.

Aramyan, L.H., van Wagenberg, C.P.A. and Backus, G.B.C. (2009). EU policy on GM soy – tolerance threshold and asynchronic approval. Accessed on 03/25/2010 at: http://edepot.wur.nl/7856.

Brundtland, G. (1987). Our Common Future: The World Commission on Environment and Development. Oxford University Press.

Dangour, A.D., Dodhia, S.K., Hayter, A., Allen, E., Lock, K. and Uauy, R. (2009). Nutritional quality of organic foods: a systematic review. American Journal of Clinical Nutrition 90, 680-685.

Denison, R.F., Bryant, D. and Kearney, T.E. (2004). Crops yields over the first nine years of LTRAS, a long term comparison of field crop systems in a Mediterranean climate. Field Crops Research 86:267-277.

Diouf, J. (2009). Agriculture to 2050: the challenges ahead. Inauguration of the High Level Forum on the future of food. Accessed on March 23rd, 2010 at: http://www.fao.org/news/story/en/item/36195/icode.

Gonsalves, D. (2004). Transgenic papaya in Hawaii and beyond. AgBioForum 7, 36-40.

Goulding, K.W.T and Trewavas, A. (2009). Can organic agriculture feed the world? Accessed on March 23rd, 2010, at: http://www.agbioworld.org/newsletter_wm/index.php?caseid=archive&newside=2894.
Huang, J., Hu, R., Rozelle, S. and Pray, C. (2005). Insect resistant GM rice in farmers' fields: assessing productivity and health effects in China. Science 308, 688-690.
James, C. (2010). Global status of Commercialised Biotech/GM Crops in 2009. Report accessed on March 23rd, at http://www.isaaa.org.
Jonas, H. (1985). The imperative of responsibility. University of Chicago Press.
The Keystone Center. (2009). Field to market: the Keystone Alliance for sustainable Agriculture. Report accessed on March 23rd, 2010 at: http://www.keystone.org/spp/environment/sustainability/field-to-market.
Mäder, P., Fliessbach, A., Dubois, D., Gunst, L., Fried, P. and Niggli, U. (2002). Soil fertility and biodiversity in organic farming. Science 296, 1694-1697.
Nuffield Council on Bioethics (2003). The use of genetically modified crops in developing countries: a follow-up discussion paper. Accessed on March 25th, 2010, at: http://www.nuffieldbioethics.org/go/ourwork/gmcrops/publication_313.html.
Paarlberg, R. (2008). Starved for science – how biotechnology is being kept out of Africa. Harvard University Press.
Raney, T. and Pingali, P. (2007). Agricultura sostenible. Investigación y Ciencia, Nov. 2007, 60-67.
Ronald, PC. and Adamchack, RW. (2008). Tomorrow's Table. Oxford University Press.
Zwart, H. (2000), A short story of food ethics. Journal AAE 12/2000, 123-124.

Nanotechnology and trust in aquaculture

B.K. Myskja
Norwegian University of Science and Technology,Department of Philosophy, NO7491 Trondheim, Norway; bjorn.myskja@hf.ntnu.no

Abstract

Salmon aquaculture has proven a valuable source of nutritious food production, but it is also a source of considerable controversy. One major challenge of fish farming concerns spread of diseases in dense salmon populations. The current crisis in Chilean salmon farming due to the ISA virus and the recurrent problems of salmon lice and other diseases in Norwegian and Canadian salmon farming, indicates that disease control is crucial for a trustworthy salmon production. Vaccination has proven to solve some of the disease problems in fish farming, but creating others, such as serious malformations, probably caused by the oil emulsion used in vaccination. This presentation considers a Norwegian research project seeking to avoid oil emulsion related problems by developing a vaccine using nanoparticles as vaccine vector. Such technological solutions to problems created by a technologically framed fish farming industry appear to raise the issue of public trust. A controversial industry known to induce a number of serious environmental and fish welfare problems may experience increased public distrust due to such attempts to solve its problems by means of yet another technological solution. The central concern in this presentation will be to discuss communicative strategies of scientific researchers developing the nanoparticle vaccine. It is often taken for granted that openness is essential for trust building. In cases like this there is a close connection between trustworthiness and transparency, as obscuring or downplaying the problems is morally unacceptable. But transparency can, intentionally or not, lead to misconceptions of the issues at stake. In addition, transparency may together with other forms of public engagement in science lead to a transferral of responsibility from science to society, because the public is given the task of accepting or rejecting applications of the new technology. The presentation will suggest some rules for responsible trust-building communication of scientific uncertainty based on this case.

Keywords: nanovaccines, salmon farming, trustworthiness, transparency, responsibility

Nanovaccines as solution and problem

Aquaculture promises nutritious and healthy food produced under environmentally sound conditions paying heed to fish welfare. Parts of the industry, such as salmon farming has been a commercial success, but also criticised on a number of accounts. The overarching criticism is that it does not represent sustainable food production. The salmon is a carnivore, and producing 1 kilo farmed salmon requires somewhere between 1.5 and 4 kilo wild fish, depending on whose estimate one trusts. Salmon escape threatens wild salmon stock in rivers in Norway, Britain, Canada and Alaska, with increasing numbers of wild salmon stock becoming extinct. This is due to escape followed by interbreeding with wild stock or the escapees superseding the wild salmon in the rivers and spawning grounds. Another factor that is argued to contribute to this decimation of wild fish stock but is also considered a major issue for fish welfare is the spread of diseases. Due to the narrow confines and rapid growth of the farmed salmon it is particularly susceptible to diseases. In the early years of farming, the use of antibiotics was high but still diseases were source of great loss. This is not only a grave concern for the industry, as the spread of diseases such as salmon lice in wild salmon is correlated with the proximity of salmon farms (Krkošek *et al.,* 2007).

In recent years the Chilean salmon farming industry has grown to become one of the major producers in the world, but this development came to a sudden halt when the ISA virus struck in 2007, probably imported via Norwegian salmon embryos, halving the production (Vike *et al.*, 2008). Clearly, diseases are problematic both to the survival of wild salmon and to the profits of salmon farming. High levels of antibiotics use pose a problem both for the industry, the environment and in the long run even for human health. The solution has been vaccination. Although vaccination cannot prevent the major problems such as sea lice, it has lead to significant reductions in antibiotics use and reductions in some of the major farmed salmon diseases. However, this is merely a partial solution as some bacterial and significant viral diseases are not prevented by today's oil-adjuvanted vaccines. In addition, it is documented that this vaccination has some negative side effects, the more serious being inner organ adherations and effects on growth and skeletal deformations leading to reduced product quality and, perhaps more significantly, to reduced fish welfare (Berg *et al.,* 2006).

A research project attempts to avoid these vaccination side-effects and increase the efficacy by replacing the current vaccination strategies with using PLGA nanoparticles as carriers of vaccine antigens. We still have little knowledge of the probability of success for this project, but the project illustrates some general problems concerning the use of new and unfamiliar technologies on important social issues such as food production. Salmon farming is of great economic significance in Norway and provides jobs for people in many coastal communities else threatened by depopulation. However, the industry is also controversial due to issues of fish welfare, pollution and its impact on the economically and culturally significant wild salmon – an iconic species in Norway. Solving one of these problems with an unknown technology associated with hope as well as fear, may cause serious problems with public trust.

Technology and society: a relation of trust

The relation between science and society has earlier been regarded as a contract, or a trust-based relation, where science is given resources and liberty to develop new solutions and improvements that are beneficial for society. Gibbons (1999) argues that this contract has been replaced by a new, demanding not only reliable but also socially robust knowledge. He holds that this new approach where science and society is jointly producing knowledge can be seen in some large technology projects. It is reasonable to argue that this kind of co-production perspective has been part of the establishment of nanotechnology as a field, exemplified in the Royal Society and Royal Academy of Engineering (2004) report. Gibbons points out that this new approach to knowledge production means a closer interaction between science, technology and society, including greater transparency regarding the potentials and risks involved in developing and applying the research. It is reasonable to say that the old authority based trust is gradually been replaced with a trust based on a combination of scientific authority and interaction or co-production.

The picture drawn by Gibbons and others arguing for this new perspective on the contract between science, technology and society can be considered a paradigm or a normative ideal rather than a description of reality. Many scientists conceive of their relation with society according to the old contract, insisting on the value of distinguishing clearly between producing knowledge and evaluating its applications. Even if one accepts that society must be involved in deciding what kind of research we should fund and under which circumstances new technologies should be used, it is reasonable to hold that the scientific work should be conducted in peace, assuming that scientific autonomy at this stage will yield the best results. Adding to this is the fact that most scientific and technological projects are fairly uninteresting to outsiders, and involving the public in these stages will be difficult and of little use in any respect. Thus there is still a stage of technology development that is conducted in relative peace, and the relation between science, technology and society can be described as one of trust. This trust-based relation is however precarious and may easily be eroded if some technologically related harm should occur or if the communication between technology spokespeople and the public fails. A paradigmatic example of the first case is how trust in nuclear energy production was greatly affected by the 1986 Chernobyl disaster, and how the BSE scandal in the UK affected trust in politicians and expertise on food production. Many biotechnology proponents have argued that the European resistance to GM food production is also an example of distrust due to communicative failures. Some, as Norman Borlaug (2000), argue that the public is misled by activists, others think that the public is ignorant but place at least some of the blame on the scientific community's inability to communicate their knowledge in a good way (Cook *et al.,* 2004). In both cases the dominant idea of communication is as a one-way exercise, where scientists teach the public about the realities of the technology. This idea of trust-building information from science to the public about technology is incompatible with the ideas of a new contract between science and society suggested by Gibbons, which sees two-way communication as an essential part of making socially robust science. Trust must be established by scientists listening to the arguments of the public in addition to presenting their own knowledge and views on the matters.

Trustworthiness as a moral obligation

Trust is necessary for a well-functioning relation between science and society, as long as we hold that we need new and better knowledge improving our lives, and we believe technology development is a crucial part of reaching that goal. But trust is more than a necessary element in a well-functioning modern technological society, as is the focus of functionalistic trust accounts (Luhmann, 2000). Trust is fundamentally descriptive of a moral relationship, where the trustor explicitly or implicitly gives the trustee responsibility in some particular regard. The trustee lives up to these expectations when she acts in a trustworthy way in this regard. If the public trusts that nanotechnologists will not release toxic particles into the environment, then nanotechnologists are trustworthy when they take all precautionary measures needed to prevent the release of toxic particles. Onora O'Neill (2002) argues that the Kantian idea of autonomy as expressed in the Categorical Imperative entails that the essence of morality is to avoid deception and coercion in the relation with other human beings. This is also the essence of trustworthy behaviour. Even if the sociological literature regards trust as a particular aspect of life in modern technological society because it is a necessary means for reduction of complexities and a way to deal with risks, trust is also a fundamental aspect of human morality. When someone is trusted they are expected to live up to the moral demands connected to having power over other people. We live up to the trust of others by behaving in a trustworthy way, which means not using the power we have to deceive or coerce others, but ensure that they retain their autonomy.

Now we are in a position to see the moral significance of the trust-based relation between science and society. The contract is not one of reciprocity, as the scientists are given a certain privilege through resources and liberty to pursue a certain line of research. In this way they, at least potentially, are given power over society while society expects to reap the potential benefits of the research. Although the political community has some sanctions to use against scientists who abuse the power, this possibility is seldom used and usually comes into play long after the abuse of power has taken place. For all practical purposes, this is a real trust-based relation, where the trustors to some extent are in the power of the trustee, and the trustee is under a moral obligation not to abuse this power. This picture is complicated by the fact that most research is not of a kind that give scientists power to harm members of society due to their research, regardless of its utility in the world beyond the laboratories. The only possible harm for many projects is that they are of poor quality and may take funding from more worthy recipients, as the project in itself only has impact as part of a collective research effort. But in principle every research project is potentially both beneficial and harmful, and the researchers are under the moral obligation of living up to society's trust.

Trustworthiness and transparency

If we return to Gibbons' idea of the new contract between science and society, one of the implications of this idea is that the old paternalistic role of science is replaced by the ideals of interaction and two-way communication. As stated above, this does not mean that members of the public are involved in all aspects of the technology development, but that they ought to be informed about all significant matters. '[R]eliable knowledge can only become socially robust if society sees the process of knowledge production as transparent and participative. The old image of science working autonomously will not suffice' (Gibbons, 1999: C83). For Gibbons this is primarily descriptive, capturing a necessary way to ensure a stable relation between science and society, but we should be aware that transparency is mandatory from a moral perspective on this relation. If trustworthiness means avoiding deception, transparency seems to be part of the moral requirements of the one who accepts the trust of others. Thus, the group developing nanovaccines to combat fish diseases and the fish welfare problems following conventional vaccination, are morally obliged to report all, including negative, results of their research.

There is however a paradoxical trust effect in connection with transparency. Being trustworthy means not deceiving the trustor, and this demand is perfectly fulfilled when the trustee is totally open regarding relevant information to the relevant audience. But transparency often leads to distrust, as the information provides an inadequate basis for public judgement. Only through second-order judgements about the first-order controls that are exercised by experts, can they have a basis for trust (O'Neill, 2006). Transparency in itself cannot provide basis for trust unless the trustors know and understand the real risks of the activity. In a highly specialised field as nanotechnology in fish health research, this is not possible. Even if the researchers attempt to live up to the standards of trustworthiness by being transparent about their research in all respects this may result in distrust in case the public fails to understand their communication or choose to weigh the information differently from the scientists.

Distrust based on poor communication or misunderstandings is not merely a theoretical possibility, but reality in many technology related issues. One of the standard examples is the fact that people believed and still to some extent believe that only genetically modified foods contain genes (Gaskell *et al.,* 2003) and presumably that this makes such food potentially dangerous. Communicating and understanding risk information related to new technology is furthermore difficult due to its complexities and the ideal of information neutrality (Schwennsen *et al.,* 2008), an ideal that can be connected to the new understanding of the relation between science and society. If we assume that trustworthiness entails complete transparency on all potential risks and uncertainties, it is reasonable to assume that this will increase rather than decrease trust in the scientific activity, even if the trust in the veracity of the technologists is unaffected.

Nanotechnology, transparency and trust

When the nanovaccine researchers present their project, they focus mainly on the technical challenges and the potential benefits of the research. They also see the need for transparency concerning possible risks and scientific uncertainties, as they employ a technology that still is novel and dealing in substances that are not easily controllable and known to have detrimental effects on health and environment (The Royal Society and the Royal Academy of Engineering, 2004). Under the new paradigm for the relationship between science and society, research in materials with these kinds of properties is a matter where all affected parties have a right to participate in regulatory debates. The particles employed in this project are degradable and no harmful effects are known, but lack of experience with new materials on the nanoscale means that unforeseen harm cannot be excluded. The potential benefits of nanomaterials are connected to the altered characteristics due to their size, and these characteristics may also be sources of harm. The particles may interfere with other natural processes than intended, have other biological properties than their larger counterparts, and may more easily be taken up by cells and cross natural barriers (Myhr and Dalmo, 2007). The issues on how to communicate these uncertainties about possible harm is complicated by the fact that they already exist in nature. We are exposed to nanoparticles already, some of them quite harmful (*ibid.*). But the fact that these are natural substances does not diminish our moral responsibilities when we intentionally introduce new particles of this kind, perhaps into new contexts. When humans interfere with nature we cannot hide behind nature's own risks. We carry responsibility for our deeds, even when they replicate natural phenomena. As long as we lack sufficient experience with these materials released in nature, we cannot reject the possibility of unforeseen harm.

The nanotechnologists may believe and communicate that unexpected harm to nature due to the release of these degradable organic nanoparticles is highly unlikely, but they have a duty to communicate their knowledge, not their assumptions only. Then they have to admit that the effects of the altered properties of nanoparticles are not fully explored and that harmful consequences to the environment are at least theoretically possible. This transparency may lead to reduced public trust, despite the fact that it makes the scientists more trustworthy. The reason is simply that it is reasonable to relate the potential harm to the benefits. As long as the benefits are negligible or perceived as irrelevant for most people, the harm is unacceptable. This is especially important when the harm is a matter of uncertainty, because then it is possible that the harm is more significant than what is considered scientifically plausible – even if we know that the opposite usually is the case. This situation poses a dilemma for the nanotechnology researchers: Should they downplay the risks and uncertainties or should they accept the possible public rejection of their research due to a – in their eyes – exaggerated risk perception? In the first case they more or less fail to fulfil the transparency obligation and become less trustworthy while making the technology more trusted. In the second case they are trustworthy but may fail to ascertain the kind of trust in the technology they hold to be appropriate.

This dilemma could be solved in several ways. One is to engage with the public in dialogue through different forums such as consensus conferences, arguing that the relation between benefits and potential harm is appropriate, and meeting their concerns and counterarguments. Although this is the politically correct approach given the renegotiated contract between science and society, it is a time-consuming exercise that has no guarantee of success. Another approach would be to acknowledge the risks and uncertainties in relation with the nanovaccines, and then present ways to deal with these problems. First they must present the benefits of the use of the nanoparticles as vaccine vectors. Are the harmful consequences of traditional vaccination avoided, and do these new vaccines provide better disease protection? The new vaccines should not be introduced unless they have fewer negative side-effects and provide better protection compared with the traditional vaccines. Then they should present the current knowledge concerning the potential harm from releasing these particles into the environment. Are there forms of harm that can be subject to standard risk assessments, and if so, what are the risk estimates? Furthermore they should present plausible worst case scenarios given harmful effect of release, and suggest how to deal with these potential harm scenarios. They should furthermore say what kind of research would be needed in order to ascertain that we can control the negative consequences. Then they should open a public discussion through the media engaging expertise and ethicists as well as the general public. The issues should be whether the benefits are such that these measures to prevent the harm are acceptable. Usually such debates fail to achieve consensus, but some arguments tend to be better than others. These arguments should be basis for decisions on how to deal with the scientific uncertainties concerning harm from using the nanovaccine in open fish farms. In this way the researchers remain trustworthy and may still be trusted by significant parts of the public. And they receive advice on how to utilise their research. As such debates tend towards conservative conclusions, the advice will probably be of a precautionary kind. This kind of conservatism is usually more in keeping with the principle of sustainability, although one can ask to what extent a sustainable strategy for nanovaccines in fish farming is a rational goal for these discussions. The more important question is perhaps whether fish farming in itself can become sustainable.

References

Berg, A., Bergh, Ø., Fjelldal, P.G., Hansen, T., Juell, J.E. and Nerland, A. (2006). Dyrevelferdsmessige konsekvenser av vaksinasjon av fisk – effekter og bivirkninger. Animal welfare and fish vaccination – effects and side-effects. Havforskningsinstituttet, Bergen. Available at: http://www.imr.no/filarkiv/2003/12/Nr.9_2006_Dyrevelferdsmessige_konsekvenser_vaksinasjon_av_fisk_effekter_og_bivirkninger.pdf/nb-no. Accessed 24 March 2010.

Borlaug, N. (2000). The Promise of Biotechnology and the Threat of Antiscience Zealotry. Plant Physiology 124: 487-490.

Cook, G., Pieri, E. and Robbins, P.T. (2004). 'The Scientists Think and the Public Feels': Expert Perceptions of the Discourse of GM Food. Discourse & Society 15: 433-449.

Gaskell, G., Allum, N. and Stares, S. (2003). Europeans and Biotechnology in 2002. Eurobarometer 58.0. Available at: http://ec.europa.eu/public_opinion/archives/ebs/ebs_177_en.pdf. Accessed 24 March 2010.

Gibbons, M. (1999). Science's new social contract with society. Nature 402: C81-C84.

Krkošek, M., Ford, J.S., Morton, A., Lele, S., Myers, R.A. and Lewis, M.A. (2007). Declining Wild Salmon Populations in Relation to Parasites from Farm Salmon. Science 318: 1772-1775.

Luhmann, N. (2000). Vertrauen. Ein Mechanismus der reduction sozialer Komplexität. Lucius & Lucius, Stuttgart, 140 pp.

Myhr, A.I. and Dalmo, R.A. (2007). Nanotechnology and Risk: What are the Issues? In: Allhof, F., Lin, P., Moor, J. and Weckert, J. (eds.) Nanoethics. The Ethical and Social Implications of Nanotechnology. Wiley, Hoboken, N.J., pp. 149-159.

O'Neill, O. (2002). Autonomy and Trust in Bioethics. Cambridge University Press, Cambridge, 213 pp.

O'Neill, O. (2006). Transparency and the Ethics of Communication. In: Hood, C. and Heald, D. (eds.) Transparency: The Key to Better Governance? Oxford University Press, Oxford, pp. 75-90.

The Royal Society and The Royal Academy of Engineering (2004). Nanoscience and nanotechnologies: opportunities and uncertainties. The Royal Society, London, 116 pp.

Schwennsen, N., Svendsen, M.N. and Koch, L. (2008). Beyond informed choice: Prenatal risk assessment, decision-making and trust. Etikk i praksis. Nordic Journal of Applied Ethics 2 (1): 11-31.

Vike, S., Nylund, S. and Nylund, A. (2009). ISA virus in Chile: evidence of vertical transmission. Archives of Virology 154 (1): 1-8.

BENEFISH: benefits and costs of welfare interventions in aquaculture

C.M. Mejdell[1], S. Kadri[2] and B. Damsgård[3]
[1]National Veterinary Institute, P.O. Box 750 Sentrum, 0106 Oslo, Norway; [2]Trans-National Consulting Partnership, 34 Lawrence St. 1/1, Glasgow, G11 5HD, United Kingdom; [3]Nofima Marine, P.O. Box 6122, 9291 Tromsø, Norway

Abstract

There is increasing agreement in Europe that a high level of animal welfare, including fish welfare, is an ethical value that can be regarded as a public good. Improved welfare will often however involve extra costs for fish farmers, whereas economic benefits may be less visible. The BENEFISH project has developed a bio-economic model, which provides a tool for estimating the overall monetary outcome of actions taken to improve fish welfare. The model can handle all costs, including capital investments and operational costs, and benefits including direct effects on production as well as any added value through increased willingness for business partners and consumers to buy or to pay a premium. Degree of uncertainty can be integrated, and analysis can be done for single farms or the whole industry. Visualisation of the economy of welfare improvements may motivate fish farmers to prioritize welfare. Further, the tool may be used to select the more cost-effective welfare actions, and as a decision support tool for policy makers and legislators.

Keywords: fish welfare, welfare actions, bio-economic modelling, consumer

Introduction

There is a growing concern among citizens, scientists, farmers, and legislators for the welfare of production animals in general. There is also increasing agreement that a high level of animal welfare is an ethical value that can be regarded as a public good. Over recent years, these reflections have been expanded to include farmed fish in Europe, such as Atlantic salmon, rainbow trout, sea bass, sea bream, eel and carp. This is illustrated by the pan-European work of the Scientific Animal Health and Animal Welfare Committee (AHAW) under the European Food Safety Authority (e.g. EFSA, 2008), and also by national legislations on fish welfare.

In 2007, the European Union funded the research project BENEFISH: *Evaluation and Modelling of Benefits and Costs of Fish Welfare Interventions in European Aquaculture*. The aim of the project was to develop a bio-economic analysis model demonstrating how actions taken to improve fish welfare may influence factors such as production costs, productivity and consumer behaviour, and hence predict the economic consequences of improved fish welfare in aquaculture (Berrill *et al.*, 2009).

The concept of animal welfare

Animal welfare describes the status of individual animals, but there is up till now no universal agreement on a *single* welfare definition. Suggested definitions can be coarsely grouped into three categories according to their focus: Biological function; a natural life; or the animal's affective state, or various combinations of these categories (see for example the review by Huntingford *et al.*, 2006). These views can be seen as expressions of underlying values (Fraser, 2003). Among citizens, it seems that the association between animal welfare and a natural life is widespread, whereas among farmers, scientists and veterinarians the biological function view is more common. In practice, welfare researchers should be aware of these differences and measure welfare by using various welfare indicators, preferably covering more than one aspect of the welfare concept. Such indicators may include health parameters (e.g. disease, injury), behavioural parameters (e.g. movement, abnormal behaviour), physiological parameters (e.g. stress hormone levels, respiration) and environmental parameters (e.g. water quality, management factors).

All welfare indicators should provide a valid reflection of welfare, be repeatable and comparable between farms. For on-farm surveillance of fish welfare, the welfare indicators should be relatively simple and inexpensive to monitor, and thus, a welfare assessment will often concentrate on a few key welfare indicators. These are often called Operational Welfare Indicators (OWI). Examples of OWIs include mortality, deformities, fin damage, swimming behaviour, feed efficiency and oxygen levels, and OWIs will differ between for example rearing systems, fish species, life stage and the degree of production intensity.

The BENEFISH project

BENEFISH aimed to produce a scientific basis for comparing the biological effects, and the direct and indirect monetary consequences, of various welfare measures through the development of high level bio-economical decision analysis models.

Welfare indicators and welfare interventions

The first step to improve fish welfare is to identify and quantify potential welfare threats. Whilst some threats to welfare are broadly generic and relate to more than one species, rearing system or life stage, in most cases threats to welfare will be species and system specific. For example, a low concentration of dissolved oxygen in water is a threat to welfare for most species of fish, but the actual levels that are harmful will vary between species and life stages. Consequently, any action taken to prevent unwanted low oxygen levels must be specific, and it is thus not possible to provide a generic welfare action.

In BENEFISH, four specific OWIs were selected: mortality, fin damage, water CO_2-levels, and deviations from expected patterns of feed intake. These cover different aspects of welfare, i.e. animal health, behaviour, and their environment. They are also widely applicable across European aquaculture species and systems, and reflect data which is often routinely monitored by producers. A range of case studies using these OWIs was selected, covering different European farmed species and systems. A welfare intervention or action can be defined as any technological or methodological step that is taken to improve fish welfare, and for each case one or several interventions were chosen and analysed. For example, increased water flow rate in flow-through rearing systems can be an intervention to alleviate poor water quality, which might be negatively impacting welfare, as identified by e.g. high mortality, low appetite or high incidence of gill infections. Alternative interventions to improve water quality can be to treat the water or to reduce fish density. In BENEFISH the initial case studies included for example: interventions to improve fin condition, where fin damage in Atlantic salmon may be due to a range of farming conditions. Furthermore, to reduce water CO_2-levels for juvenile Atlantic salmon in Norway, the CO_2 was actively removed from the water, using a stripping machine. To reduce mortality in rainbow trout farming in UK, diploid fish were replaced by triploids. To improve feed intake of flatfish kept in recirculation systems in the Netherlands, water quality was improved by extended water treatment (protein skimming and ozone treatment) for sole and increased specific flow rate for turbot (Noble *et al.*, 2010).

Change in OWIs and production measures

The next step in BENEFISH was to assess changes in the OWIs as a consequence of the welfare actions, e.g. less fin damage or lower concentration of CO_2. In addition to this, the direct effects upon farm productivity was estimated, measured as potential improvement in production. Three widely used and recognised productivity measures were used, which have a direct impact on the economical performance of the farm: growth rate, feed conversion ratio (FCR) and mortality. Poor growth reduces the farms' profitability primarily by extending the length of time required to grow fish to a market size. FCR is a measure of the efficiency with which feed is turned into body tissue. Feed costs account for up to 60% of a farms total cost outlay (Sveier and Lied, 1998), and inefficiency in feed use will rapidly reduce the economical outcome of a farm. Mortality affects overall production biomass and leads to direct economic losses for the farmer. However, the impacts on productivity depend on when in the production cycle mortalities occur. Losses of fish in the early stages of the life cycle have a relatively low impact, while when fish which are close to being harvested die, considerable effort and financial resources (e.g. feed, labour) have been spent and mortality will then have a considerable effect on productivity.

There are also costs to the farm connected to the welfare interventions and which must be included in this type of modelling exercise. Such costs typically include technical investments, changes in running costs, and increased costs for monitoring the OWIs.

Product quality and added value

Product quality can be directly influenced by welfare related factors such as disease or stress, which can lead to effects upon intrinsic quality traits such as shelf life, fillet texture, colour, and taste. In addition, the perceived or extrinsic quality may be affected, i.e. where the consumer prefers fish reared under welfare friendly conditions, which may lead to a willingness to pay a premium, or result in more people willing to buy fish (e.g. Olesen *et al.*, 2009). These mechanisms may enhance the value of the fish, at both the business-to-business and consumer levels. In BENEFISH, industry partners in the project were consulted on these issues, and in addition, interviews with producers along with tests of consumers' willingness to buy and pay in restaurants were used to estimate added value of a more welfare friendly fish product.

Welfare actions may not be value free

Not all welfare interventions are necessarily viewed as good or even value neutral. The action itself could be regarded as either morally right or wrong, independent of whether the OWI is improved. This is well known from the laying hen case, were the public does not consider battery cages as an appropriate action to prevent cannibalism, as cages may prevent birds to display natural behaviours. Similar concerns may arise for the triploid strategy to reduce mortality in farmed fish (Kankainen *et al.*, 2010). Because the procedure of making triploid fish can be regarded as unnatural or artificial, implicitly considered to be morally wrong, some people will object to it. On the other hand, others may welcome this particular action due to a side effect that they consider beneficial: triploids are sterile and thus triploid escapees do not represent any danger to the genetics of the wild population. Variations in public opinion are likely to have a significant impact on consumers' willingness to buy or pay for such a product. This has implications for bio-economic modelling of the consequences of specific interventions. In the BENEFISH project, consumers were asked about improved fish welfare in general as well as in relation to the welfare indicators, without detailed information on specific interventions.

Utility

The total utility of a welfare intervention is the sum of all costs and benefits that result from the intervention, i.e. from implementation of a welfare action, productivity factors, costs regarding monitoring and auditing changes in welfare, and any added value in the supply chain from retailers to consumers. This total utility was first calculated for an average company within the respective sector (i.e. the farm, micro level) and then multiplied up to the sectoral (macro) level.

There will always be a certain degree of uncertainty for some costs and outcomes, and these can be fed into the model as probability distributions. The probabilities can be defined according to the data source, varying from experimental data or statistical standard errors to expert statement uncertainty.

Using the model in practice

Modelling the bio-economic consequences of welfare interventions requires an understanding of the causes and consequences of relevant welfare threats and quantification of any relationships between them. This in turn allows quantification of any improvement in welfare (measured as change in one or several OWIs) which results from welfare interventions proposed to address such threats. To be effectively and successfully modelled, the identified threats and their corresponding welfare interventions must be clearly defined and unequivocal. It is necessary to establish suitable datasets, sufficiently robust for analysis. Such datasets can be either experimental or commercial in origin, but they must be applicable to on-farm conditions.

Once a suitable intervention has been established, it is important to be able to quantify the efficiency with which that action improves welfare, as measured through a change to the OWI, in order to allow for incorporation into the bio-economic model. This information may be available from the dataset used to identify the welfare action, or it may be available elsewhere, but it is vital for bio-economic modelling. The development of effective and practical welfare actions involves scientific interpretation of the identified welfare threats, often combined with consultation with farmers and welfare experts.

The bio-economical model is a tool to visualise the economy of welfare improvements. This will hopefully motivate farmers to prioritize fish welfare and to select the more cost-effective welfare actions. The model may also provide a valuable tool for policy makers and legislators to gauge the potential economic consequences of a change in welfare regulation.

Acknowledgements

This study has been carried out with financial support from the Commission of the European Communities, specific RTD programme 'Specific Support to Policies', SSP-5-FISH 'BENEFISH'. It does not necessarily reflect its views and in no way anticipates the Commission's future policy in this area.

References

Berrill, I.K., Kadri, S., Ruohonen, K., Kankainen, M., Damsgård, B., Toften, H., Noble, C., Schneider, O., and Turnbull, J.F. (2009). BENEFISH: A European project to put a cost on fish welfare actions. Fish Veterinary Journal 11: 23-38.

EFSA (2008). Scientific report on animal welfare aspects of husbandry systems for farmed Atlantic salmon. The EFSA Journal 736: 2-122.

Fraser, D. (2003). Assessing animal welfare at the farm and group level: the interplay of science and values. Animal Welfare 12: 433-443

Huntingford, F.A., Adams, C., Braithwaite, V.A., Kadri, S., Pottinger, T.G., Sandøe, P. and Turnbull, J.F. (2006). Current issues in fish welfare. Review paper. Journal of Fish Biology 68: 332-372

Kankainen, M., Berrill, I.K., Noble, C., Ruohonen, K., Setälä, J., Kole, A., Mejdell, C.M., Kadri, S. and Turnbull, J.F. (2010). Modeling the economical utility of improving welfare in fish farming – a case study from the UK rainbow trout industry. Aquaculture Economics and Management, submitted.

Noble, C., Berrill, I.K., Waller, B., Kankainen, M., Setälä, J., Honkanen, P., Mejdell, C.M., Turnbull, J.F., Damsgård, B., Schneider, O., Toften, H., Kole, A.P.W. and Kadri, S. (2010). A multi-disciplinary framework for bio-economic modelling in aquaculture: a welfare case study. Aquaculture Economics and Management, submitted.

Olesen, I., Alfnes, F., Røra, M.B. and Kolstad, K. (2009). Eliciting consumers' willingness to pay for organic and welfare-labelled salmon in a non-hypothetical choice experiment. Livestock Science, in press. doi:10.1016/j.livsci.2009.10.001.

Sveier, H. and Lied, E. (1998). The effect of feeding regime on growth, feed utilisation and weight dispersion in large Atlantic salmon (*Salmo salar*) reared in seawater. Aquaculture 165: 333-345.

Food sovereignty, a right to life: the case of the Peruvian Andes

L.A. Hallazi Méndez
Universidad Autónoma de Madrid, Spain

Abstract

It is time for the community that defends human rights to take note of a pattern that has been developing the world over: diverse communities that live in certain harmony now have to live in fear of socioeconomic and political threats arising from governments that are partial to transnational enterprises. Governments that try to generate dependency in the various agricultural stages: production, distribution and consumption, whether it be from the management of seeds to the final consumption of genetically modified organisms (GMO) called transgenic organisms. In this vein, this article will briefly analyse the case of the Peruvian Andes (Puno), where agriculture is characterised by peasant farming, and the resultant food insecurity that would be generated by the recent governmental policies such as free trade agreements (FTAs) with the United States and China. The analysis further ties in such considerations as intellectual property, the control of imports and exports, the production and distribution of food, and the protection of biodiversity. Therefore, what is intended in these brief analyses is to search for mechanisms for defending the right to food sovereignty, like the right of the population to define its own sustainable policies, guaranteeing the population's right to food using small and medium producers as its foundation, thereby respecting both culture and the diverse indigenous and peasant methods of production, agribusiness commercialisation and management in rural spaces.

Keywords: free trade agreement, intellectual property, biofuels, agrofuels, transgenic organisms

Regional antecedents

When we refer to that part of the Peruvian territory, reference is made to one of the most difficult zones of the Andes in terms of liveability. That said, despite the climatic conditions, populations have established themselves in this area from time immemorial, managing to develop themselves in harmony with the assigned space. The southern Andes of Peru – which contains the territories of Puno and others, such as Cusco and Arequipa – is principally located in the region known as Quechua and Aymara, a region of cold, temperate climate. Here exist abundant natural pastures where stockbreeding is extensive. Medium-sized estates predominated, surrounded by communities and rural plots where agriculture, livestock farming and mining are the principal economic activities. For this reason, it is deduced that, in addition to being a region of rural agricultural production, agriculture is the mainly the product of two of the most numerous indigenous populations, namely the Quechua and Aymaras. It is from this point of view that we can measure the impact that the imposition of FTAs will have.

In summary, we can indicate the following as defining elements of the southern Andean region of Puno: it is a region of long historical formation with the predominant indigenous populations being the Quechua-Aymara-speaking groups; of equal predominance is the agribusiness sector which contrasts with the high Andean altiplane, a terrain that in large part characterises the department of Puno. Further, we can add that, from the 19th century, the southern region of the Andean Puno passed through four great periods: the first, between 1885 and 1930, characterized by the expansion of the 'commercial houses' of Arequipa and the large Cuzco and Puno estates, whose foundation was the extensive exploitation of the pastures for livestock, using manpower sourced from the indigenous populations from the interior; the second period, between 1930 and 1945, was characterized by attempts at capitalist modernisation based upon the introduction of technical improvements at the level of the large estates, and the attempt to proletarise the agricultural peasantry, the majority of which was indigenous. A third period, between 1945 and

1980, is defined by increasing urbanisation, the breakdown of the large-scale landholder property and of the chieftain society, which has its climax the application of agrarian reform in 1969 under the government of General Juan Velasco Alvarado and, with the reform, the different 'claims' on land on the part of the peasant farmers. A fourth period, which extends from 1980 to 2000, is characterised by social de-structuring at various levels and has to do with political violence unleashed by the Partito Socialista de Peru Sendero Luminoso and the repression of the Armed Forces which placed the society at its most unprotected state. Finally, a fifth period may be included and it is understood as this latest decade. This latest period must be observed from a perspective that is integrated and multidimensional i.e. regional, national and international, and implies visualising a basically economic globalisation, where regulated markets from the Northern Hemisphere try to induce market liberalisation in Latin America, thereby consolidating asymmetric trade relationships.

Free trade agreements and human rights

When the United States failed to see World Trade Organisation (WTO) negotiations concretised with the rapidity they hoped for, the US promoted Free trade agreements (FTAs) as the way of forcing trade along the 'fast track'. Hence, they began to prompt negotiations for the formation of regional free trade blocs. The FTAs form part of a complicated story of international and regional economic and financial bilateral conventions and treaties that have subordinated or supplanted the basic instruments of international law and regional human rights, national constitutions, legislation directed towards national development and the social and labour laws tending to mitigate inequality and exclusion. In the public's opinion, bilateral treaties are characterised as having low visibility with many final agreements celebrated in secret. Consequently, they are even more prejudicial to the rights of the populace than existent and prospective regional and international treaties. We must keep in mind that bilateral trade agreements include treaties of promotion and protection of foreign investment (TPPI), free trade treaties, intellectual property rights, import-export treaties, and cooperation in science and technology with one clear objective: to benefit the stronger economy.

The States are those that celebrate the treaties, but the rights that they afford are conferred by others; if they guarantee anything, it is the private firms' investment including the relative dispositions of the same to resolve controversies, such as the international responsibility of the receiving State for damages incurred.

If one remembers the Zapatista uprising in 1994, it will be recalled that it coincided with the launch of the North American Free Trade Agreement (NAFTA). The current consequences are disastrous. Two years after the agreement, Mexico imported 43 per cent of its food, reducing consumption of the same to 30 per cent. This is related too to the mounting boom in agricultural biofuels where 80 per cent of the world's corn is controlled by US firms – which in turn explains the corn crisis in Mexico where increases reached as high as 400 per cent. These effects of trade liberalisation are clear. Here production is controlled by the same firms, adding to this the commercialisation of corn for ethanol production. Fifteen years after the FTA in Mexico, the small and medium-sized agriculturists have not been able to withstand transnational companies or compete with subsidised US producers and have decided to abandon the field.

Perhaps there is something that makes us believe that the same will not occur in Peru. Peru's buying into the FTA with the United States will have different, significant costs for the country's economy and society. Costs will be even higher for a region like Puno where 60 per cent of the population is rural, with 66 per cent of the population in poverty and 30 per cent in extreme poverty; where there is small scale agricultural production, where 81 per cent of the agribusiness units own less then 10 hectares, the majority being villagers, plot owners and small producers. If the organisation of the production here is usually referred to as nascent, developing or rudimentary, because it maintains traditional methods of production and distribution (such as tools for cultivation, communal fairs for barter, etc.) we ought to be aware that this does not constitute a disadvantage, given that the principal objective of these farmers is food autonomy. Therefore, the problematic issue is not in the low indices of profitability and competitiveness, but in the possibility of providing minimal conditions for facing the sustainability of production and consumption of our own products, and that, furthermore, allow for improvement of the life of the rural producer, considering that he is the subject of and principal actor in what relates to his own wellbeing; where the State must protect his right to his own self-sustenance and must promote the public policies necessary for such assignment.

The FTAs, therefore, have effects that are particularly serious for the case of the agriculturist from the regions of the Andean range, who has been one of the greatest sacrifices in FTA negotiation. FTAs disallow price bands, the absence of which leaves the local farmer on the fringes without protection, in that he directly faces North American and Chinese producers who are the subsidised beneficiaries of internal assistance; these conditions will introduce unfavourable competition with artificially reduces prices that leave national producers without possibilities, and force them into consuming imported products. It is absurd to think of some sort of compensation be offered by the government and it is impossible to offer compensation to all the agriculturists.

The region of the South Andes also contains an important flora biodiversity and traditional knowledge. However, this is matters little to the negotiating government and the result is the vulnerability of the same. By this we mean that the FTA will permit these practices to increase with many of these plants and the associated traditional knowledge being patented by transnational firms. These patent rights configure an important risk for the sovereign food security of the region, such as has occurred in other agricultural communities and regions where they could not continue with the control and management of seeds. The management of seeds is the first link to be taken over in controlling the food system and patents are the legal way of doing so, given that they challenge rights to keep, to exchange and to multiply seeds in a free and autonomous way, not only of agriculturists but of everyone who wants to secure his own food. The right of farmers in lessened in two ways, according to Vandana Shiva: one, via legislation regarding seeds that obliges the use of 'registered' varieties and, two, through 'intellectual property rights' that obliges farmers to renounce their rights to maintain, exchange and improve seeds – finally the small farmers cannot assume the cost of registration and, little by little, see themselves grow dependent upon the large industry.

From here, genetically engineered or genetically modified organisms (GMO) come into play. They flood the market both in production and consumption to such an extent that our traditional methods of cultivation will be supplanted by these kinds of products, resulting in a truncating of our biodiversity in favour of the boom in the demand for agrofuels so well publicised by the current president of the government. In the Peruvian case, the boom is accompanied by growing infrastructural integration, which implies transport and exportation, a mega agricultural fuel project which forms the axis of Peru-Brazil-Bolivia integration, and the construction of a hydroelectric dam on the Brazil-Bolivia border, the production of biofuels (like soy) and access routes facing the Pacific (such as a highway and ports) in Southern Peru. Brazil has been the worst example of sustainable agriculture, in that, in 2007, the sugarcane harvest grew 7 per cent and the prices fell more than 35 per cent because there is no market to absorb the excess demand; the excess crop in turn is converted into a way of making the situation precarious for indigenous and rural labour who lose their means of production and livelihood.

Regarding measures of environmental protection in free trade agreements, it remains clear that, apart from good intentions and rhetorical declarations, no manifest commitment exists for applying and improving environmental standards, with the possible exception of those that directly affect trade.

Finally, in the departments of Arequipa, Puno or Apurimac, the production of sensitive crops is markedly superior to the production of exportable crops.

Therefore, this regional analysis reveals that the departments of the mountain range and forest zones would be those most negatively affected with the signing of the treaty. By this, the segments of the rural mountain-forest population remain highly vulnerable. Regardless, we hear enthusiastic and demagogic declarations from current President Alan Garcia with regard to FTAs with the United States, China and the European Union – declarations such as, 'Peru, with these treaties, reaches 2,100 million people throughout the world, who will now receive our products free of tariffs and taxes which are an obstacle to the capacity of Peruvians'. There is nothing sadder and more demagogic.

In search of food sovereignty

It is clearly shown that the problem of agricultural and free trade agreements is not a priority in Peru: rather, it is evidently *not* important and, therefore, it will not strengthen the large scale production of food. By contrast, the cultivation of agrofuels and food importation is proposed and the market will be flooded with foreign products. There are no great plans or programmes for food sovereignty in Peru. Enthusiasm revolves around celebrating larger numbers of free trade agreements while, surely, attention is centred on areas of high biodiversity and access to water.

Concretely, our practices in the area of food should be aimed at questioning the methods of rural production and the methods that are linked to trade. The history of Peru has not recognised rural/peasant groups and their productive contribution – not only to themselves – but to the entire country. The peasant farmer in this area is isolated, and may be separated from the national dynamic, silenced and extinguished. Our theme of sovereignty is linked to the theme of peasant production. It should be noted that, the peasant has always produced food, but not in the romantic way of feeding everyone; rather, he produces under forms of domination from owners to whom he submits his produce, not as an owner of the soil but as an exploited worker. Solidarity with peasant farmers delves into redefining the ownership of the means and factors of production; passing through the market and producing food outside of systems of domination, outside of close and closed circuits involving those who currently manage the circulation of produce. Sovereignty not only implicates those who buy from farmers and how but it also looks at the problem globally, including infrastructure and production means.

In order to achieve popular food sovereignty, it is necessary to work for: the integral autonomy of producer communities, networks of free trade, placing consumers and peasants in direct relationship, ecological agriculture, respect for natural cycles, recovery of seeds, strengthening of local economies and solidarity without being joined to the State and the policies that it promotes, attempting to break the concept of food as a merchandise, among other things. To construct food sovereignty, it will be necessary to recognise 'how' what we eat is produced, thereby drawing us close to other circles – the very producers themselves – who discuss the theme of autonomy.

The first fundamental right is the right to life. This implicitly signifies the right to not go hungry and the right to earn a living. Building upon this premise, the search for food sovereignty will have to establish mechanisms of protection for land-use in favour of small agriculturalists and peasant farmers, the free circulation of seeds, the protection of ecological diversity (including cultural diversity), the management of imports and exports propelled by FTAs, the destruction of production and distribution monopolies in food, the regularisation of patents in favour of the citizen, the fight against transgenic organisms and the denouncement of personal deterioration (i.e. illnesses caused by chemical pollution) and denouncement of deterioration of the natural environment (i.e. the physical and biological degradation of soils against the mandatory use of unnecessary technologies, etc.). It is therefore the task of the Peruvian state to guarantee these conditions in a world where globalization plays in favour of the economic capacity of actors with greater economic power.

References

Anonymous, undated. Documento Oficial Tratado de Libre Comercio http://www.tlcperu-eeuu.gob.pe/index.php.
Calle, A. (2005). Nuevos movimientos globales: Hacia la radicalidad democrática. Ed Popular Madrid.
Desmarais, A.A. (2007). La Vía Campesina. La globalización y el poder del campesinado. ed. Icaria. Madrid.
MAELA (2000). Perspectivas del movimiento agroecológico latinoamericano en el nuevo milenio. Cochabamba. Bolivia: AGRUCO.
Martinez Alier, J. and Ropke, Y. (2008). Recent Developments in Ecological Economics, 2vols., Edward Elgar.
Shiva, V. (1996). Future of our seeds, future of our farmers: Agricultural biodiversity, Intellectual Property Rights and Farmers' Rigthts, p. 170.
Shiva, V. (2003). ¿Proteger o expoliar? Los derechos de propiedad intelectual. ed. Octaedro. Madrid.138p.
Vargas, M. (2009). Agrocombustibles: ¿Otro negocio es posible?. ed. Icaria, p. 130.

Part 3. The role of local breeds and traditions in sustainable agriculture and aquaculture

Traditional part-time grazing: a more sustainable sheep milk and cheese production in the Basque country

J.R. Mauleón[1], L.J.R. Barrón[1], M. Virto[1], N. Mandaluniz[2], G. Amores[1], F.J. Pérez Elortondo[1], J. Arranz[2], E. Abilleira[1], I. Beltrán de Heredia[2], J.C. Ruiz de Gordoa[1], A.I. Nájera[1], R. Ruiz[2], M. Albisu[1] and M. de Renobales[1]
[1]Faculty of Pharmacy, University of the Basque Country UPV/EHU, Spain; [2]Neiker Tecnalia, 01080 Vitoria-Gasteiz, Spain; mertxe.derenobales@ehu.es

Abstract

The concept of sustainability when referring to food production rests on three main aspects: (1) respect for the environment; (2) economic and social benefits for all involved in production; and (3) production of sufficient quantity of healthy food at an affordable price. We focus herein on those aspects of the traditional sheep milk and cheese production (under the Denomination of Origin Idiazabal Cheese) in the Basque Country of primary relevance for the sustainability of this production. It is based on the local *latxa* or *carranzana* breeds, adapted to mountainous terrain. Sheepherders take advantage of local resources to reduce management costs by combining indoor dry forage and concentrates with outdoor grazing throughout lactation, according to local pasture availability. This system facilitates recycling of manure, fertilising pastures and forests at the same time. Cheese is produced within an Industrial Food System (51.7% of the total cheese produced in 2008), or within an Artisanal Food System (48.3%; by the sheepherders using the milk from their own flocks or by a sheep farmer association). The Artisanal System is advantageous for the following reasons: (1) there are higher economic returns because the cheese price directly sold to consumers is more competitive than industrial cheese sold in supermarkets; (2) it increases the value of women's work (over 80% of the cheese makers are women) in the community and their self-esteem; (3) it creates rural jobs and contributes to rural development; (4) we have demonstrated (experimental and commercial flocks) that part-time grazing allows the sheep farmer to obtain high yields of milk, and cheese, of high nutritional and functional quality. Currently, a less sustainable, intensive sheep milk production with foreign imported breeds kept constantly indoors is gaining favour among milk producers because of its perceived higher economic profitability

Keywords: outdoor feeding, social sustainability, healthier food product, food systems

Introduction

The best known definition of 'sustainability' is perhaps that of the Brundtland Report (1987) which extended the concept of sustainability to all activities that lead to the development of a society by stating that it *consists in meeting the needs of the present without compromising the ability of future generations to meet their own needs.* References regarding the concept of sustainable development of agricultural food producing systems are diverse. Paniagua and Moyano (1998) propose the definition accepted in the present work. An agricultural system is sustainable when in the long run it guarantees the following:

1. respect for, or equilibrium with, the environment, maintaining its quality (this is the environmental dimension);
2. economic and social benefits for all those involved in production (socio-economic dimension); and
3. access to sufficient quantity of healthy food at an affordable price, respecting cultural preferences for certain foods (consumer dimension).

In this paper we study the sustainability of sheep cheese producing and marketing systems in the Basque Country region of Northern Spain, focusing on the second and third dimensions of our concept of sustainability because the environmental and animal welfare aspects of outdoor grazing systems have been more thoroughly investigated.

Sheep cheese production systems within the Idiazabal Protected Denomination of Origin (PDO)

Sheep herding is the traditional sheep production system, linked to the local *Latxa* breed and to outdoor part-time grazing during the lactation period in late winter and spring. The sheepherder combines the available fresh grass with indoor feeding (dried forage, concentrate) to provide the animals with the necessary nutrients. Around 90% of the sheep milk produced is transformed into Idiazabal PDO Cheese. In 2008, the DO comprised 435 sheep milk producing farms (sheepherder is farm owner), producing a total of 7.7 million litres, and 1.3 million kg cheese. Cheese producers (111 in 2008) can be categorized into 3 groups:

- large dairies (a total of 3) produce more than 100 Tm per year (44.5% of the total Idiazabal PDO cheese in 2008). These dairies buy all their milk.
- small dairies (a total of 13) produce less than 100 Tm per year (17.2% of the 2008 cheese production). Some dairies buy part of the milk they use, and others are associations of sheep farmers which transform the milk of their members.
- individual farmhouses (a total of 95) transform the milk of their individual flocks (38.3% of the 2008 total Idiazabal cheese).

A *food system* comprises raw materials producers, food industries which transform them, distributing and marketing outfits and consumers who buy a given food product (Renting *et al.*, 2003). In the case of the Idiazabal PDO Cheese, our research clearly differentiated two food systems on the basis of the role of the sheep farmer: industrial and artisanal. If the sheep farmer produces only milk, (s)he is ascribed to the industrial system, whereas if the sheepherder makes cheese (s)he is ascribed to the artisanal system. Thus, the industrial system comprises large and small dairies (about 52% of the PDO cheese produced). By contrast, the artisanal system is comprised by the 95 individual farmhouses, and those associations of sheepherders which transform the milk of their members. In addition, outside any PDO production, there is a third food system for sheep cheese. It comprises sheepherders who produce milk from breeds of foreign sheep (Assaf or Lacaune). They sell their milk to cheese factories which make 'Idiazabal type' cheese, with no PDO seal, and market it through supermarket chains.

Our working hypothesis is that the artisanal food system for PDO cheese production is the most sustainable of the three food systems.

Methodology

The following social analytical techniques were used: (1) analysis of secondary sources of information; (2) in-depth interviews to four expert agents in sheep milk production; (3) semi-structured interviews with 8 sheepherders. For the assessment of milk and cheese quality, 11 commercial flocks, members of the DO, were selected. Milk from each flock was sampled on 2 consecutive weeks, before and after turning the sheep out to pasture. Cheeses made with these milks were sampled after ripening for 120 days.

Results

Considering the second, socio-economic dimension of our definition of a sustainable food system, the most advantageous food system is that in which sheepherders make cheese with their own milk.

Their gross profit for each Annual Working Unit was 27,617 Euros in 2007 as compared to 19,665 Euros for those who sell their milk to cheese factories. In addition, the former food system requires the contribution of another family member, which in almost all cases is a woman. On average, there are 1.94 AWU in farms that make their own cheese as compared to 1.2 AWU in farms that sell their milk. Establishing direct relationships with customers significantly increases farmers' self-esteem because they see themselves as capable of preparing a highly appreciated food instead of merely providing a raw material for cheese factories. This self appreciation of the value of their own work is probably as important for the long-term continuity of sheep farming as their higher level of income. In fact, in the last few years an increase in the number of sheep farmers who make their own cheese has been observed.

Regarding the third dimension, food quality, of a sustainable food producing system, that in which sheep farmers make their own cheese is also the most appropriate. Factories market their cheese through supermarket chains at around 20 Euros/kg, after paying producers an average 8-10 Euros/kg. By contrast, most sheepherders sell their cheese directly to consumers, at an average 14-16 Euros/kg. The cheese made with milk obtained during the outdoor period had higher concentrations of healthier fatty acids (Pariza, 2004), a 35% lower atherogenicity index, 14% lower cholesterol, and 10% higher antioxidant capacity than milk obtained during the indoor period (Abilleira *et al.*, 2009).

Conclusions

Results presented in this work confirm our initial hypothesis. An important aspect to promote the continuity of this cheese producing system is that consumers be aware of its benefits to help them make buying decisions which will potentiate and strength it. In addition, institutions should also support this type of food producing systems which benefits such widely different aspects.

Acknowledgements

The authors thank the Idiazabal Denomination of Origin, and Nekazal Ikerketa eta Teknologia (IKT), S.A., for the use of their databases, and the sheepherders Association Artzai-Gazta. Financial support: University of the Basque Country (UNESCO 02/05) and the Spanish INIA (RTA2006-00100-C02-02).

References

Abilleira, E., Collomb, M., Schlichtherle-Cerny, H., Virto, M., de Renobales, M. and Barron, L.J.R. (2009). Winter/Spring changes in fatty acid composition of farmhouse Idiazabal cheese due to different flock management systems. Journal of Agricultural and Food Chemistry 57, 4746-4753.

Brundtland, G. (1987). Our Common Future: The World Commission on the Environment and Development. Oxford University Press.

Paniagua, A. and Moyano, E. (1998). Medio ambiente, desarrollo sostenible y escalas de sustentabilidad. Revista Española de Investigaciones Sociológicas 83:151-175.

Pariza, M.W. (2004). Perspective on the safety and effectiveness of conjugated linoleic acid. American Journal of Clinical Nutrition 79: 1132S-1136S.

Renting, H., T.K. Marsden and J. Banks (2003). 'Understanding alternative food networks: exploring the role of short food supply chains in rural development'. Environment and Planning A 35: 393-411.

Sambucana sheep: ethical and socio-economical successful outcomes of a 20-years *in situ* preservation program

P. Cornale, M. Renna, R. Fortina and L.M. Battaglini
University of Torino, Dep. Animal Science, v. L. da Vinci 44, 10095 Grugliasco (TO), Italy; paolo.cornale@unito.it

Abstract

A study on the breeding evolution of the Sambucana sheep in the last 20 years has been carried out. This dual-purpose sheep (meat and wool) is bred since the 18th century in the Stura di Demonte valley (Piemonte, N-W Italian Alps). In this area the Sambucana has played a major economical, social, environmental and cultural role thanks to its adaptation to the harsh climatic conditions and the poor feeding resources of the native territory. For many decades, cross-breeding with Biellese rams was performed to increase the birth weight and the growth rate of lambs. This practice has progressively worsened quality and yield of meat and wool, as well as the ability of exploiting marginal pastures. Moreover, the number of purebreds drastically decreased: in the early '80s only a hundred Sambucana purebreds out of a total population of 5,000 sheep survived in the Stura di Demonte valley. For these reasons and with the purpose of recovering the Sambucana breed's heritage, local initiatives have been carried out since 1985. The local authority of the valley founded the consortium *Escaroun*, gathering local Sambucana sheep breeders. The association aimed at safeguarding the breed and economically valorising the derived meat and wool productions. In 1988, the Consortium started a genetic selection program and created a centre for the performance test of rams. The morphometric evolution of the Sambucana breed in the last two decades was investigated by body measurements on some selected animals of the rams centre. Measurements of 2007 were compared to those collected at the beginning of the recovery program, and significant variations ($P<0.001$) were observed. Adult rams (≥18 months) showed reductions in height at withers, height at rump, and chest girth, together with an increase in rump width, confirming the effectiveness of the backcross breeding program. Considering the wide loss of genetic diversity in livestock, the results point out the importance of *in situ* preservation programs able to promote a sustainable management of local animal genetic resources.

Keywords: local breed, recovery program, biodiversity, rural traditions, Alpine valley

The territory

The Piedmont region (N-W Italy) is mainly constituted of mountains (43.1%) and hills (22%) with forests covering approximately 670,000 ha (26%) of the regional area. Agriculture is mostly characterized by family farms (97% of total farms and 80% of farmland) with an average area of 13.3 ha per farm (ISTAT, 2000). Among Piedmonts' Alpine valleys the longest (60 km) and widest is the Stura di Demonte valley, located in S-W part of the region and running East-West between Maritime and Cottian Alps. It ranges from Borgo San Dalmazzo plain (515 m.a.s.l.) to the Maddalena pass (1996 m.a.s.l.) on the French border. The valley has always been a strategic link between Italy and France and it has historically represented one of the ancient *'salt routes'*. Agriculture has always played a key role in the local economy. Grasslands and extensive rangelands of these marginal areas represented a feed source for low-intensity livestock systems only; sheep breeding has always been one of the most important rural activities of this valley.

Even if several Alpine valleys shared the above mentioned characteristics, the Stura di Demonte valley is an exception to the mountains' massive abandonment trend and its relative facing decline. Which factors have enabled the survival of this valley?

The valley and its sheep breed

The autochthonous sheep breed of the Stura di Demonte valley is the Sambucana (Figure 1). Locally known as '*Demontina*', its name origins from *Sambuco*, a small village located in the upper part of the valley. Its origin is still unclear (Di Stasio *et al.*, 2001) but it is known that the Sambucana sheep has been introduced in the 18th century. Rearing this sheep hundreds of years in poor, rocky, and marginal pastures with harsh climate conditions contributed to make it well adapted to the unfavourable territory (Brignone and Martini, 2001). Nowadays, shepherds appreciate Sambucana especially for its rusticity, agility, and hardiness. From spring to late autumn sheep flocks are reared in alpine pastures (1,500-2,500 m.a.s.l.) without shelters for night or frequent off-season snowfalls. This sheep is able to pass steep slopes through paths carved into the rock and to reach the highest pastures. During winter they are stabled and fed with local hay only.

After the second half of 20th century, some breeders have introduced in the valley other more improved breeds (e.g. Biellese sheep) aiming at increasing birth weight of the lambs and consequently being able to obtain a higher income since the animals were sold by their weight. Actually the cross-breeding practice resulted in an increase of weight of the lambs due to larger and heavier bone structure rather than to major growth rate and meat yield. Furthermore, the use of Biellese rams progressively resulted in: (1) a worsening in meat and wool quality, (2) a decrease in cross-breds adaptability to native marginal areas, and (3) a dramatic decrease in the number of Sambucana purebreds (Massimino, 1989). In the early '80s, following the reporting of this negative trend, the Food and Agriculture Organisation (FAO) inserted the Sambucana sheep in its extinction-threatened breeds list (FAO, 1980). But it was thanks to the willingness of some local inhabitants that Sambucana recovery began and took shape (Luparia, 2000).

Figure 1. A Sambucana ewe with its lamb (Cristina Baudino, 2007).

The Sambucana preservation program

Since 1985 the *Comunità Montana* (local authority) has started to safeguard the breed involving the whole valley and especially the aged shepherds, witnesses of the original characteristics of the purebreds. The discovery of few flocks, in which Sambucana ewes have never been crossbred with Biellese rams, gave the *Comunità Montana* the opportunity to realize the protection program. It was estimated that only a hundred Sambucana purebreds out of a total population of 5,000 sheep survived in the valley. Local authority and Sambucana sheep breeders founded the consortium *Escaroun*, referring in Occitan language to a small group of sheep that leave the flock to find the highest and best pastures. Supported by a specific European financing (Reg. 2078/92/EEC) in 1988 the consortium started a genetic selection program and created a centre for the performance test of rams, with the primary objective of recovering the genetic heritage of Sambucana sheep.

The Rams Centre buys young male lambs from farmers. Each lamb is measured and checked (performance, morphometric and sanitary parameters) during its growth. During the breeding season, rams that reached sexual maturity are randomly lent to the associated shepherds from spring to late autumn. At the end of this period, they are returned to the Rams Centre where they are checked again and evaluated for their suitability for the following breeding season.

'Breeding' results

In order to highlight the results obtained in twenty years of recovery program, the population data and the morphometric evolution of the Sambucana breed were investigated.

As mentioned above, in the early '80s only hundreds of purebred subjects were reared in the valley. With the creation of the consortium *Escaroun* and with the successful activities of the Rams Centre the population has shown a positive trend. The first and most important result was the recognition by FAO: after few years from the beginning of the recovery program the Sambucana sheep was cancelled from the list of breeds at risk of extinction, and it was classified as 'breed at limited diffusion' (FAO-UNEP, 1993). An Official Register of the breed, managed by Italian Association of Pastoralism, was set a few years later. The number of the registered animals constantly increased with a peak of 4,206 subjects divided in 70 farms in 2006 (ASSONAPA, 2010).

Rams body measurements collected at the beginning of the recovery program (1988) were compared to those collected two decades later (2007) from rams bred in the centre. Significant evolutions were observed from crossbreds to recovered purebreds and these results are shown in Table 1.

Adult rams (≥18 months) showed reductions in height at withers, height at rump, and chest girth, together with an increase in rump width, confirming the effectiveness of the backcross breeding project.

The obtained results are a remarkable success of the genetic recovery program and saving an autochthonous breed from extinction corresponds to the best measure in conservation of livestock biodiversity (El Aich and Waterhouse, 1999; Köhler-Rollefson, 2001). But this result is only one of the several outcomes deriving from the preservation program (Figure 2) which contributed to save people, traditions, landscape and economy of the valley (Petrini, 2008).

Successful outcomes in the valley

The cross-breeding practice with Biellese rams decreased the rusticity and the adaptability of sheep to exploit marginal areas of their native territory. Not recovering the breed and its characteristics

Table 1. Morphometric evolution of Sambucana adult rams from 1988 to 2007 (mean±std dev).

		Morphometric data		Evolution[1] [+/-(*P*)]
		1988	2007	
Age	(months)	26.12±7.532	26.55±5.882	n.s.
Height at withers	(cm)	81.25±2.405	78.52±3.178	- ($P<0.001$)
Height at rump	(cm)	81.66±2.387	78.21±3.279	- ($P<0.001$)
Chest height	(cm)	35.09±1.674	34.32±3.400	n.s.
Chest width	(cm)	22.83±1.485	23.21±2.685	n.s.
Rump width	(cm)	23.00±1.118	25.96±2.502	+ ($P<0.001$)
Trunk length	(cm)	82.71±3.205	82.39±6.855	n.s.
Chest girth	(cm)	98.92±4.957	93.57±8.694	- ($P<0.001$)
Weight	(kg)	85.5±9.624	83.99±7.783	n.s.

[1] n.s. = not significant.

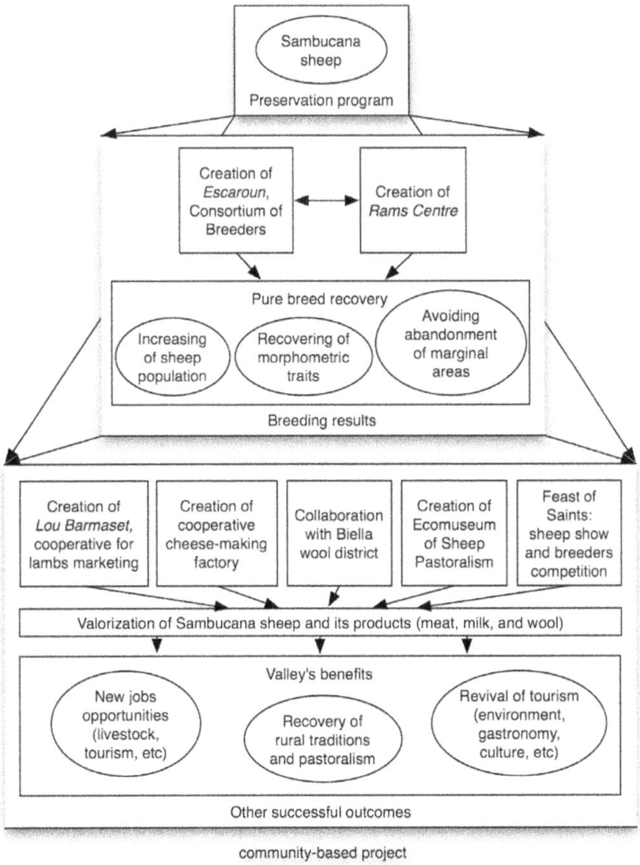

Figure 2. Scheme of the successful outcomes of the Sambucana preservation program.

would almost certainly have caused the abandonment of grasslands and the deterioration of pastures qualities. Locally adapted livestock breeds have been proven to be the best (or in some cases the only) feasible opportunity for farming systems in marginal areas (Gandini and Villa, 2003; Rege and Gibson, 2003) and they contribute to the appropriate management and preservation of semi-natural habitats (Ostermann, 1998).

The pure-breed return and the associated recovered rusticity of Sambucana sheep, together with correct financial supports of pastoral activities, avoided the depopulation of the valley. These conditions allowed additional possibilities for employment of young people. The local authorities realized that the validity of the entire project would depend on a correct promotion of animal products. The cooperative '*Lou Barmaset*' (referring in Occitan language to a small shelter in the rock) was founded in 1992. It gathers today almost fifty shepherds and is responsible for the marketing of lambs bred in valley (approximately 2,000 per year). Two official awards further valorised lamb meat. The '*Agnello Sambucano Garantito*' (Guaranteed Sambucano Lamb) brand, recognized by the Italian Ministry of Industry (1996), was created. Subsequently, in 2001 the Sambucano lamb was inserted into the 'Presidia' list of Slow Food. Some researches investigated also its meat quality (Battaglini *et al.*, 2004, Cugno and Cavallero, 2000) showing high tenderness and low fat content.

Additionally several initiatives were carried out to safeguard the other Sambucana products. In Europe local sheep wool has been severely neglected and it represents more often a cost instead of a resource. In fact, the introduction of fine wool from extra-European countries almost completely excluded the autochthonous breeds from the production of textiles. Thanks to relatively good fleece quality of Sambucana, textiles (sweaters, scarves, and plaids) made with its wool have been produced in collaboration with a firm of the Biella wool district.

Milk has traditionally been a secondary product mainly used for lamb feeding. The derived typical product is the heavy lamb (locally known as '*tardoun*') which remains with the ewe throughout the lactation. However, if lambs are slaughtered at an earlier age (18-20 kg weight) the remaining milk is used for '*touma*' production, a typical cheese with a delicate taste. This product is economically valorised by a cooperative cheese-making factory, created in 2000 together with the Ecomuseum of Sheep Farming, in Pietraporzio village (Martini and Pianezzola, 2001).

The Ecomuseum was born to rediscover and preserve the culture and the traditions of pastoralism after the effective safeguarding of the breed. The numerous cultural and educational activities of the Ecomuseum, together with the other local initiatives (e.g. Fest of Saints with sheep show and breeders competitions) are able to catch several thousands of tourist in the valley.

This community-based project clearly demonstrates that rescuing a local sheep and its traditional extensive breeding systems in a marginal area can really help recovering a valley, its traditions and its economy.

Acknowledgments

The authors gratefully thank Antonio Brignone (Comunità Montana 'Valle Stura'), Stefano Martini (Ecomuseum of Sheep Farming), and Crisina Baudino.

References

ASSONAPA (2004). Banca Dati Asso.Na.Pa. – Consistenze Provinciali della Razza Sambucana – Anno 2006. Available at: http://www.assonapa.it/Consistenze/Cons_raz-za.asp?Anno=2006&Cod_razza=R1. Accessed 17 November 2009.

Battaglini, L.M., Tassone, S., Cugno, D. and Lussiana, C. (2004). Sambucana sheep breeding in Valle Stura di Demonte and meat characteristics: present situation and outlooks on future. Options Méditerranéennes, Série A, 61: 195-199.

Brignone, A. and Martini, S. (2001). La razza ovina Sambucana – Rusticità e adattamento all'ambiente. In: Guillame Lebaudy – Dionigi Albera (ed.) La routo. Sulle vie della transumanza tra le Alpi e il mare. Ass. Cult. PRIMALPE, Cuneo, Italy, pp. 133-136.

Cugno, D. and Cavallero, A. (2000). Nouvelle orientation de la production de l'agneau de race 'Sambucana' en alpage: caractérisation de sa viande et transformations des exploitations provoquées par les prédations des loups et chiens errants. 11th FAO Meeting of the FAO working group of Mountain Grasslands. Quality and valorisations of mountain animal production. 13-17 September 2000. Luz St. Sauver (France).

Di Stasio, L., Fiandra, P. and Rasero, R. (2001). 'Patrimonio zootecnico del Piemonte- Iniziative per la salvaguardia delle razze autoctone', Suppl. al n.28/01 dei Quaderni della Regione Piemonte-Agricoltura.

EEC (1992). Regulation 2078/92/EEC of 30 June 1992. Official Journal of the European Communities 215: 85-90.

El Aich, A. and Waterhouse, A. (1999). Small ruminats in environmental conservation. Small Ruminant Research 34: 271-287.

FAO (1980). Animal genetic resources, conservation and management. FAO Technical Consultation, 2-6 June 1980. Rome, Italy.

FAO-UNEP (1993). World watch list for domestic animal diversity. 1st ed. Loftus, R. and Scherf, B., FAO, Rome, Italy, 376 pp.

Gandini, G.C. and Villa, E. (2003). Analysis of the cultural value of local livestock breeds: a methodology. Journal of Animal Breeding Genetics 120: 1-11.

ISTAT (2000). 5° Censimento Generale dell'Agricoltura – Presentazione dei dati definitivi – Piemonte. Available at http://censagr.istat.it/piemonte.pdf. Accessed 10 March 2010.

Köhler-Rollefson, I. (2001). Community-based management of animal genetic resources with special references to pastoralist. In: FAO (ed.) Proceedings of Community-based management of animal genetic resources workshop. May 7-11, 2001. Mbabane, Swaziland.

Luparia, S. (2000). The Sambucana sheep: a project to save a valley. Animal Genetic Resources Information 27: 27-33.

Martini, S. and Pianezzola, M. (2001). L'ecomuseo della pastorizia – Strumento di valorizzazione di una valle. In: Guillame Lebaudy – Dionigi Albera (ed.) La routo. Sulle vie della transumanza tra le Alpi e il mare. Ass. Cult. PRIMALPE, Cuneo, Italy, pp. 117-131.

Massimino, R. (1989). La razza ovina sambucana: indagini biometriche e rilievi sull'attitudine alla produzione carnea (rese alla macellazione). Degree thesis. University of Torino.

Ostermann, O.P. (1998). The need for management of nature conservation sites designated under Natura 2000. Journal of Applied Ecology 35: 968-973.

Petrini, C. (2008). La vittoria del 'mondo dei vinti'. La Repubblica, 5 ottobre 2008.

Rege, J.E.O. and Gibson, J.P. (2003). Animal genetic resources and economic development: issues in relation to economic valuation. Ecological Economics 45: 319-330.

The Latxa dairy sheep in the Basque Country: importance, challenges and opportunities for a traditional livestock activity

R. Ruiz, B. Díez-Unquera, I. Beltrán de Heredia, N. Mandaluniz, J. Arranz and E. Ugarte
Environmental Research Institute Neiker, P.O. Box 46, Vitoria, Spain; rruiz@neiker.net

Abstract

Livestock farming has played a key role for the livelihoods of people in the Basque Country since ancient times. In fact, the Latxa and Carranzana sheep breeds are native of this area. The features of these breeds have been shaped in order to fit to the environmental conditions through the management practices and production system carried out by shepherds. Initially, these sheep could be considered as dual-purpose breeds and were reared basically upon the base of natural resources by means of a transhumant system (seasonal movement of flocks). Since the beginning of a breeding programme for dairy production in the middle 80's, a certain level of intensification has been observed in the management practices, the standard of living conditions of the farmers have improved significantly and the products obtained are protected under quality and denomination of origin labels. Therefore, the implications and importance of this system are unquestionable in relation to landscape configuration, rural development in mountain areas, diversity of products, cultural heritage, etc. However, some of the flocks disappear without generational turnover despite being economically profitable, showing that sustainability does not depend only upon productivity parameters. Indeed, the sustainability of some of the farms is seriously endangered nowadays, and the remaining ones will have to face adaptation strategies to get adapted to changing policies, consumer demands, social, economic and environmental conditions. Several pressures have been identified in this system: low meat and milk price, standards of living expectations, access to land, growing dependency on inputs (purchased feedstuff, energy, labour, etc.), wolf predation, etc. But there are also interesting tools, actors involved and opportunities to be developed in order to improve sustainability at the farm level: cheese making, closer access to consumers, environmental services, training and education initiatives (shepherds school), agro-ecology approaches for farming practices, growing sensibilization to quality food consumption, etc. Therefore, social and environmental indicators have to be considered as they can be crucial for adaptability, resilience, self-sufficiency and equity attributes.

Keywords: sustainability, farming, local breeds

Importance of the grassland based Latxa dairy system

In order to assess the real importance of the production system of Latxa dairy sheep in the Autonomous Community of the Basque Country (BC), it is important to understand the multifunctional character of dairy farming in mountain areas and its the cultural, social, environmental and economical implications. The Latxa breed is recognized as one of the most ancient in the Iberian Peninsula, and those mountain areas are still managed by means of communal structures (*parzonerías*, communal pastures, etc.) and grazing livestock basically during late-spring, summer and autumn, which exert a crucial role in landscape modelling. There are also evident implications of the shepherding activities in the development of the Basque language and the cultural heritage (local markets, gastronomy, typical food products, etc.).

Regarding environmental issues, many authors agree on the generic importance of grasslands for the conservation and management of biodiversity (Gibon, 2005; Henle *et al.*, 2008), although the relationships between grazing animals, pasture management and the richness of plant and animal species are complex, with various spatial and temporal scales concerned (McCracken and Bignal, 1998). The importance of keeping suitable livestock pressures for the ecosystems in the mountain areas of the BC has been demonstrated (Amezaga *et al.*, 2004; Mandaluniz *et al.,* 2009). This is particularly relevant for nature preservation purposes taking into account that semi natural pastures, meadows and shrublands (including heathlands protected by the European Habitat Directive 97/62/EU) cover around 20% of the territory, and the fact that 20% of the BC is declared part of the Natura 2000 Network.

In terms of social and economical importance, the primary sector and related industries provide less than 1 percent of nominal gross domestic product of the Basque Country, which is assessed in 31,110 €/per capita in 2008. However, grassland based systems valorise non-marketable natural resources by converting them into high quality food products; in this case 9.2 M. litres of sheep milk per year, 1,400-1,500 tons of Idiazabal cheese, and around 150,000 young lambs.

As this production takes place mainly in economically less favoured areas, where there are few alternative economic activities; it plays a crucial role to maintaining human populations, as well as generating labour and richness. In this sense, around 83% of the BC is considered as mountain agricultural areas. Therefore, the interest of keeping local dairy sheep linked to the territory in these areas is clear for rural development purposes, as they can provide the economic basis for leisure activities and tourism, as well as recreation areas for urban populations.

To improve the conditions, efficiency and sustainability of this activity, a wide variety of technical and advisory programmes have been put into practice since the mid-eighties and continue with successful results: the breeding programme of the Latxa sheep, improvement of the quality of forages and feeding advisory, milk quality, economic advisory, programs to control some diseases, etc. However, some of the flocks disappear without generational turnover despite being economically profitable, showing that sustainability does not depend only upon productivity parameters.

Challenges

The following challenges are perceived as the most important ones in terms of the sustainability of the system:

- Low price of products:
  - Lamb: the price paid to farmers shows a seasonal variation depending on the market supply, but overall levels have remained unchanged for many years.
  - Wool: whereas the cost of shearing a sheep is around 1.5 € per animal, the incomes generated do not reach 0.5 €.
  - Raw milk: farmers claim that the price offered by the dairy industry is not fair enough taking into account the final price of the Idiazabal cheese branded with Protected Denomination of Origin (PDO).
- Uncertainty regarding the continuous revisions of the Common Agricultural Policy (i.e. application of environmental objectives and measures into agriculture activities), and disappointment due to high levels of exigencies in terms of administrative requirements and bureaucracy.
- Growing concern regarding livestock – wolf conflicts: the reappearance of wolfs in the western areas of the BC and the periodical sheep losses due to predation have arisen social concern within the professional livestock sector, who claim that grazing livestock and predators is not compatible. Meanwhile, the administration is trying to set the rules that make it possible, in an attempt to meet also the demands of ecological organisations.
- Increasing dependence on inputs related to intensification strategies, which mean higher costs (protein and energy sources, etc.) and could drive to the substitution by highly specialized foreign breeds (Ugarte *et al.*, 2001).

- Lack of interest among young people for this activity as a professional opportunity and problems for generational turnover.
- Increasing pressures over the agricultural land to build communication infrastructures (roads, high speed train), residential or industrial areas, and limited access for new settlers.
- Adaptation to new constraints regarding the expected impact of global warm, in particular to modification in the patterns of availability of resources and emerging of new infectious diseases (new serotypes of blue tongue).

Opportunities

Nevertheless, there are also interesting tools and opportunities to be developed in order to improve the sustainability of dairy sheep farms. On the one hand, there is a whole structure of agents, technicians and programmes working in favour of the conservation and improvement of dairy sheep. In particular, the successful experiences of the breeding programme of the Latxa sheep (nowadays the annual average genetic improvement is assessed in 2.5-3%), the PDO Idiazabal cheese (prestigious product and consolidated market) and the Shepherds' School (more than 80 young farmers established in 12 years) are just an example that should encourage new approaches.

The development of added value activities either individually at the farm level or through the joint transformation and commercialisation of milk, cheese, lamb or wool, have to be envisaged as an interesting option in an attempt to improve the price perceived by farmers. In fact, the number of farmers that have begun transforming into cheese the milk produced in their own farms, has increased continuous and significantly since 1995. Strategies to develop closer commercialisation channels have recently also been carried out even for lamb meat.

Regarding the generation of public goods and externalities related to landscape preservation, it is crucial to develop innovative approaches so that society can valorise economically the role of grassland based systems. The figure of farming contracts in response to environmental services should be explored. Finally, social and environmental indicators have to be considered together with technical and economic data in the design and application of policy measures aimed to guarantee the sustainability of the Latxa dairy system.

References

Amezaga I., Mendarte S., Albizu I., Besga G., Garbisu C. and Onaindia M. (2004). Grazing Intensity, Aspect, and Slope Effects on Limestone Grassland Structure. Rangeland Ecology & Management 57, 606-612.

Aranburu A., Macía H., Díez J., Fernández J.M. and Francisco, M. de. (2006). The Natura 2000 Network in the Autonomous Community of the Basque Country. XVI Congreso de Estudios Vascos: Desarrollo Sostenible-IT. Donostia: Eusko Ikaskuntza, (2006), pp. 111-118.

Gibon, A. (2005). Managing grassland for production, the environment and the landscape. Challenges at the farm and the landscape level. Livestock Production Science 96, 11-31.

Henle, K., Alard, D., Clitherow, J., Cobb, P., Firbank, L., Kull, T., McCracken, D., Moritz, R.F.A., Niemelä, J., Rebane, M., Wascher, D., Watt, A. and Young, J., (2008). Identifying and managing the conflicts between agriculture and biodiversity conservation in Europe–A review. Agriculture, Ecosystems & Environment 124, 60-71.

McCracken, D.I. and Bignal, E.M. (1998). Applying the results of ecological studies to land-use policies and practices. Journal of Applied Ecology 35, 961-967.

Mandaluniz N., Aldezabal A. and Oregui L.M. (2009). Atlantic mountain grassland-heathlands: structure and feeding value. Spanish Journal of Agricultural Research 2009 7(1), 129-136.

Ugarte E., Ruiz R. and Gabiña, Beltrán de Heredia I. (2001). Impact of high-yielding foreign breeds on the Spanish dairy sheep industry. Livestock Production Science 71, 3-10.

A bio-integrated model of food production and knowledge: a study case in Ciudad del Carmen, México

L. Domínguez-Trejo[1], M.R. Morales-Garza[2] and W. Cano[3]
[1]Centro Regional de Investigaciones Pesqueras (CRIP-Carmen) Héroes del 21 de abril s/n, Playa Norte Cd. del Carmen, C.P. , Campeche, México; [2]Centro de Investigaciones sobre Desertificación-CIDE, Cami de la Marjal s/n Albal, Valencia, Spain; [3]University of the Basque Country (UPV/EHU), Departamento de Lógica y Filosofía de la Ciencia, Av. Tolosa 70, 20008 Donostia, Spain; miremo@uv.es

Abstract

In 1946, shrimp fishery provided employment and income for the population of Ciudad del Carmen, Mexico. For forty years overfishing was rampant affecting the natural populations and changing the traditional way of living of native people. Also, pollution produced by the Mexican state oil company diminished the shrimp population. In those days, there was no concern on a precautionary and ecosystem approach to fisheries (EAF). Nowadays, new generations of fishers do not want to suffer again the shrimp experience and are now interested in an EAF and having a more actively participation. We developed a bio-integrated model system that links recirculating aquaculture with hydroponic vegetable production (aquaponic system) as a different way to integrate fish and plants in a polyculture that increases diversity of food products. To develop this project we integrated scientific and technological knowledge to set up the system as well as traditional knowledge of women to decide the production of plants. As a result, they are improving their nutrition and incomes with the sale of the products. Cooperation between experts and communities can be possible when recognize that the value and legitimacy of traditional knowledge is a first step to reduce social inequality. We expect these experiences provide the community an empowerment to participate in decision making actions.

Keywords: aquaculture, hydroponic, aquaponic, polyculture

From the imposition to the integration

Populations that directly depend on a specific natural resource for a high proportion of their income tend to be very vulnerable to the availability of the resource, market valuations, competition, and also local policies (West, 1994). In Ciudad del Carmen the overfishing of shrimps was rampant, affecting the natural populations and creating many social conflicts.

The research center CRIP-Carmen had been working on the investigation of fisheries and the biology of regional species with economic importance. Even though transference of scientific results to society is problematic. We thought it was due to scientists' linear vision of knowledge and technology transference. Thus, it is quite difficult for the community to adapt an unknown technological practice to its own context or needs. As a result, when they do not find sense or the conditions to incorporate a new technology to their own lives, sometimes they simply reject the project or abandon it. Moreover, it is more complicated if the transfer of new technology has to be implemented using an ecosystem approach to fisheries (EAF), because this linear model does not consider the cultural and traditional dimensions that EAF emphasizes (FAO, 1998).

Considering the potential knowledge held in local communities of Ciudad del Carmen, we tried to develop mechanisms to use this knowledge appropriately. Instead of transferring scientific knowledge to the community or impose the scientific model as a superior or better than other knowledge, we tried to integrate new forms to share information with and within the community. We decided that the community was not going to be any more the object of study, instead, it became an active subject of change. In this way we can interact, investigate, and understand their needs and start working in a different environment of mutual respect and understanding. The production of knowledge was determined since the beginning by the interests and needs of the community. This transfer of knowledge is bidirectional and more interactive because the transference answers the social demands inside a specific sociocultural context (Gibbons *et al.*, 1997).

Aquaponics, scientists, and community

For several years we were working in close contact with fishers. From this experiences we learnt that the scientific information produced at the CRIP-Carmen where only for other scientists and never reach local people. It was then when we decided to apply a human approach based on common values and interests. We learnt that the participation of the fishermen also recognized their economic, social, and cultural importance in the community.

To develop the model, we first considered to set up real objectives based on our human and economic resources. We always considered the qualities of the project but we also recognized that the use of both knowledge, even in a very specific context, could have limitations and be in some cases problematic. Fortunately, interacting with the community allowed us to exchange perspectives and knowledge. This mutual learning helped us to build strong bonds between participants and prepare us to start the new project of aquaponics in Ciudad el Carmen.

Since the mid 1980s aquaponic has become a viable system of food production, however requires intensive management. Aquaponic is the combination of aquaculture and hydroponics in an integrated system (Rakocy *et al.*, 2004). In the aquaponic laboratory of the CRIP-Carmen, we decided to design our system using mainly recycled materials. The bio-integrate model that we designed were thought to answer the needs of the community, where households in Ciudad del Carmen are mainly supported by women. Considering this, we believed that a gender aspect of this local knowledge would be more appropriate when developing the bio-integrated model because they have more traditional knowledge of the local resources. These skills and special knowledge are very important to get a bidirectional share of knowledge. Once the system is installed, the activities to maintain it are easy to do. The system allows a healthy functioning ecosystem inside households that not only sustains itself, but also the family, local communities or even regional economies. When you grow fish and plants together in one integrated soilless system, the benefits for the households are many because it can increase their food supply and improve nutrition. Also the household economy benefits through the sale of fish and plants.

A learning experience

The combination of traditional and scientific knowledge is not easy and requires commitment, a process of learning and time. But the results are beneficial for both, scientists and community. The products obtained from our bio-integrated model were first used in households that were effectively engaged in sustainable development, and had a shared sense of purpose. Women where the ones who decided the vegetables and fruits that had to be grown in the system. Also, they taught some traditional cooking recipes using the production of the aquaponic system. To integrate woman and make them feel more comfortable most of the interviews and future plans were made in their kitchens.

The model system worked well at a household level, and now we expect to expand the transference of the bio-integrated model in more households in different areas of the island. But we need to involve decision makers, scientists, managers and the local population. These actors need to cooperate in the elaboration of a pattern program that will meet both the development requests of local communities, and the protection and conservation needs of the natural environment. Unluckily, it is not the case in Ciudad del Carmen, there are economic and political interests that for many years have segregated local communities. However, we consider that if we can obtained more support from our authorities, we could have an extended program for public sensitizing and collaborative working, to encourage the local population participation, as well as the integration of other scientists to work in an interdisciplinary way. By now, we can not carry out all our plans with such a limited budget, inadequate facilities, lack of human resources and in some cases no basic resources such as water and light. Even though we have faced many problems, we will continue to work with this community because we think that communication, a permanent dialogue, and the engagement of householders were and still are a priority reason to go on.

We believe that active participation of the community can train its members and even empower them to define in an autonomous way the course of their own personal and collective development.

References

FAO (1998). Social issues in fisheries. By Phillip Townsley. FAO Fisheries Technical Paper, No. 375, Rome. 93 pp.

Gibbons, M., Limoges, C., Nowotny, H., Schwartzman, S., Scott, P., and Trow, M. (1997). La nueva producción del conocimiento: la dinamica de la ciencia y la investigación en las sociedades contemporáneas, Ediciones Pomares-Corredor, España, 235 pp.

Rakocy, J.E., Donald S., Bailey, R., Shultz, C., and Thoman, E.S. (2004). Update on tilapia and vegetable production in the UVI aquaponic system. In: New Dimensions on Farmed Tilapia: Proceedings of the Sixth International Symposium on Tilapia in Aquaculture. September 12-16, 2004. Manila, Philippines, pp. 676-690.

West, P. (1994). Natural Resources and the Persistence of Rural Poverty in America: A Weberian Perspective on the Role of Power, Domination, and Natural Resource Dependency. Society and Natural Resources 7: 415-427.

Autochthonous breeds, sustainable local exploitation, and quality of animal products: the case study of Piemontese cattle

M. Renna, P. Cornale, C. Lussiana, A. Mimosi and L.M. Battaglini
University of Torino, Dep. Animal Science, v. L. da Vinci 44, 10095 Grugliasco (TO), Italy; luca.battaglini@unito.it

Abstract

Piemontese cattle, autochthonous of Piedmont region, are bred in many Alpine valleys of North-Western Italy due to their high environmental adaptation degree and ability of exploiting marginal mountain areas. In the last forty years the genetic selection of this breed has been mainly oriented to meat production. Piemontese breed is now primarily known and appreciated for its precocious growth, high slaughter yield, and meat quality. Although Piemontese milk production recently underwent to a substantial decline, it historically represented the base for some Italian Protected Designation of Origin (PDO) cheeses (e.g. Castelmagno and Raschera). Milk and dairy products obtained from this breed are now gaining importance as a consequence of consumers' increasing attention towards traditional food. A study was conducted to investigate the nutritional characteristics (i.e. fatty acid profile) of milk obtained from Piemontese cattle bred on mountain pastures and fed with fresh grass and hay, only. A comparison with the fatty acid (FA) profile of milk types obtained in the same territory from different cattle breeds, stabled and fed with hay and concentrates due to their reduced rusticity, was also performed. Both regular milk and high quality milk (characterised by peculiar hygienic and chemical features according to the Italian Law no 169/89) have been considered in the comparison. Piemontese milk revealed the healthiest FA profile among the considered milk types, showing the lowest level of hypercholesterolemic saturated FA ($P<0.001$), and the highest level of total monounsaturated FA ($P<0.001$), rumenic acid (C18:2 *c9t*11, which has been proven to hold many beneficial biological activities for human health; $P<0.001$), and n3/n6 fatty acids ratio ($P<0.01$). These results suggest the existence of a strong relation among autochthonous breeds, their ability of exploiting local resources, and the quality of animal-derived food products. Given the above-mentioned results, a return to a double aptitude genetic type for the Piemontese breed should be taken into account for its further valorisation.

Keywords: local cows, dual-purpose, milk, Alpine areas, fatty acids

Piemontese breed: past and present history

The Piemontese cattle breed takes its name from Piedmont, its native region located in North-Western Italy. Twenty-five thousand years of selection under the local ecological conditions of this geographical area have made this breed well adapted to the Alpine environment and highly capable to exploit marginal mountain areas. In 1886, a natural mutation in the myostatin gene resulted in the appearance of an over-development of muscle (muscular hypertrophy, commonly known as double-muscling) in this breed. Many breeders recognized the possibility to obtain economic benefits from this feature (increased income due to increased retail product yield and positive effects on dressing percentage, carcass conformation and meat quality) and decided to hand it down by means of a genetic selection. Despite of the remarkable commercial interest linked to the above mentioned traits, some disadvantages associated with the double-muscle phenotype were later revealed. In particular, antagonistic genetic associations exist between muscularity and calving ease, determining an increase in the incidence of dystocia (birth difficulties), which can determine pain and suffering of animals (Carnier *et al.*, 2000; Lips *et al.*, 2001).

Until the half of the twentieth century, Piemontese cattle were triple-purpose, providing such different outputs as meat, milk and draught power. With the introduction of agriculture mechanisation, the use of this cattle as draught animals declined and milk and meat productions became the main purposes. Milk from Piemontese cows had traditionally been used as one of the basic ingredients for dairy transformation into widely well-known and appreciated Italian Protected Designation of Origin (PDO) cheeses (e.g. Castelmagno and Raschera). However, through the last forty years of genetic selection, the dual-purpose breed have been transformed into primarly meat producing animals. Milk production consequently underwent to a substantial decline, being mainly used for the feeding of calves only. Nowadays the Piemontese breed is the most important beef cattle in Italy, accounting for about 300,000 cattle registered in the Italian Genealogical Book. Its meat has high quality, with high percentage of protein and unsaturated fatty acids, and it is greatly appreciated by the Italian consumers especially for its very low amount of fat.

The nutritional quality of milk obtained from Piemontese cattle

The recent increasing attention by the consumers towards high-quality, locally- and traditionally-made food products (e.g. PDO cheeses) is giving a gaining importance to milk and dairy products obtainable from Piemontese cattle. We carried out a research study in order to investigate the nutritional characteristics (lipidic composition determined according to Renna *et al.*, 2009) of Piemontese cows' milk. The fatty acid (FA) profile of raw milk obtained from Piemontese cattle was compared with that of raw milk types obtained in the same territory (Stura and Maira Valleys, Cuneo province) from different Italian cow breeds (Italian Holstein 70%, Piemontese 15% and the remaining 15% from Aosta Red Pied, Alpine Brown and Italian Red Pied). Piemontese cattle were bred on mountain pastures and fed with fresh grass and hay only (low-input farming model, high environmental sustainability of the production system) while the other breeds were stabled and fed with hay and concentrates due to their reduced rusticity. Both regular milk and high quality milk have been considered in the comparison. According to the Italian Law no 169/89, high quality milk is characterised by peculiar hygienic (fixed maximum levels of total mesophilic aerobic and somatic cell counts) and chemical features (fixed minimum amounts of total fat, total protein and not denaturated soluble whey proteins). Data were submitted to one-way ANOVA (SPSS, 2004). Piemontese milk revealed important differences in its FA profile relative to the other considered milk types (Table 1), showing the lowest level of saturated FA characterized by hypercholesterolemic effects in humans ($P<0.001$), and the highest levels of total monounsaturated FA ($P<0.001$), rumenic acid (C18:2 *c9t11*, which has been proven to hold many beneficial biological activities for human health; $P<0.001$), and n3/n6 fatty acids ratio ($P<0.01$).

Table 1. Some remarkable differences in the fatty acid composition (% of total fatty acids) of three milk types produced in Piedmont (mean±std dev).

Fatty acid composition[1]	Piemontese milk[2]	Regular milk	High quality milk	*P*
HSFA	42.71 ± 1.67^{c}	47.41 ± 2.92^{b}	50.52 ± 2.09^{a}	<0.001
MUFA	35.35 ± 1.16^{a}	31.01 ± 1.66^{b}	29.58 ± 1.35^{b}	<0.001
RA	0.92 ± 0.05^{a}	0.35 ± 0.05^{b}	0.38 ± 0.06^{b}	<0.001
n3/n6	0.13 ± 0.06^{a}	0.03 ± 0.01^{b}	0.05 ± 0.01^{b}	<0.01

[1] HSFA = hypercholesterolaemic saturated fatty acids; MUFA = monounsaturated fatty acids, RA = rumenic acid; n3/n6 = n3/n6 fatty acids ratio.

[2] Different letters within rows indicate statistically significant differences among milk types.

The future of the Piemontese breed: an ethically, economically and socially-desirable return to a dual-purpose aptitude

Several recent studies (Battaglini *et al.*, 2008a,b) have pointed out an increasing interest in the Piemontese cattle for its exemplary sustainable qualities (adaptability to harsh environments, ability to make good use of roughage, fertility and longevity) and for its remarkable added value assets (i.e. good nutritional characteristics of milk and good cheese yield and taste). Some breeders have recently decided to start milking their Piemontese cows in order to obtain milk and derived products for self-consumption, direct retail sale in their farms and/or at local markets.

Animal-derived food products of high nutritional quality are often the result of a blend of local traditions and autochthonism. The latter is strictly linked to the ability of domestic animals to survive and produce in an economically and environmentally sustainable manner (frugality, resiliency and ability to thrive on forage-based diets). Piemontese cattle have been bred for centuries for their ability to graze and produce both meat and milk used for outstanding cheese manufacturing. The genetic selection towards the mono-purpose specialisation for meat with consequent disadvantageous effects on milk production risks leading to the loss of important heritage, both in terms of nutritional aspects of the milk itself (excellent chemical characteristics and good taste) and traditional aspects of its use (PDO cheeses which are now produced by using milk from other cow breeds).

To conserve this heritage, we argue that Piemontese milk production should be preserved through specific recovering programs. These are necessary in order to obtain: (1) the valorisation of economic activities in marginal mountain areas, (2) the maintenance of dairy farming in these disadvantaged regions where more specialized dairy cow breeds make less efficient use of locally available resources being consequently unable to realise their full genetic potential, (3) the rural traditions upturn, and (4) touristic and gastronomic opportunities.

References

Battaglini, L.M., Burdese, S., Mimosi, A. and Renna, M. (2008a). Osservazioni sul benessere di bovine di razza Piemontese allevate in aziende montane della provincia di Cuneo. In: Quaderni SoZooAlp no 5 'Benessere animale e sistemi zootecnici alpini'. Nuove Arti Grafiche, Trento, Italy, pp. 154-170.

Battaglini, L.M., Bosco, S., Malfatto, V., Lussiana, C. and Renna, M. (2008b). Relazioni tra composizione lipidica della dieta e dei prodotti lattiero-caseari di bovine di razza Piemontese allevate in ambiente alpino. Scienza e Tecnica Lattiero-Casearia 59 (6): 487-490.

Carnier, P., Albera, A., Dal Zotto, R., Groen, A.F., Bona, M. and Bittante, G. (2000). Genetic parameters for direct and maternal calving ability over parities in Piemontese cattle. Journal of Animal Science 78: 2532-2539.

Lips, D., De Tavernier, J., Decuypere, E. and Van Outryve, J. (2001). Ethical objections to caesareans: implications on the future of the Belgian White Blue. In: Pasquali, M. (ed.) Food safety, food quality and food ethics. Univ. press, Milan, Italy, pp. 291-294.

Renna, M., Garda, A., Lussiana, C., Ambrosoli, R. and Battaglini, L.M. (2009). Chemical, nutritional and microbiological characterization of organic Fontina PDO cheese. Italian Journal of Food Science 21 (3): 287-303.

SPSS (2004). User's Guide, Base 13.0 for Windows. Statistical Package for Social Sciences Inc., Chicago, Illinois, USA.

Part 4. The role of biotechnology in food security and ethical and legal aspects of bioscience patenting

Synthetic biology, ethical challenges and its impact on food security

M. Engelhard
Europäische Akademie zur Erforschung von Folgen wissenschaftlich-technischer Entwicklungen, Bad Neuenahr-Ahrweiler GmbH, Wilhelmstraße 56, 53474 Bad Neuenahr-Ahrweiler, Germany; margret.engelhard@ea-aw.de

Abstract

Synthetic biology is a young and very dynamic discipline. Its high goals reach from attempts to create artificial life in the test tube to the search for new effective medicines, cheap biofuels and new methods to produce petrochemicals. With its systemic approach and its attempts to generate biological systems that might not have any counterparts in existing living systems it goes far beyond mere genetic engineering. Compared with other emerging technologies synthetic biology has prompted concomitant ethical research at a remarkably early stage. One important discussion is that synthetic biology brings along challenges to social justice and food security. Synthetic biology will add to the rising demand for agricultural raw material and might lead to competition with food production and promote forest destruction. In addition, synthetic production of otherwise naturally derived molecules are likely to have huge effects on trade, particularly in the developing countries.

Keywords: *Organismus syntheticus*, carbon sources, food security

Synthetic biology goes beyond genetic engineering

In the last six decades biology has been fast-paced from a descriptive to a predominantly applied discipline. The description of the genetic code set the foundation for the first artificial introduction of a gene into an organism, and in 1977 the first human gene was cloned into a microorganism in order to produce recombinant human proteins. This experiment is often referred to as the advent of biotechnology. Since then genetic engineering has generated numerous applications reaching from the production of single molecules like biopharmaceuticals to the development of transgenic crops that are for example resistant to herbicides. Synthetic biology now extends beyond mere genetic engineering: Instead of modifying existing biological systems by adding a limited number of genes taken from another species, it seeks to generate new systems from the scratch.

Synthetic biology envisions the assembly of novel – usually microbial – genomes from a set of standardized genetic parts. These genetic parts may be natural genomic sequences that are being applied to a new purpose, natural genomic sequences that have been redesigned to function more effectively or artificial genomic sequences that have been designed and synthesized from scratch. In some cases even the genetic code itself is redesigned. In theory, these systems could have new features that have never been found in existing natural living systems, and they would serve the specified purposes and vision of their creators. In short, synthetic biology moves from 'manipulatio' to 'creatio'. Like genetic engineering synthetic biology has its conceptual roots in the 1970s, but only the new achievements of system biology and automated synthesis of DNA have enabled synthetic biology to flourish in recent years. There is no generally accepted definition of synthetic biology yet, and one aim of the accompanying philosophical research is to provide a problem-oriented definition in order to facilitate discourse. In the meantime the following working definition may serve its purposes: 'Synthetic biology is the engineering of biology: the synthesis of complex, biologically based (or inspired) systems which display functions that do not exist in nature. This engineering perspective may be applied at all levels of the hierarchy of biological structures

– from individual molecules to whole cells, tissues and organisms. In essence, synthetic biology will enable the design of 'biological systems' in a rational and systematic way.' ('Synthetic Biology: Applying Engineering to Biology', report of an EU NEST High Level Expert Group). The systemic approach of synthetic biology emphasized in this definition is the second main differentiating factor in comparison with genetic engineering. Also, despite having appropriated the term 'engineering', genetic engineering is by far not as influenced by the engineering disciplines as synthetic biology is; in fact, the engineering perspective in synthetic biology is currently changing the way of working in biotechnology significantly. At the moment scientists are developing a minimal cell that is stripped down to a few basic features onto which standardized genomic sequences can be mounted expressing the desired function. Some of the most prominent applications hoped for are the creation of bioengineered microorganisms that can produce new medicines and diagnostics and generate petrochemicals, climate-friendly biofuels and hydrogen for the post-petroleum economy. In addition, synthetic biology will generate a vast amount of new information about life processes – from the role of specific genes or of genetic circuits to the metabolism and the functioning of a whole organism.

Ethical framework of synthetic biology

Compared with other emerging technologies, synthetic biology has prompted concomitant ethical discourse and research at a very early stage. Right from the beginning when synthetic biology emerged as a new field, ethical considerations (mainly risk issues) were taken into account by many synthetic biologists themselves and, on a more general level, by NGOs. Furthermore, funding institutions and several academies of sciences are currently getting involved in the discussion. Finally, among academic ethicists, the ethics of synthetic biology is a vivid emerging field. Most ethical challenges in synthetic biology are not ingenious to the discipline though. They have their parallels in ethics of other new technologies such as genetic engineering, nano-technology or neuroprosthetics, but the specification in the context of synthetic biology is an important task. In addition, some authors argue that there are specific ethical challenges in synthetic biology, referring for example to the creation of new life. Most authors propose a risk-benefit and case-to-case evaluation of synthetic biology. Ethical challenges can be grouped into biosafety risks, biosecurity risks, and challenges to concepts of life, to human self-conceptions and to social justice.

So far, in most cases statements and discussions on the ethics of synthetic biology have focused on the evaluation and management of direct biosafety and biosecurity risks. While biosafety risks comprise unintended risks, bio-security risks are characterized by deliberate misuse. The biggest problem in the assessment of biosafety risks of synthetic organisms is their strong divergence from well-known natural systems. When there is no natural reference system, many proven methods of technology assessment are not applicable. Another challenge is that the precise characteristics of synthetic organisms remain 'noisy' and hard to predict. In addition, genetic circuits tend to mutate rapidly and become non-functional or develop new unknown characteristics. In response to these problems an augmented utilisation of organisms that are not expected to survive outside the laboratory or organisms with an artificial genetic code are often cited as a good way of preventing synthetic organisms to intermingle with their environment. However, these scenarios have not been experimentally validated so far. Therefore, in summary, the rules of the precautionary principle are suggested to be applied, particularly in cases of high complexity and uncertainty. This includes, above all, the principle of containment and in some cases even a moratorium.

One main aim of synthetic biology is to make biology easier to engineer by the development of a toolbox to design biological systems. It is expected that this toolbox will spark a wave of innovation, as more and more people outside the traditional biotechnology community have the necessary skill and currently also access to material and information to engineer synthetic organisms. However, experimenters not sufficiently educated in biology may lack the biological foresight to judge risks scenarios and may never have attended biosafety training. In the case of biohackery, garage and do-it-yourself biology which are upcoming fashions, guidance from scientific supervisors is even missing altogether. The de-skilling agenda of synthetic biology and the possibilities of dual use also lead to enormous biosecurity risks, since outlaw states, terrorist organisations, or individuals may exploit synthetic biology for hostile or malicious purposes.

Besides these biosafety and biosecurity risks, ethical problems of synthetic biology arise from conceptual questions. Synthetic biology has made the first steps to cross the borderline to create new life from small parts, and it will be only a matter of time before the first *Organismus syntheticus* will come to existence. That the creation of life is a realistic option is new to philosophy. A number of concepts are challenged by this new development: the concept of life, our understanding of nature and the human self-conception (as creator) – to name only a few. In many ethical theories, entities that are described as alive are assigned different values or moral standing than artefacts and machines. Consequently, the way in which newly created organisms are conceptualized has an ethical impact on how life in general is understood and valued. Therefore, a properly managed debate with understandable and reliable communications that addresses the challenges of synthetic biology requires a problem-orientated and sufficiently consistent definition of living matter with the clearest possible borderline to non-living matter. Semantic problems currently arise, because many representatives of synthetic biology use terms and metaphors (for example 'living machines') that appear to blur the boundary between living and non-living matter. Similar to this example our understanding of nature and, therefore, our way of dealing with nature is likely to be changed. When we assume a dichotomy of nature and culture, one can expect that a change in technical prospects will influence the conceptualization of nature, too.

Potential impact of synthetic biology to social justice and food security

Finally, synthetic biology brings along challenges to social justice. Synthetic biology will add to the rising demand for agricultural raw material and might lead to competition with food production. This could cause an accelerated forest destruction, thus threatening both biodiversity and climate. Agroenergy monocultures have already led to harm to biodiverse agricultural systems and caused land grabbing. One huge problem in the assessment of technologies that utilize biomass is the lack of a realistic estimation of how much biomass exists, how much energy and food respectively can be derived from it, and what the ecological consequences of biomass removal on such a large scale could be. In addition, synthetic production of otherwise naturally derived molecules is likely to have huge effects on trade, particularly in the developing countries. This could destabilize southern economies and employment. Both rubber in Malaysia and indigo dyes in India provide historic examples of natural molecules that faced marked dislocation from synthetic competitors. While rubber maintained marked shares, natural indigo vanished from the marked. Meanwhile, the shift from natural to synthetic production of molecules could lead to a higher independence of countries where the production has been dislocated to, which is often regarded as a factor of national security. In this context the questions of access to the benefits of the technology and intellectual property rights need to be discussed. However, in most cases these factors are hard to assess and difficult to predict. Thus, in a risk-benefit analysis they can currently only be integrated on a very generalized level.

References

Ball, P. (2004). Synthetic biology: Starting from scratch. Nature 431, 624-626.

Carole, L., J.I. Glass, N. Alperovich, R. Pieper, P.P. Parmar, C.A. Hutchison, III, H.O. Smith, and J. Venter (2007). Genome Transplantation in Bacteria: Changing One Species to Another. Science 317 (5838) 632-638.

Check, E., (2005). Synthetic biologists face up to security issues. Nature. 436(7053):894-5.

Cho, M.K., D. Magnus, A.L. Caplan, D. McGee, and Ethics of Genomics Group (1999) Ethical Considerations in Synthesizing a Minimal Genome Science 286 (5447), 2087-2090.

Gibson, B., C.A. Hutchison, C. Pfannkoch, E.A. Denisova, H. Baden-Tillson, J. Zaveri, T.B. Stockwell, A. Brownley, D.W. Thomas, M.A. Algire, C. Merryman, L. Young, V.N. Noskov, J.I. Glass, J.C. Venter, C.A. Hutchison III and H.O. Smith (2008). Complete Chemical Synthesis, Assembly, and Cloning of a *Mycoplasma genitalium* Genome. Science 319: 1215.

Glass, J.I., N. Assad-Garcia, N. Alperovich, S. Yooseph, M.R. Lewis, M. Maruf, C.A. Hutchison, H.O. Smith and J.C. Venter (2006). Essential genes of a minimal bacterium. PNAS 103 (2): 425-430.

Highfield, R. (2007). 'Man-made microbe 'to create endless biofuel'', Telegraph, June 8.

Keasling, J.D. (2008). Synthetic biology for synthetic chemistry. ACS Chem Biol. 2008 18;3(1): 64-76.

Piccirilli. J.A., T. Krauch, S.E. Moroney and S.A. Benner. (1990) Enzymatic incorporation of a new base pair into DNA and RNA extends the genetic alphabet. Nature 4;343(6253):33-7.

Pilkington. E. (2007). I am creating artificial life, declares US gene pioneer. The Guardian, October 6.

Rutz, B. (2007). Synthetic biology thought the prism of scenarios. Biotechnol. J. 2. 1072-1075

Smith, H.O., C.A. Hutchison, C. Pfannkoch and J.C.Venter (2003). Generating a synthetic genome by whole genome assembly: {phi}X174 bacteriophage from synthetic oligonucleotides. PNAS 100 (26): 15440-15445.

Terrence, M.T., C.F. Basler, P.V. Aguilar, H. Zeng, A. Solórzano, D.E. Swayne, N.J. Cox, J.M. Katz, J.K. Taubenberger, P. Palese, and García-Sastre (2005) Characterization of the Reconstructed 1918 Spanish Influenza Pandemic Virus. Science 310: 5745:77-80.

Climate-ready GM crops, intellectual property and global justice

C.A. Timmermann, H. van den Belt and M.J.J.A. Korthals
Wageningen University, Department of Social Sciences, Applied Philosophy Group, Hollandseweg 1, 6706 KN Wageningen, the Netherlands; Centre for Society and Genomics, P.O. Box 9010, 6500 GL Nijmegen, the Netherlands; cristian.timmermann@wur.nl

Abstract

So-called climate-ready GM crops can be of great help in adapting to a changing climate. Climate change, caused in great part by anthropogenic greenhouse gases released in the atmosphere since the industrial revolution by the developed world, is felt much stronger in the developing world, causing unexpected droughts and floods that will cause large harvest loss, leading to more hunger and malnutrition, rising death tolls and disease vulnerability. The current intellectual property regime (IPR) strikes an unfair balance between profit oriented seed industry giants and low-acreage famers. The biotechnology industry, mainly headquartered in the developed world, has started to patent new seed varieties, taking biomaterials from developing countries, with the claim, that they will be more tolerant to flood, drought, heat, and cold. Special sales contracts prohibit saving seeds from the harvest for the next season, thus forcing the farmers to buy them every season anew, which goes against traditional farming. Using a widely accepted concept of global justice, that tackles the fact that around 10% of the world's population profits from 90% of the earth's resources, and by taking into account the feasibility of a fair IPR, we will discuss three issues. First, an ethically acceptable IPR should prevent unjust and unfair assignments of property rights (e.g. patents) that completely ignore small-scale farmers inventiveness and efforts to save agrobiodiversity. As a second task, this regime should encourage globally a just distribution of the objects of innovation that are covered by patents. Third, such a regime should encourage innovations at a rate that is effective to cope with climate change.

Keywords: development ethics, climate change adaptation, biotechnology

Introduction

The changing climate is striking the tropics particularly hard. Many river deltas that feed millions of people, especially in Egypt and Bangladesh, have had an increase in salinity over the last years. A temperature rise of 3 °C could reduce harvest yields of some crops by 15 to 25% in North Africa (FAO, 2008). Seeds that have traditionally been planted in those areas do not promise a prosperous harvest anymore. Since the poorest countries are located around the equator, an area where the consequences of climate change are very unfavourable, those changes have become an additional burden for development. Most of those countries don't only have difficulties feeding a growing population, but also adapting their agriculture to the new climate conditions. Here biotechnology could give a hand by developing new crop varieties that are able to better withstand abiotic stress like droughts, floods, saline water, heat and cold. By manipulating the genome of food crops, specific traits could be incorporated at a much faster speed than by traditional plant breeding. To make such improvements on seeds most of the developing world lack the required expensive infrastructure.

Taking into account the proportion of arable land that is affected by the changing climate a huge demand for those crop varieties can be expected. One could talk about a splendid business opportunity for the seed industry, if only those land areas where not destined to feed the poorest people of our planet. And here lays the major problem in the way biotechnology innovations nowadays have to emerge. Research and development cost a huge amount of money and the people who could benefit most from the new seed varieties are not able to pay for the improvements. Therefore research efforts aim to more profitable enterprises and not to investigations that could benefit the largest number of people. Although this problem is well-known in biotechnology research and development, a new element is brought into the debate, when the case is analysed taking climate-ready seeds into account. Climate change has been caused by greenhouse gases emitted mainly by developed countries.

A reason to act: global justice

Especially poor countries in the tropics, that received little benefits from the industrialization of the north, have to spend a huge amount of money for climate change adaptation. Global justice demands from the countries that have emitted most of the greenhouse gases to assist the developing world in adapting to a changing climate. If no solution is offered to overcome the problems caused by the new climate conditions, those countries will lose large harvests, with the consequence of hunger and malnutrition, making the population much more vulnerable to diseases, thus causing massive suffering and slowing down their development.

The UN member states have agreed with the Millennium Development Goals to halve the proportion of people who suffer from hunger by 2015. To achieve this goal, the additional challenge of decreased harvest yields due to climate change has to be taken into consideration. As this goal is ambitious on its own, improved infrastructure might not be enough to reach it due to the additional abiotic stress imposed on crop fields. Biotechnology research that is specifically oriented to crop development for the tropics has to be promoted, as it currently is severely underfunded. Privately funded agriculture research that is focused on the developing world amounts only to 6% of the total (De Schutter, 2009), showing a disproportion of resource dedication that is even higher than the already severely criticized 10/90 gap in pharmaceutical research and development along its failure to provide for medicines for tropical diseases at an adequate rate (Pogge, 2005). Since the right to development is recognized around the world as such, this unfair allocation of research efforts cannot be defended on any ethical grounds.

Another point in favour of aid is that the situation in which poor countries are finding themselves is not self-inflicted. Strong evidence shows that climate change is and has been largely caused by anthropogenic greenhouse emissions of countries in the north. This does not just leave the poor countries in a situation where they were accidentally placed in a situation of harm, in which altruism demands the affluent to aid. The countries that flourished with an early industrialization have unintentionally put those countries in this situation, as they ignored the consequences of massively releasing greenhouse gases in our atmosphere, and nowadays consciously worsen this situation, as they fail to take the necessary measures to limit emissions. Under this argument those countries should not only be helped out of charity, but also from an obligation of retributive justice.

To overcome this research gap a fair intellectual property regime has to come in place. After a brief introduction to the biotechnology research and development landscape, three issues will be discussed. Those are the unfair and unjust distribution of intellectual property rights, the problem of access to objects of innovation and the importance of a higher rate of innovation to cope with climate change.

The research and development landscape

The most common way biotechnology companies recoup their research and development expenses is by bringing a new product onto the market. For this step they also need to spend a considerable amount of money for biosafety regulations. In order to make sure that nobody copies their product, they apply for intellectual property rights, most commonly patents. Once granted, they have a monopoly right for a given time, thus being able to charge prices as they see fit. This method to recoup expenses has a series of problems. It is very expensive, many resources have to be relocated in legal expertise and fees for patents.

It is slow, it takes years to get a patent granted. It can be misused to block other competitors out, since once an intellectual property right is recognized, in practice there is no mechanism to force the patent holder to satisfy the market demand with products. When using a series of features patented by different institutions for the development of one product your expenses for licenses increase drastically because you will have to negotiate with every involved party.

To be able to put a product on the market, special task-driven alliances between a private and public part have been called for. They share much of their patented and traditional knowledge, are more open to share genetic materials, get better prices for licenses and more willingness to be assisted in biosafety testing, due to their not-for-profit orientation, as experience with private-public partnerships for neglected diseases has shown (Moran, 2005). Since they are promising, a brief introduction follows below.

A path for a fairer distribution of property rights: private-public partnerships

The poverty of the developing world is mainly pecuniary. However, those countries harbour the areas with the richest biodiversity. A late or still not occurring industrialisation of farming, has helped many small scale farmers survive, who continue to utilize in their fields a broad diversity of seeds and are still involved in breeding. Therefore the rich biodiversity of the developing world includes also an impressive amount of food crop varieties.

Seed industry giants have lost the trust of many local governments and communities, due to biopiracy and charging them unaffordable prices for their new products. As a result local governments have denied them access to genetic materials and traditional knowledge, slowing the rate of innovation down. To reverse this situation, past practices have to be changed. Philanthropic institutions have to mediate between biotechnology corporations and local governments and construct an atmosphere of trust. This can be done with clear contracts, that give a fair share of the benefits of intellectual property to all parties. Once local communities start to feel that they can benefit from such a cooperation, the news will spread out and more communities will be willing to share their traditional knowledge and own-bred varieties. If the world society manages to capture the inventiveness of small-scale farmers for the globalized system of innovation, a huge acceleration in the rate new crop varieties come out can be achieved. The saving of agrobiodiversity cannot be seen as a minor undertaking. The tremendous fund we now have is fruit of the effort to improve harvests done by millions of small farmers worldwide who have continued to use a wide variety of seeds (Altieri and Koohafkan, 2008). This fund is supplemented by the dedication of developing world's governments to conserve nature intact in wildlife reserves, thus rescuing wild relatives of crops varieties from disappearing.

Recognizing the inventive step made by indigenous communities does not only mean that they should be granted a part of the profits, the moral right that is customarily assured to the inventor by international intellectual property regimes, that is the right to have his name mentioned in the merits, if so he chose, should also be safeguarded. This is particularly important for communities that are neglected by their home governments, they can regain a sense of pride for their people's achievements.

The custom of charging excessive prices has to be changed too. The seed industry is at risk to be seen as one single player, as the foreign giant that is allied with an oppressive government, therefore abuses by one company can jeopardize the image of the whole industry, because many people will fail to distinguish among the different companies. It will be difficult to keep track of fair and unfair players, as companies continue to merge and develop alliances.

An aspect of the intellectual property regime that is often ignored in the discussion is the importance of trade secrecy. Here the private part of the partnership has to have the certainty to be able to hire scientists from abroad without running the risk, that they will sell the information gained inside to competitors. At a national level this situation is controlled by special contracts, in which the employee assures secrecy. Contract violations may conduce to criminal offenses that might only be enforced nationally. International laws to punish violations of contracts for trade secret protection are important to enforce and enact, only that way companies will not be afraid to hire at a higher rate scientists from developing countries. This is key to professional development in a globalized world, it facilitates communication and companies could ease their patenting habits up to a certain degree, because they will gain a new tool to protect intellectual property at a much lower cost.

Fair access to improvements

In the case of climate-ready seeds the problem of access is quite severe. The developing countries feel a double injustice: they have collaborated with the genetic materials for a solution to a problem that has been caused by the countries in the north, and now they have to pay for this solution a price that they can hardly (if at all) afford. Individual farmers are tempted to borrow money for improved seeds from local money lenders, paying enormous interest rates, with no guarantee that the harvest will have as high of a yield as promised by the seed industry. It is clear that this is a situation of injustice, but accusing the agricultural biotechnology industry alone of unfairness is also not right. The biotechnology industry is profiting from the situation, while offering something, that can be considered as the solution for one of the problems caused by climate change. They succeed making those huge profits because as a private industry they have virtually a monopoly. The lack of political will, that is connected with the refusal to recognize one's responsibility for the effects of climate change, to strive for a solution that helps the countries around the equator to adapt to a changing climate, has given rise to a situation, where the profit-driven seed industry is the only provider for a relief.

As it is unfair to let alone the poorer countries pay for climate change adaptation, it is equally unjust to make the seed industry solely responsible for providing an affordable solution, since the seed industry is only one of the many actors who have benefited from industrialization.

To make a solution for climate change adaptation accessible there has to be a generous subsidy from the countries who have profited from industrialization during the longest time, but for that, a clear inventory of the research and development costs of climate-ready seeds has to be available.

After this, there is still a major problem that has to be solved. Farmers have saved seeds from their harvest for the next season for millennia, but now, in some geographic locations, where the climate has changed without their fault, those farmers are forced to pay for every season anew for their improved seeds. Justice demands some sort of remuneration for the additional burden faced by those farmers.

Catalyzing and prioritizing innovation

An effort has to be made to keep innovation systems open. Facilitating access to working communication networks at least to mid-scale farmers around the world is an important goal. This is particularly important, since the inventiveness of millions of farmers could be added to the system of innovation. Observations about varieties withstanding effectively a particular abiotic stress could be thus easily shared. This information can help institutions and industry around the world provide solutions to adapt to the new climate conditions.

Research institutes and industry have to realize that patenting is not always profitable and is very time consuming. There are other alternatives, such as open source approaches, that allow freedom to operate and keep the technology from being monopolized by competitors, while putting the invention in the public domain and, depending on the conditions stated in the technology access agreement, even the built-up improvements.

Time is a factor that has to be taken into account. Lack of seeds improvements accessible at the right moment can mean large-scale famine. It is thus important to financially and materially aid partnerships that have the aim to develop certain crops that are urgently needed to feed the affected people. Crops that feed the largest number of people have to be prioritized. Those crops should also have a preference position in the waiting list for biosafety testing.

Conclusion

Key to a new start in a relationship of trust is to depart from old customs. Private industry in particular has to make a fresh start. They have to condemn employers that keep being unfair with indigenous communities and small farmers. Being consistent with this new conception of doing business, and showing publicly that they are condemning past practices, will aid them in establishing an atmosphere of trust. Communities in the developing world have to learn to distinguish among fair and unfair foreign and national biotechnology companies. As they blame the developed world for unfair prices and for not helping them to develop solutions for their particular problems, they should also be aware that a blanket distrust from their side slows down innovation that could benefit and even save millions of people from adversity and hunger, too. The appeal to share fairly is addressed to both parties, the developing and the developed world. Further, as they demand from the developed world recognition of rights for traditional knowledge, they should not underestimate the amount of new knowledge that is put continuously in the public domain and the utility of information that becomes open for public use after the intellectual property protection time-lapse expires.

To ensure food security under changing climate conditions, the developed world has to recognize its responsibility for the consequences of elevated CO_2 levels in our atmosphere. This responsibility demands finding solutions for climate change adaptation. Developing new seed varieties that can withstand better abiotic stress is an urgent task for adaptation. To accomplish this goal there has to be an increased expenditure in the agricultural research for developing world agriculture. As any delay in development threatens food security, countries with rich agrobiodiversity should be more open to share and be aware that they are not just ruining a business possibility for the biotechnology industry, but also blocking an opportunity for agricultural development to other countries in need around the world.

References

Altieri, M.A. and Koohafkan, P. (2008). Enduring farms: climate change, smallholders and traditional farming communities. Environment & Development Series 6. Third World Network, Penang, Malaysia, 62 pp.

De Schutter, O. (2009). The right to food. Seed policies and the right to food: enhancing agrobiodiversity and encouraging innovation. Available at: http://www.srfood.org/images/stories/pdf/officialreports/20091021_report-ga64_seed-policies-and-the-right-to-food_en.pdf. Accessed 23 March 2010.

Food and Agriculture Organisation (FAO) (2008). Twenty-ninth FAO regional conference for the Near East. Climate Change: Implications for agriculture in the Near East. Available at: ftp://ftp.fao.org/docrep/fao/meeting/012/k1470e.pdf. Accessed 23 March 2010.

Giddens, A. (2009). The politics of climate change. Polity, Cambridge, UK, 264 pp.

Kindelerer, J. (2008). The Cartagena protocol on biosafety. Collection of Biosafety Reviews 4. ICGEB Biosafety Unit, Trieste, Italy, pp. 12-65.

Moran, M. (2005). A breakthrough in R&D for neglected diseases: new ways to get the drugs we need. PLoS Medicine 2(9): 828-832.

Oldham, P. (2009). An access and benefit-sharing commons? The role of commons/open source licenses in the international regime on access to genetic resources and benefit sharing. Initiative for the Prevention of Biopiracy, Research Documents, Year IV, No. 11. Available at: http://ssrn.com/abstract=1438027. Accessed 23 March 2010.
Pogge, T.W. (2005). Human rights and global health: a research program. Metaphilosophy 36 (1/2): 182-209.

Transgenic livestock: a contribution to global food security?

R. Huber[1], C.B.A. Whitelaw[2] and I.A.S. Olsson[1]
[1]Instituto de Biologia Molecular e Celular (IBMC), Laboratory Animal Science – Universidade do Porto, Portugal; [2]Division of Developmental Biology, The Roslin Institute, University of Edinburgh, United Kingdom; rhuber@ibmc.up.pt

Abstract

According to the Food and Agriculture Organization (FAO), worldwide demand and production of animal derived protein is growing massively, due to demographic and societal developments especially in developing and newly industrialized countries. One of the key factors limiting the growth of livestock production in these countries is the climate: the highly productive breeds used in Europe and North America do not do well in the tropical, subtropical and desert climates of many developing countries and the native, locally adapted breeds often have poor production traits. It has been argued that genetically modified (GM) livestock may contribute to food security and food safety, by improving both production traits (e.g. growth performance) and disease resistance. Unlike the present situation with crops, no transgenic animal products have been approved for human food consumption so far, but currently regulatory bodies (e.g. the Food and Drug Administration – FDA) are assessing related food security and biosafety issues. However, such applications have been met with some scepticism – in surveys conducted across the developed world applications related to food production are less accepted by the public than biomedical applications of transgenic animals. Within the scientific community there are also doubts about the introduction of transgenic animals into the food chain at present. In the present paper/presentation, we want to give an overview of (potential) contributions of transgenic livestock to food security and safety, to pinpoint important issues of discussion/controversy related to the technology and to identify areas of research that should be undertaken before considering the use of transgenic animals in food production.

Keywords: biotechnology, animal husbandry, developing countries

Introduction

'Food security exists when all people, at all times, have access to sufficient, safe and nutritious food to meet their dietary needs and food preferences for an active and healthy life.' World Food Summit (1996)

Food security and the role of livestock

The EC–FAO Food Security Programme identifies 4 dimensions of food security: (1) the physical availability of food, (2) the economical and physical access to food, (3) adequate utilization of food resources and (4) the stability over time of points 1-3. Sub-Saharan Africa has faced the worst situation in terms of food security: 13 of the 17 countries listed in the EC-FAO Food Security Country Briefs are located there (FAO, 2010).

Livestock plays an important role for food security in developing countries, both because of the socioeconomic importance animals have for many people and households and in terms of protein supply. While high meat consumption is now associated with health problems in affluent populations, for low-income countries health gains are expected from moderately increasing the present low consumption, (McMichael *et al.*, 2007). As a follow up to the 'green revolution' in the 1960's and 1970's, the 'livestock revolution' reflects an exponential growth of production and demand of meat and dairy products worldwide since the early 1980's. The developing world account for 80% of the increase in meat production between 1980 and 2004, doubling per capita consumption and trebling production over the same period. The trend in production and demand for animal derived products is likely to continue over the next decades with extensive socioeconomic and ecological consequences (FAO, 2005). Three major 'players' – Brazil, India and China – have emerged and are responsible for most of this increase, while regions like Sub Saharan Africa have seen a comparatively moderate rise in production. They are often characterized by low economic growth and unfavourable production conditions regarding climate and incidence of diseases.

Breeding and biotechnology in developing countries

Despite an ongoing intensification process, small scale producers, the use of indigenous breeds and traditional local market structures are predominant. Indigenous breeds kept in low input – low output systems are well suited to these constraints while high-performance exotic breeds do not thrive in unimproved production environments (FAO, 2006). There has been continuous effort to improve animal genetic resources. Apart from traditional breeding techniques, biotechnological approaches like marker assisted selection, artificial insemination (AI), embryo transfer (ET) and multiple ovulation combined with embryo transfer (MOET), *in vitro* production of embryos (IVEP) have been used to improve breeding success. Gene based technologies have also been employed for the development of vaccines, diagnostics and epidemiology. Presently genetic engineering only influences the livestock sector indirectly via the use of genetically modified (GM) crops as animal feed (Madan, 2005).

Transgenesis in livestock: promises and applications in food production

Transgenesis involves the integration of exogenous genetic information into the genome of animals. A number of potential applications of transgenesis to production animals have been attempted with varying success. The most advanced studies targeted classical production parameters such as animal growth, carcass composition, meat quality and milk composition and yield. Moreover, successful attempts to improve the animal's resistance to diseases (e.g. mastitis) and decrease the environmental impact (Enviropig™) have been reported (Laible, 2009). Presently, growth enhanced GM fish is on the verge of obtaining regulatory approval by the FDA. Although no genetic modifications in livestock targeting parameters specifically relevant for developing countries have been reported yet, such applications have been discussed (Thomson *et al.*, 2009).

Transgenic livestock: discussion

In industrialised countries transgenic technology has been the focus of widespread discussion both in the public domain and within the scientific community. Public perception, ethics, risk assessment, decision making and regulatory mechanisms are important factors determining whether or not a technology will be implemented.

Ethics, public opinion and scientific opinion

Major topics in the public and ethical debate on transgenic animals are: (1) the relation of (potential) harms and benefits for humans, animals and the environment (2) the purpose of the modification (merely economical vs. wider benefit), (3) the position on the phylogenetic scale of the organism to be modified and (4) the level of trust in regulatory mechanisms and transparency. Public acceptance of biomedical applications of transgenesis has remained higher than other applications, e.g. food production or creation of transgenic companion animals, especially in Europe (Einsiedel, 2005; Lassen *et al.*, 2005). However, available data on these issues is largely based on surveys performed in western societies, empirical data on risks and benefits is limited and almost nothing is known about attitudes of farmers and consumers in developing countries.

It has been argued that too much emphasis on the risk of using GM animals can lead to a 'risk of not using GM animals' (Murray and Maga, in press). However, economical, technical (e.g. efficiency of gene transfer) and biological (e.g. long generation intervals in livestock, high influence of the environment on the phenotype) considerations also limit the perspective of imminent introduction of transgenic technology in food production. Overstatement of the potential benefits of transgenesis has also been criticized (Blasco, 2008).

Regulation and risk assessment

A framework for the introduction of the technique into the livestock production chain has been devised that incorporates (potential) benefits/costs for the major stakeholders involved (i.e. producers and consumers) and highlights the decisive role of regulation in the process (Novoselova, 2007). It is likely, however, that regulatory bodies in different regions of the world will adopt differing approaches to transgenic livestock production similar to the 'transatlantic rift' that has existed between the US and European policies towards GM foods and crops (Aerni and Bernauer, 2006; Ramjoue, 2007).

Transgenic livestock: relevance for developing countries

Why use transgenic livestock in developing countries?

It is unlikely that application of transgenesis in livestock will replace functioning conventional measures like selective breeding, the use of efficient vaccines, etc. The great allure of transgenesis lies in the possibility of introducing changes that are difficult or even impossible to obtain through more conventional measures of management, breeding and health care. By introducing genetic information from another species, traits can be targeted that could not be improved at all or only after long periods of selective breeding. Apart from species specific absence of genetic information on a desirable trait making breeding for it impossible, the success of conventional breeding may be limited by low allele frequency of desirable traits in a population, thus complicating selection. Similarly, genes with desirable effects that cannot be introduced from other breeds by introgression are potential candidates for a transgenic approach (Blasco, 2008).

Ideal candidate traits for improvement of livestock through transgenesis are monogenic traits or traits controlled by few loci that are not very susceptible to environmental conditions. Gene based disease resistance is an example of a promising field that could be targeted by transgenic technology and is highly relevant for developing countries (Whitelaw and Sang, 2005). It is possible that ongoing efforts in genome mapping will reveal additional candidates for application of transgenesis specifically in developing countries (Thomson *et al.*, 2009).

Lessons from biotechnology applications in livestock breeding and GM crops

In extremely poor parts of the developing world it is likely that transgenesis will face similar obstacles as other biotechnological methods used in livestock breeding. These include the lack of qualified human resources, insufficient cooperation between industry, universities and institutions, failure of developing risk assessment schemes, economical limitations and the delicate equilibrium of the local ecosystems. Also technical limitations (e.g. mistakes in processing, storing and thawing of semen for AI) under field conditions have been identified as inhibiting factors (Madan, 2005).

Genetically modified livestock would not be the first application of transgenesis in developing countries and experiences from the discussion surrounding the introduction of GM plants may provide a background perspective. There is a great deal of disagreement on the benefits of GM crops in developing countries, often connected to the aggressive marketing policies of multinational companies, overstating the potential benefits of their technology, and putting countries and producers under pressure (Glover, 2008).

However, stakeholders in some developing countries seem to show rather pragmatic attitudes towards genetically modified crops, seeing the potential of solving important problems in agriculture (e.g. drought, pests) without ignoring the potential risks and obstacles.

In the case of GM crops, researchers emphasize the need for an adequate estimation of the time and money necessary, prudent design of regulatory frameworks and the support of crossdisciplinary research to cope with the diversity in agro-ecological and socio-economical systems. Strong institutional infrastructure coupled with functioning cooperation between public and private stakeholders to secure financing and the building and maintenance of (local) capacities and human resources to support the technology are also regarded as crucial (Eicher *et al.*, 2006; Raney, 2006).

Transgenic livestock in developing countries: closer than we think?

Scenarios resulting from an introduction of transgenic livestock into production systems in developing countries are hard to predict and depend on many factors. However, it is possible to think of a variety of desirable applications, some requiring substantial changes to the way farmers have to manage their livestock, others compatible with present production systems and conditions. As an example of the latter, one could imagine the introduction of genetic disease resistance into a local breed where no efficient vaccine is known. Theoretically, this would not change greatly the requirements to the care and management while improving production and resource economy by reducing loss of animals. The level of production per animal would remain within the biological limits of the local breed, therefore not increasing pressure on the environment through nutritional requirements. Fewer animals would have to be reared for the same production and disease specific veterinary treatments would be obsolete. Of course, this is the case only if the genetic modification does not interfere with characteristics other than the targeted ones.

On the other end of the spectrum, animals with an increased output in milk and/or meat may prove more challenging for the agro-ecological system due to the increased nutritional requirements needed to grow faster and secrete more milk. Highly productive animals are comparatively sensitive towards environmental conditions, regular and sufficient availability of food and water and susceptible to diseases. This may be challenging in itself for the farmer and it would become even more challenging in the case of use of exotic transgenic breeds.

The often extreme production environments found in developing countries cannot be neglected in the context of breeding, as the phenotype is a result of genotype and environment. These environmental conditions also 'contribute' to traditional breeding by eliminating unfit animals by natural selection, and it has been suggested that this process would have to be anticipated and mimicked by humans in the case of transgenic livestock (Blasco, 2008).

Apart from market and management relevant aspects (Jabbar, 2003) consumer and producer preferences of livestock pheno/genotypes in developing countries often incorporate cultural values and culturally valued traits (e.g. horn shape in cattle) more than in most industrialized countries (Zander, 2008). Including these traits, whatever technique used, may be crucial for acceptance by livestock keepers in developing countries.

Lessons learned from the already used biotechnological methods (GM crops and livestock breeding) point towards technical and management limitations rather than lack of public and political support; however very little is known about how the different stakeholders view the ethical issues surrounding transgenic livestock.

Although technical difficulties have been and will be overcome in the future, transgenesis remains a technology requiring resources and research infrastructure which may not be available in most developing countries. If a participatory approach is selected, local research and infrastructure has first to be improved before jointly developing strategies to implement transgenesis into the food chain. A participatory approach is also valuable in seeking to determine the most relevant genotypes for the respective region and avoiding economical and political dependence.

Transparent communication of risks and benefits associated with the technique and a transparent and comprehensive layout of regulations concerning large animal transgenesis will be a particular challenge to countries with serious problems in terms of democracy and stability.

Conclusions

At the moment, transgenic livestock are not contributing towards global food security as no applications have entered the market yet. It is difficult to predict if/when this is going to happen; however theoretically transgenesis seems promising for a number of reasons. It allows a desirable phenotype to be obtained within few generations. Depending on which that phenotype is, it may be possible to improve quality and quantity of food produced and reduce loss of animals, thus contributing towards increased sustainability. However, a number of technical, institutional, societal and political considerations will determine whether transgenesis will be implemented as to serve the populations most affected by food security problems. Here research is urgent to understand the specific situation of developing countries as regards stakeholder views, risks and regulation, taking into account the variety of agro-ecological, cultural and socio-economical needs. Case-based surveys may help to better understand what people need and what they would be willing to accept in terms of transgenic livestock.

References

Aerni, P. and Bernauer, T. (2006). Stakeholder attitudes toward GMOs in the Philippines, Mexico, and South Africa: The issue of public trust. World Development 34, 557-575.

Blasco, A. (2008). The role of genetic engineering in livestock production. Livest. Sci. 113, 191-201.

Eicher, C.K., Maredia, K. and Sithole-Niang, I. (2006). Crop biotechnology and the African farmer. Food Policy 31, 504-527.

Einsiedel, E.F. (2005). Public perceptions of transgenic animals. Rev. Sci. Tech. Off. Int. Epizoot. 24, 149-157.

FAO (2005). Livestock policy brief. Responding to the 'Livestock Revolution' – the case for livestock public policies. ftp://ftp.fao.org/docrep/fao/010/a0260e/a0260e00.pdf last accessed 24.03.2010.

FAO (2006). Livestock Report 2006. http://www.fao.org/docrep/009/a0255e/a0255e00.htm last accessed 24.03.2010.

FAO (2010). EC-FAO Food Security Programme. http://www.foodsec.org/programme.htm last acessed 24.03.2010.

Glover, D. (2008). The corporate shaping of GM crops as a technology for the poor. J. Peasant Stud. 37, 67-90.

Jabbar, M.A. and Diedhiou, M.L. (2003). Does breed matter to cattle farmers and buyers? Evidence from West Africa. Ecological Economics 45, 461-472.

Laible, G. (2009). Enhancing livestock through genetic engineering-Recent advances and future prospects. Comp. Immunol. Microbiol. Infect. Dis. 32, 123-137.

Lassen, J., Gjerris, M. and Sandoe, P. (2005). After Dolly – Ethical limits to the use of biotechnology on farm animals. In: Symposium on Agricultural and Societal Implications of Contemporary Embryo-Technologies in Farm Animals, Copenhagen, DENMARK, Jan 12, 992-1004.

Madan, M.L. (2005). Animal biotechnology: applications and economic implications in developing countries. Rev. Sci. Tech. Off. Int. Epizoot. 24, 127-139.

McMichael, A.J., Powles, J.W., Butler, C.D. and Uaoy, R. (2007). Food, livestock production, energy, climate change, and health. Lancet 370, 1253-1263.

Murray, J.D. and Maga, E.A. (2010). Is there a risk from not using GE animals? Transgenic Res., in press.

Novoselova, T.A., Meuwissen, M.P.M., Huirne, R.B.M., (2007). Adoption of GM technology in livestock production chains: an integrating framework. Trends Food Sci. Technol. 18, 175-188.

Ramjoue, C. (2007). The transatlantic rift in genetically modified food policy. J. Agric. Environ. Ethics 20, 419-436.

Raney, T. (2006). Economic impact of transgenic crops in developing countries. Curr. Opin. Biotechnol. 17, 174-178.

Thomson, R., Molina-Portela, P., Mott, H., Carrington, M. and Raper, J. (2009). Hydrodynamic gene delivery of baboon trypanosome lytic factor eliminates both animal and human-infective African trypanosomes. Proc. Natl. Acad. Sci. U.S.A. 106, 19509-19514.

Whitelaw, C.B.A. and Sang, H.M. (2005). Disease-resistant genetically modified animals. Rev. Sci. Tech. Off. Int. Epizoot. 24, 275-283.

Zander, K.K. and Drucker, A.G. (2008). Conserving what's important: Using choice model scenarios to value local cattle breeds in East Africa. Ecological Economics 68, 34-45.

Communication and biotechnology: a decade of misunderstandings

E.F. Guiral
Executive Secretary of the Asociación de Periodistas Agroalimentarios de España (APAE), University School of Agricultural Technology Engineering, 28040 Madrid, Spain; elena.f.guiral@apae.es

Abstract

Biotechnology and its applications, especially in the areas of food and farming, have constantly been in the media over the last ten years. Reports, interviews and opinion columns have regularly referred to this technology. The media's approach to the controversial and scientifically complicated subject of genetically modified seeds is a typical example worthy of further study. It reflects the enormous challenge involved in communicating scientific innovation, within the life sciences, to the public. Journalists have been duly faced with diverse sources of information, many of which contradict one another. Therefore on many occasions journalists have been unable to meaningfully satisfy either the needs of their professional sources, or the interests and concerns of their readers. Furthermore, communication professionals have attributed the same level of importance to technical and scientific arguments as they have to emotional or social ones. As a result the public has been bombarded with confusing arguments that have provided more 'misinformation' than information. The role of scientists and the media should be to close the gap between the overabundance of data and the lack of understanding that assails us every day. As mediators in the public debate on biotechnology, the media have a great responsibility. A good science journalist should not limit herself to presenting two opposing arguments and attach the same validity to both. Instead she should give greater priority to scientific rigour and truth, and this can only be done by consulting the appropriate scientific sources. Journalists have an ethical duty to objectively inform the public so that it in turn can exercise its rights and responsibilities. This is the pending task of agro-food biotechnology, following ten years of bad communication and misunderstandings.

Keywords: biotechnology, science journalism, ethical responsibility

Introduction

Over the last decade, biotechnology – and its applications in health, agriculture, food and the environment – has featured in innumerable press articles of all sorts. Features, interviews, opinion pieces and even humorous vignettes have been written about the subject, mainly related to agriculture and food. It is not advisable to cling to the old idea of the media as the 'fourth estate' – a metaphor for their influence on the public, opinion formers and, consequently, political decision making. However, the influence they have had on the development of this technology in Europe and Spain is undeniable, which is what concerns us here. The lack of communication between scientists, companies and journalists, together with the barrage of contradictory or polarized information between different sectors, has made the media's informative role more difficult.

One of most prominent characteristics of information regarding biotechnology is that it has a wide range of implications: scientific, sanitary, environmental, economic, societal, ethical and even political. This is why nearly all media, written and audiovisual, have published, sooner or later, information about the subject. Science journals that have published the latest findings, publications aimed at housewives concerned about food safety, environmental media sensitive to the criticisms of ecologists... they have all treated the issue of biotechnology from very different perspectives, using completely different sources and resources.

The complexity of science journalism: a definition and problems

Manuel Calvo Hernando, an authority on the popularisation of science, defines science journalism as 'the specialisation of popularising science and technology through the mass media' (Calvo Hernando, 1997). For Calvo Hernando, hope and fear go hand in hand in science journalism, a reflection of what happens in science itself. These days our lives are dominated by technology, so we have a certain apprehension about an uncertain future full of risks: new types of food, nuclear energy, toxic residues, overpopulation...

Strangely enough, our current society is increasingly less prepared to take risks and craves more and more certainty – but science cannot guarantee this certainty with the methods at its disposal. But at the same time scientific progress means we experience new risks and uncertainties, such as new modes of transport or new foods. On the other hand, while we are a lot more sceptical than at the beginning of the 20th Century, humanity continues to place many of its hopes in scientific development, above all in two areas: a solution to global overpopulation and a cure for serious diseases such as AIDS and cancer. In these cases journalists play a crucial role as intermediaries between scientists and the public regarding technological advances, but also, as has already been commented, as a guide regarding the uncertainties and possibilities that these new technologies represent.

Agro-food biotechnology: a fascinating but controversial subject

Why has biotechnology, and more specifically the issue of food derived from genetically modified seeds, aroused such interest among journalists and the public? Although it is difficult to give a comprehensive explanation, it is quite likely because biotechnology is concerned with the fundamental issue of where life itself comes from. The ethical and philosophical repercussions associated with biotechnology are undeniable.

From the moment that we ask whether or not it is 'natural' to modify the genome of a bacteria, virus, animal, plant or even a human being, we are faced with huge uncertainty, precisely because it is about such a new, and potentially infinite, possibility. Genetics, the second big revolution of the 20th Century after information technology, awakens doubt as much as hope, and this makes it an especially attractive subject for readers.

Added to this is the fact that it is such a relatively young technology, so consumers are largely worried about the health repercussions, above all in the long term, given that scientists are still not sure. Biotechnological applications related to agriculture and the environment also interest many people who are increasingly concerned with protecting the planet. In this regard, biotechnology represents both sides of the coin. On the one hand, bioremediation (the elimination of heavy metals from the soil using transgenic plants designed with this function) opens up huge possibilities to improve the environment in the future. On the other hand, there are renewed doubts concerning the influence of genetically modified crops on the ecosystem and biodiversity.

Added to all these factors, more or less related to the popularisation of science, is one more: the potential for biotechnology to generate economic wealth, skilled jobs and stock market success. These things have all turned biotechnology into a 'new value' for the economic press. Despite being a subject of the utmost interest which has generated a veritable avalanche of information and tens and tens of media outlets, today's public does not feel adequately informed. Or, to be more precise, the public feels 'minsinformed.' In the latest Euromonitor survey on Biotechnology (2006), 80% of those interviewed said they had not received enough information – presumably clear and correct – about the subject.

The reason for this situation is the constant stream of contradictions that reach journalists, whose work is to receive, evaluate, select and offer up clear and straightforward information to the reader, radio listener or television viewer. On the one hand, as Victoria Camps says in her study *The Perception of Biotechnology in Spain* (Camps, 2002), published by the Victor Grifols Foundation, 'environmental organizations have produced alarmist information which is rarely based on scientifically rigorous data.' Added to this problem, the companies that produce the genetically modified seeds at the heart of the controversy have taken too long in satisfying the media's requirements for information.

Apart from not being a reliable source of information because of their obvious interest in the issue, to make money from biotechnology, these companies were also surprised by the avalanche of news stories about a sector that had traditionally only been written about by the specialist agricultural press. Furthermore, until that point companies in the agro-fishing sector had hardly invested at all in their communication departments, which meant they virtually had to start from zero.

The lack of contact between the companies and journalists meant that when the controversy broke over genetically modified organisms, the relationship between the two sides was 'momentarily paralyzed.'Another problem was that the journalists that first covered stories on biotechnology in the general media were the same ones that had previously been assigned to environmental stories. Therefore their first instinct was to turn to their trusty sources, environmental organizations, which initially influenced their point of view and the way in which they dealt with the news. Nor did environmental scandals involving multinationals as Bolidén in Alnazcóllar help much to improve relations between journalists and companies.

Communication between scientists, the main source of information on biotechnology, and journalists has not been as fluid as it might have been for several reasons. Firstly, scientists find it very hard to explain a subject as complicated as genetics in an accessible way, not only to readers but also to general news journalists. Then there is the difficulty they have of expressing things categorically, given that scientific hypotheses can rarely be guaranteed to be one hundred percent valid.

Another concern is that sometimes the scientific community does not approve if one of its members participates in certain media, such as magazines aimed at housewives, because these types of publications do not contain authorized opinions on particular issues. Finally, one must not forget that journalists have to work 'against the clock', which means that often there is not enough time or resources to find countervailing information to that provided by primary sources, or to check that what is published is scientifically rigorous.

As Mariano Belenguer explains, referring to specialist journalists in his book *Journalism: research proposals* (Belenguer, 2001), 'science journalism is more dependant on groups of journalists than it is on groups of scientists. Science journalists should go to the places where science is generated, but those that ultimately decide what is and what is not science are the scientists themselves.' In terms of biotechnology, the role of the journalist should be 'to distinguish in each specific case the genuine scientific sources, as opposed to the pseudo-scientific ones, and collaborate with scientists to unmask false science. Web 2.0 revolution through social networks, blogs, webs and Google news offer interesting communication channels, but could be a little valuable and scientific source of information. How distinguish true and false science? Applying common sense, interest and going to the best scientific info providers. As an example, some public scientific informations platforms as Cordis, SINC and Alphagalileo.

Biotechnology, ethics and communication: conclusion

Journalists play a crucial role in the development of Global Bioethics which, as described by Marcelo Palacios (Palacios, 2000), 'should guide the use of science and technology to promote the development of society, combining respect for human dignity with respect for the environment and all living creatures.' Communicators have the attributes and intellectual capacity to take on this key role: to contribute to the sustained and sustainable development of society, paying special attention to those communities that need help in order to survive. Providing information in an ethical and rigorous way regarding the possibilities and challenges presented by agricultural biotechnology, without being manipulated by biased or sensationalist views, is essential.

The main aim of the Asociación de Periodistas Agroalimentarios de España (APAE), founded in 1933, is to promote the professional and ethical development of agro-food journalism in Spain, in order to guarantee the recognition of its key role in the sector, as well as its goals and aspirations. It is part of the International Federation of Agriculture Journalists (IFAJ) that brings together agro-food journalists from around the world.

References

Alphagalileo, (undated). Agencia de información europea sobre ciencia, medicina y tecnología. Available at: http://www.alphagalileo.org/.

APAE, (undated). Asociación Española de Periodistas Agroalimentarios: www.apae.es.

Belenguer, M. (2001). El periodismo científico, objeto de investigación. Padilla Libros Editores y Libreros, Sevilla, 2001.

Calvo Hernando, M. (1997). Manual de periodismo científico, Editorial Bosch, Barcelona, 1997

Camps, V. (2002). Percepción social de la Biotecnología, Fundación Víctor Grifols i Lucas, Barcelona, 2002.

Cordis, (undated). Servicio de Información Comunitario sobre Investigación y Desarrollo. available at: http://cordis.europa.eu/home_es.html.

Eurobarometer 64.3: 'Europeans and Biotechnology in 2005: Patterns and Trends' , Comisión Europea, 2006.

Palacios, M. (2000). 'Bioética 2000', Sociedad Internacional de Bioética, 2000.

Renobales Schleifer, M. (2010). 'Alimentos más sostenibles: las semillas transgénicas en la agricultura ecológica', Sociedad Internacional de Bioética, 2010.

SINC, (undated). Servicio de información y noticias científicas de la Fecyt. Available at; http://www.plataformasinc.es/.

Part 5. Eating, nutrition and health: ethical and legal considerations

Conflicting concerns in fighting obesity

K.K. Jensen
Danish Centre for Bioethics and Risk Assessment, Faculty of Life Sciences, University of Copenhagen, Rolighedsvej 25, 1958 Frederiksberg C., Denmark; kkje@life.ku.dk

Abstract

Fighting the obesity epidemic gives rise to conflict between several major ethical concerns. From the point of view of total welfare in society, it seems justified to seek to minimise the prevalence of obesity through restrictions on consumption believed to increase obesity and to seek to change peoples' life style away from obesity through positive and negative incentives. It may be compatible with liberal principles to invoke various restrictions on consumption or to put incentives for a healthy life style in place. Still, these strategies might involve a tendency to undermine the autonomy by not respecting individuals as persons with responsibility for their own life. This tendency is strengthened by public health care, which generally lets the community bear the health care costs for people with unhealthy life styles. This paper argues that, if the principle of responsibility is carried to its logical conclusion, then health insurance would have to be exclusively a personal matter. But the principal reason for having a public health system is that private health insurance would be unfair. Moreover, the complicated causal mechanisms underlying obesity makes it difficult to apply the concept of personal responsibility. Hence, once public health care is in place, a principal ethical question is how to weigh up the concern for efficiency of resources spent, and the concern for fairness. Still, the problem with responsibility remains. Clearly, it would be better, if obese people would become motivated to take more responsibility for their own health. Presumably, however, obese people are victims of negative attitudes in the general public and also to some extent of hidden discrimination in health care. In order to respect autonomy, and also to be effective, offers of help must be free of condemning attitudes.

Keywords: efficiency, fairness, responsibility

Introduction

Worldwide, an obesity epidemic is evolving (WHO, 2010). This development represents a huge challenge for health care budgets since, as a consequence of the epidemic, the prevalence of obesity related diseases are increasing and is expected to be increasing dramatically in the future.

From the point of view of total welfare in society, it seems justified that governments seek to minimise the prevalence of obesity through restrictions on consumption believed to increase obesity and seek to change peoples' life style away from obesity through positive and negative incentives.

This is an example of governments trying to influence a (large) minority's life styles in the direction which the majority finds desirable. Is this a threat to the freedom of people in a liberal society to choose their own life style? As John Stuart Mill (1859) famously stated the principle of liberalism, 'That the only purpose for which power can be rightfully exercised over any member of a civilised community, against his will, is to prevent harm to others', it seems compatible with liberal principles to invoke various restrictions on consumption or to put incentives for a healthy life style in place, since incentives are not coercion.

However, on a deeper level, these strategies might involve a tendency to undermine the freedom of people by considering them as objects of social engineering rather than addressing them as autonomous subjects with responsibility for their own life. This tendency is strengthened by public health care, which generally lets the community bear the health care costs for people with unhealthy life styles, rather than letting them experience the consequences of their choice of life style themselves without interference.

Hence, to promote the autonomy of people, there is reason to hold them responsible for their choice of life style. And for reasons of fairness, the costs from an unhealthy life style should not be borne by the community, but rather by the person choosing to live unhealthy.

I shall demonstrate that, if this reasoning is carried to its logical conclusion, then health insurance would have to be exclusively a personal matter. But the principal reason for having a public health system is that private health insurance would be unfair, because people do not have access to a fair share of resources.

However, even within a public health care system, it is not clear what fairness more precisely implies as regards treatment of obese people. There might be a right to equal access to treatment, regardless of life style; but there is also a concern for cost efficiency, requiring resources to be spent where they produce most health per Euro. And obese people might not benefit as much as people living more healthilyfrom a specific treatment. I shall discuss how these concerns should be weighed against each other.

Moreover, if the concern for fairness only can be met by a public health insurance system, there seems to be conflict between the concern for autonomy and the concern for fairness. I conclude by arguing that this dilemma can be solved to a certain extent, if public health care abstains from condemning obese people.

Responsibility for one's own health

Every now and then, suggestions are made to the effect that people with unhealthy life styles, e.g. smokers, should be kept responsible for diseases resulting from their life style by not getting treatment with the same priority in public health care as people living more healthily. The obesity epidemic has made such suggestions no less frequent.

Often, these suggestions reveal a condemning attitude in that only risky life styles with a negative image are mentioned, e.g. smoking, drinking, over eating and lack of exercising; however, other risky life styles like going in for extreme sports, skiing or mountain climbing are not mentioned even though injuries resulting from these can be said equally well to be a matter of personal responsibility. Clearly, if not all forms of responsibility are taken into account, the condemned groups are being unjustly discriminated against.

A widespread argument against holding people responsible for certain diseases observes that many diseases do not have well defined single causes; rather, multiple risk factors are known. Under these circumstances, it seems impossible to ascertain exactly when a person is responsible for a disease because of his life style. The Danish Ethics Council (Etisk Råd, 1996), for one, uses this observation to reject priorities in health care based on responsibility.

However, from the perspective of autonomy, this conclusion is perhaps too hasty in dropping the concern for responsibility. From the fact that, in most cases, there is uncertainty about the exact consequences for a person of his choice of life style, it does not follow that it does not matter what he chooses. We can all be said to be responsible for how we are dealing with the uncertainty about our future health.

Since dealing with uncertainty is a matter of one's attitude to risk, and since people diverge in their attitudes to risk, it seems fairer if also the costs of health insurance becomes peoples' own responsibility. Thus, people with risky life styles cannot just invoke the costs of their life style on the community; they have to decide themselves, how much insurance they are prepared to pay for.

But to many, private health insurance would only be fair, if people had a fair share of resources at their disposal. Some people are born in very poor families, and some people are born with innate handicaps or diseases. A private health insurance system without some form of compensation in the amount of resources available for these groups would appear unfair.

However, even in theory it is very difficult to determine what would be a fair share of resources. And even if it could be determined, it would be very difficult to point out the practical implications. Finally, it would probably not be politically feasible to ensure the necessary schemes of compensation.

Both in theory and in practice, however, it is feasible to ensure an equal right to treatment in public health care. This might be less fair than equal shares of all relevant resources, but this is how far fairness is feasible in most societies. And in my view, this is the main argument underlying systems of public health care. But if the concern for fairness is met in this way, a conflict between the concern for fairness and the concern for autonomy has been opened. The problem remains, that public health care might cause people to become more disinterested in taking responsibility for their own health.

Efficiency and equal right

However, there are further questions about fairness in health care. Apart from the right to equal access to treatment in relevantly similar cases, there is a concern for cost effectiveness in public health care. If resources are not spent efficiently, welfare is lost through since fewer people than feasible are getting treated. This might be considered an aspect of fairness. But even if it is not, there is conflict between the concern for cost effectiveness and the concern for equal access to treatment which might affect obese people.

Thus, in many cases, an obese person continuing an unhealthy life style after treatment for a given disease, e.g. diabetes II, will not benefit as much as a more healthily living person. Hence, from the perspective of cost effectiveness, resources are better spent on the more healthily living person than on the more unhealthily living.

How should the concern for cost effectiveness be weighed up against the concern for fairness in the sense of equal access to treatment? John Broome (1990-91) has pointed out that a natural and probably widespread answer to this question does not take fairness into account in the right way.

Consider the case of diabetes II outlined above and assume that just one treatment is available. Fairness gives the more healthily living and more unhealthily living equally strong reason to be treated. However, cost effectiveness gives stronger reason to treat the more healthily living. When we weigh up these reasons, it is very hard to avoid the conclusion that overall, the strongest reason are in favour of treating the healthy. But then, Broome says, the right to equal access has no content. The unhealthy's right to equal access to treatment is bound to be overridden. He could make the complaint that this would be unfair to him.

The right way to take fairness into account, according to Broome, is first to decide which concern has more weight, cost effectiveness or right to equal access. If the latter is the case, then both have the same right to be treated. If resources are indivisible, i.e. if there is only one treatment available as I assumed, the only way to respect right to equal access is to hold a lottery. This at least gives the more unhealthily living a chance of treatment. If, on the other hand, cost effectiveness has more weight, then the more healthily living person should get the treatment. This is unfair to the more unhealthily living, but in this case the unfairness is overridden by the concern for cost effectiveness.

The important question is, of course, when exactly fairness has more weight. I shall not attempt a detailed answer to this question. I shall only suggest that the closer we get to matters of life and death, the more weight the concern for fairness seems to have.

Autonomy and condemnation

The way the concern for fairness is taken into account by public health care might not promote autonomy. Equal access to treatment regardless of life style does not let people experience fully the costs of their choice of life style. This might be an obstacle to motivating obese people for a healthier life style.

It might be objected that more responsibility not necessarily leads to less obesity. Thus, obesity appears more prevalent in the U.S. than in Europe, even though public health insurance in the U.S. is covering less than in Europe. Assuming that being obese is not in the self interest of anyone, it seems clear that part of the problem with the obesity epidemic is that even in the face of relevant information about the consequences of obesity many obese people appear unable to change their life style. Life style habits are often formed during childhood, and once a person happens to be trapped in obesity, it is very difficult to lose weight even if the ambition to do so is present.

Exactly in these cases, personal motivation is probably most important in order to achieve positive results. It is hard to imagine deep going changes in life style and personality without a personal recognition of the problem and a strong wish for a change. And for children, it appears necessary to engage their families if a change of life style is going to have lasting effect. And most severely obese people do appear to want a change, but find it very difficult to achieve it.

It seems clear that if people are motivated to change to a healthier life style in order to fight their obesity, but find it hard to accomplish this task, public health care should offer help. However, the general condemnation of obese people in many societies clearly makes it more difficult for them to keep up the spirit and self esteem necessary to accomplish an efficient change of life style. Moreover, there appear to be both open and more hidden discrimination against obese people in public health care.

It is in my view, as argued above, open discrimination if obese people in cases of serious diseases get less treatment as compared with more healthily living people. Apparently, there is also widespread, but more hidden discrimination against obese people in public health care (Hebi and Xu, 2001). In both cases, the result is of course increased difficulties for obese people in forming a wish for a change of life style, sticking to it and carrying it to success.

The psychological challenge here is to take autonomy seriously. Roughly, it might well promote autonomy to insist on personal choices being able to make a difference; however, it does not promote autonomy to condemn people. In other words, in order for public health care servants to meet obese people as autonomous agents, they should become able to not condemn them, but to meet them as people who have difficulties in living as they want. This probably involves a form of therapeutic neutrality.

Conclusion

Increased autonomy appears to be a necessary precondition of fighting severe obesity. However, the concern for fairness in health care is probably best met in public health care systems where autonomy cannot be fully promoted because people in these systems by definition are protected against bearing the full cost of their choices.

Autonomy as a precondition of fighting obesity is also threatened by general condemnation and open and hidden discrimination in public health care. If autonomy is to be promoted within a public health care system, discrimination as well as condemnation must be avoided. Public health care servants should meet obese people in a neutral way in order to promote their autonomy.

References

Broome, J. (1990-91). Fairness. Proceedings of the Aristotelian Society 91: 87-102.

Etisk Råd (1996). Prioritering i sundhedsvæsenet – en redegørelse. København: Etisk Råd [In Danish].

Hebi, M.R. and Xu, J. (2001). Weighing the care: physicians' reactions to the size of a patient. International Journal of Obesity 25: 1246-1252.

Mill, J.S. (1859). On Liberty, in Mill, J.S. (1993), Utilitarianism, On Liberty, Considerations on Representative Government. London: Dent, pp. 69-185.

World Health Organisation (2010). Obesity and owerweight. Available at: http://www.who.int/dietphysicalactivity/publications/facts/obesity/en/. Accessed 26 March 2010.

Morbid obesity in children and the law

S. Romeo[1,2] and P. Nicolás[1]
[1]Interuniversity Chair in Law and the Human Genome, University of Deusto, University of the Basque Country, Avda. de las Universidades, 24, 48007 Bilbao, Spain; [2]University of Las Palmas de Gran Canaria, Spain; pnicolas@genomelaw.deusto.es

Abstract

The meaning of the proclamation of the Rights of the Childhood in relation to the diet is connected primarily with the problem of malnutrition, but other situations may arise where the health of children is not affected by food shortages, but by a damaging diet. In this regard, the increase of childhood obesity in Spain is spectacular and disturbing: 15 years ago, 5% of Spanish children were obese; this proportion is 16% nowadays, showing an obvious public health problem that requires the adoption of immediate measures of health, educational and legal nature. Actually, perceptions of parental blame have become so widespread, that some child protection agencies have gone as far to suggest removing obese children from their families. In this sense, in Spain, in 2009, the Public Administration decided to remove an obese boy from his family by social workers because of his weight. This case, and other of similar nature in other countries, brings up fundamental issues about our perceptions of obesity and how to improve public health, but it has also opened a debate on many other issues, including that about a certain similarity between malnutrition and abuse or, at least, neglect. To address this problem is important: (a) to know the duties of those in custody of a child in relation to diet; (b) to establish some criteria to evaluate the fulfilment of those duties and to identify a situation of helplessness; (c) to assess the child's interests in order to take the appropriate decisions by the authorities; (d) to analyse the proportionality of the measures and their possible extension, by analogy, to other situations; (e) and finally, to analyse the extent of the legal liability (especially civil and criminal) of those who neglect their duties.

Keywords: obesity, overweight, minors, legal guardianship, neglect, criminal liability

Introduction

The meaning of the proclamation of the Rights of the Childhood in relation to the diet is connected primarily with the problem of malnutrition, but other situations may arise where the health of children is not affected by food shortages, but by a damaging diet. Morbid obesity is an unfortunate problem that is only becoming worse everyday.

According to the World Health Organization (WHO) obesity and overweight have become a global epidemic which means that it is a problem that affects all regions of the world. The alarming aspect of it is that it is affecting people at a much earlier age; thus, young children are becoming morbidly obese and are experiencing the same health problems as middle-aged adults. In this regard, the increase of childhood obesity in Spain is spectacular and disturbing: 15 years ago, 5% of Spanish children were obese; this proportion is 16% nowadays, showing an obvious public health problem that requires the adoption of immediate measures of health, educational and legal nature.

According to experts, obesity is a complex disease involving many factors. It is a complex, multifactorial chronic disease that develops from an interaction of genotype and environment. It is difficult to determine the implication of each factor in the development of this illness. Obesity appears to be the result of biological, genetic, environmental and lifestyle factors, and the unbalanced relationship between diet and caloric requirements. Although there are powerful genetic factors affecting individual families who have genetic susceptibility, the overwhelming influence for 99% of the population is environmental (abundance of 'energy dense' foods and drinks and sedentary lifestyle).

Public Health policies for the control and prevention of obesity must take all of these factors into consideration and not just some of them, as they sometimes do. However, diet and sedentary lifestyle are the two most important factors because both of them can be easily changed. For this reason, public health policies to fight against obesity should be based not only in 'health education' (improving public knowledge on the relationship between diet and health, energy intake and output, sport, physical activity and health, etc.), but also in other relevant factors such as transportation, urban planning, children advertisement, sports, menus at schools cafeterias, school environments, food labelling, and others.

However, the failure to remedy the epidemic of childhood obesity through interventions such as school-based programs, nutritional counselling, and behaviour therapy can result in a drastic form of intervention – state intervention. State intervention in childhood-obesity cases is a controversial topic because it necessarily implicates parents' fundamental liberty interest; parents have a general right to raise their children free from interference. Nevertheless, state intervention is permissible when it is necessary to protect the physical and emotional health of a child. Because obesity threatens the physical and emotional health of children, states must determine when and how to intervene. However, current Case Law limits intervention to situations where there is a compelling interest in preventing harm to the child. In cases of morbid childhood obesity, there is clearly such a compelling interest because the child's life is in imminent danger.

Privacy vs. public intervention

The parents, or those who exercise the legal guardianship, are the ones in charge of providing a favourable environment to the minor for his/her suitable psychical and physical development. Therefore, in case of a medical problem originated by obesity, like in the rest of the cases of treatment of pathologies in minors, the health professionals should inform the parents about the measures to relieve the situation, both in the sanitary and in the domestic area; as well as of the consequences that a certain system of nourishment can have for the health of the minor.

But if the parents do not comply with their duty of care, that is to say, if the problem is not solved inside the family context, and this conduct affects seriously the health of the child, the public administration must take actions, which will be graduated depending on the seriousness of the situation and of the verification of possibilities of improvement of the minor environment.

The purpose of these administrative and civil measures is not the sanction for a neglectful conduct (as it would be in case of a criminal process, as it is explained later, but simply the fulfilment of the duty of protection to the infancy that comes entrusted to the states by the international law and also by the national legislations. In this way, according to Articles 3-1 and 9-1 of the Convention on the Rights of the Child, of the UN, of 20 November 1989: 'In all actions concerning children, whether undertaken by public or private social welfare institutions, courts of law, administrative authorities or legislative bodies, the best interests of the child shall be a primary consideration.' 'States Parties shall ensure that a child shall not be separated from his or her parents against their will, except when competent authorities subject to judicial review determine, in accordance with applicable law and procedures, that such separation is necessary for the best interests of the child. Such determination may be necessary in a particular case such as one involving abuse or neglect of the child by the parents, or one where the parents are living separately and a decision must be made as to the child's place of residence'.

The actions that the public administration shall take, can be of educational or preventive nature, but also of a more forceful character in concrete cases. In England, for example, it was well known the case of Connor McCreaddie, a child of eight years that weighed almost ninety kilos. His mother faced in February 2007 the possible retreat of the custody on the part of the British authorities when the paediatricians assured that she had not followed the advices received for a long time to improve the health of the minor. In 2009 in Galicia, Spain, the Public Administration assumed the guardianship of a 9-year-old child with morbid obesity, and decided his revenue in a foster home. The child had been 48 times in the ICU (intensive care unit) because of her health problems. The social services had developed a follow-up of his life and habits for two years, trying to modify the habits in the family. A specific centre was controlling the boy from Monday until Friday, but the weekends, at home, the strict diet was skipping. The parents disagreed with the administrative decision but a Court of Justice of Galicia ratified it. The Court emphasized that it was a matter of a serious risk for the life of the minor based on paediatric reports. The attitude of the parents was negatively valued in the court decision. Likewise, the court thought that the administrative decision was not going to break the familiar relation of the minor, because the parents were able to visit him and to communicate with him according to the rules of the minors' center.

In fact, in these situations, the State can design educative programmes, as it has been done in the United States, as Deena Patel describes in her article 'Super-sized kids: Using the Law to Combat Morbid Obesity in Children'. These programmes have tried to relieve the lack of awareness on the prejudice of the obesity in the childhood, which remained stated in a survey (*2003 Childhood Health Survey*). The findings indicate that parents have skewed viewpoints of their children and weight: almost half of the parents surveyed have obese children but do not believe that their children are obese, a third of the parents feel that their obese children have a 'healthy enough or as healthy as it could possibly be', diet, half of these obese children think likewise and almost half of the parents are likely to agree that good health is mostly determined by people's genes and not by things like diet, exercise, and drinking habits. Evidently, lack of awareness is yet another contributing factor to the morbid obesity crisis. While the government needs to implement education for children and parents of all weights and ages, there is an imminent need to educate morbidly obese children and their parents today. Although it would be beneficial to implement a general education program for everyone, that should be considered a long-term goal, whereas a short-term goal would use the court system to target morbidly obese children and their parents now. This might single out morbidly obese children and their parents, but the children are at risk of dying and require immediate attention. In these programs, parents would learn how to provide healthy food options for their children and encourage more active lifestyles, while children would learn how to make smart food choices and how to engage in more physical activity.

But one more step in the intervention of the State in a situation of risk for the minor caused by his/her obesity, can affect in a more forceful way the family sphere that in principle, is a sphere of the life that the citizens have right to keep reserved from the interventions of the public administration. Once a general framework of a regulation of the familiar relationships is established, the parents exercise the legal authority on their child, which implies obligations and rights with regard to them, which they exercise according to their personal criteria.

This set of duties and rights has experienced an evolution that has determined the modification of its legal nature, since it has changed from being an absolute right of the father, as it was conceived in the original Roman law, into a set of power directed to fulfil a few duties and obligations that the law imposes to the parents.

The limitation of the faculties of the parents have been legally established, since the function of the guardianship takes as a limit the best interest of the child and his/her benefit, staying in hands of the public administration the possibility of depriving of the legal authority to the progenitors. The *favor filii* is the criterion that justifies the intervention of the public administration in the parental relations. In effect, a neglectful conduct by a legal guardian may derive in administrative and civil consequences in order to protect the minor.

That is why the State must take charge of the minors in situation of abandonment and give them assistance. In addition, the parents who don't comply with their duties, can be deprived of the guardianship (totally or partially) by a Judge.

The basis from which this intervention remains valid is the verification of a situation of abandonment of the minor. The intervention can imply the guardianship of the minor on the part of the competent organs of the Administration, which replace the parents in the exercise of their functions, and the consistent suspension of the guardianship for them. It may also manage to suppose its definitive privation.

In any case, the privation of the right – duty that endures this relation of kinship, cannot be adopted but for the concurrence of a reason of well-known seriousness, from which dangerous prejudices stem for the minor.

Nevertheless, it has been said (Alexander *et al.*, 2009) that in cases of failure to thrive and in cases of severe obesity, clinicians may observe immediate health risks as well as longer-term risks, including psychological harm. The 'immediate' or 'urgent' risks more often associated with failure to thrive will – for that reason alone – are seen to be important. However, risks to a child's health that are progressive and have lifelong consequences are also important, even if there is no particular point in time at which they constitute an 'urgent' risk of 'immediate harm'. In cases of *severe* paediatric obesity, clinicians will need to decide at what point the longer-term risks are sufficiently important that notifying the case to child protection services is the *best thing to do now,* even if the circumstances are not 'urgent'.

As a matter of fact, it is not infrequent that the State takes charge of the minors in a situation of abandonment caused by the disregard of the parents who do not attend their duties for different reasons (economic or of another type). But this question has already appeared in relation with the suitable nourishment that the parents must provide to the children as a part of their duty of care. For example, in Spain, the Community of Madrid withdrew the guardianship from vegetarian parents because it was proved that their son's physical and psychical health had been affected because of the nourishment.

The protection of the minor justifies the intervention of the public administration in a situation of vulnerability, independently of the reasons of the conduct of the parents (cultural, religious, etc.). The freedom of the parents in the exercise of the *patria potestas* remains limited by the best interest of the minor. On the other hand, the evaluation of the situation of abandonment refers to a geographical certain context, in which it could demand a certain well-being in the childhood and in which, in addition, the public Administration has means to take the necessary actions.

To estimate the situation of vulnerability, the specific circumstances of the case have to be examined meticulously, taking into account, as the most important objective, the best interest of the minor, and also the necessary protection of the family. So it becomes necessary to estimate that the moral and material assistance of the minors in order the declaration of abandonment, has to deserve a restrictive interpretation, looking for a balance between the benefit of the minor and the protection of the family relationships, in such a way that only the existence of the abandonment is estimated when it means really exactly the lack of a the minimum attention to the minor, demanded by the most common social values. Even if the best interest of the minor is basic and preferential, it is necessary to emphasize the extraordinary importance of other rights and interest at stake, that is to say, those of the biological parents and those of the remaining persons involved in this situation. This is connected directly with the principle of priority of the natural family proclaimed in the Declaration of the General Assembly of the United Nations of December 30, 1986 in which art. 9, is affirmed the interest of the child to being educated by his/her natural parents. Definitively, the situation of abandonment, be voluntary or wanted by the progenitors, has to be always estimated restrictively. Neither the poverty nor the lack of education are reasons that imply disregard and abandonment of the offspring; and in any case, when a situation is based on those causes, the duty of the public administration is to provide and adequate help to the family to avoid the separation of the child of his/her biological family (Supreme Spanish Court, July 30, 2009).

Now then, when the educational measures have not worked and it has been appreciated a serious risk for the minor, there have been cases when the public administration have decided to control and to assume the guardianship of the minor, because of the situation of vulnerability. These cases are based in that facts that the action or omission of the parents has been the reason of the situation of danger for the physical or psychic integrity of the minor, and that the separation of the environment was indispensable for the improvement of the minor health.

Deena Patel, in the above mentioned article, refers some cases in the United States providing the temporary removal of a child from his or parent's custody for this reason. In some of them, the minors got an important improvement during the period in public centres, and when they returned home, the parents aroused also to the need to establish a few more suitable nourishing habits. On the contrary, in other cases, it was decided finally the definitive removal of the custody.

Child obesity and criminal law

So far we have analyzed some issues related to the Child Obesity under Civil and Administrative Law. In this way, we have seen that the State can impose some measures for Obesity Prevention through education and good nutrition. We have also explained how parents can lose custody of their children in some circumstances when some requirements are not fulfilled. That can happen when we are dealing with cases of severe or extreme childhood obesity, which can have adverse consequences, such as psychosocial problems, increased risk for coronary heart disease, abnormal glucose metabolism and the persistence of obesity into adulthood with resultant long-term consequences. There is no question that morbid obesity poses an enormous health risk on children.

In these cases we could also consider the possibility, in addition to the civil measures adopted, of charging the parents with a criminal offence. The crimes that could be applied would depend on different issues, such as the severity and the nature of the obesity, or the specific national legislation in each country. Depending of these factors, we could consider, in principle, the following offences (that can be felonies or misdemeanours, depending of the country's legislation): neglect, abuse and injuries.

The image most associated with parental neglect is an undernourished child. However, as we mentioned above, a child's obesity related health concerns can be considered neglect and result in a parent's loss of custody. In many countries, neglect can also be a criminal offence. The problem here would be determine when neglect can pass from be considered a situation where social services should intervene, to a situation that requires the intervention of the Criminal Justice System.

No doubt that 'neglect' is a very complicated concept. Definitions of 'neglect' and 'medical neglect' (parental neglect of a child's medical needs) are not universally consistent. The word 'neglect' implies that someone has given up caring. There are generalised concepts and standards that are expected of parents towards their child. Failure to meet these standards may legitimately be considered neglect.

Charging someone with 'neglect' cannot be based on a child's weight alone. Clearly not all obese children are neglected – indeed, parents of obese children may be very devoted to their child. It should be limited to cases where a child is at high risk of harm because a parent is neglecting to provide necessary care.

An example of what can be considered neglect could be the following one: A child has cancer that is 90 percent curable with a certain treatment, and his parents refuse that care. It's neglect if a mother isn't giving her sick child proper medication.

Unfortunately for obesity cases, however, the responsibility of the parent is much less clearly defined than in cases of outright refusal to give a child desperately needed medicine. Not feeding a child is neglect, but what if a father only feeds his son pizza and soft drinks?

This is why Criminal charges should be a last resort. Charges of neglect should be related to a morbid obesity caused by poor nutrition. In these cases, before applying this offence we should wait until the Public Administration or a Court has already ordered the parent to comply with some sort of treatment plan for the child (such as exercise, keeping a food log, etc.) and he/she has refused to comply.

Actually, the cases of neglect that can be considered to be treated as a criminal offence are closer to the child abuse, what can also be a criminal offence in many jurisdictions. In effect, according to some experts over feeding a child, especially to the point of obesity or morbid obesity, is a type of neglect that constitutes child abuse. It is up to a parent or guardian to ensure the basic needs of a child are met. Parents must provide many basic needs. One of these is food, but it is food for nourishment, not food to the point of excess and obesity. Nor is it food that puts the child in a malnourished state. One must consider what the child is being fed, not just the quantity.

Unlike measures of civil or administrative nature (which includes removing a child from their home as a last resort), the use of Criminal Law in such situations has just been started to be explored. Back in June 2009 a woman of South Carolina (USA) was arrested and charged with criminal neglect. The reason was that her son weighed 555 pounds (250 kg) by age 14. This case attracted national and international attention and it could open the door to more criminal action against parents whose children have become dangerously overweight.

State courts in Texas, Pennsylvania, New York, New Mexico, Indiana and California have grappled with the question in recent years. In all of those cases, except the one in California, courts expanded their state's legal definition of medical neglect to include morbid obesity and ruled that the children were victims of neglect. Criminal charges were filed only in the California and Indiana cases, but the parents weren't sentenced to jail time in either.

Apart from a few cases in which extreme child obesity has attracted public attention, the intervention of the authorities only takes place when the neglect results in death or serious harm. And this can happen because, as said above, obese young people are more likely than children of normal weight to become overweight or obese adults, and therefore more at risk for associated adult health problems. Actually obesity increases the risk for high blood pressure, stroke, cardiovascular disease, Gallbladder disease, diabetes, respiratory problems, arthritis, cancer and emotional problems such as depression and anxiety.

As we have seen, although the health problems tied to childhood obesity don't become chronic until adulthood, it doesn't make impossible to charge parents with child abuse. But what happens if the high risk of health problems becomes real? What happens if a childs suffers a stroke or a heart attack as a consequence of the obesity?

The question here is if we can take the criminal liability further and charge parents with injuries or homicide (usually manslaughter). This could be possible if we can find and prove a link between the poor nutrition or the neglect and the obesity, and between the obesity and the illness or the injury. However this won't be something easy to do most of the times.

We must take into account that both acts and omissions may trigger the imposition of criminal liability. This means that a person can be made responsible of a result both because he/she has done something that he/she shouldn't have done and because he/she hasn't done something that he/she should have done.

In this way, we have to consider if an act (for instance, the bad quality of food the parents have been providing to their child) has produced a result (the injury or the death). To do this, we need to prove 'causation', that is to say, liability for a crime of harmful consequences may only be imposed when it has been determined that the defendant's conduct was the cause of the harm. Case law usually follows the 'but for' rule to determine causation. According to this rule, an actor's conduct is the cause of the victim's harm if it constitutes a condition without which the harm would have not taken place. In addition to this, it is also required that the conduct created an unjustifiable risk of producing a criminal harm, and that he harm was produced as a result of the unjustifiable risk created by the defendant's conduct.

In the cases we are dealing with, prove of causation will be something really difficult to do. We will never be able to be completely sure of the real cause of the morbid obesity, and we might not be sure either about the real cause of the injury. As it was mentioned above, many other factors apart from poor quality food may have had some influence in the obesity. In consequence, it doesn't seem a good way to seek criminal responsible through a positive act.

But what about a negative act, an omission to act? Would it be possible to charge a parent with injuries or manslaughter due to his/her passivity because he/she didn't anything to control the overweight of his/her child? This way can be more effective to attribute criminal liability to the parents.

Generically, an omission can be defined as the failure to perform an act that is required by law. There are two distinct types of punishable omissions. On the one hand, an actor may be held liable when he fails to perform an act that the criminal law requires to be performed regardless of whether the failure to act produces a harmful result (for instance, failing to pay child support). These failures to act are punishable even when no one is harmed as a result of the omission.

On the other hand, sometimes an actor is held liable for an omission only when his failure to act produces a harmful result. In these cases, the failure to act generates liability for the harm caused as if it had been committed by way of a positive act. An example of this would be the case of the parent who kills his child by failing to feed him or because he hasn't taken his/her child to the hospital when he is ill. Could this also be the case when he/she doesn't care about an evident overweight (extreme obesity) and the harm is caused as a consequence of it?

Generally speaking, omissions should give rise to liability for a crime of harmful consequences only when the defendant's failure to act has brought about the wrongful result as if he had caused it by way of a positive act. Sometimes it is difficult to determine whether the omission caused the result in a way that is equivalent to it having been produced by way of positive conduct.

When determining if the result caused by an omission should be deemed equivalent to the result having been caused by an act, courts must take into account whether the actor had a special legal or contractual duty to act. A special duty to act may arise as a result of the existence of close familial ties. Thus, a parent has a special duty to take care of his/her child. If the breach of this duty results in the death of the child (i.e. if a positive act – such as getting professional help to lose weight – could have very probably avoided the result), the parent would be held liable for homicide (manslaughter). Since parents can be criminally prosecuted if they starve their children to death, they should also be at risk for prosecution if their children die from morbid obesity, and they neglected to take steps to prevent or counteract it.

References

Alexander, S.M., Bauer, L.A., Magnusson, R. and Tobin, B. (2009). 'When does severe childhood obesity become a child protection issue?' Medical Journal of Australia, Vol. 190 No 3, 2 February 2009, pp. 136-139.

Gómez-Jara Díez, C. and L.E. Chiesa, (2009). 'Spanish criminal law'. In: K.J. Heller (ed.), Comparative Criminal Law, Stanford University Press, 2009.

O'Dea, J.A., (2004). 'Prevention of child obesity: 'First, do no harm'', in Health Education Research, Oxford University Press, 2004.

Patel, D. (2005). 'Super-sized kids: Using the Law to Combat Morbid Obesity in Children', *Family Court Review*, Vol. 43 No. 1, January 2005, pp. 164-177.

Parental views on public-policy options regarding healthy eating

D. Wickins-Drazilova, S. Nicholls, G. Williams, C. Börnhorst, L. Grafström, S. De Henauw, S. Marild, D. Molnar, L.A. Moreno Aznar, I. Pigeot, A. Siani, M. Tornaritis, T. Veidebaum and W. Ahrens, on behalf of the IDEFICS consortium
Department of Philosophy, Lancaster University, Lancaster, LA1 4YG, United Kingdom; Bremen Institute for Prevention Research and Social Medicine, University of Bremen, Linzer Str. 10, Bremen, 28359, Germany; d.wickins-drazilova@lancaster.ac.uk

Abstract

As part of the IDEFICS obesity intervention study, baseline opinions about key issues were elicited from the parents of children about to take part in the intervention in eight European countries. This was done by a questionnaire, in which a set of questions was specifically designed to investigate attitudes towards ethical and public-policy aspects of interventions on healthy lifestyle and diet. The aim of these questions was partly to elicit parents' views concerning the ethics and effectiveness of interventions, responsibilities of schools and parents themselves, and also to gain some indication as to how parents might view some of the policy options available. Here we present analysis of three of the questions on public-policy regarding healthy eating: (1) I feel that the authorities don't do enough to support healthy eating; (2) There should be a controlled restriction on advertising of high-fat and high-sugar foods; (3) Very unhealthy foods should be highly taxed while healthy foods should be subsidised. We report here that there are statistically significant differences between the respondents both by country, and by gender. There is strong support among parents for controlled restriction on advertising of high-fat and high-sugar foods, as well as supporting of taxation of 'unhealthy' foods and subsidising of 'healthy' foods. In most countries men are more critical of authorities' support of healthy eating than women, and they are also more in favour of taxation and subsidising of certain types of food.

Keywords: food, interventions, survey questionnaire, children, obesity

Background

The prevalence of obesity and associated diseases (such as type 2 diabetes and heart problems) is currently rising in Europe. This increase is related to wider societal changes in people's diet, physical activity, lifestyle and other factors. There is insufficient evidence to make direct recommendations on what policy changes and interventions should be adopted to reverse the trend of rising obesity and obesity-related diseases at the population-wide level. It is, therefore, necessary to conduct further research to gather reliable evidence assessing interventions aimed at promoting healthy diets and living.

The IDEFICS Study (Identification and prevention of Dietary- and lifestyle-induced health EFfects In Children and infants), that started in 2006 and will run until 2011, is funded by the European Commission to deliver reliable data from which international assessment of obesity in children will be possible. Beyond gathering data on a large sample of European children, its main research aim is to conduct and evaluate health promotion and prevention interventions in kindergartens, schools and communities, comparing these against data gathered in control regions.

As part of the study, several questionnaires were developed and administered to over 16,000 parents of children between 2- and 10-years-old in eight countries: Belgium, Cyprus, Estonia, Germany, Hungary, Italy, Spain and Sweden. The questionnaires included questions on: the dietary patterns of children and their families; medical history; physical activity; social environment; media consumption; and children's financial opportunities, independent buying power, actual spending patterns and children's influence on family purchase decisions.

Aim

As part of the parental questionnaire, a set of questions was specifically designed to investigate the ethical and public-policy aspects of interventions on healthy lifestyle and diet. The aim of these questions was partly to elicit parents' views concerning the ethics and effectiveness of the intervention, responsibilities of schools and parents themselves, and to gain some indication as to how parents might view some of the policy options available.

This paper draws on data gathered from the baseline questionnaire of the IDEFICS Study. The analysis in this paper will be limited to three questions that concerned the responsibilities of government – specifically parental responses to the statements:
1. 'I feel that the authorities don't do enough to support healthy eating'.
2. 'There should be a controlled restriction on advertising of high-fat and high-sugar foods'.
3. 'Very unhealthy foods should be highly taxed while healthy foods should be subsidised'.

These questions were designed to help us get a clearer sense of how respondents (as parents, but also stakeholders) view governmental responsibilities for providing healthy diet and lifestyle for children. Since dietary habits and preferences are formed in early life, and many of those before children even enter schools and kindergartens, it is extremely important to look at opinions of those that choose diets for today's children. We decided to conduct initial exploratory evaluation of parental responses by country of residence and gender, as did similar studies (European Commission, 2010).

The discussion over whether governments do enough to support healthy eating is longstanding. Marion Nestle has been one of the most influential advocates of stronger regulation of the food industry, in terms of clearer food labelling, banning certain ingredients, and introduction of taxes for 'unhealthy' foods and drinks (Nestle, 2002). Others have criticised the European Union for significant interventions affecting the price of foods by subsidising 'unhealthy' sugar, meat and dairy products, while failing to sufficiently subsidise and support beneficial increase in consumption of fruits and vegetables (The Faculty of Public Health, 2007).

Regarding advertising, there is currently no European legislation in place restricting promotion of food and drink high in fat and/or sugar. However, some countries have already adopted restrictive measures at the national level. For example, in 2007 the UK introduced a total ban on advertising of foods and drinks high in fat, sugar and salt in and around all programmes of particular appeal to children under the age of sixteen (Ofcom, 2010). Nevertheless, doubts remain as to whether food and drink advertising is harmful for children's health, as well as whether restrictions have a significant effect on children's diets (Livingstone, 2006).

Methods

Cross sectional surveys were conducted within each of the eight participating countries in both control and intervention regions, which were matched on key demographic variables. Children and parents were identified and recruited via institution-based (kindergarten, pre-school, primary school) approaches. Approval from local ethical committees and data protection officials was obtained by each survey centre prior to the start of the surveys. Parents were approached to give consent for both themselves and their child to take part in the study. Once they had consented, they were provided with two self-completion questionnaires: the first containing the core questions and the second focusing specifically on diet. To enhance completion, parents were offered help in filling out the parental questionnaires. This included the option to complete the questionnaire either face-to-face, through telephone interviews or in a group session. Responses to the three questions of interest were coded as 'Agree', 'Moderately Agree', 'Moderately Disagree', and 'Disagree'. All analyses were conducted using SPSS v.17. Parental responses were compared using the chi-squared (χ^2). A *P*-value of 0.05 or less was used to define a statistically significant difference between comparison groups.

Results

In each country, baseline surveys were conducted from September 2007 to February 2008 in both control and intervention regions. A total of 16,569 parents completed the questionnaire of which 16,224 fulfilled the general IDEFICS inclusion criteria. Of these 16,112 have data available for comparison with the other questionnaires and so comprise the final sample for analysis.

Country level analysis

Table 1 shows the percentage of parents who agreed with each statement by country. Whilst support varied by country and by statement, it can be seen that parents in Cyprus were most supportive in each case.

1. *'I feel that the authorities don't do enough to support healthy eating'.* Results significantly differed by country (N=14,575, chi-sq 1,507.617 on 21 df, $P<0.001$). Whilst parents in most countries were relatively evenly split in terms of broad agreement or disagreement with the statement, parents in Cyprus and Hungary were far more critical with 46.1% and 31.6% of parents respectively agreeing that authorities did not do enough to support health eating.
2. *'There should be a controlled restriction on advertising of high-fat and high-sugar foods'.* In comparison to the more mixed picture regarding the authorities, parents were largely in favour of more restricted advertising. There was, however, a significant difference between countries (N=14,724, chi-sq 1,941.359 and 21 df, $P<0.001$). Whilst parents in most countries agreed to some extent that there should be a controlled restriction on advertising, parents in some countries expressed stronger opinions than others. Parents in Cyprus again were of the strongest opinion for the statement with 81.4% agreeing and only 1.9% disagreeing, but also parents in Italy showed a high level of agreement. Parents in Estonia were least supportive of the statement with 33% either disagreeing or moderately disagreeing.

Table 1. Country level analysis.

Country	1. 'I feel that authorities don't do enough to support healthy eating' % agree (N)	2. 'There should a controlled restriction on advertising of high-fat and high-sugar foods' % agree (N)	3. 'Very unhealthy foods should be highly taxed while healthy foods should be subsidised' % agree (N)
Sweden	10.8 (186)	41.3 (720)	39.8 (691)
Belgium	13.5 (249)	36.4 (674)	35.9 (669)
Estonia	14.9 (245)	26.7 (440)	33.3 (549)
Germany	20.9 (398)	42.8 (828)	30.2 (581)
Spain	22.2 (325)	52.2 (768)	31.8 (467)
Italy	22.5 (475)	70.9 (1,549)	57.5 (1,239)
Hungary	31.6 (765)	62.7 (1,524)	52.2 (1,266)
Cyprus	46.1 (669)	81.4 (1,189)	63.3 (919)
Total	22.7 (3,312)	52.2 (7,692)	43.5 (6,381)

3. *'Very unhealthy foods should be highly taxed while healthy foods should be subsidised'.* Parents in all countries were broadly in favour of the statement, although parents in some countries were more supportive than others. As before, responses differed significantly by country (N=14,674, chi-sq 977.251 on 21 df, $P<0.001$). Again, parents in Cyprus were most supportive with 63.3% agreeing and 85% agreeing to some degree. Almost equally supportive were parents in Italy where 82.2% of parents agreed to some degree. Least supportive were parents in Germany with only 30.2% agreeing.

Responses by gender within countries

Parental responses were highly skewed, with female parents providing the majority of responses (range 80.9-94.1% women). Analysis of the aggregated data suggests that responses to each of the questions differed significantly by respondents' gender. However, once the country of residence was taken into consideration, variation in the responses was seen between countries and between questions.

1. *'I feel that the authorities don't do enough to support healthy eating.'* Men were broadly in agreement to a greater extent than women. In Sweden and Cyprus, however, this trend was reversed although gender differences were not significant ($P=0.254$ and $P=0.148$ respectively). Only in Italy was the difference significant with women being more supportive of the statement (30.7% of men agreeing compared to 21.4% of women, $P=0.002$).
2. *'There should be a controlled restriction on advertising of high-fat and high-sugar foods.'* In most countries there was little difference between responses on the basis of parental gender. However, parents in Italy differed in their responses between the genders ($P=0.001$). In Italy 80.5% men agree with the statement compared to 69.6% of women.
3. *'Very unhealthy foods should be highly taxed while healthy foods should be subsidised.'* This question appeared to generate the largest number of between gender differences. In comparison to the previous questions women were in agreement to a greater extent than men.

Discussion

The analysis here suggests that responses to the statements consistently differed by the parents' country of residence and gender. Cyprus, Italy and Hungary stand out as countries with similar response profiles, being more in agreement with the three statements. However, the two post-communist states (Hungary and Estonia) do not seem to stand out in their response profiles. Men in all countries (with exception of Sweden) are generally more critical of authorities' support of healthy eating, yet females are more in favour of taxation and subsidising of certain types of food. The finding that there is strong support among parents for controlled restriction on advertising of high-fat and high-sugar foods suggests that introduction of restrictions of advertising at the European level would be viewed positively by the public. The reason for country variation in responses to this question may in part be due to the existing regulations on marketing to children. Sweden, for example, already bans broadcast advertising of certain types of food aimed at children (Matthews, 2007), and this may be reflected in the lower levels of agreement by Swedish parents, given that they already live in an environment with restricted advertising.

Of particular interest are the positive responses of parents to the statement 'Very unhealthy foods should be highly taxed while healthy foods should be subsidised.' This is contrary to the findings of the PorGrow project (Policy options for responding to the Growing challenge from obesity), which conducted interviews with food producers, manufacturers and retailers, pharmaceutical companies, teachers, public-health professionals and public servants. PorGrow stakeholders showed low reliance on higher food taxation (Millstone and Lobstein, 2007). The results are also in contrast to our survey of the opinions of IDEFICS scientists, which showed that most are also sceptical of taxing 'unhealthy' foods and banning advertising (Wickins-Drazilova and Williams, 2010).

Conclusion

Further analysis of the questionnaire will be conducted, for example by age of parents, health of children, and socioeconomic status of families. It will be also interesting to compare the IDEFICS results with other studies that asked parents and other stakeholders for opinions on governmental interventions promoting healthier diets. These results will be also compared with follow-up parental questionnaire, which will evaluate the impact and effects of the IDEFICS intervention. This second questionnaire is being administered two years after the baseline questionnaire, and the data will be available for comparison in late 2010.

References

European Commission (2010). Special Eurobarometer: Health and Food. Available at: http://ec.europa.eu/health/ph_publication/eb_food_en.pdf. Accessed 23 March 2010.

The Faculty of Public Health (2007). A CAP on health? The impact of the EU CAP on public health. Available at: http://www.fphm.org.uk/resources/AtoZ/r_CAP.pdf.

Livingstone, S. (2006). Does TV advertising make children fat?: What the evidence tells us. Public Policy Research 13/1: 54-61.

Matthews A.E. (2007). Children and obesity: a pan-European project examining the role of food marketing. European Journal of Public Health 18/1: 7-11.

Millstone, E. and Lobstein, T. (2007). The PorGrow Project: overall cross-national results, comparisons and implications. Obesity Reviews Supplement 8/2: 29-36.

Nestle, M. (2002). Food Politics: How the Food Industry Influences Nutrition and Health, University of California Press, 469 pp.

Ofcom (2010). Television Advertising of Food and Drink Products to Children. Available at: http://www.ofcom.org.uk/consult/condocs/foodads_new/statement. Accessed 23 March 2010.

Wickins-Drazilova, D. and Williams, G. (2010). Ethical and public policy aspects of childhood obesity: Opinions of scientists working on an obesity study. Obesity Reviews (in press, accepted in February 2010).

Human rights, development and hunger in the world: ethical and legal proposals

F.J. Blázquez Ruiz
Faculty of Law, Public University of Navarra, 31007 Pamplona, Spain; javier.blazquez@unavarra.es

Abstract

One of the most imperative goals for Human Rights today is to deal with the hunger and poverty that desolate hundreds of thousands people in the world. However, both Equality and Dignity, in order to be really effective, are required to leave Declarations and International Conventions field. It is no longer the time to perpetuate speeches and proposals fed on ideas and provided with principles, but with their binding force weakened once they touch reality. Because in the meantime, the alimentary and sanitary needs do not wait and keep destroying human lives. As Mr. Mohamed Yunus, Nobel Prix of Economy insists, hunger should change its place and be part of museums only. That should be the best suitable space to accommodate it: by being part of mankind's history. The purpose is obvious. Why do thousands of people keep dying while colossal amounts of food exist that could avoid starvation and endemic poverty? Perhaps the technical and scientific achievements are not able to create and ensure the production needed to purge hunger? The 2009 World Food Summit held in Rome last November revealed that big international companies have enough means and technical resources to feed the entire world. However, the problem is quite different. The main point is not solved by questioning the technological capacity or by changing the agricultural production levels. The key is basically a matter of politics, that is, a matter of willpower and commitment along with the adequate precaution and provision. Despite of all this, no head of State of the G8 group attended the summit. In other words, we talk about planning, coordination and execution in the field of a food distribution plan that should be controlled and sanctioned not following only economic issues but also from a legal and political point of view. Only by doing that it would be possible to establish accurate objectives, to join forces efficiently and to keep track of its achievement. All this with the final purpose of turning hunger into a museum piece.

Keywords: human rights, equality, human dignity, hunger

The absence of hunger, poverty, and basic needs is a necessary and unavoidable requirement to promote a proper human development, both on personal and social levels. In fact, malnutrition, poverty, and misery are the major causes of illnesses and deaths that happen every day in the world. Specifically, each day more than one billion people lack food and more than twelve hundred million (1.2 billion) people lack access to clean water. Around eigthy percent of illnesses and over 30% of deaths in the southern hemisphere are related to water pollution.

There is no doubt that the persistence of poverty, malnutrition, and misery, sometimes in extreme terms, makes life exceptionally difficult. People cannot live their daily lives without being affected by harmful conditions. Homelessness injures millions with wounds and scars, not just by material deprivation and economic potential, but mainly because it reduces and constrains the possibilities to enhance their capacities and abilities. This could be seen as undermining personal growth.

However, it is obvious that the strategic battle against poverty and hunger goes beyond the geographical and political frame, related to specific areas and located in concrete zones. Hence, to address and tackle this enormous problem, it is necessary to adopt a broader perspective, one that is supranational and global.

In this regard, for some authors such as economist A. Sen, we must delve beyond the immediate reality, and we must penetrate into the underlying root causes. Some of the most significant causes lie in the ability of participation and cultivation of individual freedoms, in the promotion of skills, as well as the in permanent empowerment of various social manifestations.

As a close reference in political terms, we may remember that in the recent history of humanity, there have never been famine situations in the countries with democratic systems, whether they are economically rich such as Western Europe and North America, or relatively poor such as India after independence, or the cases of Botswana and Zimbabwe.

The basic motivation is to explain such glaring facts: despotic and authoritarian rulers, who will never aid during these food crises, nor do they typically influence hunger epidemics, are hardly competitive and largely incompetent. Besides, they lack political incentives, such as sufficient motivation and tension to anticipate, to prevent and to take proactive measures in good time. They are casually accustomed to lament again and again, and to blame the unpredictable and uncontrollable force of nature, to maliciously excuse their apparent incompetence and irresponsibility.

That activity itself is intrinsic to individual liberty, economic freedom, and political freedom. Then the overlap, interaction, and feedback between them are inexcusable.

However, as seen above, it is not possible to ignore the current situation experienced by the economy from a broader perspective, i.e. from the international perspective. We cannot escape the continued and expansive process of globalization which we have been witnesses in recent years.

Globalization is characterized by the prevalence of economic aspects, in particular by the hegemony of capital and the hypostatized market, in other dimensions of social reality, including the areas of policy and the decision-making power of nations.

It is well known that the increasingly intangible nature of production and activity, the permanent transfer of knowledge, and the management of financial flows mainly characterize a central theme, financial globalization. Capella states that 'finance capital becomes pure spirit' (Capella, 1999).

It in fact operates without the intervention of other classic factors of production, which had been in place historically, such as land and labour. It takes as a single regulatory approach the venerable and majestic market.

However, that approach is not without a certain vagueness and ambiguity that can cause a unique conceptual confusion. In fact, the phenomenon of globalization seems to promote and generate a process of interaction, communication skills, and interdependence. And there even seems to be a certain degree of integration or cohesion between the countries involved. To some extent it appears as if it were an advancing reality, with unavoidable dynamics. Namely as if it were the logical result of a process and direction, the course of which becomes inexcusable.

Whereupon, without further delay we might ask what relation can be established between the phenomenon of globalization as it is manifested, on the one hand, and the model of human development which in turn we have been defending.

Firstly it is clear that before the appearance of a certain relationship of global economic interdependence, we find a very different reality, one that is in no way considered symmetrical. The degree of connection and relationship and hierarchy of real dependence between countries is clearly growing.

And that trend of the globalization process is unmistakable. And its consequences are incontestable, namely: increasing inequalities and increasingly over imbalances, which allow us to deduce that the economy is not just 'global'. In this context the relationship between freedoms, human rights and the market is rather weak, if not antagonistic. Free market and free movement of capital without borders, are the specific traits that have basically enabled the process of globalization. But it does not include for example the free movement of people.

In this respect A. Touraine warned repeatedly about the need to distinguish and to clarify conceptually the terms *description* and *interpretation*, when we refer to the phenomenon of globalization.

From this interpretation it is possible to argue, justify, and legitimize the whole process of economic expansion. Among the key proposals of his speech defending the neoliberal mentors as one of its nuclear arguments, is that the free market is able to coordinate the efforts and individual actions of an economic nature.

The other objective is not to reach a quantity of goods and wealth greater than could be achieved through other means of organizing the economy, as it could be through certain models of planning or state intervention, a model reviled by the recent experiences of the communist countries. They stubbornly insist only as an axiom, that the intervention of the authorities on the development of the economy must be as small as possible, both in frequency and intensity.

The neoliberal movement, which has emerged since the late seventies and early eighties, defends the goodness and priority of the solutions offered by the market to all economic and social problems. And it strongly encourages and promotes the economic activity of individuals, businesses and large corporations in their respective markets. It is underlined as a prerequisite and essential freedom of action. However, the results remain eloquent when reducing world hunger, malnutrition, and infant mortality.

Therefore, instead of uncritically accepting its principles as proposed, inexcusable, or inexorable processes, we should be analytical and methodological. We should examine the course or dynamics followed by the macrophenomenon of globalization, as well as uncover the discourse designed to legitimize it. Finally we should evaluate its obvious effects and consequences

If we look back on this phenomenon with some historical perspective, it is easy to see how the globalization of the economy is clearly biased, and has greatly benefited the developed North, and eventually other countries, such as the development in Southeast Asia. But a small reflection from the framework of Human Rights on the reality experienced and suffered by the citizens of those countries, for the sake of vaunted economic progress, would unfortunately be very significant.

It is also clear that the consequences of globalization, once again, have not been the same for all countries and areas. In fact, there is a vast part of the world where hundreds of millions of people who are not able to participate in and to receive the potential benefits of globalization. And they suffer from being far removed from the circuits of economic or financial interest.

As an example it should be mentioned some states or areas on the periphery, namely outside the prevailing economic orbit, whose experience could be illustrated with a profusion of data and statistics. This can include sub-Saharan Africa, far from the reach of decision-making centres and capitals. That has reduced foreign trade, which was the only movement of people that would allow citizens to escape poverty, disease and exclusion, causing the expansion of social and political fracture as a result of progressive increase in inequality.

Detailed studies on the distribution of wealth and poverty in the world for millions of people have attested that globalization has not improved but worsened living situations considerably. The prices of agricultural products and raw materials have been reduced to unseen levels. Jobs have disappeared, and the unemployed have had to emigrate to survive.

As a result, people have felt increasingly powerless against forces that lie outside their control. They have seen their democracies undermined and their cultures gradually eroded. And for many, emigration stands as the most imminent and hopeful response to a future dark und uncertain. Therefore, the geopolitical dimension is no longer omnipresent in any way.

In fact, as warned Stiglitz, winner of the Nobel Prize in Economics, former President of the World Bank, globalization will continue on the path undertaken without the correct direction.Then globalization will not only fail in promoting development, but will continue to generate poverty and instability. If there is no reaction to reforms already underway, discomfort to wariness of globalization will continue to increase.

Hence, instead of using economic terms to describe and assess the reality of a country or geographical and political environment, it would be more appropriate to study development, whose purpose is basically to transform societies, improve the lives of the poor, allow everyone the chance to succeed, and to give access to health services and education.

Similarly, the globalist project of presumed universality is easily distorted when you look at other planes of reality. Development requires a pre-condition, and between them, as noted A. Sen, we must remove the factors that are opposed to freedom: poverty and tyranny, lack of economic opportunities, poor social conditions, and lack of public services, as well as intolerance or systematic repression exercised by authoritarian states.

The same could be said if we refer to the field of health. When we find that deadly diseases like AIDS or malaria, which have afflicted many third world countries for many years with unusual force, whose pandemic and mortality rates have not been diminished, because the economic gain does not allow medicines and health remedies reach to these latitudes.

In fact, the usual attitude of the international pharmaceutical companies and laboratories, and their repeated refusal to commercialize low-cost, effective products, in the case of AIDS, or generic products at lower prices is a nagging but unequivocal proof of this perverse desire of collaboration with universal progress and development. Between those the more directly related to educational projects and training, or in health programs and infrastructures in those countries seeking to channel their steps along the development path.

In other words, the battle against world hunger must be fought on several fronts, fleeing the discourse that seeks to legitimize the benefits of economic globalization. In fact, the struggle or fight against underdevelopment and subsequent poverty must be addressed not only reflecting the effects and consequences that generates unfear, as we have been discussing. This battle must be fought as well, taking into account the field of early departures, perversely, those strategies.

And if we talk about principles, not just verbally, we will also have to act and speak from the principle of solidarity. First we must address the inequalities and discrimination that traditionally accompany the fight against injustice. That and no other is precisely the spirit that permeates through the Universal Declaration of Human Rights since 1948.

Therefore as noted by A. Senn the fight against poverty and underdevelopment must be fought on several fronts. One should look not only at how to improve income and welfare of individuals, but also must provide the necessary means to realize their full autonomy; to be able to establish their own life plans and to find appropriate ways to carry them out. And finally, they must reach and improve the social bases of self-respect (Senn, 1999).

Undoubtedly, the will and subsequent determination and commitment to solidarity are inexcusable. But that decision has to be accompanied by the relentless competition of the principle of political accountability. And it is insisted that, as N. Bobbio says in his work *The Problem of War and the Ways of Peace*, 'Man cannot relinquish the fight against oppression, to fight for freedom, justice, and independence'.

Of course, from a more global aspect and in terms of international politics, it is obvious that beyond transnational economic groups, the responsibility of the governments of most developed countries is also necessarily challenged. In particular by adopting punitive measures through taxes and tariffs levied on imports of agricultural products from poorer countries.

Unfortunately there are measures that facilitate and lead to a double perverse effect. First, because the cost of imported product is considerably more expensive. And as a result, this generates unfair competition that hinders the economic development process of the source countries. In order to avoid unfair competition, the producer foments subsidies in developed countries, generating an artificial and false price on the market.

However, all these surreptitious measures are part of the daily agendas of commissioners and political representatives who advocate models of liberal societies, free markets, and little regulation or control by the corresponding administrations. The motivation of a political nature leaves no room for doubt. Voters and parties, who should and who propose and vote for them are there in its immediate environment, not abroad. This is probably one reason why in the last World Summit on Hunger held in Rome, (XI-2010) none of the Presidents of OECD countries attended.

In conclusion, we might recall the famous phrase of I. Kant referring to the political dynamics, when he recalled the maxim that often prevails in its actions: 'si fecisti negat.' For indeed one of the most blatant paradoxes of this problem is the lack of consistency and huge permanent contradiction between concepts, categories, and projects that are an inherent part of political discourse. On the other hand, the reality is that there are millions of people who suffer ignominious hunger, malnutrition, and constant exposure to a damaging disease that threaten their health and who often fall victim to these maladies.

An antagonistic situation endures as a result of implementing political and economic approaches of unacknowledged but spurious interests, and pregnant with expectations that have nothing to do with the values that make up democratic and liberal societies. Societies in which principles such as equality, freedom or dignity, supposedly non-restricted rather open and universal regulatory criteria, are put forward and provide the basis and foundation on which is found a coexistent society.

Hence the words of Pablo Neruda clearly remain in full force when he said, 'Earth, who measured you and put up walls, wires, and clasps? Were you born divided? When the meteors cross you and your face grew seas and crumbling rocks, who distributed your gifts for a few people?'

References

Capella, J.M. (1999). Las transformaciones del derecho, Madrid, Consejo Genral del Poder Judicial, p. 97.
Senn, A. (1999). Development as Freedom, Oxford, Oxford University Press.

Dietary disorders and mental health: a contemporaneous debate

M.F. Freire de Sá[1] and A.C. Brochado Teixeira[2]
[1]PUC Minas. Belo Horizonte, Brazil; [2] Centro Universitário UNA. Belo Horizonte, Brazil

Abstract

The multidisciplinary debate on health – relate issues is essential. What happens is that certain mental disorders may influence on the subject self determination capacity, concerning his health, although, many times, this does not affect his wisdom in other existential and property areas. This is the case of dietary disorders, such as anorexia and bulimia, which may cause harm on the patient's competence. In Brazil, the Act 10.216, from April, 6th, 2001, treats the possibility of non consented internment for people holding mental disorders, which comprises the dietary disorders holders.

Keywords: health mental, capacity, anorexia, boulimia

Introduction

Once the psychiatrics has been concerned with the cultural influences on the mental health, aiming at inquiring at up to what extent habits, customs and social values may influence the development of mental diseases, the Law, on the other hand, may inquire at up to what extent certain mental disorders influence the person's capacity.

Concerning that the object of psychiatrics has already been more specific or punctual relatively to mental insanity diseases, hordienally, due to the flow of information about health as a whole, higher divulgation of dietary disorders related diseases which have as their background originated from, many times, a pre-established beauty standard or an obsession for the perfect body. 'Dietary disorders are characterized by disturbances on the dietary behaviour, which may lead to extreme emaciation (cachexia – due to inadequate food reduction), to obesity (due to the ingestion of large amounts of food), or other physical problems. The main types of dietary disorders are the nervous anorexia and the nervous boulimia, and both are commonly characterized by high concern about weight and the exacerbated fear of getting fat, a distorted perception of the body shape and self evaluation based on the weight and body shape' (Anonymous, undated).

'Some young ladies take it for granted that it is impossible to be happy, and at the same time show a little fold on the belly when seated. Some lads sacrifice a great portion of the capacity of acculturation provided by nature, in endless and obsessive hours 'pulling iron' in the muscling gym centers. Not mentioning the ease which some people enjoy to attribute a little fat excess in the waist, a slightly more profuse nose or a less exuberant breast, as being responsible for their failure in attracting the opposite sex. It seems as if there is an anatomic allocation of happiness, sometimes on the breast, other times on the nose or on the scales' (Ballone, undated).

It is with this background that we will discourse on the reflexes of the dietary disorders in Law.

Dietary disorders and physical and mental health

Nervous boulimia is a dietary disfunction, a mental disorder, which is characterized by excessive food ingestion, followed by a guilty feeling which leads the person to take some dangerous attitudes towards his/her health, such as throwing out induction, laxative ingestion, appetite moderators and diuretics, or even fast or excess of physical exercises, to make for the compulsive eating.

Another dietary disorder is nervous anorexia, which is characterized by a poor diet. It is considered as being a complex disease, as it involves social, psychological and physiological aspects. The anorexic subject may also be boulimic. It is estimated that the death risk ranges between 5 to 15% of the cases.

Finally and more recently, it comes the figure of orthorexia, characteristic of people who hold fixation, obsession for a healthy diet. Such condition may incur in deficiencies in essential nutrients, anaemias, avitamisosis, besides social isolation yet. Studies have shown that many patients must be interned, because they search suicide and self mutilation.

'Patients holding dietary disorders and personality disturbances, are more liable to develop chroniphication. Even if the word 'chroniphication ' shows odd connotation and it is not well seen by some groups who are dedicated to mental diseases, it can be understood as an adaptable form in these cases. Dietary disorders are severe highly-rated morbidity diseases. This morbidity, depending on the type and severity of the associated personality disorder, may reach such high standards that the fact of a patient surviving along ten years of treatment, should be celebrated as some sort of victory. Becoming chronic may mean success. A deeply abandonment-feared female patient and not feeling able to live by herself, tends to chroniphy herself, remaining, for several years, in institutions for the treatment of dietary disorder patients. They assume a boulimic identity or one of the anorexic patient. This is very clear, as the therapist perspectives are anxiously lived by the patient. Frequently, the patient shows worse dietary symptoms or even threatens or tries suicide and self-mutilation, in certain moments when she feels about to be abandoned due to any sort of improvement in her clinical state, which rebounds on a lower attention form the therapists' (Martins and Sassi, 2004).

Through the transcription above, we can conclude that people holding dietary disorders, may not have at some extreme situations, condition to elaborate a free and conscious will. In face of nervous anorexia, for example, at first, someone is apt for exercising all the acts of civil life, but even though, does not hold the mental conditions to make the right decisions about the treatment. This person, despite desiring the cure, rejects the treatment, due to the belief that this treatment would make her even 'fatter'. In such situation, underweight people may refuse food, putting their own lives at risk. The misjudgement is clear. Then, we ask: would there be the anorexic patient's (or boulimic's, in other cases) competence to decide about the medical treatment prescribed?

Dietary disorders and the competence for the decision making on health

The 1988 Federal Constitution ensures it as fundamental right, in its art. 5°, *caput*. Besides that, it also treats the health as a social right, which shall be rendered by the State, in the absence of the own person do it. There are several judicial decisions which determine the payment of some sort of medication to the federation's beings, considered as the hypo sufficient patient., including in case of the dietary disorders as in the following decision: 'Injunction. Reexamination necessary. Patient holding severe depression with past anorexia and self – extermination ideas.need of medication for indetermined period. Right to health undissociable to the life right. Constitutional duty, together and solidary of all the beings. It is assured to the chronic and severe disease holder the constitutional right to the medical-hospital treatment, through the supply of the necessary medication, on the SUS expenses during undetermined period, as he can not afford such expenses. To all citizens it is assured the right to health – fundamental right indissociable to the right of life, being the duty of the State, together and solidarily with the institutional spheres of the federative organization, put into effect social economic policies for his promotion, protection and recovery. The health protection which implies the State' s main objectives. Federal Constitution Intelligence Federal, arts. 5°, *caput*, and 196' (TJMG, Injunction n. 1.0395.06.014266-2/001, Rel. Des. Nepomuceno Silva, J. 5/7/2007, DJMG 18/7/2007).

Although the Constitution does not expressly treat the person's autonomy to take decisions about his health – as it is the case of the Spanish and Italian Magna Carta – it is perfectly possible to as being constitutional the body autonomy in Brazil, once the person owns psychiatric wisdom to understand the act and its consequences.

This is a fundamental question to be treated by us: doublecheck whether the dietary disturbed patient is able to take a decision for not eating, not being treated, or whether the order would determine the treatment to be obligatory. All this reasoning goes around the fact of whether a person holding such disorders be competent.

It is competent the one who autonomously decides about his/her own health care. It means that it is necessary to inquire the person's cognitive and emotional resources and the reach of information which the physician discloses about this person's health.

The competence elements are as follows: (1) patient's ability to communicate a decision about health care; (2) understand the information provided by the physician who assists him/her; (3) manipulate the received information, building up an authentic decision, by weighing the risks and benefits (Ribeiro, 2010: 120-121).

If someone holds such dietary disorders – which are all considered as emotional, and so mental problems – this person has his/her competence jeopardized by them, it is possible the non-consented through:

1. *Involuntary internment:* Art. 8 by Act no. 10.216, from 6th April, 2001. The voluntary or involuntary internment Will only be authorized by physician duly registered in the Medicine Regional Council – CRM of the State where the hospitals are located. §1° The involuntary psychiatric internment shall, within the seventy-two hour term, be communicated to the State Public Ministry by the hospital's technical responsible, where it occurred, and this procedure be adopted when the certificate of discharge is released. §2° The end of the involuntary internment will occur by the written request of a relative or legal responsible, or when set by the specialist responsible for the treatment.
2. *Compulsory internment*: Art. 9° by Act no. 10.216, from 6th April, 2001. The compulsory internment is determined according to the legislation in force, by the competent judge, who will take into account the hospital's security conditions, relating to the patient's safeguard, as well as the other interns and staff.

This is a juridical resource juridical review which aims at protecting the patient against himself, once his will is viced due to the absence of psychiatric wisdom. Therefore, the law authorizes the internment against the patient's will. Studies have shown that anorexia is one of the most frequent causes of involuntary internment in face of the difficulty of the patient to join the treatment (Barros and Serafim, 2009: 175-177).

In such cases, the family would hold the standing to sue to claim for the patient's internment, so that his decision is not fruit of a vice act of autonomy. There are cases which there is lack of discernment and also incompetence, which would fall into guardianship; but there are cases which there is no competence, but there is discernment, which could lead to involuntary internment, without the patient's loss of autonomy for the other acts of life. Thus, if the lack of mental sanity is punctual, referring just to the health issues, the guardianship will be an excessive measure, as it will imply the sick person's subjectivity commitment. All in all, it does not make sense protection in the spaces where there is no vulnerability.

What must be double checked is whether the dietary disorder impedes the awareness, by the patient's side, about the cognitive and emotional resources, for the reach and assessment of the information which is provided to him. Then, there are cases which he just the competence for medical decision making, which is possible to be attested by the professional. In such cases, the patient's autonomy may- and shall be relativelized, but always within the boundaries which the concrete case determines, claiming the benefficience principle..

'This way, although the autonomy is one of the pillars of ethical actuation in health assistance, there are cases in Psychiatrics, in which the person's capacity to decide autonomously is jeopardized. The states of law recognize this and foresee specific laws for such circumstances, and it is the health professionals' responsibility to know them.' (Barros and Serafim, 2009: 175-177).

References

Anonymous (undated). Available at: www.ambulim.org.br/transtornos_alimentares.php. Accessed 8 June, 2010.

Ballone, G.J. (undated). Anorexia nervosa. In: PsiqWeb, Internet. Available at http://www.psiqweb.med.br/. Acessed 8 june 2010.

Barros, D.M. and Serafim, A.P. (2009) Parâmetros Legais para a internação involuntária no Brasil. Rev Psiq Clín. 36(4):175-177. Available at http://www.hcnet.usp.br/ipq/revista/vol36/n4/pdfs/175.pdf. Acessed 8 june 2010.

Brasil (undated). Act *n. 10.216*, from 6 April. 2001. It disposes on the protection and rights of the mental disorders holders and redirects the assistance model in mental health. Available at http://e-legis.anvisa.gov.br/leisref/public/showAct.php?id=547. Accessed 9 June 2010.

Lima, T.M.M. and Sa, M.F.F. (in press) Autonomia privada e internação não consentida. Revista na Universidade Federal de Minas Gerais (in press).

Martins, F.C.O. and Sassis Jr., E. (2004). A morbidade entre transtornos alimentares e de personalidade e suas implicações clínicas. Revista de Psiquiatria Clínica, vol. 31, n. 4. Available at em www.scielo.br/scielo.php?script=sci_arttext&pid=S0101-60832004000400005&lng=en&nrm=iso. Acessed 8 june 2010.

Ribeiro, G.P.L. (2010). A competência da criança e do adolescente para o exercício do direito a recusa de tratamento médico. Tese de doutorado. PUC Minas. 2010.

Obesity in ethnic minorities and low-income populations: social and public policy considerations from a comparative study

A. Poveda, A. Jelenkovic, M.E. Ibañez and E. Rebato
Department of Genetics, Physical Anthropology and Animal Physiology, Faculty of Science and Technology, University of the Basque Country, Bilbao, Spain; esther.rebato@ehu.es

Abstract

Over the last decades, the high prevalence of obesity in developed countries has lead to a rising interest in the nutritional status and health of the populations. Accordingly, several strategies have been developed to prevent excess body fat accumulation on the different segments of the population (e.g. The NAOS strategy (2005) through the Spanish Agency for Food Security and Nutrition (AESAN); the comprehensive plan to prevent childhood obesity in the CAV (2004); and the strategies followed by the World Health Organization (WHO) and the International Obesity Task Force (IOTF)). The Roma people, due to their particular social and ethnic backgrounds, are particularly vulnerable to developing obesity and related diseases and may be considered as a risk group. Indeed, considering living standards and key health indicators from an epidemiological point of view, the Roma are still much closer to developing populations than to their European neighbours. It is in this context that the current Roma population has been able to develop higher frequencies of overweight and obesity than the surrounding groups. This fact may be explained, at least in part, by the decline of the manual jobs that carry higher energy expenditure, summed to the increasing rates of sedentariness observed over the last years in many populations. This comparative study analyses the extent to which disparities related to socioeconomic status, membership of a minority ethnic and lifestyle contribute to the prevalence of overweight and obesity, with it possible application to specific policies for the Roma people.

Keywords: Roma people, SES, health policy, nutrition

Introduction

During the past 20 years, obesity levels in adults and young children have increased in developed and developing countries. In order to avoid the obesity increasing prevalence throughout the population, the experts advise people to do a little exercise and to eat plenty of fresh, healthy, and unprocessed foods every day. However, all individuals do not have equal access to these recommended habits. Low-income people and ethnic minorities suffer the highest rates of preventable, diet-related diseases linked to restrictions in time and opportunity to make healthy eating and activity choices. But little is known, however, about what promotes or hinders engagement with prevention programmes. Therefore, in recent years, there has been a considerable interest in studying the socioeconomic and ethnic variations in relation to health indicators, as the prevalence of obesity, showing clear evidences of large disparities between different population groups (Wang and Beydoun, 2007). The Roma people due to their low socioeconomic status (SES) and the recently experienced rapid transition from a traditional to a Westernized lifestyle are considered a 'high-risk' group (Hujová *et al.,* 2009).

From an ethical point of view, however, that kind of comparative studies raise significant ethical and legal questions. Some research programs in the past have involved the mistreatment of minority research subjects and contemporary research programs have to contend with that legacy (Castle *et al.*, 2007). But it is accepted that only focusing the populations groups that are specially suffering some illnesses, health inequalities could be reduced; we mean, those differences in health between populations that are believed to be unfair, avoidable and therefore, reducible (Braveman *et al.,* 2001; Whitehead, 1992).

Governments (among others US, Spain) admit that serious studies on this inequality provide a sound basis for political decision making.

Council of Europe's Convention on Human Rights and Biomedicine (1997), Helsinki Declaration and Spanish Law on Biomedical Research (2007) have been taken into account in this comparative study, whose mayor aim is to examine the relationship of SES, ethnicity and/or sex with the prevalence of overweight and obesity. All subjects have given free and informed consent and their associations agreed with the goal of helping to reduce health inequalities. Several measures were taken for guaranteeing their fundamental rights, including privacy and data protection.

Materials and methods

The present research was conducted on 119 (54 men and 65 women) Roma people subjects living in the great Bilbao (Basque Country, Spain). Height and weight were measured following standard anthropometric techniques (Lohman *et al.,* 1988). From these anthropometric measurements body mass index (BMI) was computed as weight (kg)/height (m^2) for each individual. Underweight was defined as a BMI <18.5, normal-weight as a BMI of 18.5 to 25.0, overweight as a BMI of 25.0 to 30.0 and obesity, 30 or higher. The present results (Roma people settled in Basque Country: BCRP) have been compared with some other samples of different ethnic, SES and/or origin (Table 1).

Results and discussion

The main results of this research are summarized in Table 2. BCRP population shows significantly greater rates of overweight and obesity than BCP, particularly in women (73.45% vs. 25%). The same fact is observed between SRP and SP (64.1% vs. 59.6%, for men and 56.7% vs. 50.4%, for women, respectively) pointing to a disproportional prevalence of overweight and obesity based on ethnic group. In relation to international Roma people (SLRP) the same trend is presented as shown in the study of Dolinska *et al.,* (2007). Although in this sample women obesity and overweight rates decline to a 31.5%, the differences in obesity prevalence in comparison with SLP (7.5%) remain substantial. Regardless of Roma population, this tendency has been already confirmed in a number of international researches, such as, the study of Ogden *et al.,* (2006) in which three American populations were compared. They concluded that there were significant differences in obesity prevalence among ethnic groups and especially among women, being the highest obesity rate of 45.0% for Non-Hispanic Blacks, followed by Mexican-American (36.8%) and finally Non-Hispanic Whites (30%).

Table 1. Sample size and measures characteristics of the samples used in the comparisons.

Sample	Individuals	Males	Females	Measures
Basque country general population (BCP)	855	346	509	direct
Spanish Roma people (SRP)	993	-	-	self-reported
Spanish general population (SP)	14,616	6,796	7,280	direct
Slovak Roma people (SLRP)	396	0	396	direct
Slovak general population (SLP)	444	0	444	direct
American non-Hispanic white population (AWP)	2,357	1,183	1,174	direct
American non-Hispanic black population (ABP)	876	432	444	direct
Mexican-American population (AMP)	873	458	415	direct

Table 2. Prevalence of underweight, normal-weight, overweight, obesity and overweight & obesity by sex and ethnic/socioeconomic group.

Sex	Samples[1]	Underweight	Normal-weight	Overweight	Obesity	Overweight and obesity
Males	BCRP	0	22.2	27.8	50	77.8
	BCP	0	36.2	51.9	12	63.9
	SRP	0.9	35	45.9	18.2	64.1
	SP			46.4	13.2	59.6
	AWP			39.5	31.1	70.6
	ABP			35.1	34	69.1
	AMP			44.5	31.6	76.1
Females	BCRP	1.6	25	21.9	51.5	73.4
	BCP	1.4	73.5	20.4	4.6	25
	SRP	1.9	40.4	31.3	25.4	56.7
	SP			32.9	17.5	50.4
	SLRP	9.3	59.2	21.4	10.1	31.5
	SLP	20.7	71.9	4.8	2.7	7.5
	AWP			27.8	30.2	58
	ABP			27.7	53.9	81.6
	AMP			33.1	42.3	75.4

[1] See Table 1.

The higher differences in overweight and obesity prevalence between BCRP and BCP than between SRP and SP can be explained by the larger differences in SES between the two Basque Country population samples; since BCP belongs to a medium-high SES and BCRP have a much lower SES. It also must be taken into account that while the data from the samples of the Basque Country were obtained by direct measures, SRP study was carried out by interview. This could lead to a bias between self-declared values and real measures, which may also influence the amount of difference observed in obesity rates. In order to go into socioeconomic differences in depth, we have also compared two Spanish populations with different SES but of the same ethnic background, BCP and SP (representing medium-high SES and medium SES, respectively). The comparison shows clear differences in obesity prevalence rates related to SES, especially in women. Higher overweight and obesity rates are associated with lower SES (50.4% vs. 25% for SP and BCP, respectively) and vice versa (only 4.6% of the BCP women are obese). Furthermore, the negative association between SES and obesity, resembles the pattern observed in other female population in developed regions and could be partially explained by the expected associations between SES and diet and nutrition knowledge, standards of physical attractiveness and concern with weight control (diet and exercise).

In conclusion, obesity and overweight disproportionately affect some ethnic and socioeconomic groups. The negative association between obesity rates and socioeconomic status is more consistent among women. Minority groups have more limited access to health care and preventive medicine and also are severely underrepresented among biomedical and behavioural researches. Change in these issues could be achieved through information, health policy and educational interventions, which would increase their knowledge on some basic aspects of health, as diet and activity choices, without changing its idiosyncrasy.

References

Braveman, P., Starfield, B. and Geiger, H.J. (2001). World Health Report 2000: How it removes equity form the agenda for public health monitoring and policy commentary: comprehensive approaches are needed for full understaning. British Medical Journal 323: 678-681.

Castle, D., Cline, C., Daar, A.S., Tsamis, C. and Singer, P.A. (2007) Science, Society and the Supermarket.

Dolinska, S., Kudlackova, M. and Ginter, E. (2007). The prevalence of female obesity in the world and in the Slovak Gypsy women. Bratisl Lek Listy 108: 207-211.

Hujová, Z., Desatnikova, J. and Gabor, D. (2009). Some cardiovascular risk factors in Gypsy children and adolescents from Central Slovakia. Bratisl Lek Listy 110: 233-239.

Lohman, T.G., Roche, A.F., Martorell, R. (eds.) (1988). Anthropometric standardization reference manual. Champaign, Illinois: Human Kinetics Books.

Ogden, C.L., Carroll, M.D., Curtin, L.R., McDowell, M.A., Tabak, C.J. and Flegal, K.M. (2006). Prevalence of Overweight and Obesity in the United States, 1999-2004. American Medical Association 295: 1549-1555.

Wang, Y. and Beydoun, M.A. (2007). The Obesity Epidemic in the United States – Gender, Age, Socioeconomic, Racial/ Ethnic, and Geographic Characteristics: A Systematic Review and Meta-Regression Analysis. Epidemiologic Reviews 1-23.

Whitehead, M. (1992). The concepts and principles of equity and health. International Journal of Health Services 22: 429-445.

Eating and drinking: challenges for global bioethics

T. Ausín
Institute of Philosophy, CCHS, Spanish National Research Council, CSIC, Madrid, Spain;
txetxu.ausin@cchs.csic.es

Abstract

The lack of food and drinking water does nothing but contribute to the chronic circle of poverty and disease that afflict millions of people worldwide. Then, the effective treatments and sanitary aid are useless since, inevitably, people fall in reinfection caused by nutritional deficiency, poor sanitation and hygiene, and lack of drinking water supplies. We characterize food and drinking water as basic and fundamental human rights ('human rights begin with breakfast', said Leopold Senghor), against some liberal conceptions that reduce them to something purely welfare and subject to the rules (marked, not invisible) of markets. In fact, the causes of the famine situation in much of humanity have their source in economic and trade policies: programs of the International Monetary Fund and the World Trade Organization, production of biofuels, the abandonment of consumption agriculture, and speculation with food. This scenario requires a deep rethinking of public policy on food and drinking water, departing from its conceptualization as public goods, and not as mere commodities. Finally, the satisfaction of the right to eat and drink is the minimum that is necessary precondition for the exercise of autonomy. It is this basic need that requires an individual to achieve sufficient control over their abilities and skills – both from the standpoint of needs, from the point of view of capabilities, or from the rational conditions for human agency. Therefore, eating and drinking are undoubtedly unavoidable claims for a global bioethics that has put the emphasis on interdependence and vulnerability as fundamental categories that actually define the human condition. This entails a necessary 'politicization' of bioethics not only in terms of global justice but of participation, deliberation and public policy; a bioethics at the bridge between the sciences and the humanities would not refuse either to economy, ecology, political science, or international relations.

Keywords: poverty, famine, public policies on food, public policies on drinking water

Introduction: the human face of data

Bioethics, which had been conceived by V.R. Potter as a science of survival, a bridge or a third culture between the sciences and the humanities, has ended today limited as a kind of biomedical ethics, very focused on issues concerning clinical relations in health systems of developed countries and linked to the debates provoked by the new biomedical technologies (cloning, biobanks, gene therapy, hybridization, etc.). However, if we look at health care from a global perspective, it is impossible not to consider that 100,000 people die every day from hunger and lack of drinking water or their immediate consequences.

The scourge of hunger is an old companion of humankind, ancient and persistent, which has plagued humans at various times in their history (Bessis, 1992). However, dramatically today, in the twenty-first century, in the context of some of the most incredible scientific advances as communication technologies, gene therapies, tissue regeneration from stem cells, nanotechnology, etc., hunger increases: More than 1000 million people suffer from food shortages (until 2007 the figure was 824 million and forty years ago were around 400 million). Within this huge group of people are 20 million children under 5 annually affected by severe acute malnutrition and each year 9 million people die due to health reasons in which malnutrition is the key factor of the disease (hunger-related diseases). This lack of food also plays a role in the emergence and spread of infectious diseases because the immune system is weak, like the AIDS pandemic in Africa:

> Returning from a trip to southern Africa, Peter Piot, head of UNAIDS, United Nations organization responsible for the global fight against AIDS, wrote: 'I was in Malawi and met with a group of women living with HIV. As I always do when I meet people with HIV/AIDS and the other community groups, I asked them what their highest priority was. Their answer was clear and unanimous: food. Not care, not drugs for treatment, not relief from stigma, but food' (Ziegler, 2006: 114-115).

About drinking water, over 1,100 million people lack access to safe water and over 2,600 million people still lack adequate sanitation systems and domestic wastewater treatment (United Nations Program for Development – UN Human Development Report, 2006: 33).

Elements of the global food crisis

We just indicate that famine, rather than fall, is ever increasing in recent years. It is outrageous that agricultural production has tripled in less than a century, on par with the world's population also multiplied by three. Moreover, according to the World Food Report of the FAO, global agriculture, with the ongoing development of its power production, could feed at a rate of 2,700 calories per adult per day, to 12,000 million people, i.e. almost double of humanity.

Consequently, starvation is not the product of an objective lack of goods (*der objektive Mangel*), such as Marx assumed, nor is a Malthusian regulatory mechanism of the population; the cause is a radically unequal distribution of these goods, food and drinking water. Hunger is not inevitable, nor a problem of resources but an enormous injustice, so that 20% of the population holds 80% of resources. This means that, on a par with that comes to more than 1000 million people are undernourished, nearly as many are overweight or obese (Patel, 2008; de Sebastián, 2009; Duch, 2010).

But beyond this terrible paradox, we should clarify some structural elements that contribute to the advancement of world hunger.

Abandonment of consumption agriculture

Unlike the boom in wheat and corn, often produced from hybrids, millet, sorghum and tubers are still cultivated in much of the Third World for millions of farming families, being the basic livelihood of their diet, but its production has stagnated due to lack of economic incentives and technical and financial means. Similarly, it has reduced production of legumes, rich in protein, and supply base in many countries of the South: peas, beans, lentils, peanuts.

Moreover, international aid to indigenous agriculture has fallen sharply over the past 20 years, although 75% of the poor in developing countries live in rural areas and depend on agricultural production.

Unequal production

20% of world population, rich countries, is both the largest producer of food, leading consumer and the biggest exporters of food products. Besides, Northern agricultural production is heavily subsidized and protectionism is applied to not suffer too much foreign competition (agricultural dumping). By subsidizing agricultural production in rich countries, international markets filled with cheap food during the seventies and eighties of last century, causing many developing countries were dedicated to the import, because it was cheaper to buy than to produce at home, contributing to the mentioned neglect of indigenous agriculture.

Structural adjustment policies

We include under this heading three major lines of action of the international economic institutions like the IMF, the World Bank and the World Trade Organization.

First, the 'scandal of the debt': poor countries pay annually to the ruling classes of the rich countries much more money than they receive from them in form of cooperative loans, humanitarian aid investments, or development funds.

Secondly, it has been applied ultra-liberal agricultural policies on Third World countries, while applying protectionism, as noted, on agricultural production in rich countries, an incredible exercise in cynicism and double standards. In this vein, international economic institutions have restricted banks of seed, state organizations for the purchase of grain storage, price support policies, etc. That is, socialism for the rich, liberalism for the poor.

Thirdly, these international economic organizations have prioritized and encouraged a kind of agriculture focused on export (coffee, cocoa, cotton, rubber) and controlled by multinational agribusiness companies, which has done extremely poor countries dependent on imported food up to 50%, resulting in the abandonment of the land, the enormous increase in urban population, and vulnerability to the price increase.

Price rises

We have already referred to the dramatic increase in prices that have undergone basic foods in recent years. If the average has been from 54% in the past year and a half, rice has risen by 74%, soybeans 84%, corn 31% and wheat by 130%.

Besides the decline in global food supply, there has been an increasing consumer demand since world population increases at a rate of 75 million people a year. And there has been an increase in per capita food demand, especially in emerging countries like China and India, where the increased purchasing power of the population and increasing urban population compared to rural, has grown significantly the number of net buyers of food in front of the producers.

Moreover, in recent years, the non-food use of agriculture has dramatically increased, mainly for the production of biodiesel and bio-ethanol, which has led to increased demand for cereals. Experts do not agree on the impact that bio-fuel production has on the rising price of food. For some economists like Jeffrey Sachs and Paul Krugmann, bio-fuel, also called 'dead-fuel', is a huge mistake and, even, would not save significant emissions of CO_2, as noted by Hartmut Michel, Nobel Laureate in Chemistry. OECD and the International Food Policy Research Institute indicate that the impact of bio-fuel demand on inflation experienced in food prices is over 30% (Steenblink, 2007).

Finally, the culture of finance capitalism is installed in the food market so that, increasingly, with the upheavals in international financial markets, there has been a change in priorities for large investments to the purchase of agricultural commodities and food core. In 2000, the total amount of trade involving agricultural products between all the stock exchanges (Chicago Commoditiy Stock Change, Agriculture Parworld Classic, etc.) numbering 10,000 million dollars. In May 2008 amounted to about 175,000 million. In January 2008, 38% of U.S. investment funds were invested in food markets that offer, next to atrocious famine map, a tremendous return on capital (the variation June 2007 to June 2008 was a 43.70%). World Bank economists estimate the role of speculation in the rising price foods is around 37% (World Bank Report, 14 April 2008).

The right to eat and drink

The right to food is recognized in Article 25.1 of the Universal Declaration of Human Rights in and ratified in the Vienna Conference on Human Rights (1993). It has also reached an explicit recognition by the UN Committee on Economic, Social and Cultural Rights interpreting Article 11 of International Covenant on Economic, Social and Cultural Rights of 1966 which recognizes the right of everyone to an adequate level of life, including food, clothing and housing, and the fundamental right of everyone to be free from hunger and malnutrition. Thus, the Committee has established that the human right to food and drinking water is essential to decent living (General Comments Ns. 12 [1999-05-12] and 15 [2002-11-29]).

The right to eat and drink is one of those rights that give and preserve life. Beyond the difficulties of grounding other rules considered basic and universal human rights, in the case of food supplies and drinking water is no doubt that these are needs of every human being and in this sense, are essential and urgent. These are elements required for survival and psychological and physical integrity of human beings, and whose absence causes serious and objectives damages, leading to vulnerability and the impossibility of social participation and the pursuit of the own aims. Doyal and Gough (1994) have referred to physical survival and personal autonomy as the preconditions for any individual action in any culture, so similar to Gewirth (1982) saw freedom and wellbeing (physical, mental balance) as the necessary conditions for human action (agency). From this perspective, certainly food and drinking water are needed material resources sufficient for the psychophysical integrity of all human beings, beyond cultural differences, being objective and universal and, therefore, valuable and necessary for everyone, which leads to moral universal demands (human rights).

In this sense, the *Ad hoc* Group on Poverty and Human Rights of the UN Human Rights Commission has reiterated the indissoluble link between the right to life and the satisfaction of basic human needs (document E/CN.4/Sub.2/2002/15 of 25 June 2002, paragraphs 3, 15 and 17): The most essential level of the right to life is the maintenance of the biological level of human beings and to that extent the right to life encompasses four basic rights: (1) The right to adequate food; (2) The right to have drinking water; (3) The right to housing; and (4) The right to health. The lack of one or more of these elements constitutes a violation of the dignity of human beings as well as torture, restriction of freedom, or killing. Hunger (and thirst) poses extreme violence on individuals, as damaging as the worst torture, and so psychologically paralyzing as the worst repression.

Food and drinking water: commodities or public goods?

Despite being, as we analyzed, key elements necessary for human life and, therefore, universal human rights, water and food are considered by international agencies like the World Trade Organization – through its General Agreement for Tariffs and Trade (GATT) – as tradable and open to profit operation. This is the case with great agribusiness corporations. In this sense, food and drinking water are compared to other commodities and consumer goods such as furniture, vehicles or appliances, and undergone to the criteria of 'free' market speculation and the stock market.

Exclusive submission of water and food to global market forces results in a distribution of those resources based on ability to pay, where the citizen is reduced to the category of user or consumer of goods and services far from a citizen as owner of the goods necessary for life, health, bodily integrity and, thus, for its possibility of agency and freedom.

Departing from previous conceptualization of food and drinking water, we label them as public and social goods – as do Human Rights Committees of the United Nations in its General Comments –, as essential resources to life, and as the stuff of the universal human right to eat and drink – the right to be free from hunger. Just stating food and drinking water as public goods is possible to extend the protection on them and to ensure universal access to them, regulating its production, financing and trade.

So, it should be reform the legal framework for international trade relations, since, as recalled Joachim von Braum, Chief of the International Food Policy Research Institute and a participant in the high-level meeting on food safety (Madrid, January 2009), it is not possible to ensure food security without fair trade. And that includes to put on the same level tariff across countries, the reformulation of agricultural subsidies in rich countries, and especially the regulation and control of the entire food production system, now controlled by a group of large agribusiness companies operating in the market, almost as a monopoly, with the consent of a legal framework agreed by governments and international economic organizations.

Challenges for global bioethics

The lack of food and drinking water does nothing but contribute to the chronic cycle of poverty and disease that afflict millions of people worldwide. Then, the effective treatments and sanitary aid are useless since, inevitably, people fall in re-infection caused by nutritional deficiency, poor sanitation and hygiene, and lack of drinking water supplies.

The satisfaction of the right to eat and drink is the minimum that is necessary precondition for the exercise of autonomy. It is the basic need that requires an individual to achieve sufficient control over their abilities and skills – both from the standpoint of needs (Doyal and Gough, 1994), and from the point of view of capabilities (Nussbaum and Sen, 1993) or the rational conditions for human agency (Gewirth, 1982).

So, we should lay emphasis on interdependence and vulnerability as fundamental categories that actually define the human condition. This lead us to consider the dimensions of care and responsibility, linked to the above categories, from a global viewpoint, when we recognize the impossibility of an actual exercise of autonomy if not met certain minimum life-elements – like access to food and drinking water. Thus, we stand that poverty is the major bioethical issue.

This entails a necessary 'politicization' of bioethics not only in terms of global justice but of participation, deliberation and public policy; a bioethics at the bridge between the sciences and the humanities would not refuse either to economy, ecology, political science, or international relations.

References

Bessis, S. (1992). El hambre en el mundo. Madrid: Talasa.

de Sebastián, L. (2009). Un planeta de gordos y hambrientos. (La industria alimentaria al desnudo). Barcelona: Ariel.

Doyal, L. and Gough, I. (1994). Teoría de las necesidades humanas. Barcelona: Icaria.

Duch, G. (2010). Lo que hay que tragar. Minienciclopedia de política y alimentación. Madrid: Los Libros del Lince.

Gewirth, A. (1982). Human Rights. Essays on Justification and Application. Chicago: The University of Chicago Press.

Patel, R. (2008). Obesos y famélicos. (El impacto de la globalización en el sistema alimentario mundial). Madrid: Los Libros del Lince.

Nussbaum, M. and Sen, A. (1993). The quality of life. Clarendon Press.

Shiva, Vandana (2003). 'Food Rights, Free Trade, and Fascism', en Matthew J. Gibney (ed.). Globalizing Rights. Oxford: Oxford University Press, pp. 87-108.

Steenblink, R. (2007). Biofuels. At What Cost? Government Support for Ethanol and Biodiesel in Selected OECD Countries: http://www.iisd.org/pdf/2007/biofuels_oecd_synthesis_report.pdf [download 2010-03-26].

Ziegler, J. (2006). El imperio de la vergüenza. Madrid: Taurus.

Food policy and institutional conflict between food and health: a sociological approach to obesity in Spain

C. Díaz-Méndez
Dpto.Sociología, Universidad de Oviedo, 33006 Oviedo, Asturias, Spain; cecilia@uniovi.es

Abstract

The figures for obesity continue to rise, especially among the lower age groups both in Spain and the rest of Europe. Moreover, intervention by public bodies to halt this rise does not achieve satisfactory results. Consequently, analysts from different scientific fields have pondered this scant success of institutional intervention in the problem of obesity. Underlying all the criticisms is the lack of evaluation in many of the measures adopted, but in-depth analysis is needed of the form of intervention and the consequences of the different strategies adopted to stem the epidemic. In this work we review the different solutions adopted and comment on each one. This review leads us to state that it is impossible to halt obesity without coordination between the sectors to solve the problem. Relationships between the agents involved in food problems, the state, businesses and consumers are conflictive, giving rise to uncoordinated actions to deal with food problems and offer possible solutions. We, therefore, maintain that, at least in Spain, there is no institutional environment favorable to halting the rise of obesity. The source of the problem lies in ascribing responsibility.

Keywords: food obesity policies, health policies, food safety, diet

Introduction

Studies carried out in developed countries point to a steady increase in overweight and obesity in Europe. For this reason obesity is today the subject of academic and institutional debate throughout Europe and has been called by WHO 'the non-infectious epidemic of the 21st century' WHO (2004, 2006); EC (2007). All factors would seem to point to Spain having one of the highest rates of child obesity in Europe. From their first survey in 1987 to their latest in 2006, the Spanish Health Survey has confirmed that the problems of overweight and obesity have been on the rise. According to this source 18.7% of the age group of 2 to 17-year-olds is overweight and 8.9% obese. In the age group 18+ 15.6% of the population is obese (National Health Survey, 2006). Data from the International Association for the Study of Obesity show that the percentage of 7 to 17 year-olds, especially boys and adolescent youths, places Spain among the top positions for these problems.

The figures for obesity continue to rise, especially among the lower age groups both in Spain and the rest of Europe. Moreover, intervention by public bodies to halt this rise does not achieve satisfactory results (Costa-Font and Gil Trasfi, 2008). Consequently, analysts from different scientific fields have pondered this scant success of institutional intervention in the problem of obesity. Underlying all the criticisms is the lack of evaluation in many of the measures adopted, but in-depth analysis is needed of the form of intervention and the consequences of the different strategies adopted to stem the epidemic. Existing studies would indicate that the scant success of the intervention measures to contain child obesity is directly related to their partial nature. For this reason we hold the opinion that only a global view, both of the problem and its solution, can help to reduce the current rates of obesity.

In this work we review the different solutions adopted and comment on each one. This review leads us to state that it is impossible to halt obesity without coordination between the sectors to solve the problem. However, relationships between the agents involved in food problems, the state, businesses and consumers are conflictive, giving rise to uncoordinated actions to deal with food problems and offer possible solutions. We, therefore, maintain that, at least in Spain, there is no institutional environment favorable to halting the rise of obesity. The source of the problem lies in ascribing responsibility.

Critical theoretical views on obesity: different ways to understand and halt obesity

The policy and analytical treatment obesity has received has undergone important changes over the last twenty years. In this time obesity has gone from being a health matter of an individual nature, dealt with in the field of medicine, to a social and public health problem with the intervention of government bodies. We are going to carry out a brief review of the different theoretical standpoints from which this problem has been analyzed.

Obesity as a biological, medical and individual problem

Food problems, and in particular, obesity, have been treated mainly in a health environment. This viewpoint has its basis in the fact that problems linked to eating are considered to have their origin in biological factors (Neel, 1962). These single-cause and biologically oriented views were dominant for a long period of time before obesity was looked at a social problem and explain the individual, medical treatment of the problem.

This strictly medical- health intervention, just as other interventions for specific illnesses or epidemics, needs to be performed on the persons affected by the illness. In the case of obesity and overweight the medical intervention aims at getting the person affected to reduce their weight and change their eating and physical habits. The sufferers are asked to change their state of health by changing their eating habits and their sedentary life style (calorie reduction and physical exercise) and at times, some kind of medication is added.

There has been a two-fold questioning of this view. Firstly, medical intervention for obesity does not show the expected individual results, as the person does not always achieve a steady weight loss. (Cuevas and Reyes, 2005; Vázquez Martínez *et al.*, 2008). Some analysts maintain that while success obtained from therapies is frequently written about, failures are not (Jain, 2005). Be that as it may, the patients do not always reduce their weight significantly and the problem is recurring. Secondly, and above all, the intervention, being of an individual nature, has no impact whatsoever on the current epidemic in the population (Jain, 2005). Indeed, the medical-health type of intervention makes obesity into an illness whose main cause lies in the excess of calorie-rich foods and lack of physical exercise, forming part of the affected individual's life-style.

Obesity as a socio-educational problem analyses

The medical viewpoint, although it coexists with other kinds of interventions, is no longer the norm in the current climate where obesity is a social problem. Doctors who have examined nutritional changes in the population agree in asserting the necessity of dealing with the matter from a nutritional policy perspective that encourages eating habits in accord with nutritional recommendations and allows obesity to be treated as a public matter (Serra Majen *et al.*, 2007). This view is shared today by a large number of scientists in the health and nutrition fields and by policy management officials in institutions related to food. Nutritional studies favor various external factors of a socio-educational nature to understand and analyze the causes of obesity. This trend has meant variables of a social nature being incorporated into 'classical' nutritional studies. Among these, Varela's studies stand out: they result from institutional collaboration between the Surveys of Family Budgets and Nutritional Surveys carried out by the National

Institute of Statistics. There has been a spate of surveys favoring the integration of variables from one field or another offering the possibility of linking the nutritional health of the Spanish people to family incomes, occupations, or household composition (Varela, 1995, 1999). Other works have rounded off this social trend by including a study on dietary beliefs (Vioque and Quiles, 2003). In general surveys on nutrition now include questions on practising sport, leisure time physical activity or ,as in the Survey on Health by the Ministry of Health, people's perceptions of their own state of health. A step forward has been made towards the understanding of obesity and overweight in modern societies by including together with the nutritional aspects others of a social nature (Contreras, 2009).

The step away from a medical and individual view of the problem to one more social and public in nature has led intervention against obesity into the educational field, endowing the solution with a preventive aspect. This type of proposal is upheld by the verification that the bad eating habits of children giving rise to overweight and obesity such as excessive consumption of sugars, lack of physical activity or bad breakfasts is confirmed in nationwide studies (Study ENKID, Aranceta and Serra Majen, 2001) and in national health surveys. Intervention in the youngest section of the population is justified by the close relationship between child obesity and obesity in adult life and emphasis is laid on the future illnesses that can be caused by overweight and obesity. The financial consequences for the state are also mentioned as these kinds of illness entail an increase in health costs because of their treatment (Rodríguez Caro and González López-Valcárcel, 2009). Hence an intervention is chosen with training and information programs, directed particularly at young people and with special intervention in educational centers, although also directed at the general population by way of information campaigns. Thus, for example, the Ministry of Health's program PERSEO is directed at 6 to 10 year-olds and takes place in schools.

Generally in any intervention of an educational nature there are great difficulties in verifying its success. On the one hand it is because its results cannot always be measured and on the other, long periods of time are needed to corroborate its effectiveness. It is perhaps because of these difficulties the programs carried out have not been fully evaluated and their effects on the targeted population are unknown. In Spain, especially, there are no academic articles describing the impact of the educational strategies aimed at reducing child obesity. In other countries it has been found that this kind of intervention is more effective with sections of the population with high educational and income levels (Swinburn and Egger, 2002). This finding would call into question the impact on less favored social groups who are more affected by the epidemic, in effect, these are the ones with least qualifications and lower income levels (Aranceta *et al.*, 2003; Lhuissier, and Regnier, 2005).

Another of the criticisms usually made about this kind of socio-educational intervention refers to the ascribing of responsibilities. If obesity is analyzed as a question of a lack of information or education, the bad results are the responsibility of educators and parents, incapable of keeping the eating behavior of the children in order. From the individual responsibility of the first medical measures we now pass to group responsibility, in which the family especially is the main factor responsible for the problem by being incapable of guiding the behavior of its younger members towards healthy eating habits.

Obesity as a policy problem

There is a third sphere of intervention against obesity that merits attention, the ascribing of responsibility to the so-called obesogenic environments. From this point of view it is argued that inappropriate eating norms, leading to unhealthy life-styles, producing obesity and overweight are behaviors imposed from outside the individual and beyond their control. Eating healthily means facing daily commercial centers where consumption decisions are made and in these consumer environments foodstuffs favoring obesity are what is offered (Swinburn and Egger, 2002). The lack of healthy offerings in the market lies behind the inadequate decisions. On one hand the food industry, with strong publicity backing, offers foodstuffs at low prices with few or no healthy ingredients. On the other hand, healthier products cannot count on vigorous marketing strategies to encourage their consumption and, moreover, their prices are not always low and they are often introduced by local distribution channels or minor networks (such is the case in fresh fruit). The result is that healthy products are competing against the less healthy ones at a disadvantage and their consumption does not spread among consumers. Indeed, the analysis of obesity cannot be considered apart from an environment of economic and industrial sets of reasons decisive in the production and the maintaining of bad eating habits (Pollan, 2008; Luque, 2008; Jain, 2005; Lang and Rayner, 2007).

From this viewpoint it is understood that these obesogenic environments encourage consumers to choose foodstuffs following criteria that draw them away from reasonable nutritional advice and hinder them from adopting healthy life habits. Responsibility here does not belong to the individual but to the group, mainly businesses, and the subject is but a victim of economic agents that impose far from healthy behaviors which are difficult to resist.

Some authors have added a policy dimension to the analysis of obesity (Nestle, 2003; Lang, 2006; Lang and Rayner, 2007). These views show the complex framework of relations between the agents involved in food policies and especially the inborn complexity of the search for solutions to obesity in the developed countries (Nestlé, 2003).

Some analysts consider the triangle of interests between the state, business and civilian society is an ideal relationship model guaranteeing a good social and policy background for putting national food policies into effect (Lang, 2007). However, in practice this network of relationships between the agents involved in nutrition has given place to what some have called food cacophony. Lang (2007) gives the term a social sense while Fischler (1995) analyzes it from the individual standpoint and refers to a large amount of confusing information about how to feed oneself properly. Working with this assumption it is maintained that the lack of results in the policies aimed at halting obesity is due to the complex framework of relationships between the agents involved in solving the problem. Policy intervention carried out to tackle obesity is sectoral and the failure of programs to combat obesity is due to the multiplicity of standpoints and the difficulty in achieving a comprehensive intervention. This cacophonic situation, 'noise drowning out symphony effort', causes the actors involved in solving food problems to fail to act in concert sharing the policy field of action and thus achieving results in reducing obesity (Lang and Rayner, 2007).

Conclusions

The backbone of the relationship between state, food industry businesses and consumers is their shared interest in food and health. What causes this relationship not to be the right one is a matter of the ascribing of responsibilities in the deterioration in health linked to diet and especially in obesity figures that continue to rise.

The ideal agent relationship model could be the triangulation of interests proposed by Lang (2006) in which the state, the citizens and the food industry share interests and responsibilities, but this does not seem very probable. In a society such as the Spanish one where the relationships between the food agents pivot round the state, with a not highly developed civilian society and a food industry performing in a highly regulated environment would lead to the following opinion: the principal responsibility in the lack of effectiveness in the measures to stem child obesity lies in the role adopted by the state with regard to the other agents. The theatre of action is therefore less complex than it would seem to be. Even so, relationships are not free from conflict.

The scenario for relations between the state and the food industry companies is set in the planning of regulations as to how food should be produced, changed, distributed and sold. (Leon Arce, 2008). It does not seem that these relations are going to run more smoothly in the future. Against a background of globalization of the international food business regulation will be more necessary and therefore it will be more difficult to reach agreement with respect to what extent businesses are in favor of a greater state intervention in their actions.

The scenario for relations between consumers and companies is mainly though not exclusively in the field of publicity as there are few new interaction scenarios. Perhaps a more thoughtful and critical consumer might not make this scenario to be one of conflict, but under-age young people continue to be a vulnerable group in need of protection. Consequently the state will continue to intervene in these relationships as long as support and criticism is needed.

The scenario for relations between consumers and the state is more characterized by its asymmetry than by conflict. (Ramiro Pérez *et al.*, 2002). In a society in which citizen consumer participation is weak (even more so in the case of food), the state takes on the role of guardian and mediator between the consumer and the other agents. The consumer does not play a very active part and is possibly under-represented in the policy field. This supposes a lack of initiative to participate in the governance of food policies.

In view of what we have examined here, we do not think there is an appropriate environment for halting the growth in obesity because the responsibility is multiple but not all the agents are equally ready to assume their share. The result is a weakening in responsibility or rather preference is given to partial interventions that avoid confrontation between agents but which prevent a comprehensive intervention in the problem.

References

Aranceta, J., C. Pérez Rodrigo, L. Serra Majem, L. Ribas Barba, J. Quiles Izquierdo, J. Vioque, J. Tur Marí, J. Mataix Verdú, J. Llopis González, R. Tojo and M. Foz Sala (2003). Prevalencia de la obesidad en España. Resultados del estudio SEEDO 2000. Revista de Medicina Clínica, vol 120, 16.

Aranceta, J. and Serra Majem, J. (eds) (2001). Obesidad infantil y juvenil. Estudio ENKID. Barcelona, Editorial Masson.

CE (2007). Libro blanco sobre la obesidad.

Contreras, J. (2009). Dimensiones socioculturales de la obesidad. Humanitas humanidades médicas. Tema del mes on-line, 43.

Costa-Font, J. and Gil Trasfi, J. (2008). ¿Por qué ha aumentado la obesidad en España? Temas de debate. Catedra Fedea.

Cuevas, M. and Reyes, M.A. (2005). Lo último en diagnóstico y tratamiento de la obesidad. ¿Hay lugar aún para la terapia conservadora?. Rev.Med.Chile, 1333: 713-722.

Díaz Méndez, C. and Gómez Benito, C. (2008). Evolución de las recomendaciones institucionales sobre salud y alimentación. Alimentación, Consumo y salud, 24, 25-50.

Encuesta Nacional de Salud (2006). Ministerio de Sanidad y Consumo.

Jain, A. (2005). Treating obesity individuals and populations. Clinical Review, 331, 1387-1390.

Lang, T. (2006). Food, the law and public health: three models of the relationships. Public Health, Journal of the Royal Institute of Public Health, 120, 30-41.

Lang, T. (2007). Overcoming policy cacophony on obesity: an ecological public health framework for policy makers. Obesity Reviews, 8, 1, 165-181.

Lang, T. and Rayner, C. (2007). Overcoming policy cacophony on obesity: an ecological public health framework for policymakers. Obesity Reviws, 8, 1, 165-181.

Lhuissier, A. and Regnier, F. (2005). Obesity and food in working classes. An approach of the femanle body. CORELA, INRA. No. 3-4. December.

Luque, E. (2008). La obesidad más allá del consumidor. Raíces estructurales de los entornos alimentarios. Alimentación, Consumo y salud, 24, 130-149.
Leon Arce, A. (2008). 'Políticas alimentarias y seguridad del consumidor'. Alimentación, consumo y salud, 24 81-104.
Lobo, F. (2007). Políticas públicas para la promoción de la alimentación saludable y la promoción de la obesidad. Revista Española de Salud Pública, 2007, 5.
Neel, J.V. (1962). *Diabetes mellitus*: a thrifty genotype rendered detrimental by progress. Am J. Hum Genet, 14, 353-362.
Nestle, M. (2003). Food policy. How the food industry influences nutrition and health. University of California Press.
Pollan, M. (2008). El detective en el supermercado. Temas de hoy. Madrid.
Ramirez Pérez, Navarro and Trujillo, 2002. Consumerismo y movimiento de los consumidores. Valores, actitudes y comportamiento consumeristas de la población andaluza. Revista Española de Investigaciones Sociológicas, 202, 145-176.
Rodríguez Caro, A. and González López-Valcarcel, B. (2009). El trasfondo económico de las intervenciones sanitarias en la prevención de la obesidad. Revista Española de Salud Pública (online), 83, 1.
Serra Majem, L., Ribas Barba, L., Salvador Castell, G., Roman Vinas, B., Castell Abat, C., Cabezas Pena, C., Cruz Pastor Ferrer, M., Raido Quintana, B., Ngo De La Cruz, J., Garcia Alvarez, A., Serra Farro, J., Salleras Sanmartf, L. and Plasencia Taradach, A. (2007). Tendencias del estado nutricional de la población española: resultados del sistema de monitorización nutricional de Cataluña (1992-2003). Revista Española de Salud Pública (online), 81, 5.
Swinburn, B. and Egger, G. (2002). Preventive strategies against weight gain and obesity, The international association for the study of obesity. Obesity reviews, 3, 289-301.
OMS (2004). Estrategias globales sobre dieta, actividad física y salud.
OMS (2006). Carta europea contra la obesidad.
Varela Mosquera, F. (ed.) (1995). Encuesta de presupuestos familiares 1990-91. Estudio nacional de nutrición y alimentación 1991. Madrid. INE.
Vioque, J. and Quiles, J. (2003). Encuesta de Nutrición y Salud de la Comunidad Valenciana, 1994. Alicante.

Part 6. Food labelling and consumer protection

Certification and ethical standards in aquaculture

A.S. Haugen[1], M. Kaiser and V. Almaas
University of Bergen, Centre for the Study of the Sciences and Humanities, Bergen, Norway; arne.sveinson@svt.uib.no

Abstract

The diner in Darden restaurant will have to pay 5 $ extra in order to serve organic farmed salmon from Scotland. The fish thus on offer is certified by the U.K. Organic Food Federation. This is a newsline in Fish Farming International January 2010. Apart from the question whether it is worth to spend extra money for organic food, one may also ask what the certification amounts to and how good a certification it actually is. After all, is organic food really compatible with the farming of fish high in the food web? There is already a plethora of existing certification schemes available, which in addition to certifying organic production include such as eco-labelling, sustainable production, food safety, health and nutrition, food quality, animal welfare, fair trade, food security and internal retail standards. The United Nations Conference on Trade and Development estimates the number of schemes at 400 and rising. Explicit guidelines (e.g. FAOs guidelines for eco-labelling, responsible fisheries and aquaculture, the FAO *Codex Alimentarius*, FAO/NACA Aquaculture certification systems), industrial standards and regulatory frameworks (e.g. ISO 9000, 14000, 22000 standards), the major private/voluntary standards and certification systems (e.g. Global GAP, Global Aquaculture Alliance (GAA) standards, IFOAM, MSC certification and WWF-coordinated Aquaculture Dialogue standards). The World Wide Fund for Nature in 2007 analysed standards and certification schemes in aquaculture and found most of them wanting in many respects. Given the consumer demand, and in part the consumer scepticism, it seems that the work on better labelling and good certification schemes in this area is set to continue. As part of the EU project SEAT (Sustainable Ethical Aquaculture Trade) we examine these standards and ask whether ethical issues are reflected in them and what the prospects are for supplying transparent and reliable information to the ethically minded consumer of aquaculture products. Are voluntary schemes really satisfactory or should one develop mandatory ones? In this paper we present our first approach towards handling these kinds of questions.

Keywords: certification, guideline, code of conduct, fair trade, sustainability, food ethics, ethical standard, ethical index, ethical matrix, stakeholders

Introduction

Aquaculture products make up the fastest growing food sector world-wide. Production and trade is truly international, and markets are diverse. Expectations from European consumers are comprehensive, covering not only product quality, but also issues like traceability and sustainable production methods. In particular the expectations relating to ethical issues in aquaculture production and trade are far from clear. In this paper the authors present the ideas behind a search on how and to what extent ethical standards can be found in existing certifications schemes, guidelines and codes of conduct. This is part of a process towards developing an ethical aquaculture food index (EAFI) as part of a larger EC funded project on Sustaining Ethical Aquaculture Trade (SEAT; coordinated by the University of Stirling; http://seatglobal.eu/). Participatory stakeholder processes in Europe and four Asian countries will probe deeper into a realistic set of possible index standards.

The task of developing an EAFI will be done through an in-depth study of four particular cases of cultured species. These are tilapia, pangasius catfish, peneid shrimp and macrobrachium prawns. This will be done by looking at their production in the four different Asian countries of China, Vietnam,

Thailand and Bangladesh (Little *et al.*, 2009). Given the cultural and socio-economic diversity between producing and consuming countries, a particular challenge will be to not let the best stand in the way for the good. This means not to impose top-down standards merely anchored in the affluence and cultural values of Europe.

Values and goals, not moral judgements

Part of the work plan to develop an EAFI is a critical review of current ethical standards for aquaculture. This includes identification of costs, benefits and risks against livelihood contexts associated with compliance to such standards for different stakeholders in the food chain, from producers to retailers to consumers. One basic and necessary step in this process is to identify relevant ethical principles that might be relevant and useful for the adaptation of an EAFI as an ethical decision-making tool incorporating participatory approaches. Important here is also identification and description of relevant stakeholders, both primary and secondary.

The development of an EAFI has a lot to do with clarifying how aquaculture production and trade can fit in with global values and goals, such as poverty alleviation and environmental sustainability. It has not in any way to do with passing moral judgements on aquaculture or parts thereof, neither positively nor negatively.

One of the basic research assumptions is that aquaculture production and trade is embedded in a web of social values in the sense of both affecting these values and being affected by them. These might be values of fairness, equity, well being and sustainability. The further assumption of the project is that these values matter vitally in regard to the consumer acceptability of seafood products, as they eventually also may matter to governmentally imposed standards or certifications for trade.

Ethics is in the SEAT project seen as a clarification of values underlying social and economic practices. For this clarification it is necessary to include value perceptions from various stakeholders and affected parties. For example, some forms of shrimp cultures have contributed greatly to creating workplaces in poor communities, especially perhaps among women who otherwise had small chances of income. This value is not often visible from the consumer end. On the other hand, some forms of production practices have proven a great burden on the environment by threatening biodiversity generally and degrading mangroves specifically. The SEAT project shall chart all concerns by not only focusing on the worries, but just as much also on the positive effects along the value chain. Consumers are not merely following food habits and preferences, but are making food choices that they see as part of their vision of a good, just and sustainable society. One may say that they see food as an arena for political action and are 'voting with the fork'.

Ethical considerations

The term 'ethics' shall be used in a non-technical sense, which means that it shall not imply adherence to any specific ethical theory or body of specific ethical principles (Kaiser, 2010). Non-technical thus refers to ethical issues concerning deeply embedded values and value-conflicts, and to human choices where different courses of action are possible. Foremost among those values and value-conflicts is the concept of sustainability, which may be regarded of a wider scope than purely ethical. Sustainability linked to the livelihood and welfare of human beings refers to ethical traits.

Sustainability for whom?

A core challenge when developing the EAFI is to decide who should be regarded as stakeholders to care about with regard to their sustainability. By this it might be said that the first ethics related question to be investigated will be: Sustainability for whom? This means not only to find which humans or human bodies might be regarded as stakeholders, but also whether nonhuman organisms or entities might 'enjoy some moral standing' (Kaiser, 2010). For the SEAT project this means more specifically whether fish, shrimps, prawns or ecosystems might 'enjoy some moral standing'.

When, on the basis of the above, existing certifications schemes, guidelines and codes of conduct are examined with regard to whether ethical issues are reflected in them, the question of 'Sustainability for whom?' will have to be investigated. A preliminary sketc.h of who these stakeholders might be, taking the whole value chain related to aquaculture into consideration, is shown in Figure 1.

Sustainability of what?

The next core challenge when the question of 'Sustainability for whom?' is analysed, is to find what might be perceived as values important for securing sustainability from the perspective of the different stakeholders respectively. The second ethics related question will then be: Sustainability of what? This refers to anything perceived by the stakeholders to be of importance both for their present and their future prosperity and needs. In other words this is about the stakeholders' long-term needs and preferences.

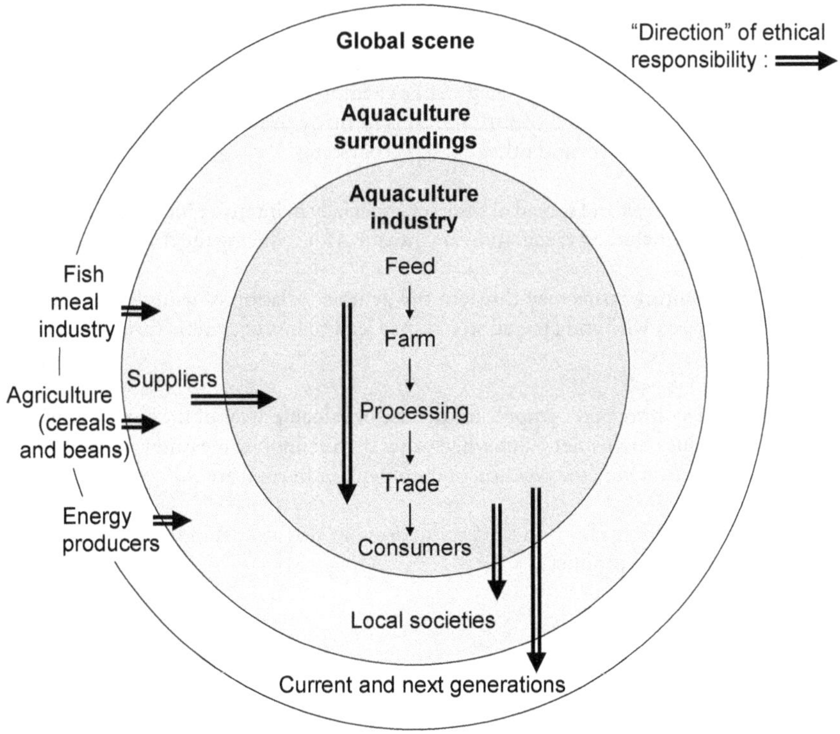

Figure 1. Sustainability for whom? Possible stakeholders.

Four ethical principles

In relation to evaluations of aquaculture from an ethical and sustainability point of view Kaiser (2010) suggests the following set of ethical principles: Welfare as eliminating negative utilities, Welfare as promoting positive utilities, Dignity and autonomy, Justice as fairness. These are principles that might be used in connection with the so-called ethical matrix used as a decision tool to include ethical criteria in decision making (Kaiser *et al.,* 2007). Since the ethical matrix is suggested as a tool also for the development of the EAFI, it is found reasonable to link the above mentioned ethical principles to the question 'Sustainability of what?' This means to see the long-term needs and preferences of all stakeholders within the context of the four ethical principles.

When crossing cultural borders as in the SEAT project, one should not take for granted that the four ethical principles communicate easily to stakeholders educated in a very different cultural framework. It is therefore an open question how such concepts and ideas fare in inter-cultural dialogues, specifically between Asia and Europe.

Possible conflict areas

Kaiser (2010) and Haugen (2008) address some of the practical challenges faced in this connection. What species to cultivate and how to feed them? Cultivation of predators, which are high on the food web, receives a high market price. They depend on feed with a high percentage of fish meal and fish oil. The risk is that this contributes to the over-exploitation of the seas (Naylor *et al.,* 2000).

Environmental sustainability

Discharge of nutrients and organic waste is inevitable in open-cage systems. Furthermore, dependent on how it is administered, various forms of medication like chemotherapeutants and antifoulants may enter the environment. Uncontrolled spread of shrimp farms in many tropical countries threatens valuable natural resources such as mangroves and other coastal ecosystems.

There is always a risk of outbreak and spread of diseases, especially in intensive industrialised aquaculture. Good primary health-care including vaccination are measures for reducing the risk of disease outbreaks.

Fish escapes form aquaculture farms may threaten the genetic variation of wild stock. The concern is that cross-breeding between wild and farmed stocks may lead to loss of genetic variation with negative effects on performance traits.

Production sites for aquaculture may compete for the use of valuable areas of the coastal zone for which alternative uses may produce higher net-value when other than economic measures are applied. Counter arguments for aquaculture include the creation of employment in rural areas.

A core issue is also the use energy both in the production but not at least in the transportation world wide of raw materials and final products.

Quality and consumer issues

This handles the question of whether good quality of food safety should be the only issue of concern in relation to food production or if other issues related to the production should be addressed.

Traceability is foremost a matter of quality assurance of the food product. This is about consumer trust and quality assurance through all links of the food chain.

The question of whether to declare or not that aquaculture products are not capture fish products. Until recently many producers wanted to slip their aquaculture products into the market by giving it names that people might associate with the wild relatives of the product.

Another question is the extent to which the consumers get reliable information on the sustainability of the product.

Animal welfare has become more important as a consumer issue. It is also true that animal welfare is viewed by many as a consideration that needs to be differentiated between different species. For instance has Singer (1975) drawn a line 'somewhere between shrimp and an oyster' when considering animal welfare and the suffering of the animal in the slaughter process. Many people see, however, a difference between a fish and a mammal in terms of their capacity to suffer. Yet, several researchers now indicate that fish most likely do feel pain and can suffer more than we would expect from its simple nervous system (Nordgreen, 2009). This seems to indicate that some animal welfare considerations need to apply also in aquaculture. The main reason for a focus on animal welfare might be about securing market acceptance. This is because the far-reaching consequences an animal welfare ethics for all kinds of fish handling, whether in farming, game fishing or commercial fishing, is currently too complicated to be handled to its full extent in the societies where the aquaculture industry operates.

The trend should be addressed that consumers in rich industrialised countries will continue to exert pressure to implement stronger standards of food safety which will eventually force the producers to reduce even the slight trace of contaminants in aquaculture products.

Choice of ethical position

The status with regard to perceptions about the moral status of ecological entities such as fish, shrimps, prawns or ecosystems should be clarified. This is to secure an upright and honest argumentation about what might be perceived as ethically good or wrong with regard to those nonhuman entities.

Equity and fairness in access to food

The question of equity and fairness in access to food is triggered by the threat that forage fisheries represent to local artisanal fisheries. There is a risk that forage fisheries for raw materials for feed production reduce the fish stocks which the local artisanal fisheries are dependent on, and by that threaten the local catches and local access to fish.

Food safety

There are diverging perceptions in society and among consumers on what safe food is. This is especially so with regard to GMO, which tend more to be a market acceptance issue than a factual food safety issue. Therefore the aquaculture industry has a special responsibility for being transparent with regard to production processes. The aquaculture industry should be transparent in this matter and disclose all the information that is necessary for customers themselves to decide whether they regard the farmed salmon to be safe to eat or not.

Parallel to the above possible conflict areas addressed by Kaiser (2009) and Haugen (2008) the SEAT project (Little *et al.,* 2009) has highlighted environment and contamination, food safety and public health, and social and economic as core issues of the project. Taking all this together it is found that following five themes can be a synthesis of main possible conflict areas in relation to the sustainability of possible stakeholders linked to the aquaculture industry:

- Net food supply – including issues such as food security and artisanal fisheries.
- Socio-economic conditions – including issues such as profit sharing, working conditions, child work.
- Environmental controversies – including issues such as pollution, area use.
- Food quality – including issues such as food safety, nutrient quality, gastronomic experience.
- Animal welfare – including issues such as moral status perceptions, pure market demands.

These five themes will be approached in the search for ethical standards in existing certifications schemes, guidelines and codes of conduct.

Search for ethical standards

Based on the issues discussed above, reviews will be made both on current ethical frameworks for aquaculture products and on existing certification schemes applicable to aquaculture. As an input for these reviews, a search will be done on how and to what extent ethical standards can be found in existing certifications schemes, guidelines and codes of conduct.

Conclusion

The search for ethical standards in existing certifications schemes, guidelines and codes of conduct will probably reveal a great discrepancy both with regard to whether ethical standards are found and with regard to how these standards might be presented. The learning from this will be useful information for the further process towards the development of an EAFI. Particularly this will be useful for the preparation and running of planned participatory processes linked to the ethical matrix approach in each project country. All this together is meant to add in with a holistic, evidence-based perspective to the EAFU and be basic question of what constitute fair and sustainable trade.

References

Haugen A.S. (2008). Values behind biodiversity and moral duties – Development of a conceptual model for ethical environmental accounting using salmon farming as a case study, PhD Thesis, University of Oslo, Norway.

Kaiser, M. (2010). The ethics and sustainability of aquaculture. Forthcoming in: David Kaplan (ed.), The Philosophy of Food, University of California Press.

Kaiser, M., Millar, K., Thorstensen, E. and Tomkins, S. (2007), 'Developing the ethical matrix as a decision support framework: GM fish as a case study', Journal of Agricultural and Environmental Ethics, vol.20, 65-80.

Little, D.C., F.J. Murray, T.C. Telfer, J.A. Young, L.G. Ross, B. Hill, A. Dalsgard, P. Van den Brink, J. Guinée, R. Kleijn, R. Mungkung, Y. Yi, J. Min, L. Liping, L. Huanan, L. Yuan, Y. Derun, N.T. Phuong, T.N. Hai, P.T. Liem, V.N Ut, V.T. Tung, T.V. Viet, K. Satapornvanit, T. Pongthanapanich, M.A. Wahab, A.K.M. Nowsad Alam, M.M. Haque, M.A. Salam, F. Corsin, D. Pemsl, E. Allison, M.C.M. Beveridge, I. Karunasagar, R. Subasinghe, M. Kaiser, A. Sveinson Haugen and S. Ponte (2009). 'Sustaining ethical trade in farmed aquatic products between Asia and the EU', in: K. Millar, P. Hobson and B. Nerlich (eds.), Ethical Futures: Biosciences and Food Horizons. Wageningen Academic Publishers, Wageningen, pp. 226-231.

Naylor R.L., Goldburg R.J., Primavera J.H., Kautsky N., Beverldge M.C.M., Clay J., Folke C., Lubchenco J., Mooney H. and Troell M. (2000). Effect of aquaculture on world fish supplies, *Nature*, 405, 1017-1024.

Nordgreen, J. (2009), Nociception and pain in teleost fish, PhD Thesis, the Norwegian School of Veterinary Science, Oslo.

Singer P. (1975). Animal Liberation: A New Ethics for Our Treatment of Animals, A New York Review Book, New York, USA. 301p.

Buyer be-aware: the ethics of food labelling reform and 'mobilising the consumer'

X. Frohlich
MIT Science, Technology, & Society, 77 Massachusetts Ave., E51-185, Cambridge, MA 02139, USA; frohlich@mit.edu

Abstract

Growing public health concern over chronic degenerative diseases and their association with the 'nutrition transition', a shift from undernutrition to overeating, has challenged many classical governance policies with regard to food production and fostering a healthy, consuming citizenry. Among the many tools that governments are using to address this new understanding of food, diet, and risk are information labels, designed to empower and mobilize the consumer to make informed and 'correct' food purchases. This paper discusses the history of nutrition labelling in the United States in order to illustrate some of the common scientific and legal challenges, assumptions, and limitations of such 'read the label' campaigns. While nutrition labelling is often celebrated as a liberal reform method which puts the consumer in control, allowing for individual rational decision-making, ironically, such labels, in their design and implementation, depend heavily upon backstage expertise that determines what should go on the label, how they should be calibrated, and what epistemological or institutional frames are most appropriate for them. By emphasizing consumer literacy and autonomy of choice over direct consumer protections (such as product or ingredient bans), food labelling as a strategy for reform is susceptible to several critical weaknesses: labels only reform the already literate and interested reader, consumers are empowered only in so far as they have the purchase power to choose designated healthier foods, labels are always reductions of a fuller food information disclosure and therefore reflect foremost the values and biases of the experts that design them, and the consumer's interpretation of labels depends on an informational environment or context which is easily shaped (and manipulated) by the substantial marketing resources of the food industry. This paper explores these tensions between the perception of food labelling as more democratic and transparent, and the the practice of labelling as a transformation in the relationship between consumer-citizens, scientific and legal experts, and the state.

Keywords: food labelling, informational environment, empowered consumers

Introduction

Growing public health concern over chronic degenerative diseases and their association with overeating has challenged many classical governance policies on food production and fostering a healthy, productive citizenry. Among the tools that governments are using to address new understandings of food, diet, and risk are information labels. Regulators designed these to empower and 'mobilise the consumer' (Miller and Rose, 1997) to make informed and 'correct' food purchases. This paper discusses the history of nutrition labelling in the United States to illustrate some common scientific and legal challenges, assumptions, and limitations of 'read the label' campaigns. While food labelling efforts may open up opportunities to cultivate new understandings of food and risk, labelling alone does not offer a solution to enduring problems of ensuring food quality nor settle more recent questions concerning the proper role of expertise in democracies.

What can a *longue durée* history of food labelling tell us?

Food labelling is situated within a much longer history of efforts by manufacturers, distributors, retailers, consumers, and the state to establish mechanisms for quality assurance and trust between buyer and seller. Early efforts to regulate food quality took advantage of the physical space of market*places*, rationalising the food supply through the control of vendors' physical location within town markets. While popular protests over food often centred around negotiating a fair price, another common problem was adulteration or trickery that disguised poor quality or unsafe food (Hutt and Hutt, 1984). The emergence of food labelling in the late 19th century was a response to the rise of packaging and the lengthening of the supply chain between producer and customer following the industrial revolution. Manufacturers embraced labelling as an important tool to establish market segmentation, product positioning, and brand loyalty, and to regain some control over 'substitution evil', where grocers substituted one brand for an equivalent absent one. Manufacturers also used labels to address a central concern with packaged foods, that '[c]ans and boxes concealed colours and odours and prevented shoppers from tasting food before they bought it' (Strasser, 1989: 35). Branding ideally functioned as a private tool for accountability, encouraging consumers to develop product loyalty with producers whom they could no longer meet directly at the marketplace. The state, however, played an important third-party role in policing these technologies of trust, particularly owing to the importance that food provisioning plays in maintaining public order. In the United States, early federal labelling laws focused on preventing 'unfair' and deceptive packaging practices through slack-fill container laws and setting product standards. Only later, beginning around the 1920s with the popularisation of vitamins, did rules on nutrition and health claims become a regulatory concern. In this early period, government regulation focused on preventing outright nutrition quackery, pseudoscientific claims, and what was known as 'economic adulteration', the use of cheap substitutes which cheated consumers out of the more wholesome authentic product.

This brief history illustrates several important points to bear in mind when considering the ethical implications of nutrition labelling. First, a wide heterogeneity of actors and constantly shifting relationships are in play. While scholars and policymakers try to make sense of this heterogeneity with heuristic concepts such as 'commodity chains', 'food networks', 'from farm to fork', or even 'agribusiness', such efforts to trace 'linkages' between different actors should not overshadow the fact that such links are highly contingent and often fraught with power struggles (Hamilton, 2009). Consumers weigh a variety of issues when choosing their food, including price, flavour, environment, community, religion, status, familiarity, and health. While in contemporary times we tend to divide these attributes between different epistemologies, historically they have been commonly interlinked (e.g. price or place as an indication of quality, quality an indication of safety or nutritiousness), a fact that seems to motivate many organic and locavore movements to seek more holistic labels of food quality. A long history of trickery in the marketplace has meant concomitant state interest in ensuring the safety and quality of its citizens' food supply. Debates about informational labelling, we are reminded, are really debates about trust between consumer and producer. What has changed is not this underlying anxiety, but rather the institutions and mechanisms by which that trust is established. In the last century there has been a shift from focus on consumers *knowing* foods (because foods were self-evident) to consumers *reading* foods (because they were now packaged and processed), what amounts to an informational turn. In many respects this turn reflects the abstraction of the marketplace, but it also grows out of a particularly progressive belief in the rationalisation of the market, with an emphasis on transparency and consumer choice.

A brief institutional history of United States nutrition labelling

The history of U.S. nutritional labelling carries two key lessons. First, food health claims have long posed classificatory conundrums over labelling products as either 'food' or 'drugs', which can be understood to really be a debate over whether to establish stricter or more lenient standards for determining efficacy, or instead place that burden upon the consumer. Second, the Food and Drug Administration's (FDA) nutrition labels should be understood as a 'government brand', intended to standardise and legitimate information, whose effectiveness as an intervention is directly affected by consumers' trust in their government.

Nutrition labelling in the United States grew out of two intersecting institutional movements: the FDA's shift to informational regulation, beginning in the 1970s, and increasing public health attention to what nutritionist Marion Nestle calls the 'nutrition transition', the dietary shift from societal problems with malnutrition and underconsumption to an increase in diseases associated with overeating. Much has been written about the nutrition transition and how the rising prevalence of 'diseases of affluence' are transforming people's understanding of food. I mention it to underscore two points. Over the last fifty years there has been a tension inside the nutrition profession between the old regime, focused on curing deficiencies amenable to specific technological fixes, and a new regime concerned with over-consumption and overall healthiness of diet. While at the same time, there has been a broader cultural transformation from a pre-WWII society of consumers who generally saw themselves as healthy, to a post-1960s society of consumers who imagine themselves on a continuum of healthiness, where everyone has some degree of disease risk.

The FDA's turn to information labelling reflected the agency's gradual recognition of this transformation. Up until the 1960s, the FDA took a conservative, restrictive approach to food health claims. Any reference on a label or in any advertisement to a food's ability to cure or prevent a disease would prompt the FDA to declare the product not a 'food' as defined by statutory law, and to remove it from the market as a misbranded 'drug' lacking adequate FDA approval. Similarly, the FDA established 'standards of identity', a set of standard recipes with fixed common name labels (such as 'peanut butter' or 'tomato soup') for all mass-produced foods. Any foods that contained substitute ingredients (e.g. filled milk or low fat substitutes) would be labelled 'imitation', implying they were inferior to the authentic standard product. Nonstandard 'imitation' products also had to carry an ingredient label, while standard foods did not. Food companies' first efforts, in the 1960s, to market unsaturated fats such as vegetable oils and margarines for their health-promoting properties in preventing heart disease were initially rejected by the FDA. According to the FDA, such health claims were an inappropriate message for mass marketing. Under the agency's regulatory logic, most consumers were healthy and thus did not benefit from such alarming food risk information. Those consumers who did require special dietary foods were best directed to their personal physician.

Growing pressure from medical professionals, eager to put new diet science to use in preventive public health measures, helped lead to a shift in the FDA's policy. In 1973, the FDA introduced the voluntary 'nutrition information' panel, allowing producers to make nutrient disclosures so long as they included a standard panel specifying quantities of vitamins and certain macronutrients, as well as an ingredients panel. The agency also dropped the 'imitation' label for all products deemed to be nutritionally equivalent or superior to an existing standard food product, effectively freeing up industry to introduce a whole host of new diet recipes without seeking FDA approval. While this move by the FDA reflected its belief that nutrition information itself was objective, the agency continued to prohibit so-called disease-claims that linked eating foods to preventing specific diseases, because these claims blurred the legal distinction between 'foods' and 'drugs'. In the 1980s, nutrition and health claims began to proliferate. The introduction in the early 1990s of the mandatory 'Nutrition Facts' panel reflected an effort by the agency to standardise nutrition information by linking all nutrient declarations and specific FDA-approved health claims back to the label. A key innovation was the inclusion of '% Daily Values' for nutrient declarations, introduced to meet Congress's mandate to present information 'in the context of the total daily diet'.

Ethical problems with food labelling

The turn to food labelling, and information disclosures more generally, as a tactic for social reform raises a number of ethical problems. Perhaps the most fundamental problem is that labelling movements presuppose consumer choice and by extension personal responsibility for those choices. In the title of this paper, I play on the classic property law concept of *caveat emptor*, or 'buyer beware'. This doctrine places the burden on the buyer to determine any defects in quality of a purchased good, waiving her right to protest after the purchase unless she can prove deceit on the part of the seller. Over the course of the twentieth century, a variety of regulatory frameworks utilised labelling, particularly warning statements, as one way to 'level the playing field' for the consumer. Food labelling in this respect fits within a broader sociological shift often referred to as the emergence of the 'risk society'. This focus on the consumer's 'right to know', while popular, is problematic in that it doesn't resolve a central ambiguity: once the buyer is informed of the stated risk, does the responsibility for the risk now fall fully on her shoulders? Policymakers may debate this by asking, how informed is the 'ordinary' consumer?

Critics of excessive food labelling, as well as proponents of 'better' labelling systems, have pointed out the following tensions with using information labels to make the buyer be aware of risky products. For one, how coercive is information on a label? Is it merely information (available for the interested reader), or is it education or advertisement (intended to reach and change the minds of uninterested readers)? These questions reach to one of the major problems with labels: do people (and in particular the 'target audience') even read them? Or better, *who* reads them? Advertisers have historically used labelling for market segmentation, for example, targeting certain niche audiences who read nutrition labels avidly, not worrying that other consumers ignore them. When the U.S. FDA introduced the 'Nutrition Facts' panel in the 1990s, for example, the food industry rushed to change the ingredients of many of its food products in order to create a more nutritious profile, even though only a minority of American consumers were likely to change their purchase decisions based on the pre-existing (but now visible) unhealthy ingredients. This poses a difficulty for reformers who believe that nutrition labels are educational. Labelling is meant to empower consumers by helping them to 'vote' for healthier options through their purchases, but ultimately it only enables the purchase-power of those who have the luxury to be discriminating shoppers. Labelling reforms therefore do not redress existing structural social disparities, which more directly shape point-of-purchase decisions about food risk.

The use of food labelling to reform also presupposes a rational, literate, and somewhat already informed decision-maker, the kind of perfect reader who may not exist. Critiques of whether labels truly create transparent markets range from those that accept Enlightenment ideals of improving educated decision-making to those that reject the liberal notion of the individual decision-maker and question the underlying models of the rational-actor. In the former category are at least two main concerns: information overload and information noise or distortion. The hazard of overwarning is that consumers, if presented with too many labels or too much information, will become saturated and lose interest in or ignore the valid information. This concern manifests as arguments over consumers' attention spans and the economy of label space. A slightly different criticism focuses on whether the flood of unrelated or distorted nutrition 'noise' in the marketplace distorts the label's intended message. One version of this criticism is the 'capture thesis', which contends that the nutrition profession's or government's health messages have been appropriated by industry (Nestle, 2002). Some go so far as to argue that nutrition advertising has become an extension of the food industry's fabrication of choice and novelty so as to sell more products (Pollan, 2006). These critiques suggest that real reform would involve an institutional cleansing of 'corrupted science' with stricter controls on the kinds of information allowed in advertising.

The other camp of label detractors describe a more intractable kind of 'consumer confusion'. Since at least the 1970s, some scholars in risk studies have raised concerns about excessive risk-aversion and its consequences when labelling alarming yet presumably safe products such as 'irradiated' foods. This concern has gained renewed traction in recent years because of behavioural economics and psychology studies demonstrating that, under certain experimental settings, people may act against their economic rational interests, either because of preexisting biases (irrational actor) or misleading heuristics (limited-rationality actor). Critics relying on these studies do not reject labelling efforts, but rather advocate a more conservative and arguably 'libertarian paternalistic' kind of governance, whereby experts improve consumers' 'choice architecture' by taking into account predictable patterns of irrational choices and building those into the design of labels (Thaler and Sunstein, 2008).

In such accounts of consumer decision-making it is imperative to ask how self-reflective policymakers or experts are when they use the term 'consumer confusion'. By characterising 'incorrect' consumer food purchases as evidence of 'consumer confusion', policymakers risk privileging expert understandings of food and underestimating consumers, misinterpreting consumers' priorities because they are different from their own. Simplifying labels has the virtue of clarity, but the drawback is wresting decisions about what information is important away from consumers.

Ethical problems specific to nutrition labelling

Nutrition labels that appear on nearly all food products in America (and increasingly a majority of food products in Europe) draw upon a scientific epistemology for their legitimacy. This raises additional policy challenges. One such debate centres on whether nutrition labels ought to or in fact do mark foods as 'good' or 'bad' to eat. Proponents of so-called street-light labelling systems, which mark healthy foods with a green light and unhealthy foods with a red one, often dismiss this concern as industry obfuscation. Yet among nutrition scientists there exists a legitimate debate over whether it is more expedient to single out specific foods, or more accurate and therefore more effective to keep the focus on one's entire diet in the context of his or her lifestyle. Here nutrition labelling, particularly in a saturated advertising and diet advice culture, runs up against the philosophical is-ought quandary. A striking recent example of where a simple positive statement of nutrient content could end up functioning like a normative prohibition is that of 'trans fats' labelling in the U.S. Popular concern was so great about the 'toxic' effects of eating trans fats that as soon as it was included on the Nutrition Facts panel, in 2003, most companies had to remove them from their recipes or risk losing their customers.

A less obvious ethical difficulty arises from the fact that nutrition labelling is a mass medium read by a diverse audience with vastly different nutritional needs. Science-based advice, especially diet advice, rests upon expert constructions of an imagined consumer. The architects of the U.S. Nutrition Facts panel targeted the label to an *averaged* consumer. One example is that they calibrated the percent daily values to a daily caloric intake of 2,000 kilocalories. The number was below the average recommended male intake of 2,350 kilocalories, but slightly above the recommended female daily intake level. The decision to use this lower-leaning average reflected a victory of pragmatics over scientific accuracy in public food labelling. By aiming low (for men) on daily recommended calories, the label's architects believed they would offset the American tendency to eat too much, and by choosing a round number, 2,000, they both simplified calculations for consumers and sent the message that the number represented an estimate rather than an exact science. The label averages out a heterogenous mix of special interest groups: the elderly (for whom the FDA put in a minimum type font requirement); mothers (for whom the vitamin thresholds were set high); ageing men (the imagined target population for concern over heart-disease and fat consumption); and diabetics and hypertensives (sodium and sugars). In each of these instances, regulators grappled with the consequences of what sociologist Wendy Espeland calls 'commensuration' – scientific and legal rationalisations of social disputes which transform participants' positions into measurable, quantifiable equivalences (Espeland and Sauder, 2007).

Moreover, regulators were using population-level data as the foundation to design a tool for individualised recommendations. This leads to a significant ethical problem, what epidemiologists call the 'paradox of prevention' – when a public intervention 'brings large benefits to the community [but] offers little to each participating individual'. For example, encouraging U.S. citizens to eat less saturated fats might reduce an individual's risk of cardiovascular disease very little, while at the population level many fewer cases of heart disease would probably occur. The '% Daily Values' of the Nutrition Facts panel embody the epidemiologist's preoccupation with collective risk – 'treating sick populations not sick individuals' – and reflects a utilitarian approach to the problem of how to move from aggregate statistical data about population risk to an inference about individual risk (Rose, 1985). The perfect reader of the Nutrition Facts panel is, therefore, a population of readers, not an individual.

Conclusions

I have tried to underscore a central irony of food labelling movements, that they seek to empower individual consumers to make choices for themselves, but they rely heavily upon a backstage of expertise that determines what should go on the label and how they should be framed. Labels offer the potential for facilitating the cultivation of new food-identity communities. Yet given their historical use for market segmentation, as well as the problem that food labels may ultimately reform only the already literate and interested reader, they are likely to reinforce existing social and consumer disparities. The history of U.S. nutrition labelling also suggests that labels cannot function as a 'technical fix' for reformers who believe that simply placing the information on the food is adequate to educating consumers. The Nutrition Facts panel was initially released as part of a much broader education campaign, situating the label at the centre of a whole host of tools intended to both personalise nutrition advice and to ensure that consumers could adequately and correctly interpret nutrition information (Taylor and Wilkening, 2008). As that public education campaign has disappeared, or more correctly been conceded to the substantial marketing resources of the food industry, age-old problems with nutrition quackery and misleading health claims have reemerged.

A broader concern is whether nutrition labelling cultivates scientism and nutrition reductionism. Here, cultural critics refer to a kind of modern alienation occurring with industrial food, drawing upon Marxist traditions of agrarian critique (Scrinis, 2008). Packaging foods alienates consumers from the actual systems of production that create the foods, and labels always end up being reductions which reflect the values and biases of the experts (often producers) who design them. Nutrition labelling and 'nutritionism', according to these critics, compartmentalises other attributes that arguably commingle with nutrition (health): economics (price), taste (flavour), place (environment, community), status (conspicuous consumption), and tradition (familiarity and religious significance). While offering an important reminder that nutrition labelling represents a comparatively new (and untested) way of envisioning food and diet, these often nostalgic appeals ignore the underlying change which motivated labelling in the first place – the dramatic transformation that has occurred to our diets as a result of our greater affluence and abundant food resources. It is in this context – developing incentives to consume less – that nutrition labelling remains such an appealing, albeit highly problematic tool for social reform.

References

Espeland, W., and Sauder, M. (2007). Rankings and Reactivity: How Public Measures Recreate Social Worlds. American Journal of Sociology, 113(1), 1-40.

Hamilton, S. (2009). Analyzing Commodity Chains: Linkages or Restraints? in W. Belasco and R. Horowitz, eds. Food Chains: From Farmyards to Shopping Carts. Philadelphia: University of Pennsylvania Press.

Hutt, P.B. and Hutt, II, P.B. (1984). A History of Government Regulation Of Adulteration And Misbranding Of Food. Food and Drug Cosmetic Law Journal. Vol. 39, No. 2: 2-73.

Miller, P. and Rose, N. (1997). Mobilizing the consumer: Assembling the subject of consumption. Theory, Culture & Society. Vol. 14, No. 1: 1-36.
Nestle, M. (2002). Food Politics: How the Food Industry Influences Nutrition and Health. Berkeley: University of California Press.
Pollan, M. (2006). The Omnivore's Dilemma: A natural history of four meals. Berkeley: University of California Press.
Rose, G. (1985). Sick individuals and sick populations. International Journal of Epidemiology. 14: 32-38.
Scrinis, G. (2008). On the ideology of nutritionism. Gastronomica: The journal of food and culture. Vol. 8, No. 1: 39-48.
Strasser, S. (1989). Satisfaction Guaranteed: The Making of the American Mass Market. New York: Pantheon Books.
Taylor, C. and Wilkening, V. (2008). How the nutrition food label was developed, Part 2: The purpose and promise of nutrition claims. Journal of the American Dietetic Association. Vol. 108, No. 4: 618-623.
Thaler, R. and Sunstein, C. (2008). Nudge: Improving decisions about health, wealth, and happiness. New Haven: Yale University Press.

Nutrient profiling: are dairy products really unhealthy?

L. Escajedo[1], A.M. Rocandio[2] and M. de Renobales[3]
[1]Department of Constitutional Law, Faculty of Labour Sciences, and Inter-University Chair in Law and Human Genome, University of the Basque Country, Bilbao, Spain; [2]Department of Nutrition and Bromatology, University of the Basque Country, Bilbao, Spain; [3]Department of Biochemistry and Molecular Biology, Faculty of Pharmacy, Vitoria-Gasteiz, Spain; leire.escajedo@ehu.es

Abstract

In an attempt to reduce the frequency of diseases related to inadequate diets, efforts are being directed towards providing consumers with tools to help them make healthy dietary choices. A recent EU regulation resorts to the nutrient profiles as a method to estimate the healthiness of a food. However, in our opinion nutrient profiles have several short-comings: (1) there are several different definitions of a nutrient profile; (2) each definition may classify a given processed food differently; (3) foods that experts (even EFSA experts) consider important in the Mediterranean diet are often classified as 'less healthy', such as vegetable oils and dairy products – among others. This is often due to considering only the gross composition of a food, but ignoring its fine composition (i.e. types of fatty acids, minerals, quality of the protein, etc.) when determining nutrient profiles; (4) they stress the degree of healthiness of a given food while de-emphasizing the importance of a balanced diet, which should be suited to the particular needs of each individual. It also ignores the value of the diet in certain pathologies, and the fact that the components of various foods ingested in the same meal may interact to facilitate, or hinder, the absorption of important nutrients. In any case, regardless of the type of nutrient profile finally adopted, the most important aspect is not to confuse consumers. Thus, dairy products offer clear advantages because of their high content of quality protein with biologically active peptides, and important minerals such as calcium and phosphorus. Dairy lipids are negatively labelled as 'saturated', ignoring positive physiological functions of certain saturated fatty acids, or the amounts of other functional unsaturated fatty acids. These components are usually not included in food composition tables. Yet, dairy products, within a balanced diet, are particularly beneficial for certain groups such as the elderly, children, and women (including pregnant women).

Keywords: consumer information, nutrient profiling, dairy products, food classification systems

Nutritional profiles as a reference to estimate the healthiness of a given food

Until recently, government measures to help prevent cardiovascular diseases focused on *education* for a healthy diet emphasizing the presence or absence in a diet of certain categories of foods, rather than specific foods. Guidelines recommended to increase the intake of fruits, vegetables, lean meat, fish and cereals, or to reduce the intake of high sugar- and/or fat-containing foods (Lobstein and Davis, 2008). But in the last decades, new foods, 'less healthy than traditional ones', (WHO, 2009) have gained popularity at the same time overweight and obesity, even in children, have drastically increased in an alarming manner. So, considering that consumers may lack sufficient nutritional knowledge (Vairyam *et al.*, 1996), new tools are being developed to help the public select adequate products for a healthy diet. Thus, the label of a food product is not an exclusive communication channel between manufacturer and consumer any more, but an educational tool for consumers to improve their eating habits. Terms like 'healthy' or 'unhealthy' food products are being used instead of 'healthy' or 'unhealthy' diets, although the definition of a 'healthy' food product, and the criteria it must fulfil, are still controversial. There is no agreement among experts on how the degree of 'healthiness' of a food product should be determined (Arambepola *et al.*, 2007). Regardless of the method chosen to transmit information to the consumer (the easy-to-understand traffic light labelling, or the label with complex descriptive tables), the determination of the degree of healthiness of a food quite often depends on its relative energy content, amount of sugars, fat, saturated fat, salt and fibre.

Nutrient profiling is a method of categorizing the overall healthiness of a food product based on its nutrient content, providing a sort of 'healthiness indicator' of a food product which can now be compared to others (Maillot *et al.*, 2008). It is a science aimed at classifying foods based on their nutritional composition (Drewnoski and Fulgoni, 2007).

In the European Union, Regulation (EC) 1924/2006, on nutrition and health claims made on foods, regulates how functional properties or nutritional characteristics of foods may be transmitted to the consumer on the product label which the manufacturer voluntarily affixes to the package. It establishes several parameters to determine the 'healthiness' of a food Thus, it aims at preventing the manufacturer from hiding, or masking, the *inappropriate nutritional profile* of a certain food, '*the overall nutritional status of a food product, which could mislead consumers when trying to make healthy choices in the context of a balanced diet*'. It is necessary to guarantee that a food product labelled as *particularly beneficial for the consumer* does not turn out to be just the opposite. Health claims should be based on 'generally accepted scientific evidence relative to the relationship between diet and health' (Whereas Clause 11).

A method to establish the *healthiness* of a food would be extremely valuable because diet-related diseases are one of the main problems of European governments. Among other things, such a method would be an excellent social teaching tool to promote improving people's diet-related decision-making. The question is whether or not there exists a unique and solid method to determine the *healthiness* of a food product.

The Regulation presents *nutrient profiling* and, from the various Whereas Clauses, we infer that it is the *definitive* method and that the criteria on which to base the profile are clearly set. But occasionally we find important nuances. It specifically states that, for certain foods or categories of foods, depending on their importance in a population's diet, '*exceptions to the mandatory respect for the profiles may be necessary*.' In other words, certain foods of recognized importance in a population's diet may result to have questionable *healthiness* under the proposed model. In the present multidisciplinary study we consider the consequences of applying the proposed nutrient profiling methods to dairy products which are particularly important for certain population groups.

Considerations about nutrient profiles and certain nutrients

Regulation (EC) 1924/2006 states that nutrient profiles should be based on generally accepted scientific data regarding the food product composition and the relationship between diet and health. Nutrient profiling is becoming increasingly important, even from the regulatory point of view (Bryans, 2009), because official organisms emphasize the potential role of certain nutrients (some fats and fatty acids, sugar and salt) in the development of certain chronic diseases (Garsetti, 2007). In addition to efforts to promote consumption of fruits and vegetables (in spite of their very low protein content), fish and lean meats, consumers have been advised against the presence of 'negative' components (saturated fatty acids, trans fatty acids, fat content, high energy content, sugars, salt, cholesterol) in certain food products, and manufacturers have not been allowed to disguise their presence and/or concentrations. Thus, those 'negative' components are perhaps given too much weight in the formulas used to determine the nutrient profile of a processed food.

Up to 23 different methods to determine the nutrient profile of foods have been reported and analysed (Garsetti *et al.*, 2007). In 19 of them (82%) saturated fatty acids are included as negative components, whereas total fat is included in 14 of them (61%). Trans fatty acids are singled out as 'negative' components in 6 nutrient profiling methods (26%). Some of the nutrient profiling systems do take into account the presence of compounds with beneficial effects by subtracting points given for positive effects from those given for negative ones. However, not always are beneficial compounds recognized in individual food items. This is the case of some dairy products, like cheeses, for example.

If, as stated in Whereas Clause 11, nutrient profiles are to be based on generally accepted scientific data, it is necessary to recognize that not all fatty acids have negative physiological effects. Since the early 1990s, it is widely accepted that saturated fatty acids increase the concentration of total cholesterol in serum and that this is a major risk factor for coronary heart disease and cardiovascular disease. However, some recent publications, such as a meta analysis of 24 well-designed prospective epidemiological studies stated that there is no significant evidence to conclude that dietary saturated fat is associated with any of these disease conditions (Siri-Tarino *et al.*, 2010; review by Parodi, 2009). It has been known for a long time that straight-chain fatty acids with 4, 6, 8, or 10 carbon atoms, and a percentage of the 12-C fatty acid, after intestinal absorption, enter the portal circulation and are taken up by the liver where they are rapidly metabolized to obtain energy and other compounds. These fatty acids are not incorporated into chylomycrons, or adipose tissue, have no serum cholesterol raising effect, and are not considered to play a role in obesity (review by Parodi, 2009). Another case is dietary stearic acid which human tissues convert into oleic acid (an unsaturated fatty acid).

Nutrient profiling and cheese

As a consequence of maintaining a negative consideration for fats in general, and saturated fats in particular, certain nutritious foods, such as whole milk, cheeses and other dairy products, may receive a more negative classification than they should if current scientific knowledge would be taken into consideration. These foods have traditionally had, and should continue having, an important place in the diet of several population groups, albeit limiting ingested quantities. Cheese is a nutritious food with high protein and fat content (Table 1). In addition, 75 to 80% of the fat in cheeses (or whole milk) is saturated fat. These are the 2 main reasons for being signalled out in most nutrient profiling systems. However, cheese has a range of nutrients with positive physiological effects, from peptides with anti-hypertensive effect (Ricci *et al.*, 2010), to certain fatty acids (German and Dillard, 2006), to calcium and vitamins, which are not considered in most nutrient profiling systems. A 50 g portion of semi-hard or hard cheese can contain between 300 and 550 mg calcium. In addition, cheese (except for fresh cheese) has no lactose, a great advantage for lactose-intolerant (most adult) people.

The fatty acid composition of cheese fat is like that of milk fat because very small amount of fat is lost during cheesemaking. In addition, milk fat composition can be modified by the diet fed to the animals, with various supplements and/or by allowing them to graze outdoors, as shown in Table 2. This is a well known effect in the dairy industry which is behind the current interest in animal feeding studies to obtain dairy products with increased concentrations of beneficial fatty acids and other functional compounds.

Table 1. Average content (g) of nutrients in a 50 g portion of fresh, soft, semi-hard, hard, and extra-hard cheese (adapted from Walther et al.*, 2008).*

Type of cheese	Water	Protein	Fat	Minerals + vitamins
Fresh cheese	35	5.5	4	4
Soft cheese	26	10	11	3
Semi-hard cheese	20	12.5	13.5	4
Hard cheese	17	13.5	15.5	3.5
Extra-hard cheese	15	14.5	16.5	4

Table 2. Fat and protein content (g/100 g) and fatty acid composition (μmol/g fat) from cow's milk according to feeding period (adapted from Ferlay et al.*, 2008)*

	Indoors feeding (forage and concentrate)	Outdoors feeding (grazing)
Fat content	4.69[a]	4.00[b]
Protein content	3.16[a]	3.25[b]
Fatty acid		
short chain fatty acids (C4-C10)	856.4[a]	866.4[a]
lauric acid (C12:0)	175.0[a]	155.0[b]
myristic acid (C14:0)	539.5[a]	473.7[b]
palmitic acid (C16:0)	1,265.6[a]	1,007.8[b]
palmitoleic acid (C16:1)	59.0[a]	43.3[b]
stearic acid (C18:0)	295.8[a]	376.7[b]
oleic acid (C18:1c9)	627.6[a]	695.0[b]
vaccenic acid (C18:1t11)	38.3[a]	117.7[b]
linoleic acid (C18:2)	46.4[a]	48.9[a]
c9t11-CLA isomer	21.4[a]	53.6[b]
atherogenicity index	4.54[a]	3.19[b]
% saturated fatty acids	79.80	75.02
% nonatherogenic saturated fatty acids	29.35	32.38

[a,b] Different superscripts in the same row indicate statistically significant differences.

Among fatty acids with beneficial physiological effects we should mention short chain, unsaturated, and conjugated linoleic acid (CLA) fatty acids (German *et al.*, 2009). CLA isomer c9t11 has been found to have anticancer, antiatherogenic and inmuno-modulating effects, both in tumor cell and *in vivo* animal studies (Lee *et al.*, 2005; Parodi, 2006).

The atherogenicity index indicates the relative contribution of the food item to the development of cardiovascular disease and it is calculated with the formula: $(C12 + 4 \times C14 + C16)$/total unsaturated fatty acids. Therefore, neither short chain fatty acids or stearic acid are taken into account. However, considering that stearic acid is converted into oleic acid in human tissues, the amount of stearic acid should be added to the total sum of unsaturated fatty acids. If so, the atherogenicity indexes would be: 3.49 (indoor feeding) and 2.37 (outdoor feeding). These results would agree better with the data reported above that dietary saturated fat does not appear to be associated with cardiovascular disease.

Nutrition in certain groups: children, women and the elderly

Although calcium has many functions in humans, its role in bone health and prevention of osteoporosis is perhaps the most important one, particularly in women after menopause. Calcium intakes are lower than recommended levels (about 1,200 mg a day) for much of the adult population, men and women, particularly in the elderly, and assimilation decreases with age, both in men and in women. Dairy products are also an excellent source of calcium for children. In addition, intake of dairy products has been associated with significantly decreased systolic blood pressure (Walther *et al.*, 2008). The contents of minerals and vitamins in milk and dairy products make them difficult to replace in the diet of children and older adults (German *et al.*, 2009).

Conclusions

Nutrient profiling systems, particularly easy to understand ones like the *traffic light* system, are a good way to call the attention of customers to the healthier, or less healthy, nutritional composition of a given food item. Front-of-pack labelling schemes clearly communicate their message at the precise moment when consumers place their products in their shopping carts. However, this and other nutrient profiling systems in the long run may be doing society a disservice, at least, for the following reasons: (1) they consider that 400 million Europeans can be reduced to 'the consumer', whereas diet has individual, personal components; (2) these systems take food items out of the context of a healthy diet (against the philosophy of Regulation 1924/2006; (3) current scientific knowledge is not adequately taken into consideration, at least as fat and fatty acids is concerned; (4) the contribution of compounds with positive physiological effects present in certain nutritious traditional foods such as vegetal oils, some dairy products and, specially cheeses, are ignored.

Oversimplification of diets while trying to improve social education in general may also lead to important cultural and nutritional losses, neglecting products which would be difficult to bring back at a later date. It is also very difficult to change the idea that all fats are equally bad, once this erroneous concept is well entrenched.

We realise that there is no easy way to convey to different groups of consumers the complex nutrition information regarding food items needed to design balanced diets. By not separating fat components according to their physiological effects, as it is often done with cheeses and dairy products, and stressing the importance of their negative components, it is likely that certain groups of the population will avoid them, and will not benefit from their beneficial components. The validity of nutrient profiling to positively contribute to consumer education to favour the consumption of healthy diets depends on a variety of factors, the most important of which is the integrity with which scientific data are used.

References

Arambepola, C., Scarborough, P. and Rayner, M. (2007). 'Validating a nutrient profile model', Public Health Nutrition, 11(4), 371-2.

Bryans, J.A. (2009). ' Nutrient profilinf: consumer friend or foe?', The Australian Journal of Dairy Technology, vol. 64, no. 1, p142.

Drewnoski, A. and Fulgoni, V. (2007). 'Nutrient profiling of foods: creating a nutrient-rich food index', Nutrition Reviews, 66 (1): 23.

Ferlay, A., Agabriel, G., Sibra, C., Journal, C., Martin, B. and Chilliard, Y. (2008). Tanker milk variability in fatty acids according to farm feeding and husbandry practices in a French semi-mountain area. Dairy Science and Technology 88, 193-215.

Garsetti, M., De Vries, J., Smith, M., Amosse, A., and Rolf-Pedersen, N. (2007). 'Nutrient profiling schemes: overview and comparative analysis'. European Journal of Nutrition 46 (Supp. 2): 15-28.

German, B.J., Gibson, R.A., Krauss, R.M., Nestel, P., Lamarche, B., Van Staveren, W.A., Steijns, J.M., De Groot, L.C.P.G.M., Lock, A.L., and Destaillats, F. (2009). A reappraisal of the impact of dairy foods and milk fat on cardiovascular disease risk. European Journal of Nutrition 48, 191-203.

Lobstein, T. and Davies, S. (2008). Defining and labeling 'healthy' and 'unhelathy' food. Public Health Nutrition, 12(3), 331-340.

Maillot, M., Ferguson, E.L., Drewnowski, A. and Darmon, N. (2008). 'Nutrient Profiling Can Help Identify Foods of Good Nutritional Quality for their Price', Journal of Nutrition – American Society for Nutrition, 138: 1107-1113.

Parodi, P.W. (2009). Has the association between saturated fatty acids, serum cholesterol and coronary heart disease been over emphasized? International Dairy Journal 19, 345-361.

Ricci, I., Artacho, R., and Olalla, M. (2010). Milk Protein Peptides With Angiotensin I-Converting Enzyme Inhibitory (ACEI) Activity. Critical Reviews in Food Science and Nutrition 50, 390-402.

Siri-Tarino, PW., Sun, Q., Hu, FB., and Krauss, R.M. (2010). Meta-analysis of prospective cohort studies evaluating the association of saturated fat with cardiovascular disease. American Journal of Clinical Nutrition doi: 10.3945/acjn.2009.27725.

Vairyam, J.N., Blaylock, J., Smallwood, D.M. (1996). Modelling nutrition knowledge, attitudes, and diet-disease awareness: the case of dietary fibre, Statistics in Medicine 15, 23.

Walther, B., Schmid, A., Sieber, R., and Wehrmüller, K. (2008). Cheese in nutrition and health. Dairy Science and Technology. DOI: 10.1051/dst:2008012.

WHO (2009). Global Health Risks, Mortality and binden of Disease attributable to selected major risks. p 3-4.

Framing omega-3: the politics of essential fatty acids in functional foods

M. Döring[1] and S. Stirn[2]
[1]University of Hamburg, FSP-BIOGUM/FG Medicine, Lottestraße 55 (2.OG), 22529 Hamburg, Germany; [2]University of Hamburg, FSP BIOGUM, Biozentrum Klein Flottbek, Ohnhorststraße 18, 22609 Hamburg, Germany; martin.doering@uni-hamburg.de

Abstract

The study of food and society is an interdisciplinary and ever more recognised area of research providing a growing understanding about the production, regulation and consumption of food. It emphasises that food is a socially charged and politically contested field in which individual bodies, scientific innovations, legal aspects and moral concerns become materially and virtually entangled. Recent research in the area of food and society investigates areas such as dietary habits, food production or the regulation of food but seems to ignore developments in the field of functional foods. The history of this type of food goes back to the early 1900s in the US and since then scientists have discovered a good deal of compounds with so-called functional qualities which – only recently – became basic ingredients in food products: probiotics in yogurts keep the intestine balance, sterines in margarines reduce LDL-cholesterol or bread with added omega-3 acids helps to prevent cardiovascular diseases. Advocates emphasise health promoting or even disease preventing qualities of functional foods while adversaries question them. This contested boundary between food and drug is the starting point of our empirical study in which we follow omega-3 acids on their travels through regulative and policy discourses in the context of the European Regulation on nutrition and health claims made on foods. We studied frames taken from written evidence (scientific reviews, legal commentaries etc.) and expert interviews on omega-3 acids and put analytical emphasis on the structure and organisation of scientific evidence that informs EU legislation. The aim of the paper is to show and analyse how scientific evidence, governance processes and licensing procedures interact and finally result in current food labelling practices.

Keywords: study of food and society, health and nutrition claims, EU regulation of food, prevention of disease

The politics of healthy food and diet: the case of functional foods

The study of food and society is an interdisciplinary and evermore recognised area of research analysing the production, regulation and consumption of food. It emphasises that food is a socially charged and politically contested field in which individual bodies, scientific research and innovations, legal aspects, economic interests and moral concerns become materially and virtually entangled and contested (Belasco, 2008). Food and nutrition embody an area of conflict and thus represent a highly contested political topic as recurrent food scandals such as the BSE crisis, public debates over genetically modified organisms or salmonella in eggs clearly indicate. This dilemma of the industrial production of food and connected food scandals has consistently changed ideas and concepts of food and a healthy nutrition: Food is no longer just food (Lien and Nerlich, 2004: 1), and a healthy nutrition no longer is a healthy nutrition. Ideas about and concepts of the politics of food nowadays include much more than just the institutional or bureaucratic administration of food security or agricultural policy – to mention just two examples. The study of the politics of food refers to a much wider framework in which the production, consumption and distribution of food form main areas of research besides traditional institutional approaches.

This re-conceptualisation of food studies and the study of food and society under the conceptual umbrella of the politics of food opens up research avenues in which a vast array of social concepts and practices such as daily cooking, aspects of healthy diet and – only recently – ideas about prevention of diseases come to into view (Lien and Nerlich, 2004). Thus, the complex interplay of food control, food producers or state policies has to be improved by the studies of everyday practices and a closer look at the interaction of actors in a vast array of social arenas in which the socio-political relations of producing, administering, distributing and consuming food are negotiated. Food represents insofar a special case as it holds symbolic as well as biological implications that range from the literal embodiment of food in physiological terms to the enhancement of human capacities and functions via research in nutrition science. Especially the latter has become of vital interest to dieticians, medical scientists and the food industry due to its potentials in the area of preventive medicine and the exploration of new markets. Novel types of food emerge now and then and only recently so-called functional foods entered the arena of foodstuffs. This relatively new type of food covers a vast range of enriched ailments such as probiotics in yogurts that keep the intestine balance, sterines in margarines that reduce LDL-cholesterol or bread with added omega-3 acids that help to prevent cardiovascular diseases. Advocates emphasise health promoting or even disease preventing qualities of functional foods while adversaries question them.

This contested boundary between food and drug recently labelled medicinal food (Chen, 2008) is the starting point of our empirical study in which we follow omega-3 acids on their travels through regulative and policy discourses in the context of the European Regulation on nutrition and health claims. We study frames taken from written evidence (scientific reviews, legal commentaries, etc.) and expert interviews on omega-3 acids and put analytical emphasis on the structure and organisation of scientific evidence that informs EU legislation in tackling possible health effects. The aim of the paper is to show and analyse how scientific evidence, governance processes and licensing procedures interact and finally result in current food labelling practices that intend to regulate the connection between food, functional foods or medicinal food.

Framing functional food and omega-3 fatty acids: theory and methodology

Omega-3 fatty acids were chosen as a starting point for this paper due to current tendencies in the food industry to enrich a variety of foodstuffs with them: *Iglo* launched omega-3 fish fingers, the *Food Doctor* offers omega-3 bread and food production engineers currently undertake research to enhance eggs with a variety of omega-3 fatty acids. This hype goes back to positive health or preventive impacts omega-3 fatty acids are supposed to have: Their multivalent therapeutic spectrum comprises the prevention of cancer, cardiovascular disease and neurodegenerative disease plus a variety of recently discovered health impacts on eyesight, compliance and mental disorders. Some of these characteristics are known since the 1930s and were substantiated in the 1970s by Dyerberg and Jorgensen (1970) in the course of research on Inuit in the north of Canada. Dyerberg and Jorgensen encountered the paradox that members of an Inuit tribe consumed high amounts of fatty sea food but were prone to cardiovascular disease. Later on, subsequent studies corroborated Dyerberg's and Jorgensen's hypothesis that omega-3 fatty acids reduce triglycerides, reduce blood pressure, lead to a decrease in heart beat and thus prevent arteriosclerosis. Since then, scientific research on omega-3 acids intensified and finally led to the previously outlined activities in the food industry to open up new markets by launching an increasing number of functional foods that bear nutrition and health claims. The European Council and Parliament reacted to this development by adopting the *Regulation on nutrition and health claims made on foods* in December 2006. For the first time, this Regulation laid down harmonised rules across the European Union for the use of nutrition claims such as 'low fat' or health claims such as 'reducing blood cholesterol'. One of the key objectives of this Regulation was and still is to ensure that any claim made on a food label in the EU is clear and substantiated by scientific evidence.

This scientific evidence is the starting point of our research in which we study the structure and organisation of scientific evidence that informs EU legislation. We generally build on Goffman's (1974) and Schön and Rein's (1994) concept of frame analysis. Frames can be understood as interpretative schemas which are used to mobilise collective action or as a signifying activity which is covered by the verb framing. The latter represents the theoretical starting point of our study as we are interested in how meaning is produced in action (Benford and Snow, 2000) and via discursive practices. Our approach stresses the relevance of frames as the outcome of meaning-construction processes and how they draw on a range of mainly scientific evidence with the aim to convince actors in the course of assessing functional foods. We thus put emphasis on how parties in the arena of regulating functional foods deploy scientific evidence in their framing activities to create cogent and scientifically resonant frameworks to influence food labelling practices. As a result, our analysis is concerned with frames as a resource to attain approved food labels for certain health claims or benefits which omega-3 fatty acids might possess.

Acknowledging the multifaceted nature of our research object we gathered different sets of written evidence such as scientific articles, the evaluation of different health claim applications, administrative, media and legal texts connected to functional foods. The juxtaposition of the sources offered rich insight into the discourses and contexts surrounding health and nutrition claims on the EU level. After tackling general developments providing insight into contemporary scientific and regulatory debates on health and nutrition claims we decided to concentrate on one substance for pragmatic reasons. We chose Omega-3 fatty acids because a variety of health claim applications in relation to them was recently submitted and examined displaying a vast range of anticipated preventive or health improving impacts. The next step consisted in building two written corpora: One on the scientific discourse and the other on the scientific opinions on the substantiation of submitted health claims related to omega-3 fatty acids as available from the European Food Safety Authority (EFSA, 2010). We reduced the corpus of scientific texts to the scientific reviews due to the incredible amount of published research on omega-3 fatty acids. The benefit of focusing scientific reviews consisted in a reduction of scientific research to be analysed and, furthermore, offered deep insight into research trends as this type of text shapes disciplines (Hedgecoe, 2006: 725) by evaluating current research undertaken und by defining future scientific problems. Data were taken from the Medline database and checked on the ISI-citation index regarding their intra scientific dispersion. Publications from ten leading authors were included in the corpus and analysed according to the requirements outlined by leading scholars in Grounded Theory (Charmaz, 2006; Corbin and Strauss, 2008). This enabled us to trace general developments and define areas of research that overlapped with those made in the submissions to the EFSA. More importantly, the interpretative in-depth analysis used to describe and articulate different aspects in relation to omega-3 fatty acids was coupled with a frame analytical approach. This analytical tool was chosen as it assured comparability on the one hand and enabled us to conceptually investigate the underlying frame structure of the scientific discourse and the submissions made to the EFSA. Questions we addressed include the following: What frames are predominant? How do they differ among the different data sets? What impact do they have on licensing and food labelling process? The next section displays preliminary 'frame oriented' results taken from our study.

Framing omega-3 fatty acids: the scientific discourse

The analysis of the omega-3 related scientific reviews taken from our corpus provided a rich underlying conceptual framework which was achieved by relying on methodological procedures as outlined in Grounded Theory. This approach was taken as it puts emphasis on the empirical development from data. Once main themes or topics emerge during the process of analysing data, text segments are grouped under these emergent headings. These headings were re-read and analysed with the main aim to isolate lexical items, metaphors or narratives which are conceived to exert a vital impact on the scope of meaning of the emergent topic. In the case of sufficient semantic evidence, the status of a topic was transferred to a frame due to its distinctive and substantiated meaning (Entman, 1993). These frames were then conceived as organising principles or devices that enabled us to assess the extent to which a certain frame had an impact on the overall framing of health impacts on omega-3 fatty acids.

Seven frames materialised in the course of our analysis: The health at risk frame (conceptual focus on the recent increase of obesity), the disease frame (conceptual focus on lifestyle diseases), the society frame (conceptual focus on industrialised societies and social decline), the prevention frame (conceptual focus on a healthy diet as a preventive measure), the science frame (conceptual focus on science as a problem solver) and finally the frame we call 'alimentation policy' (conceptual focus on nutrition education). This conceptual framework not only informs the scientific discourse, its shapes it by focusing on certain aspects while ignoring or neglecting others. The health at risk frame, for example, puts emphasis on the rise of obesity by using widespread narrative elements such as the increase in energy intake due to the omnipresence of palatable food in everyday life. This frame links up with the society frame in which obesity is depicted as an indicator of social decline and as a typical characteristic of industrialised societies This tendency also prevails in the disease frame in which obesity is in many cases depicted a lifestyle disease caused by the omnipresence of food and the increase in caloric intake. In brief, so-called corporeal decline indicates social decline and enacts counteractive measures as outlined in the prevention frame or in 'alimentation policy'. These frames stress the relevance of a healthy diet and a need for political regulation which implicitly render preventive measures such as fat taxes or proofs of physical activity possible. Science, as depicted in the science frame has the task to develop solutions for this dilemma. Especially the prevention frame with its focus on healthy diets as a preventive or health assuring measure implicitly refers to the relevance of new foodstuffs such as functional foods that may improve health.

Considered as a whole, it appears that the frame structure of the scientific discourse is quite tight and encompasses obesity as a major health problem and at the same time offers via the assessed frames a starting point in which dietary habits and new foodstuffs at least implicitly seem to play a relevant part. Thus, the frame structure combines and focuses a number of culturally and naturally determined reasons that hallmark the overall state of health in society directing attention towards preventive or counteractive measures that include functional foods. Our due analysis of submissions and reviews of health claims made to the EFSA will tackle frames in these documents and compare them with the results taken from the analysis of the scientific discourse on omega-3 fatty acids. In doing so, it will analyse the interaction of frames deployed in the submissions and contrast them against the background of frames taken from the scientific discourse and investigate their interaction in a political context. In doing so, it adds evidence to the emerging research arena of the politics of food. The next and final section summarises our provisional results so far.

Provisional results: the importance of framing omega-3 fatty acids in functional foods and the politics of functional foods

By analysing written evidence taken from two sources, this analysis builds on the still lacking sociological interest in the structure and organisation of scientific evidence that informs EU legislation on regulating functional foods. In particular, we argue that the important efforts made to conceptualise omega-3 fatty acids in functional foods is marked by resonant or culturally embedded frames that characterise discursive activities in the area of omega-3 related science. Our study has, to some extent, provided insight into the structure and organisation of the scientific discourse that is hallmarked by a variety of frames. They blend so-called natural and cultural aspects to argue for preventive measures to be achieved in the area of health promotion. Due to our present analysis of submissions and reviews to and by the EFSA, the interaction of scientific evidence and scientifically informed regulation is outstanding. Provisional results, however, indicate that the reviewing process on health and nutrition claims is driven by scientific evidence that interacts with a variety of frames as found in the scientific reviews. The findings of this study, even in their current provisional state, add fist evidence to the still under-researched area of functional foods from a sociological and science and technology perspective. On a grander level, it aims at substantiating research on the interaction of scientific evidence and governance processes that are conceived to be a basic characteristic of policy processes in a so-called knowledge-based society.

References

Belasco, W. (2008). Food: The key concepts. Berg, Oxford, Great Britain, 158pp.

Benford, R.D., and Snow, D.A. (2000). Framing Processes and Social Movements: An Overview and Assessment. Annual Review of Sociology 26: 611-639.

Charmaz, K. (2006). Constructing Grounded Theory. A practical guide through qualitative analysis. Sage, London, Great Britain, 208pp.

Corbin, J. and Strauss, A. (2008): Basics of qualitative research. Sage, Los Angeles, United states of America, 379pp.

EFSA (European Food Safety Authority) (2010): Nutrition and Health Claims. Available at: http://www.efsa.europa.eu/en/nda/ndaclaims.htm (April 2010).

Entman, R.M. (1993). *Framing: toward a clarification of a fractured paradigm.* Journal of Communication, 43: 51-54.

Goffman, E. (1974). Frame analysis: an essay on the organisation of experience. Harper, New York: Harper, 586pp.

Hedgecoe, A. (2006). Pharmacogenetics as Alien Science. Alzheimer's Disease, Core Sets and Expectations. Social Studies of Science 36, 723-752.

Lien, M.E. and Nerlich, B. (eds.) (2004): The politics of food. Berg, Oxford, Great Britain, 244pp.

Schön, D.A. and Rein, M. (1994). Frame reflection. Toward the resolution of intractable policy controversies. Basic Books, New York, United States of America, 247pp.

Seafood safety, authenticity and analytical methodologies to detect fraud

C.G. Sotelo[1], R.I. Pérez-Martín[1], M. Aursand[2], Y. Yamashita[3], M. Yamashita[3] and I. Martinez[1]
[1]Marine Research Institute-CSIC, Eduardo Cabello 6, 36208-Vigo, Spain; [2]SINTEF Fisheries and Aquaculture, 7465-Trondheim, Norway; [3]Food Biotechnology Section, National Research Institute of Fisheries Science, 2-12-4 Fukuura, Yokohama 236-8648, Japan; iciarm@iim.csic.es

Abstract

European consumers are entitled by law to information about the species, method of production and geographic origin of commercially available seafood products. Fraudulent and mislabeled products may not only differ in its biochemical composition, but also in its safety. For example, warm-water species are more likely to carry mesophile pathogenic bacteria than cold-water fish. Wild and farmed specimens from contaminated waters may contain higher levels of environmental pollutants or metals than individuals from cleaner waters and if the strict and obligatory farming practices are not properly followed, farmed fish may contain rests of veterinary drugs which wild fish is more unlikely to be contaminated with. Wild fish, on the other hand, is more likely to be parasitized than farmed fish. This work will give some examples to illustrate the relationship between raw material and risk of contamination and therefore the relevance of correct labeling. It will also mention the most appropriate analytical methods to identify species, method of production and geographic origin. It must be stressed that international cooperation and development of common databases is essential to ensure correct product labeling and consumer protection.

Keywords: seafood safety, authenticity, geographic origin, aquaculture

Introduction

Food available to consumers should comply with the legal requirements determing the acceptable levels of contaminants it may contain. However, it is impossible to guarantee the absence of accidents and the production of completely safe food: hazards may appear in the food chain and become a risk to consumers. Which hazard any given product is likely to contain vary depending on the species, geographic origin and whether the specimen was farmed or wild. For example, persistent organic pollutants (POPs) accumulate in the trophic chain; therefore, species of carnivorous organisms, in particular if living in contaminated waters from industrial countries (Hites *et al.*, 2004), are likely to contain higher levels than herbivorous or filter-feeders. Farmed organisms may contain residues of the veterinary drugs and which drugs and their levels will depend on the practices and permissiveness of the authorities in the location of production. On the other hand parasites are more often encountered in wild than in farmed fish. It is therefore clear that when it comes to ensuring authentic and safe food, the three issues must be taken into account.

Food safety and species identification

Fraudulent species substitution usually consists in the mislabeling of a cheaper and often unknown species as a local and more expensive one. Trade globalization and transport efficiency have allowed the profitable introduction of exotic species in markets willing to pay for high quality products, such as the Spanish, French or Japanese. While a local fish species may have different names depending on its size or where it is captured or consumed, novel species lack even a name in their destination and traders have sometimes chosen the name of a known species of similar aspect for their introduction in the new market.

This type of fraud happens with increasingly more frequency due to the shift in consumers' habits to purchasing products where the discriminatory morphological characters of the specimen are removed. Examples of this are the sale of South Atlantic hake species labeled simply as hake in Spain (where it refers to *Merluccius merluccius*) or products made of Pacific cod, haddock and a variety of gadoid species sold as cod or bacalao products, when bacalao refers exclusively to *Gadus morhua*. Recently, and following the immense success in the farming of *Pangasius hypophthalmus* and *P. bocourti* in Asia, where Vietnam alone is expected to produce 1.4 million tons in 2008-2009 (Josupeit, 2009), products from these species have appeared in the USA and Europe mostly as skinned and boneless frozen fillets. These products were initially labeled only as catfish in the USA (Hsieh *et al.*, 2009), where the term catfish refers to species of the Ictaluridae family (Regulatory Fish Encyclopedia, FDA, USA) and where there is zero tolerance for antibiotics and chemicals that, while absent from their local catfish, were found relatively often in the imported *Pangasius*. This led to a ban on the import of *Pangasius* and to the development of an immunological test to identify *Pangasius* spp. (Gajewski *et al.*, 2009; Hsieh *et al.*, 2009).

There are many analytical techniques suitable for species identification (Martinez, 2009). Lipid analysis has proven useful for this purpose, but the analysis of protein and DNA are usually easier, cheaper and more reliable (Rehbein, 2009a,b). Classical protein-based identifications rely on the comparison of isoelectric point (by IEF) or molecular mass (by SDS-PAGE) profiles of protein extracts, which are always species-specific, with those of reference samples (Rehbein, 2009a). User-friendlier analyses use immunological techniques and species-specific antibody preparations, preferably monoclonal antibodies as those mentioned above for *Pangasius* (Gajewski *et al.*, 2009; Hsieh *et al.*, 2009). DNA identifications compare diagnostic sequences (Rehbein, 2009b) with those of known reference samples. DNA analyses may be based on the comparison of the fingerprint of the sample under examination with the fingerprint of authentic samples by techniques called random amplification of polymorphic DNA (RAPD) and amplified fragment length polymorphism (AFLP) analyses or on the analysis of species-specific DNA fragments, usually, but not always from the mitochondrial DNA, by single sequence conformation polymorphism (SSCP), restriction fragment length polymorphism (RFLP) or single nucleotide polymorphism (SNPs) analyses, as well as by sequencing (Martinez *et al.*, 2005; Rehbein, 2009b).

There are commercial kits available for species identification. Some examples are Biofish kits (provided by Socochim SA) that claim to be able to identify some species of gadoids, salmonids, flatfish, hakes, tuna and sardines; the Biomérieux Food Expert_ID system that claims to identify 15 species of fish; and Tepnel, recently acquired by Gen-Probe, offered a kit to identify fish species. In addition, there are biotechnology companies offering identification services to which one can send a sample and the company provides the identification. Some of these companies are Xenotechs (Vigo, Spain), Bionostra (Madrid, Spain) or Eurofins-GeneSacn (LA, USA).

Different potential hazards in farmed and wild seafood

For any given species, wild specimens are usually more expensive than farmed ones and therefore the potential for fraud exists. The production chain of farmed fish consists of several steps including: (1) selection of brood stock and ova production, (2) production of juveniles, (3) on-growing, (4) slaughtering and (5) distribution of the fish products to the market. Therefore it offers the possibility to fully control the wholesomeness of farmed organisms. Potential and emerging contaminants may be introduced at any of the above listed steps. Contaminants that accumulate in the edible part of the fish, such as dioxins or heavy metals, are usually considered a higher risk than those that can be eliminated or washed out by the time of slaughter, such as bacteria or drug residues. Undesirable contaminants that may be introduced together with the feed, include toxic metals, phosphorated and halogenated organic compounds, PCBs, dioxins and PAHs. More recently new hazards have been detected, notably melamine and cyanuric acid (Andersen *et al.*, 2008). Several other hazards can enter the chain through veterinary treatments (antibiotics and antiparasitic drugs), from the natural environment (*Listeria monocytogenes*) or from the post harvest handling (*Clostridium botulinum, Staphylococcus aureus*). Fish do not seem to be a source of foodborne viral infections (which are the most relevant hazard in farmed shellfish), although improvements in detection techniques may change this perception.

Once all the potential hazards are mapped, it is possible to design and implement farming practices to avoid or minimize their presence. The production chain of wild fish, however, starts with its harvesting from the sea, making it impossible to control potential contaminations prior to this step. Parasites of concern in wild fish include *Anisakis,* in particular *A. simplex* that may cause anisakiosis (Inoue *et al.*, 2000) and *Diphyllobotrium latum,* that causes dyphyllobothriosis (Cabello, 2007). Sporadic cases of anisakidosis and diphyllothriosis have been registered in Japan (Inoue *et al.*, 2000) and Brazil (Cabello, 2007) after the consumption of raw farmed fish. Bacteria that may contaminate fish are *L. monocytogenes, Salmonella* spp and *Vibrio* spp while the most common intoxications are due to *C. botulinum, C. perfringens, B. cereus, S. aureus* and *Vibrio* (e.g. *V. cholerae, V. parahaemolyticus*) although most often these infections have been associated with consuming raw shellfish and seafood. Other contaminants, such as methyl-mercury have been detected mostly in wild large predatory fish such as swordfish and tuna, but it is usually not a hazard in farmed fish. Veterinary treatments have always been part of the routines to prevent and treat diseases. In fish farming, veterinary drugs have usually one of three purposes: as antibacterial, antiparasitic or as anaesthetics used during handling of live fish. Unfortunately, a widespread and unrestricted use of large amounts of antibiotics for prophylactic purposes has been registered in aquaculture production especially in developing countries (Cabello, 2006).

Analytical methods suitable to discriminate farmed from wild seafood have recently been reviewed by Martinez *et al.* (2009), and they include DNA, protein and lipid analysis. Since most breeding programs select the fish based on commercially interesting traits such as growth performance, resistance to diseases or to stress, genetic analyses have allowed the differentiation of wild from farmed fish populations in a variety of species, and it has usually been observed a loss of rare alleles in the farmed populations (Pampoulie *et al.*, 2006). Hayes *et al.* (2005) suggested that it was possible to assign accurately a fish sampled from the market place to either the farmed population or the wild using either microsatellite or SNP markers. Although, in principle, any marker has the potential to be useful to differentiate farmed from wild specimens of a given species, in reality the selection of markers (which ones and how many) necessary to differentiate a wild from a farmed specimen is completely dependent on the species and the breeding stock and it needs to be examined on an individual basis.

No clear protein marker has been identified to discriminate farmed from wild seafood, but, as for DNA markers, protein markers commonly used for genetic analyses have the potential to be used as markers to identify farmed and wild. In addition, differences in the protein pattern of liver (Martin *et al.*, 2003) and muscle (Martinez *et al.*, 2007, Olsson *et al.*, 2007) tissues between farmed and wild salmon and cod have been reported. The authors attributed the altered protein expression in the examined tissues to the fact that feeds and breeding conditions are not optimized for many species, and in particular to the use of vegetable ingredients to feed carnivorous fish that may induce stress in the farmed animals.

To authenticate fatty fish, the analysis of the triglyceride (TG) fraction has often given a correct classification of farmed and wild fish (Martinez, 2009). Often, the total amount TG alone may be enough information, since farmed fish usually have a much higher content than wild ones. Given that the changes in the FA composition of the TG fraction follow changes in the composition of the diet (Jobling, 2004) and that farmed fish are increasingly more often fed with vegetable oils (Bell *et al.*, 2003) absent from their natural diet, the presence of these vegetable oils serves as a diagnostic marker for farmed fish (Axelson *et al.*, 2009).

Geographic origin and potential contaminations

As already mentioned, contamination in the waters where the fish breeds and its feed or prey are key factors to ensure its safety. Thus, it has been reported that salmonids from waters off industrial regions contain higher levels of POPs that fish from cleaner environments (Hites *et al.*, 2004) and Yamashita (2008a) reported that nori alga cultivated in China and Japan had higher levels of cadmium and arsenic respectively. The levels of the toxic inorganic arsenic were also high in the brown alga *Hijika fusiforme*, known as *hijiki* in Japan. However, the arsenic may be eliminated and the product rendered safe by increasing the length of the washing step used in the Japanese traditional cooking method, and by using sea water instead of tap water, which also improved its organoleptic properties (Yamashita, 2008b).

Methods suitable to determine the geographic origin of agricultural food products have been recently reviewed by Luykx and van Ruth (2008) and include the content of stable isotopes and of trace elements which were already proposed by Kelly *et al.* in 2005. The determination of the geographic origin of fish was reviewed by Martinez (2009). As for agricultural foodstuffs, techniques aiming at identifying the geographic origin of seafood make use of the different distribution of stable isotopes and/or trace elements in the different environments, preys and feeds. Aursand *et al.* (2000) were able to classify correctly samples of wild and farmed salmon from Norway and Scotland according to both the geographic origin and the production method using gas cromatography, isotope ratio mass spectrometry (IRMS) and high resolution(HR) ^{2}H site-specific natural isotope fractionation (SNIF)- nuclear magnetic resonance (NMR) spectroscopy on fish oils and lipids extracted from Atlantic salmon. Moltenkin *et al.* (2007) could differentiate organic from non-organic farmed salmon and their geographic origin (Norway or Ireland) using a combination of fatty acid and δ^{15}N composition of the samples. Trace element signatures, usually analyzed by Inductively Coupled Plasma (ICP) Mass spectrometry or ICP-Atomic Emission Spectrometry have been used to distinguish between minke whales (Born *et al.*, 2003), tuna (Secor *et al.*, 2002) cod (Campana *et al.*, 1994) and eels (Yamashita *et al.*, 2006) from different origins.

Conclusions and future trends

A battery of analytical techniques needs to be used in order to authenticate seafood products. The trend is on one hand is to develop user-friendly and/or high throughput analysis tests many of them for screening purposes. At the same time, research institutes and commercial companies are standardizing more sensitive and quantitative analytical procedures and offering them as a service for a fee. In any case, the key issue for any successful identification is the availability of suitable databases of authentic reference samples and of data for each species, production method and origin. Such databases need to be updated regularly, since the authentic profiles change with time: different genetic markers may be selected for breeding, the composition of the feeds and of the environment will change, etc. Therefore, it is imperative to establish a close cooperation between researchers from different countries who, using appropriate and complementary analytical techniques, establish and maintain adequate databases with the necessary authentic profiles and databanks of authentic reference materials.

References

Andersen, W.C., Turnipseed, S.B., Karbiwnyk, C.M., Clark, S.B., Madson, M.R., Gieseker, C.A., Miller, R.A., Rummel, N.G., and Reimschuessel, R. (2008). Determination and confirmation of melamine residues in catfish, trout, tilapia, salmon, and shrimp by liquid chromatography with tandem mass spectrometry. Journal of Agricultural and Food Chemistry, 56, 4340-4347.

Aursand, M., Mabon, F. and Martin, G.J. (2000). Characterization of farmed and wild salmon (*Salmo salar*) by a combined use of compositional and isotopic analyses. Journal of the American Oil Chemists Society 77: 659-666.

Axelson, D.E., Standal, I.B., Martinez, I, & Aursand, M. (2009). Classification of wild and farmed salmon using Bayesian belief networks and gas chromatography – derived fatty acid distributions. Journal of Agricultural and Food Chemistry 57: 7634-7639.
Bell, J.G., Tocher, D.R., Henderson, R.J., Dick, J.R. and Crampton. V.O. (2003). Altered fatty acid compositions in Atlantic salmon (*Salmo salar*) fed diets containing linseed and rapeseed oils can be partially restored by a subsequent fish oil finishing diet. Journal of Nutrition 133: 2793-2801.
Born, E.W., Outridge, P., Riget, F.F., Hobson, K.A., Dietz, R., Oien, N. and Haug, T. (2003). Population substructure of North Atlantic minke whales (*Balaenoptera acutorostrata*) inferred from regional variation of elemental and stable isotopic signatures in tissues. Journal of Marine Systems 43: 1-17.
Cabello, F.C. (2006). Heavy use of prophylactic antibiotics in aquaculture: a growing problem for human and animal health and for the environment. Environmental Microbiology 8: 1137-1144.
Cabello, F.C. (2007). Salmon aquaculture and the transmission of the fish tapeworm. Emerging Infectious Diseases 13, 169-171.
Campana, S.E., Fowler, A.J. and Jones, C.M. (1994). Otolith elemental fingerprinting for stock identification of Atlantic cod (*Gadus morhua*) using laser ablation ICPMS. Canadian Journal of Fisheries and Aquatic Sciences 51: 1942-1950.
Gajewski, K.G., Chen, T.Y and Hsieh, Y.H.P. (2009). Production and characterization of monoclonal antibodies specific to Pangasius catfish, basa and tra. Journal of Food Science 74: C241-C247.
Hayes, B., Sonesson, A.K., and Gjerde, B. (2005). Evaluation of three strategies using DNA markers for traceability in aquaculture species. Aquaculture 250: 70-81.
Hites, R.A., Foran, J.A., Carpenter, D.O. and Hamilton, M.C. (2004). Global assessment of organic contaminants in farmed salmon. Science 303: 226-229.
Hsieh, Y.H.P., Chen, Y.T. and Gajewski, K. (2009). Monoclonal antibody-based sandwich ELISA for reliable identification of imported Pangasius catfish. Journal of Food Science 74: C602-C607.
Inoue, K., Oshima, S.I., Hirata, T. and Kimura, I. (2000). Possibility of anisakid larvae infection in farmed salmon. Fisheries Science 66: 1049-1052.
Jobling, M. (2004). Are modifications in tissue fatty acid profiles following a change in diet the result of dilution? Test of a simple dilution model. Aquaculture 232: 551-562.
Josupeit, H. (2009). http://www.eurofish.dk/dynamiskSub.php4?id=3753. FISH INFOnetwork Market Report. FAO. Accessed 19/02/2010.
Kelly, S., Heaton, K. and Hoogewerff, J. (2005). Tracing the geographical origin of food: the application of multi-element and multi-isotope analysis. Trends in Food Science and Technology 16: 555-567.
Luykx, D.M.A.M. and van Ruth, S.M. (2008). An overview of analytical methods for determining the geographical origin of food products. Food Chemistry 107: 897-911.
Martin, S.A., Vilhelmsson, O., Médale, F., Watt, P., Kaushik, S. and Houlihan, D.F. (2003). Proteomic sensitivity to dietary manipulations in rainbow trout, Biochimica et Biophysica Acta 1651: 17-29.
Martinez, I. (2009). Other principles: analysis of lipids, stable isotopes and trace elements. In: Rehbein, H. and Oehlenschläger, J. (eds.) Fishery products, quality, safety and authenticity. Wiley-Blackwell Publishing Ltd, UK, pp. 388-410.
Martinez, I., James, D. and Loreal, H. (2005). Application of modern analytical techniques to ensure seafood safety and authenticity. FAO fisheries technical paper nr 455. Rome, Italy: FAO. 73 p.
Martinez, I., Slizyte, R. and Dauksas, E., (2007). High resolution two-dimensional electrophoresis as a tool to differentiate wild from farmed cod (*Gadus morhua*) and to assess the protein composition of klipfish, Food Chemistry, 102, 504.
Martinez, I., Standal, I.B., Aursand, M., Yamashita, Y. and Yamashita, M. (2009). Analytical Methods to differentiate farmed from wild seafood. Chapter 14th. In: Handbook of Seafood and Seafood Products Analysis (Book no. 3). eds. Nollet, L. and Toldrá, F. CRC Press, Boca Raton, USA, pp: 215-232.
Moltenkin, J., Meisel, H., Lehman, I. and Rehbein, H. (2007). Identification of organically farmed Atlantic salmon by analysis of stable isotopes and fatty acids. European Food Research and Technology 224: 535.543.
Referanse, Olsson, G.B., Friis, T.J., Jensen, E., Cooper, M. (2007) Metabolic disorders in muscle of farmed Atlantic cod (*Gadus morhua*), Aquaculture Research 38: 1223-1227.

Pampoulie, C., Jörundsdóttir, T.D., Steinarsson, A., Pétursdóttir, G., Magnús Örn Stefánsson, M.Ö., and Daníelsdóttir, A.K. (2006). Genetic comparison of experimental farmed strains and wild Icelandic populations of Atlantic cod (*Gadus morhua* L.), Aquaculture 261: 556-564.
Rehbein, H. (2009a). Protein-based methods. In: Rehbein, H. and Oehlenschläger, J. (eds.) Fishery products, quality, safety and authenticity. Wiley-Blackwell Publishing Ltd, UK, pp. 349-362.
Rehbein, H. (2009b). DNA-based methods. In: Rehbein, H. and Oehlenschläger, J. (eds.) Fishery products, quality, safety and authenticity. Wiley-Blackwell Publishing Ltd, UK, pp. 363-387.
Secor, D.H., Campana, S.E., Zdanowicz, V.S., Lam, J.W.H., Yang, L. and Rooker, J.R. (2002). Inter-laboratory comparison of Atlantic and Mediterranean bluefin tuna otolith microconstituents. ICES Journal of Marine Science 59: 1294-1304.
Yamashita, Y. (2008a). Origin identification of marine products by analysis of trace elements. Presentation to the Joint Workshop of Research Cooperation on Seafood Safety Between Norway and Japan. Augusst 2008, Bergen, Norway.
Yamashita, Y. (2008b). Removal of arsenic from seaweed hijiki by boiling process. Program and Abstracts of the 5th World Fisheries Congress, 20-24th October, Yokohama, Japan. Pg 221.
Yamashita, Y., Omura, Y. and Okazaki, E. (2006). Distinct regional profiles of trace element content in muscle of Japanese eel *Anguilla japonica* from Japan, Taiwan, and China. Fisheries Science 72: 1109-1113.

GMO, consumption and consumer vulnerability in Brazilian Consumer Law: the right to be duly informed

A.P. Myszczuk[1] *and F. Glitz*[2]
[1]*University: UniBrasil/QUIS Law School, Rua Konrad Adenauer 442, Curitiba, Paraná, CEP 82820-540, Brazil;* [2]*University: UniBrasil/QUIS Law School, Rua Carneiro Lobo 570, conj. 1501, Curitiba, Paraná, CEP 80240-240, Brazil; anapaulamy@yahoo.com.br; fglitz@gmail.com*

Abstract

The Brazilian Consumer Defense system is governed by the principle of vulnerability. The adoption of this principle involves considering the consumer as the center of the legal protection in such a way that it's impossible to reduce him to the *status* of market or suppliers' object. Among other mechanisms of equalization the consumer's right to be duly informed about presence or absence of GMO ingredients on food or feed with adequate, accurate and clear information.

Keywords: consumer, vulnerability, GMO

Initial considerations

The current contractual relations seem to demonstrate the complexity of a 'new' logic of exchange where objects, before even thought, are transformed into negotiable benefits. These 'new' issues demands legal definitions and recognitions. In this sense, one of the dilemmas that arise is the regulation on regard to GMO. The question is even more relevant when dealing with contractual relations established between 'unequal parties' in which there is no balanced bargaining power, access to knowledge, management tools or marketing strategy. This article is proposed in the context of those concerns, aiming to review cases in which the consumer may be subject to legal relationships involving biotechnology, in which he does not necessarily have full access to the due information to found a conscious choice. For that purpose, it's necessary to understand how consumer's right to be informed is settled by the Brazilian Consumer Code and, among its instruments, the mechanisms established to protect the vulnerable contractual part.

The Brazilian Consumer Code and consumer vulnerability

The Brazilian Federal Constitution of 1988, sensitive to the global trend, included the legal protection of the consumer relation, incorporating into its norms the recent tendencies of modern public law (art. 5° XXXII)[1]. It was considered an important step taken by the constitutional legislature which would later motivate the edition of the Brazilian Consumer Code (BCC – Law no. 8.078/1990). These norms were stated to have public *status* social interest, in addition to the definition of rules and principles that should be well suited to the global changing socio-economic times. The same constitutional statement, which is related to the citizen's basic rights and guarantees, mentions the need for consumer protection (art. 5°, XXXII) elevates the consumer's protection to the same status as that conferred upon national sovereignty, private property and free competition (art. 170). In the light of these concerns, it is seen the importance of the Consumer Code when it comes to the GMO food and feed production chain, especially when related to the correct and reliable identification of a GMO based product. In the case of GM food, it is evident that information on the nature of it, in the eyes of the consumer, is very important when it comes to freedom of choice.

---

[1] *In verbis*: 'According to the law, the State shall promote the protection of the consumer'.

Following the provisions in the Federal Constitution, the BCC is based on the principle of consumer protection, where the legislator chooses to protect the consumption relationship by backing it, necessarily, in the principle of objective good faith. To implement that provision, BCC sets the figure of the 'consumer' as any person or entity and/or any group of persons who acquires or uses products or services as consignee. It is necessary to clarify that in this concept remains the protection of the consumer vulnerability. The BCC also prescribes the figure of the 'supplier' as any person or entity, public or private, as well as entities without personification, carrying out activities of production, assembly, creation, construction, import, export, distribution or marketing of products or services.

To operate the consumer protection, the BCC assumes, as a main principle, that the consumer is the vulnerable part of this relationship. The protection is not taken in a paternalistic aspect but as a way to preserve the balanced negotiation, through the promotion of equalization instruments. In other words, the law plays the role of promotion of contractual justice. Legally, therefore, the protection of the consumers is the recognition of their vulnerability. Based on this, the BCC sets out a series of consumer's rights, to facilitate and ensure this protection.

The consumer is considered vulnerable because can easily have his free will limited. This is especially true on regard to the choice of priorities and needs, keeping in view all the procedures, mechanisms, methods and contractual techniques used by suppliers to, even indirectly, promote, maintain, develop and ensure the circulation of their products and services. The consumer vulnerable condition does not allow evidence against it, as a conclusive presumption; it is not questionable and cannot be remedied, since it is a principle and a *praesumptio jure et de jure*. It is an intrinsic quality, inherent and inseparable part of anyone put in the consumer position. Indeed it is admitted, even those who are not *strictu sensu* consumers (article 2°) to show their vulnerability in order to receive protection under that law, to contractual abuse (article 29). The effective result of the principle of vulnerability is the duty of incitement to any information regarding consumption. In the consumer relations involving GMO, this duty is due strictly to the principle of objective good faith as recognized by the Brazilian legislation currently in force.

It is important to note that the Brazilian consumer's protection is independent of economic, social or intellectual *status*. The consumer is considered economically vulnerable, as the suppliers have full control of the market or on the production chain and products. The technical-professional consumer vulnerability comes from the lack of specific knowledge about a particular business line or service. It should be noted, moreover, that this lack of access is even more evident when analyzing the biotechnological developments. That is especially true about food and feed based on GMO and the lack of information on the consequences of its use. In pitch, the principle of consumer vulnerability is one of the pillars of the BBC that underpins the whole system and a directive to achieve its goals. It is a material guideline for the identification of implied rights, functioning as a limiter for the suppliers activities. It is not only a guarantee that the consumer will not be subject to manipulation that reduces his condition to object, but also a duty of full development of the personality of each individual and/or the community.

The Brazilian Consumers Code, standard form contracts and GMO

Under the legal perspective, it should be noted that the consumer society is broadly subject to standard form contracts. Within this reality, it's possible to sellers to use this type of contract and its complexity and lack of clarity or transparency to inhibit the consumer's expression of free will, leading to the need for legal protection. One mechanism of such protection is to impose to the supplier the duty to provide information about the composition of a particular kind of food, for instance the requirement of special kind of labeling or the definition of the joint liability throughout the supply chain (from the producer to the distributor, who must inform the wholesaler, etc.). Information on the characteristic of the product in this case is a legal requirement which, besides being a guarantee for the consumer, is also a guarantee for the supplier, who will be able to identify any eventual informational flaws in the production chain. On the other hand, the joint liability of suppliers of products and services (as foreseen by the BCC – art. 7, single paragraph), allows the consumer to ask for compensation to anyone in the supply chain (art. 25). Therefore, if the product causes damage to the consumer, the manufacturer, producer and the importer will be held liable, unless the supplier could prove that they did not place the product on the market, that there is indeed no fault in the product or, in case of damage, that the consumer himself is solely responsible (art. 12). In addition, when dealing with liability for the bad quality or faultiness of products or services, the joint liability includes the seller. Therefore, because the manufacturer and the supplier can be held liable, it is necessary that the information about stages of the production chain to be made very clear, be it for the purposes of reimbursing losses or for dividing liability among the suppliers themselves.

The Brazilian Consumer Code and the consumer right to be duly informed about GMO products

By the Brazilian legislation, the consumer has the right to be duly informed about any product. That information must be provided in an adequate, accurate and clear way (article 6, III). The clarity of the information is not only a result of acting in good faith, but more importantly, it is the result of putting the transparence principle into action (article 4), which is also one of the objectives of the National Consumer Relations Policy adopted in Brazil. The central idea of the BCC is to establish a more sincere and less harmful relationship between consumers and suppliers. In this behalf, the BCC states that suppliers should not dispose any products that are risky to the consumer's health or safety, except those which are known to be so and that are bought at the consumers own risk. But, even then, the manufacturer must supply necessary and suitable information concerning the product and its possible risks (article 8), which is even more necessary on GMO based products. Likewise, article 9 clearly states that the supplier of potentially dangerous or harmful products and services must provide information about the risks of their products and, also, adopt any other applicable protective measures to each concrete case. Once again, the concern is that information about safety issues should be made available to the consumer, even in advertising. Effective result of the principle of vulnerability is the duty of incitement to any information regarding consumption. In relations involving GMO, this duty is due strictly to the principle of good faith and strict liability.

The current Brazilian Biosafety Law (No. 11105/2005), however, already states that the GMO presence or absence in food and food ingredients intended for human consumption or animal feed must be informed (article 40). However, it establishes that the labeling would subject of further regulation. To a large extent, indeed, the absence of labeling regulations does not relieve the supplier to provide the most complete and adequate information on the ingredients of its products. Note, for instance, that the BCC settles the obligation to provide information as fundamental principle of the National Consumer Policy (art. 4, IV) and as a basic consumer's right (article 6, III). On this last point, moreover, the legal text is wide enough to allow supposing that this duty covers the specific foods produced from GMO. It seems clear, therefore, that under any analysis, as about the percentage used in the preparation of the product or on the possible health risks, the supplier must provide the necessary information. Furthermore, the same law stipulates that the supplier shall maintain technical and scientific data to support the advertising about his products (art. 36, sole paragraph), besides that any publicity or silent advertisement can be consider misleading (article 37, § 1 and 3). The same legislation states that any information or advertising about a product in any form of communications media must be accurate and that the manufacturer must be prepared to stand by his information (art. 30). Moreover, the BCC stated that all information about products or services and their characteristics, quality, quantity, composition, price, guarantee, expiry dates and origin must be clear, correct and accurate (art. 31), even more about consumer health and safety. To that extent, if there is failure in complying such provisions, Brazilian legal system states that, along with the civil consequences (for example, compensation for damages or for breach of duty of good faith), there would be administrative penalties (fines and bans on marketing) and penalties (art. 66, arrest and fine).

About GMO labeling there is, also, a specific Decree (4680/ 2003) that makes compulsory the labeling of GM food. Therefore, it is necessary to make it clear that the GMO labeling should not be confused with safety. In Brazil, if the GMO is not considered safe for consumption, it will not be authorized for commercialization. Therefore, only GM food or feed that are considered safe will be labeled, because the labeling is part of the consumer right to be duly informed and guarantee of the possibility of freedom of choice. The product is considered to have transgenic nature, by the Brazilian law, when it contains GMO over to one percent of its composition. The Decree also states that the consumer must be informed about the type of gene used to create the GMO, which must be included in the list of ingredients. As for food and ingredients produced from animals that were fed with feed that contains GM ingredients, these must be included on the main part of the label, and must meet requirements for size and highlighting as stated in article 2. As for food and food ingredients that do not contain GMOs or were not produced from GMOs, the label will state that '(name of product or ingredient) is free of transgenic', providing there is a similar transgenic product on the Brazilian market.

Concluding notes

The Brazilian Consumer Code has the purpose of balancing contractual relationships between consumers and suppliers. It recognizes that the consumer is the weaker party in this relationship and for that reason seeks to protect him. This type of protection is extremely important in times when consumption relations are also subject to the technological innovations. It is a fact that there is separation from technological knowledge and cognitive capacity of the consumer. The consumer is not always able to understand the consequences of his consumption habits and even if a certain product or service may come to be harmful to his health. To make this protection effective, it´s assumed that any and every consumer, poor or rich, cultured or ignorant, physical person or corporate entity, individual or collective entity is vulnerable, in other words, he/it may have his/its free and conscious manifestation of will attacked by commercial practices, by the absence of knowledge and by the denial of access to information. This recognition is what founds and justifies the existence of the consumer protection system, serving as a guide also for legal developments on biotechnological matters. In this context, it may be seen that the access to a broad kind of information is a consumer right. It limits the activities of suppliers and may not be suppressed or withdrawn.

References

Amarante, M.C.N. (1997). Justiça ou equidade nas relações de consumo. Lúmen Júris Publisher, Brazil.

Bittar, C.A. (1999). Direitos do consumidor. Forense Universitária Publisher, Brazil.

Bonnato, C. and Moraes, P.V. (1999). Questões controvertidas no Código de defesa do Consumidor. Livraria do Advogado Publisher, Brazil.

De Almeida, J.B. (1997). A proteção jurídica do consumidor. Saraiva Publisher, Brazil.

De Arrua Alvim, J.M., T. Alvim, E.A. Alvim and J. Marins (1991) Código do consumidor comentado. Revista dos Tribunais Publisher, Brazil.

De Carvalho, S.V. (1997) O direito do consumidor. Sagra Luzzatto Publisher, Brazil.

De Oliveira, J. (1991). Comentários ao código de proteção ao consumidor. Saraiva Publisher, Brazil.

Filomeno, J.G.B. (2001) Manual de direitos do consumidor. Atlas Publisher, Brazil.

Grinover, A.P. Benjamin, A.H.V., Fink, D.R., Filomeno, J.G.B., Watanabe, K., Nery Jr., N., Danari, Z. (1998). Código brasileiro de defesa do consumidor comentado pelos autores do anteprojeto. Forense Universitária Publisher, Brazil.

Marques, C.L. (2002) Contratos no Código de Defesa do Consumidor. Revista dos Tribunais Publisher, Brazil.

Nunes, L.A.R. (2000) O Código de Defesa do Consumidor e sua interpretação jurisprudencial. Saraiva Publisher, Brazil.

Part 7. The role of ethical engagement and public participation in EU food governance

Food citizenship: is there a duty for responsible consumption?

J. De Tavernier
Center for Science, Technology and Ethics, Katholieke Universiteit Leuven, Sint-Michielsstraat 4, 3000 Leuven, Belgium; johan.detavernier@biw.kuleuven.be

Abstract

Regulation of food consumption is related to food safety, food quality, environmental, safety, and social concerns. Contemporary politics of food is based on a redefinition of commodity food consumption as an expression of citizenship and centers around food labelling practices. 'Citizen- consumers' realize that they could make use of their buying power in order to develop a new terrain of political action. It takes for granted kinds of moral selfhood in which human responsibility is bound into human agency based on knowledge and recognition. Citizens see their consumer's preferences as expressions of social agency. Existing research on consumer's preferences often fails to recognize the full complexity of the motivations and intentions through which the identity of the moral self is built up in relation to food consumption practices. For citizens, not only food production practices but also what we eat and the ecological foot print of what we eat matter. A major driver for this is the idea that we ourselves have to take care of our own bodies: 'We are what we eat'. Since obesity has become a major public health concern, our governments also have an interest in defining how citizens ought to behave as consumers and how retailers and producers should facilitate such responsible behavior. Indeed, much more people suffer from the impact of obesity than from the effects of food poisoning. Governments are supposed to defend the '*bonum commune*', e.g. the public health of their citizens. Therefore, they have an increasing interest in 'appropriate' consumption choices.

Keywords: food consumption, ethical consumption, food citizenship, food ethics

Introduction

Consumption of food is usually related to questions concerning food quality, food safety, sustainable food production and social concerns, like availability and accessibility of adequate food for all. Future politics of food will be based on a redefinition of commodity food consumption as an expression of citizenship. Much more than in the past, 'citizen-consumers' will use their power to direct their buying behavior not only to more sustainable and just products but also to more appropriate products. Both sustainable and responsible consumption of food seems to become a new terrain of political action, to which citizen-consumers contribute by presenting their food preferences as expressions of a new kind of social agency ('Thursday: Veggie-day'). Political consumerism takes for granted kinds of moral selfhood in which human responsibility is bound into human agency based on knowledge and recognition. Food consumption acts more and more as a key boundary maker of identity. Up till recently, it's true that culture in general defines what is called food and what will be eaten but in the nearby future no longer culture but more and more singular persons will express their unique identity through food patterns. Traditional forms of ethical consumption, such as organic food and fair trade, mainly focusing on less-than-standard working conditions, often fail to recognize the full complexity of the motivations and intentions through which the identity of the moral self is built up in relation to food consumption practices. For citizens, not only fair food production and safety practices do matter but also their ecological foot print. Food citizenship and singularity are the main drivers behind the next move towards responsible consumption.

In this paper, we will concentrate on three subthemes: (1) Food labelling: the right not to label versus the right to label and the duty to label; (2) Which impact will the introduction of the idea of food citizenship have on the existing equilibrium between the principles 'the right to know/not to know' and 'the duty to know' as expressed in present labelling? What will be the impact of nanobiosensor technology? (3) Will the present legal framework on labelling defending the individual's interest in knowing a limited amount of particular information (ingredients, nutritional value, biotechnology, animal welfare) be affected by potential interests of consumer-citizens and governments stimulating responsible consumption? Could there be a duty to know?

Food labelling: the right not to label versus the right to label and the duty to label

Since two decades, stimulated by public concerns, food safety agencies pay much more attention to issues of chemical contamination (antibiotics, growth hormones, pesticides residues, additives, packaging conditions), and food-borne micro-organisms in the food chain (*Campylobacter, Salmonella, Escherichia coli*, etc.). The public authorities were obliged to inform the public with cooking advices and proper handling in order to prevent food-borne diseases.

Traditional food labelling, focusing on ingredient labelling, ensures that consumers were not defrauded. Since the public is aware about suspected links between saturated fat, cholesterol, and coronary heart disease, nutritional guidelines have been added to the labels. Interest in nutritional labelling was shown for instance for the amount of mono-saturated, poly-saturated and saturated fats. Through labelling, additional information on nutrients and fibre that relate to high blood pressure, obesity, osteoporosis, and heart disease is provided. Food safety agencies were required to draft definitions of 'high fibre', 'light', 'low fat', the amount of calories, etc.

In the meantime, food labelling has even become more complex. With the introduction of genetic modification in plant and animal production, a new debate was introduced, *id est* the public desire to make a distinction between conventional and genetically engineered products. Some were of the opinion that there was no need at all to do this because the only criteria which really count for consumers are food safety and nutritional value. If genetically engineered products were declared safe and having the same nutritional value as conventional alternatives, no extra labelling information is needed. Restricting labelling to food ingredients and nutrition is considered as fair enough. Proponents of this position therefore defend the right not to label. A good illustration of this strategy is the rbST case in the US. The FDA determined in the last decade of the 20th century that milk products from rbST treated cows were not different from milk products from untreated dairy animals. Therefore such products must not be labelled since only 'meaningful' and 'essential' information should be mentioned on food labels.

Their opponents defend the opposite strategy while asking to move the labelling issue beyond the safety and nutritional issue. They were of the opinion that consumers have a right to know, not only about genetic modification and nutritional value but also about which particular production method is used, even if this information has no direct impact on the end quality of the final product (e.g. information on eggs about husbandry systems used for hens). In line with this reasoning, European Regulation (EC) No 1830/2003 of the European Parliament and of the Council of 22 September 2003 concerning *the traceability and labelling of genetically modified organisms and traceability of food and feed products produced from genetically modified organisms* (amending Directive 2001/18/EC) prescribes: 'Traceability requirements for food and feed produced from GMOs should be established to facilitate accurate labelling of such products, in accordance with the requirements of Regulation (EC) No 1829/2003 of the European Parliament and of the Council of 22 September 2003 on genetically modified food and feed (6), so as to ensure that *accurate information* is available to operators and consumers to enable them to exercise their *freedom of choice* in an effective manner as well as to *enable control and verification of labelling claims*. Requirements for food and feed produced from GMOs should be similar in order to avoid discontinuity of information in cases of change in end use.'

Next to the general safety and risk standards, consumers receive information about ingredients, nutritional values, presence/absence of genetically modified products (duty to label) and a non-obligatory right to label for fair trade and animal welfare conditions (except for eggs on which the husbandry system is mentioned in a code). Does it stop here? I believe that other aspects should also become subject of labelling, for instance, the environmental impact of a particular product. Due to climate change, citizen consumers desire to act more responsibly in food matters. Citizen consumers are concerned about the environmental impact of what they are eating but they could not behave in an effective way because they lack at this moment specific knowledge. Citizen-consumers are willing to consider this as a matter of informed consent because they want to feel good about their food choices but as long as the necessary information is not provided, they could not consent. Therefore, the right – or even the duty? – to label has to be extended to include a life cycle analysis measurement (also known as Life Cycle Assessment – LCA, eco-balance or cradle-to-grave analysis). The purpose of an Life Cycle Assessment is to measure the environmental and societal impact of particular products and services, in order to be able to choose the least burdensome one. It's a holistic assessment which takes into account the production process, processing, distribution, use, disposal and all kinds of transportation. In this case the life cycle impact assessment could be used to improve the environmental performance of a single food product. Although there is still disagreement on the weighing factors to each impact category, we expect that a kind of LCA-model will be welcomed as a necessary step towards more food citizenship.

Which impact will the introduction of the idea of food citizenship have on the existing equilibrium between the principles 'the right to know/not to know' and 'the duty to know' as expressed in present labelling? What will be the impact of nanobiosensor technology?

Labelling could be supported by new technology, such as chips, which will enable to provide a collection of a frankly staggering amount of individual data. Data about place of production, production method, means of transport are at present only very partly known or not known at all by consumers. For sure, if data of the environmental impact of a particular food product would be known by consumers, it would alter in particular cases the decision to buy or not to buy. Since the information could be made available, why would it be any longer hidden for consumers? Don't they have a right to know?

In our liberal democratic societies consumer's autonomy and his/her right to know is the most important ethical principle not only in biomedical ethics but also in food ethics. The core variable is the 'G(o)od citizen consumer'. The 'G(o)od citizen consumer' suggests a further empowerment of the individual consumer through use of new technology, putting pressure on or even altering the traditional retailer-consumer relationship. It is not unthinkable that citizen consumers freely accept to subordinate partly their own privacy concerns to the interest of retailers in helping them to follow appropriate individual diets. The first necessary step for this, is a much better labelling system.

What are new challenges brought about by new ways of labelling which are expected to transform consumer practices deeply? A possible major driver is the promise to treat in a more effective way obesity, indeed a treatable condition. Obesity is now a very important cause of 'welfare' diseases and even death, resulting in huge societal costs (cf. WHO Global Database on Body Mass Index). Much more people suffer from obesity-related diseases than from food contamination. More than a billion adults have overweight and one third of them could be classified as obese (BMI>30). Obesity is the result of a disproportionate and wrong food intake (junk food) and shortage of mobility. From an ethics viewpoint, we should consider food intake as an important question of responsibility since a lot of suffering could be avoided, such as chronic diseases (type 2 diabetes, high blood pressure, coronary heart disease, and certain types of cancer). Since obesity has become a major public health problem in all western countries and thus create huge extra costs for the existing health care systems, public authorities are encouraged to develop strategies in defining how citizens ought to behave as responsible consumers. The impressive societal costs could be avoided if prevention strategies, like awareness about food choices, could increase. Since governments are required to defend the common good, included the public health of their citizens, they have an increasing interest in 'more adequate' consumption choices. The Government could therefore choose to empower consumers in order to opt for more 'correct' food consumption choices and stimulate retailers and producers to facilitate such behavior. The aim is to strengthen the element of individual responsiveness.

What will be the impact of nanobiosensor technology? Nanobiosensor technology used for diagnostics is considered as a subfield of nanomedicine. Nanobiosensors are meant for *in vivo* and real time diagnosis and monitoring. These sensors use micro-electromechanical systems (or MEMS) and involve a combination of nanotechnology and microfluidics. The entire sensor could be downsized to the nanoscale, but more likely it contains one or more enabling components at this scale. A series of these sensors for health care purposes is sometimes called lab-on-a-chip, a miniaturized lab that has the capacity to analyze a variety of metabolic parameters and hence increasing the analysis speed, sensitivity and reliability. *In vivo* nanobiosensors are not yet available but they belong to Europe's short term priorities for research and development. According to the nanodiagnostics timeline of the *European Technology Platform on NanoMedicine – Nanotechnology for Health*, various 'quantum dots for molecular diagnostics and imaging', and 'point-of-care analytical devices' are already on the market. They expect 'multifunctional nanoparticles for drug release and imaging' and 'implantable devices for continuous measurement of blood markers' to become commercially available around 2010 and 2015 respectively. Nanobiosensor technology for *in vivo* monitoring will not only change the scope of the physician-patient relationship but also the role of retailers versus consumers and the role of Governments.

Nanobiosensor monitoring technology will have a broad impact because *in vivo* monitoring will be used to inform individuals about their health status, including propensity to disease/medical conditions long before they occur (for instance predicting a heart attack by continuous observation of blood pressure, observing heart rhythm, monitoring glucose of a diabetes patient). One expects soon implantable biosensors which will deliver round the clock, instant and seemingly complete information on one's entire physical being. Nanobiosensor technology is not intended to become a niche innovation but part of routine health check practice. It is believed to be used on a mass-scale for 'common' disease prevention. These diagnostic tools will be cheap and not difficult to manage.

The very essence of these nanobiosensor technologies is the acquisition of information which can be used to predict, prevent, treat or ease disease developments. Hence, the huge amount of health information data will not only be used in overall clinical practice in order to predict diseases, to observe recovering patients and to secure a more accurate diagnosis, it will also help to refine existing therapeutic diets and to define particular diets in order to prevent diseases.

The consequences of these developments for traditional food distribution have not been studied yet. The concept of the 'G(o)od consumer' has recently emerged in the biomedical literature, especially in a context of cost containment (keeping future costs down by acting now), but not in a technology driven context nor in a food context. Within that context, the citizen consumer as described above threatens to reduce the importance of the dietician/physician to a 'mere' member of a much wider technology-driven team, far increasing the role of the individual citizen consumer in managing the body/medical food-related information available. Equipped by the received information data, citizen consumers will easily interpret which food stuff suits him/her at best, if need be with the help of a computer programme or an outsourced medical support staff (e.g. in a call centre halfway across the world).

As a result, the existing legal framework for consumer rights will come under pressure by labelling developments. Traditionally this framework is characterised by protection of *rights*, whereas *obligations and responsibilities* of consumers (future patients) have been largely neglected. The traditional model of the patient-physician and retailer-consumer relationship and the regulatory relationship has to be reconfigured. *Informed consent* may become *informed decision-making* with the retailer or even the physician/dietician in the role of a consultant. Highly sensitive diagnostic nano-tools will not only provide information to create highly individualised therapies but also to prescribe individualized diets.

Will the present legal framework on labelling defending the individual's interest in knowing a limited amount of particular information (ingredients, nutrition, biotechnology, animal welfare) be affected by the potential interest of consumer-citizens and governments in stimulating responsible consumption? Could there be a duty to know?

Governments have an interest in enhancing responsible consumption. Public health is one of concerns of the common good (*bonum commune*), for which governments are supposed to take up responsibility. The fast aging population poses significant challenges to restructure European health care systems. Since more diagnostic and therapeutic techniques become available it will be impossible to reimburse everything for everybody. Hence, the government is looking for means to improve the allocation of her scarce financial resources. Since public health is part of the common good, the Government will have an increasing interest in 'correct' consumption choices. Early warning diagnostic and monitoring devices preventing diseases with a high impact factor, such as obesity, would be very cost effective. The Government is expected to have very strong incentives to promote the mass distribution of the nanobiosensor technology for reasons of cost reduction and efficacy. At this stage we expect that nanobiosensor technology will have an impact on the present equilibrium between the principles 'the right to know/not to know' and the 'duty to know'. First, nanosensor technology is expected to empower the consumer and thus favor the right to know. This empowered role will strenghten the principle of autonomy (= the 'God citizen-consumer'). At the same time we expect that the Government for cost effective reasons will develop a huge interest in asking to share the medical data and even in imposing a duty to know. Since the early diagnostic and monitoring tools are becoming so important, we expect that the use of nanobiosensors will most probably not only be facilitated by the Government but also actively encouraged (the 'good citizen consumer'). In other words, governments will likely put forth a *duty to know* on their citizens in order to reduce costs and maybe even as a prerequisite for an individual to benefit health care services.

Government intervention evidently may have to take place extremely upstream. If government is to benefit from the availability of widespread medical data of a large part of its population (cost reduction, efficacy, data mining for epidemiological studies), then in the future nanomedical implants may have to be provided to the population as a matter of course. Medical science may encourage government to deploy even nanomedical monitoring much as it currently deploys vaccines for newborns and infants.

Conclusion

Citizen consumers will use their preferences as an expression of social agency. Existing labelling information does no longer meet the full complexity of the motivations and intentions through which the identity of the moral self is built up in relation to food consumption practices. For citizens, not only information about ingredients and nutrition but also about food production practices and the life cycle assessment of food products does matter. 'We are what we eat': a major driver in this is the idea that we ourselves have to take care of our own bodies and that we are supposed to feel good about our food patterns. Therefore consumers have 'a right to know'. Since obesity has become a major public health concern in all western societies, our governments also develop an interest in defining how citizens ought to behave as consumers and how retailers and producers should facilitate such responsible behavior. The huge need for counterbalancing the health consequences of poor diets and sedentary lifestyles is not only a matter of improving the quality of personal life but also of reducing costs. Since Governments have to cut their social security expenses, they have an increasing interest in 'appropriate' consumption choices and thus in imposing a 'duty to know'.

New technology, like the use of nanobiosensors, will not only push the idea of informed consent but also the idea of 'a duty to know'. The ideal 'G(o)od consumer' suggests a further empowerment, by nanotechnology, of the individual consumer, putting pressure on or even altering the traditional producer/retailer-consumer relationship. Is it unthinkable that the good citizen-consumer subordinates his/her own privacy concerns to the interest of the retailer or even the community at large (as represented by Government), amongst others in building the databases needed to build appropriate individual diets the earliest possible?

References

Andorno, R. (2004). The right-not-to-know: An autonomy based approach. Journal of Medical Ethics: 435-439.

Bawa, R. and Johnson, S. (2008). Emerging issues in nanomedicine and ethics. In: Allhoff, F. (ed.). Nanotechnology & Society: Current and emerging ethical issues. Dordrecht: Springer.

Jotterand, F. (2007). Nanomedicine: how it could reshape clinical practice. Nanomedicine 2(4): 401-405.

Kittay, E. (2007). Beyond Autonomy and Paternalism. In: Nys, T., Denier, Y. & Vandevelde, T. (eds). Autonomy & Paternalism. Leuven: Peeters, pp. 23-70.

Kunkel, H.O. (2000). Human Issues in Animal Agriculture. College Station: Texas A&M University Press.

Micheletti, M. (2003). Political Virtue and Shopping. Individuals, Consumerism, and Collective Action. New York: Palgrave.

Nys, H. (2005). The Belgian Law of August 22, 2002 Concerning The Rights of Patients. In: Yearbook of European Medical Law. Lidingö (Sweden): The Institute of Medical Law, pp. 91-100.

Pond, W., Nichols, B.L. and Brown, D.L. (eds.) (2009). Adequate Food for All. Culture, Science, and Technology of Food in the 21st Century. London-New York: CRC Press.

Thompson, P.B. (2007). Food Biotechnology in Ethical Perspective. The International Library of Environmental, Agricultural and Food Ethics 10, Dordrecht: Springer.

Vo-Dinh, T. (2007). Nanotechnology in Biology and Medicine: Methods, Devices, and Applications. London-New York: CRC Press.

Willingness to pay for farm animal welfare: consumers' motives and expectations analysed at the example of dairy cows in Germany

M. Henseleit
Institute of Agricultural Policy and Market Research, Justus-Liebig University Giessen, Senckenbergstrasse 3, 35390 Giessen, Germany; meike.henseleit@agrar.uni-giessen.de

Abstract

Investigations in the last two decades have shown that consumers are concerned about the welfare of farm animals. It is evident that this concern is multidimensional. On the one hand, ethical motives can be a reason for concern, i.e. people want domestic animals to be handled well. On the other hand, animal welfare can be interpreted as an indicator of food safety, quality and healthiness. Consequently, consumers might expect good food standards when animal welfare standards are improved. In this case, concern about animal welfare is rather motivated by self-interests. The central question is which role do these motives play in the demand for improved husbandry standards and, accordingly, whether there might occur market failure due to a gap between expected and actual product quality. Surveying this problem should help to better understand the demand for more animal welfare friendly food products. We conducted a consumer survey in a main shopping street in central Germany in December 2009. About 207 people were questioned regarding perceived adequacy of current milk prices, attitudes towards food and shopping habits, knowledge and preferences regarding husbandry conditions of dairy cows and socio-demographics. The main part of our survey performed a contingent valuation experiment in which participants were asked to state their willingness to pay (WTP) for milk produced under improved animal welfare conditions. The average premium for one litre of fresh milk was about 0.31 €. We informed participants after this question, that no increase in product quality could be proven so far, and we again asked them to give their WTP for the same package of milk. This time, the average premium was about 0.28 €. In this way our research shows that expectations in product quality make a difference, but only a small one. We conclude that market failure might occur, but won't be exceedingly.

Keywords: consumer survey, product quality, pure ethical motives, price premium

Introduction

When thinking about bringing farm animal welfare products on the market, the following questions come into mind: First, are people willing to pay a price premium for such products, and if so, how much more are they willing to pay? Second, what do they rate most important regarding husbandry standards: Should farmers pay more attention to the feed, should they demand less productivity from the animals or should they provide more room to range or maybe something else? And third, what are peoples' expectations when they are willing to pay a premium for improved husbandry standards: Higher product quality, more product safety or is their WTP rather ethically reasoned because they want the animals to have a good life?

Research over the last twenty years shows consumers' concern about farm animal welfare has risen continuously (European Commission, 2005; Harper and Makatouni, 2007). This is a result of an increasing intensification of agricultural production, which on the one hand is often seen as a reason of food scares like the BSE crisis, for example. On the other hand, people get more and more distracted from agriculture because the rate of people working in the agriculture is decreasing steadily when farming is getting more and more industrialised. This makes it more difficult for people to build a realistic image about the living conditions of farm animals. In this way, media has a big influence on peoples' idea of farming in Germany. Since many consumers at least in industrialised countries are concerned about the welfare of farmed animals when buying animal products, in some countries there are certification systems for products from animals raised and treated humanely following certain restrictions. In Canada, for example, there is the 'SPCA Certified Labelling Program' (SPCA Certified), owned and operated by the British Columbia Society for the Prevention of Cruelty to Animals, and in the United States there are already two food labels on the market: 'Certified Humane Raised and Handled' and the 'Free Farmed Program'. Also in Europe there are such certification schemes: In the United Kingdom, for example, there is the 'Freedom Food', and in France the 'Label Rouge', which however also includes other aspects besides farm animal welfare.

In Germany consumers can differentiate between husbandry systems for laying hens when buying eggs, and sometimes meat products are labelled as 'free range'. However, there is no way to differentiate between standards for dairy cows, even though there can be big differences in the housing conditions between farms. The vast majority of people in Germany regularly buy milk or milk products. Of course, not all of them are concerned about how the milk they buy was produced, and especially younger persons usually don't think a lot about the production process. Indeed, at least a proportion of consumers might be concerned, as research in other countries indicates (e.g. Bennett and Blaney, 2002; European Comission, 2005; Gracia *et al.*, 2009; McEachern *et al.*, 2007), and therefore might prefer to purchase especially products with improved husbandry standards. However, there still is little knowledge about the motives to buy products from farms with improved husbandry standards, and not much research has been done on this issue in Germany yet.

Methods and procedure

In order to answer the questions given at the beginning of this article about the demand and motives for the demand, we conducted a consumer survey in a middle-sized City in the centre of Germany dealing with the WTP for milk from dairy cows kept under certified convenient conditions. Defining convenient conditions for farm animals is difficult, because there are a number of facets like, e.g. space, outside access, expressing normal behaviour, good conditions of transport and slaughter (Gracia *et al.*, 2009), freedom from injury, disease and incapacity as well as emotional welfare (Webster, 2001). The most important aspect in the view of consumers seems to be the possibility to behave naturally in a naturally environment for the animals (Webster, 2001). Also the quality of feed is mentioned quite often as major attribute for consumers (Harper and Makatouni, 2007). However, for our research we did not want to consider this aspect because an impact on the quality of the animal product is too obvious. We surveyed preferred husbandry characteristics by a multi-item scale where we asked participants to rate how important different aspects are to them.

Furthermore, it was interesting to know which type of consumer in terms of shopping and consumption behaviour answered the questions. Therefore, we introduced another battle of items about shopping habits, price perception, and preferences for regional products, organic products, quality labels, taste as well as attention to natural ingredients. Since at the time we conducted the survey (December, 2009) the prices for milk were just extraordinarily low, we also asked about an estimation of current milk prices because this was important to consider when interpreting the stated willingness to pay for milk certified improved welfare. As already mentioned above, knowledge about the living conditions of farm animals is rather scarce, so we also included a little test of knowledge about dairy cows.

The most important part of our questionnaire performed a purchase scenario. Here it was necessary to inform people first about differences in husbandry standards before asking them about their WTP for improved standards. For this purpose, we handed over a small table with three different husbandry standards to participants: Conventional, our own created 'improved welfare' and the organic standard. The table showed that 'improved welfare' standards are the same as organic considering the possibility to graze, to get at least 4 m^2 to range in the barn and 6 m^2 to rest per animal. Regarding feed, 'improved welfare' standards were the same as legal minimum standards, which we did not clarify more specific. We left the requirements for feed unchanged to legal minimum standards because it is hardly believable that an improvement in feed quality has no impact on milk quality. We first asked participants whether they usually consume whole milk or half fat milk because there is a price difference of approximately 0.1 €. We then showed three one-litre-packages of milk to the participant, 'normal' milk, which means conventionally produced, 'improved welfare' milk and organic. The prices were given for half-fat milk as 0.49 € for the conventional and 0.99 for the organic and accordingly for the whole milk 0.59 € and 1.09 €. The prices were set on the level of discounters because this is the lower price limit many people are aware of. People were then asked about the maximum price they would pay for one litre of 'improved welfare' milk. After that we informed people that there was no scientific proof about higher quality of milk produced under improved living conditions. We then asked participants again what they would be willing to pay for one litre of the milk considering this information. The point of interest was to observe to what extend people would decrease in their stated willingness to pay because of this information, since the reduction can be interpreted as the WTP for improved product quality only.

At the end of the questionnaire we asked people besides socio-demographics about their relationship to agriculture, i.e. whether they have farming background, live in the countryside or prefer spending holidays on farms, for example. The idea was to analyse whether people who are more familiar with farming got a different point of view on farm animal welfare and maybe also got a different willingness to pay a premium for improved standards.

Survey results

We conducted the survey in November and December 2009 in a medium sized town in the middle of Germany. About 207 people were interviewed with a standardised questionnaire. The structure of the sample is given in Table 1.

Table 1. Survey sample.

	Proportion of cases
Gender	Male: 33%
	Female: 67%
Age	Mean: 39 years
	Median: 34.5 Years
Education	University degree: 34%
	Higher education: 33%
	Other: 33%
Household size	Single household: 33%
	Two persons: 34%
	Three or more persons: 23%
Household with children	11%
Net household income	Less than 750 €: 25%
	750 – 2,250 €: 29%
	More than 2,250 €: 22%
	No answer: 24%

The sample is not completely representative for the shopping population in Germany. The age is lower and the education is higher than it is the case for average consumers in Germany. In our sample we also got on the one hand a rather low rate of workers and unemployed people, while on the other hand the quota of students is about 30%. This can be reasoned by the fact that the survey was conducted in a city where nearly every fourth inhabitant studies or works at the University.

The vast majority (about 60%) of survey participants think that current milk prices are too low, while about 35% are of the opinion that the prices are just well, and only about 5% perceive them as too high. This is an important aspect which needs to be considered when asking people about their WTP for 'improved welfare' milk later in the questionnaire. Most people in our survey are quite conscious consumers considering food since they rather agreed to pay attention to quality labels and natural ingredients on the one hand while on the other hand they rather disagreed that the region of origin is kind of unimportant to and also rather refused that price and taste are most decisive when shopping for food. Such results are not surprising because of the sample structure, but they might not be completely transferable to the average German consumer due to the discrepancy in employment, age and occupation.

Since there is little knowledge about people's preferences for different aspects of living conditions of dairy cows, we asked people to agree or disagree to an item battery about the issue. In Figure 1 you can see the rate of agreement or disagreement to the statements. Not surprising, agreement was very high in our case since people neither did have to pay for, nor to trade off between the features.

As you can see, there is not exceedingly variation in agreement between the items, but the most important point to people seems to be the possibility to graze. This is in accordance with other surveys who found out that the major aspect of animal welfare is the possibility to behave naturally (Gracia *et al.,* 2009; Webster, 2001).

After we informed participants about differences in husbandry standards by the aid of the small table, we showed them three packages of milk, 'conventional', 'improved welfare' and 'organic' while giving the appropriate prices to the two milk packages 'organic' and 'conventional' (0.49 €/0.99 € and 0.59 €/1.09 €). We asked participants about the maximum price they would be willing to pay for one litre

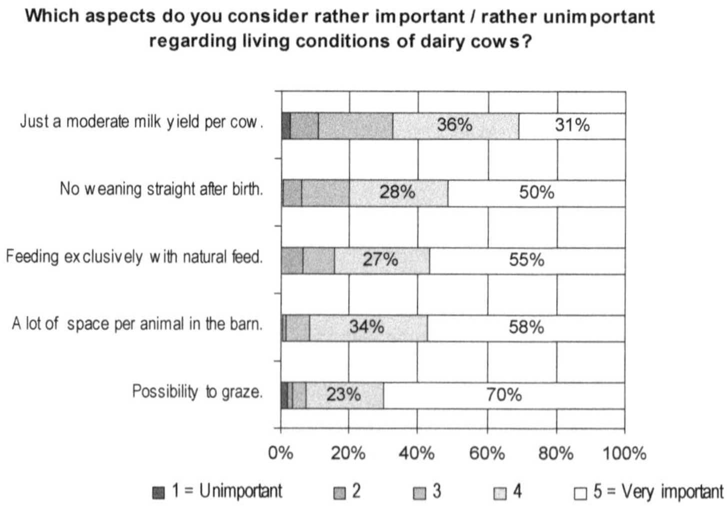

Figure 1. Preferences for different aspects of husbandry standards for dairy cows.

of 'improved welfare' milk. About 95% gave a WTP above the conventional price and in about 83% of the cases the given price was below the organic price. The average top-up to the conventional price was about 0.31 € with a median of about 0.30 €.

After this question the following text was read to participants: *'There is no impact of improved housing standards on product quality. Accordingly, 'improved welfare' milk is neither more healthy, nor in any other aspect of higher product quality than conventionally produced milk. Considering this, for which maximum price would you choose the 'improved welfare' milk out of these alternatives?'* This time the average price premium was about 0.28 €. About 22% of the participants decreased their stated willingness to pay due to the information. Figure 2 shows both, the given WTP before (squares) and after (triangles) the information. The values are reduced at the according conventional price.

As you can see, most participants gave a premium to the conventional price between 0.20 and 0.40 €. We checked the given values for plausibility and decided that any amount above 0.80 € appears rather unrealistic and therefore needs to be excluded from further analyses. The average difference between the first and the second stated WTP is about 0.038, whereas a variation seems to be more likely to occur when people gave a WTP of about 0.30 to 0.40 € above the conventional price in the first time. We tested statistically by t-test, chi-square-test and analysis of variance which factors may take influence on the one hand on the WTP and on the other hand the occurrence of a reduction between the first and the second stated WTP. In the following, the numbers in brackets indicate the level of significance.

The amount of the given WTP in the first question *is significantly higher* when:
- current milk prices are perceived as too low (5%);
- the level of income is high (1%);
- people rate improved husbandry standards as more important in total (1%).
- People are more likely *to decrease in the given WTP after the information* when they
- normally don't consume organic food (5%);
- are grown up in a rural area (5%);

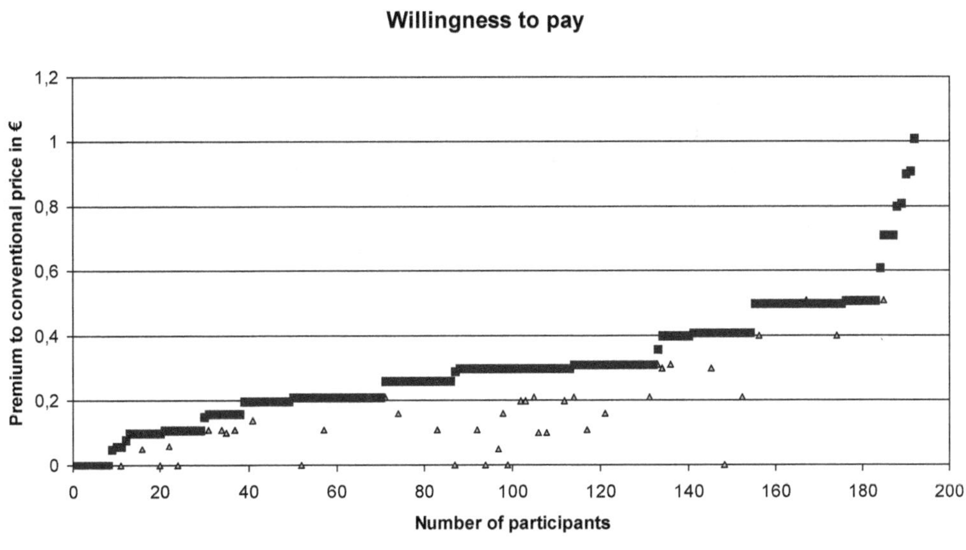

Figure 2. Willingness to pay for milk with 'improved welfare' standards.

- don't have to deal with agriculture professionally (5%);
- are rather young (5%);
- rate the regionality of food as rather less important (10%).

Nearly all of the results are plausible except the observation that people grown up in rural areas are rather likely to decrease their stated WTP after the information. The given WTP appears somewhat high, but regarding the high rate of participants perceiving milk prices as too low, the values are not necessarily unrealistic.

Discussion and conclusions

Our survey provides a first idea of the demand for improved husbandry standards for dairy cows in Germany. However, since on the one hand the sample is not representative for the German population in all parts, and on the other hand the purchase scenario in our survey is only hypothetical, the results need to be interpreted with care, even though the given amounts appear to be reasonable to far extend. Indeed, we can say that there is a broad WTP a premium for improved husbandry standards whereas the expectation of higher product quality is only a minor reason to pay more than for 'normal' milk. The perceived importance of enabling dairy cows to behave naturally by grazing and moving freely is positively connected with the WTP. Since the decrease in the WTP in consequence to the information about unchanged product quality is only small with less than 0.04 € in the average, we conclude that market failure due to maybe unmet expectations might occur, but won't be exceedingly. A bit more unclear remain factors for a decrease in the WTP. Here we only can say that people who are comparatively unconcerned about the food production process (younger people, non-organic consumers) accordingly are rather willing to pay for the product quality than for process qualities. We could not find strong a correlation between preferences for farm animal welfare and the personal relationship to farming, but at the time of writing this article our analysis is still ongoing. Further interesting results are to be expected.

References

Bennett, R. and Blaney, R. (2002). Social consensus, moral intensity and willingness to pay to address a farm animal welfare issue. Journal of Economic Psychology 23: 501-520.

European Comission (2005). Special Eurobarometer 229 Attitudes of consumers towards the welfare of farmed animals. Available at: http://ec.europa.eu/food/animal/welfare/euro_barometer25_en.pdf. Accessed 23 March 2010.

Gracia, A., Loureiro, M.M. and Nayga Jr., R.M. (2009). Valuing Animal Welfare Labels with Experimental Auctions: What do we learn from Consumers? Conference paper of the 27th conference of the International Association of Agricultural Economists. Beijing, August 16-22, 2009. Available at http://purl.umn.edu/51918. Accessed 25 February 2010.

Harper, G.C. and Makatouni, A. (2007). Consumer perception of organic food production and farm animal welfare. British Food Journal 104 No. 3/4/5: 287-299.

McEachern, M.G.; Schroeder, M.J.A.; Willock, J.; Whitelock, J. and Mason, R. (2007). Exploring ethical brand extensions and consumer buying behaviour: The RSPCA and the 'Freedom Food' brand. Journal of Product and Brand Management 16/3: 168-177.

Webster, A.J.F. (2001). Farm Animal Welfare: the Five Freedoms and the Free Market. The Veterinary Journal 161: 229–237.

The role of NGOs in food security

E. Essien
Eurogroup for Animals, 6 rue des Patriotes, 1000 Brussels, Belgium; e.essien@eurogroupforanimals.org

Abstract

Ethical consumerism is a crucial factor in the current food paradigm shift, highlighting the growing power of consumers in restructuring the mechanics of the food chain dynamics towards food security. Consumers along with other key actors in the food chain such as the government, farmers and the retailers whilst progressively united in accepting that altruism in our food system is achievable through pursuits in sustainability, do however, vary tremendously in how the principles and methods of sustainability could be effectively applied to balance all the stakes. All seem to recognise that proper knowledge, education, accountability, transparency and proper engagement of all the stakeholders are key tools towards the achievement of food security and sustainability goals. Transformation of global citizens into food citizens is therefore a dynamic process that requires a broad brush and cross sectoral and partnership approach within food security equation. Participatory tools and deliberative democracy need to be applied in designing and coordinating initiatives targeting the food chain as well as a framework for systematic assessment and monitoring of these activities as a collective. This approach also requires a new form of engagement by Non-Governmental Organisations (NGOs) to reinvigorate their collaborative role with the market and public sector. Although faced with huge opportunities created by the change in governance, the traditional central position of Non-Governmental Organisations (NGOs) as key agents of social intervention to organise this complex is now very crucial and challenging and requires a reassessment especially in relation to their role, nature and capacity to deliver a quality service that meets the needs of the community and food chain requirements.

Keywords: corporate social responsibility, food sustainability, animal welfare, civil society

Food security and sustainable development

There is a growing global realisation that food security issues are interlinked with those of sustainable food and farming; and the most effective measures in tackling them relate with avoidance of centralisation and intensification of food chain operations and greater use of stakeholders' engagement. This new approach is about the reform of food and agriculture system to install civic agriculture, relocalisation and sustainable diets and to establish articulation of responsibility for the future as well as active engagement of local community and rural economy (Alkon, 2008; DeLind, 2002; Defra, 2006; Wilkins, 2005). It also embraces and cuts across issues around carbon foot print, climate change, food miles, organic food, ecological footprint, clean technologies, animal welfare, fair trade and 'triple bottom line' (Kong *et al.*, 2002). A very important strand of this approach articulates societal change in the desire for consumers to prefer products from plant protein sources over those of animals as this particular measure could impact positively on the environment in terms of land use and utilisation of natural resources (Helms, 2004). It also confirms the assertion that 66% of global maize production as well as substantial amounts of oilseed, fish oil and fishmeal which could be used in feeding humans are routed to feeding animals which highlights the impact of animal agriculture on food security (Brown, 2001). Engagement of consumers toward step change aims at progressive transformation of food consumers into 'food citizens' (Wilkins, 2005). Sustainable food system also embeds other values linked with tackling inequalities in health, gender and expression of fundamental human rights (Heasman and Lang, 2006). It is believed that if measures are taken in this direction and food chain actors or stakeholders are assembled to work collaboratively then the balance of environmental, economic and social equity are bound to occur (Alkon, 2008).

Food sustainability as an 'ecological modernization of society' proposes policy and strategic change (Heasman and Lang, 2006). Policy reforms needed to achieve strategic targets in food sustainability relies heavily again on collaborative efforts of government and stakeholders (Defra, 2006). Achievement of sustainable food chain depends not only on a robust cross sector collaboration and dialogue but also on a balance of interests of all the stakeholders and crucially on the employment of triple bottom accounting by businesses whose interest relate with the food chain. NGOs have a crucial role in assembling all the actors in a collaborative endeavour to shape the food chain operation.

The role of NGOs as agents of change

Febrile nature of the time has influenced a shift in the concept of development, politics and governance as well as civic engagement. This implies that the traditional role of the state in providing services to communities faces a huge challenge as the modern configuration of political system and institutions has created barriers that disengage communities and raise public disaffection (Bendel, 2000). On the other hand, the emergence of the power of the market and its dominance of nation-state structure and the new role of providing the traditional services expected of the public sector portrays fundamental loss of control of the state over the market sector. This change in power relations characterised by the failure of the state to regulate the market raises disquiets in areas of corporate behaviour, conduct and legitimacy as well as creates the awareness and heightens the expectation of citizens for corporate accountability (Newel, 2000). Additionally, Murphy (2000) highlights how the impact of globalisation allied to marginalisation of economically disadvantaged has given rise to the diminution of power, trust and integrity placed on governments by people with a resultant increase in the power of citizens and the creation of civil society (Murphy, 2000) also referred to as 'deliberative democracy' advocating 'civic activism' and 'cohesive framework for sustainability' (Zompetti, 2006). Although the activities of NGOs today filter into a huge field of human endeavours, they have a very long origin and emerged from a wide diversity of philosophical histories influenced by culture, religion, politics, market and other numerous imperatives of time (Zompetti, 2006). This global phenomenon has led to a movement often referred to as 'globalisation of civil society' where local communities are engaged in formulating and challenging solutions aimed at tackling issues of social and economic significance (Murphy, 2000) as well as the emergence and prominence of NGOs under such socio-political environment.

NGOs seem to have a long record of success as key agents for social intervention and in engaging the civil society (Pyakuryal, 1989; Nicoll *et al.*, 2002; Church and Lorek, 2007; Gibson, 1993; Skobly, 1996; Wakeman, 2005). Some of the reasons are rooted on their being trusted more than the government and the governance (Gutteling *et al.*, 2006); their strategic placement to outreach people and businesses (Wakeman, 2005); advocacy work for the voiceless, powerless and socially excluded; value addition; cost effectiveness in service delivery and to achieve laudable goals; innovativeness in partnership efforts and past record of success in certain projects (Gibson, 1993; Pyakuryal, 1989; Kong *et al.,* 2002; Benson *et al.,* 2001; HM treasury, 2002). NGOs as a social system is independent and distinct from political and economic systems, empowered by the network that characterises how they operate with strong appeal to community beliefs, tradition and value system (Bendel, 2000).

In spite of their huge role, studies have hinted on the deficiencies in the utility of NGOs (Church and Lorek, 2007); efficiency issues in service delivery (HM Treasury, 2002; Zampetti; 2006) especially in areas of partnership with the government and business; as well as knowledge gap of challenges associated with new emerging issues of sustainable production and consumption. NGOs are very aware that the stake for each stakeholder is wide-ranging and often in contention with each other, which calls for systematic approach in the collective pursuit of food chain sustainability. There is a need to adopt a robust policy framework that harmonises and coordinates the conduct of food chain actors synchronously in a partnership situation to be able to balance environmental, social and economic considerations and accountability (Yakovleva, 2009).

NGOs relationship with consumers in defining the food paradigm shift

NGOs work with wider communities to inform and shape perception, values, attitudes and behaviours on issues of food production, distribution and consumption. Consumers requirements and expectations play a crucial role in food security and sustainability; and NGOs focus on consumers in creating and maintaining participatory democracy presupposes that 'if democracy intends on listening to the voice of the people, then the people need to have a voice' (Zompetti, 2006). While creating a rallying point for members of the community to 'engage each other as well as the affairs of society' (Zompetti, 2006) through campaigns, NGOs provide the information and education aimed at creating awareness, transformation and action. Campaigns to mobilise mass support of citizens are considered by NGOs as a prerequisite to successful efforts in impacting government policies. It has now become an expectation that NGOs need to succeed through robust consumer engagement and multiagency approach with businesses and partners to effectively offer real intervention in the quality of life as well as 'empower' communities to make step change in consumption behavior and lifestyle as a whole (Kong *et al.,* 2002). NGOs work with consumers aims at creating food citizens who show active engagements as actors who are inquisitive, critical, think, act and deny passivity and embrace a range of activities resulting within the system that delivers food (Wilkins, 2005). Food citizenship therefore is 'the practice of engaging in food-related behaviours that support, rather than threaten, the development of a democratic, socially and economically just, and environmentally sustainable food system' (Wilkins, 2005). Based on NGOs believe that food labels would contribute information and shape consumers actions towards animal welfare, Miele and Evans (2010) show how the use of ethical device, food labels, can act as 'subjectifiers' in creating a space of activity and 'competencies' to distinguish 'ethically competent consumer' from 'ethically non competent consumer'. Eurobarometer for instance relies on this sort of consumer space of action and competencies to extract accurate consumer information and voice to inform the state, the market and the third sector. For instance European consumers and citizens expressed their voice in Eurobarometer 2007 (EC, 2007), demanding that animals' welfare is not compromised for production efficiency and commercial benefits; they also place considerable importance to the protection of farmed animals (which is ranked on average 7.8 on a 1-10 scale); 58% of Europeans say they would like to receive more information about farming conditions in their country; the vast majority of respondents (77%) believe there is a need for further improvements to be made in their country; 62% say they would be ready to change their usual place of shopping in order to be able to buy more animal welfare friendly products. NGO's in animal welfare realm work with the market for them to internalise the cost of higher animal welfare and value addition to protect consumer's interest.

NGOs fostering partnership with government in public service delivery

Although independent, the role of NGOs places them central in working with government as strategic partners in designing, and delivering of public service. NGOs are often engaged in intergovernmental debates and consultations and, their contribution in that process serves to shape policies, values, perception and governance. NGOs have enabling position in empowering the marginalized and hard to reach groups, as well as the capacity to adapt services based on their discretion of local needs. Compared to other sectors, NGOs are unique in deploying extra resources such as volunteers and donations. They provide 'added value' through extraordinary skills and experience that they have. They are known for 'the provision of different/specialised services; cost effectiveness of provision, flexibility and innovation; advocacy; and citizen participation'; and their flexible and crossbreed structure enables them to sequestrate welfare barriers imposed upon individuals or groups with complex needs and social exclusion (HM, 2002). NGOs involvement in policy making process aims at expanding public service, articulating and enabling change through regulating the existing activities using mandatory standards. For instance, Eurogroup for Animals as a strong and effective European network of animal protection organisations with members across the EU member states, working together to achieve continuous improvements in animal welfare for all animals; aims to achieve progress through policy work and information campaigns affecting EU and national laws and supply chain standards. Eurogroup has a strong position in coordinating EU and international animal welfare organisations and participates in various advisory committee of the EU institutions to monitor new regulations and industry trends with regards to animal welfare, food safety and consumer policy. Like many other NGOs they recognise that, although mandatory standards have been used as a major baseline tool to shape organisational conduct and drive change in Europe it is not without its limitations (Blandford, 2006; Wade, 2001) as at best only minimum standards are achieved in animal welfare.

NGOs progressive engagement with the corporate sector

Retailer's position as the access controllers of the food supply chain is gaining a wider recognition by NGOs working in the food chain. Europe is a global hotspot for food retail as 60% of top 25 global food retailers (SN Supermarket News, 2009) are headquartered in Europe. As this power of the retailers continually increase through centralisation, their controls and standards become a global code, much wider encompassing, than national regulatory standards. Assigning investments in surveillance of, sometimes very stringent, private standards are one of the priorities of frontline retailers. Competitive front liners in the food chain and retailers are using voluntary standards to articulate their desires to engage with a wide range of sustainability issues. NGOs are modifying the way they lobby to engage as key partners with private sectors in driving sustainability goals and principles to impact policy agenda at international, national and regional levels. Based on these dynamics as well as the recognition of the power of voluntary standards, since 2004, Eurogroup has introduced a ground breaking initiative in *food chain sustainability and corporate social responsibility (CSR)* to work with retailers across all EU member states to support them in introducing and advancing voluntary standards in animal welfare and CSR strategy. Eurogroup's key role on this programme is to offer retailers and their subsidiaries technical consultancy towards introducing and maintaining effective supply chain that cares for animals and respects consumers' interests. The federation also inspires member organisations through collaborative work with retailers, standards and assurance scheme owners, farmers and other food chain bodies to drive change and share innovative and best practice in animal welfare. Prior to this, unique signature programmes that have characterised NGO/Private sector collaboration exist in Marine Stewardship Council (MSC), Forest Stewardship Council (FSC) and Freedom Food; all with underlying philosophy of transparency and accountability with sustainable production and consumption objectives. There are myriads of other very specific NGOs robust collaborative initiatives with private sector which underscore the scope for greater efficiency and need for strategic framework that map out a systematic approach in this direction.

Challenges for NGOs and recommendations

While this paper acknowledges successful NGO activities and points to its potential, it also recognises the barriers that they face in contributing sustainability benefits to the food chain. NGOs rooting in traditional community values and belief system means that a strong network configuration would be its hallmark; and when this is combined with attributes of globalisation that empower NGOs such as advances in technology, transport and communication leading to the shrinking of time and space they allow local issues, incidents and 'flashpoint' to be so rapidly brought to the centre of global awareness (Newell, 2000). This situation with inherent benefits could actually be perceived negatively by certain corporations, whereas strong partnership with all the sectors can benefit from using this phenomenon in tackling global food security issues. NGOs and sustainability being highly contested notions have created gaps in the understanding of each of the concepts as well as reaching a consensus on how the two are merged within the pursuits of food chain altruism. Although there is an agreement on the benefits of sustainable food system, there is still an acute lack of harmony on how its principles and methods could be effectively applied.

The triad hybrid relationship brought about by state and market partnership with NGOs may be better explored using a systematic framework to remove barriers and constraints of partnership in tackling new emerging issues; building structural capacity and reducing the skills gap existing within the sectors in understanding each other and how they could effectively work as a collective. All the sectors need to reassess the density of resource need in delivering public service; invent a measure to balance these needs; devise control of the resources as well as weigh the efficiency of use of them in competitive advantage. This might highlight the cruciality of building the third sector capacity and a more transparent accountability system to boost their delivery of quality service that meets the needs of the community and food chain requirements. As a relatively new concept, all sectors must own up to the realism that none has all the answers and non can successfully deliver food security and sustainability goals on their own. Participatory tools and deliberative democracy need to be applied in designing and coordinating initiatives targeting the food chain as well as the systematic assessment and monitoring of these activities as a collective.

References

Alkon, A.H. (2008). From value to value: sustainable consumption at farmers markets, Agric Hum Values 25:487-498.

Bendell, J. (2000). Working with stakeholder pressure for sustainable development. In: Bendell, J (ed.). In Terms for Endearment, Business, NGOs and Sustainable Development. Greenleaf Publishing, UK.

Benson, C., Twigg, J. and Myers, M. (2001). NGO Initiatives in Risk Reduction: An Overview, Disaster 25:3, 199-215.

Blandford, D. (2006). Animal welfare, Choices, the magazine of food, farming and resources 21:3, 195-198.

Brown, L. (2001). Eradicating Hunger: A Growing Challenge; State of the World 2001, The WorldWatch Institute; pp. 43-62.

Church, C. and Lorek, S. (2007). Linking policy and practice in sustainable production and consumption: an assessment of the role of NGOs, International Journal of Innovation and Sustainable Development 2: 2, 230-240.

Defra. (2006). Food Industry Sustainability Strategy. Available at: http://www.defra.gov.uk.

DeLind, L.B. (2002). Place, work and civic agriculture: Common fields for cultivation. Agriculture and Human values 19: 217-224.

European Commission. (2007). Special EUROBAROMETER 229 (2) 'Attitudes of consumers towards the welfare of farmed animals' Report.

Gibson, A. (1993). NGOs and income-generation projects: lessons from the Joint. Funding Scheme, Development in Practice, Vol. 3 No. 3, 184-195.

Elkington, J. (2004). Enter the Triple Bottom Line In: Henriques, A. and Richardson, J. (eds.). Triple Bottom Line: Does it all add up? Earthscan Ltd.

Gutteling, J., Hanssen, L., van der Veer, N. and Seydel, E. (2006). Trust in governance and the acceptance of genetically modified food in the Netherland 15:1,103-112.

Heasman, M. and Lang, T. (2006). Plotting the future of food – Putting ecologically driven, community-based policy at the heart of Canada's food economy, Making Waves 17:2, 12-17.

Helms, M. (2004). Food Sustainability, food security and the environment, British Food Journal 106: 5, 380-387.

HM Treasury (2002). The role of the voluntary and community sector in service delivery, A cross cutting review.

Kong, N., Salzmann, O., Steger, U. and Ionescu-Somers, A. (2002). Moving Business/ Industry Towards Sustainable Consumption: The Role of NGOs, European Management Journal 20: 2,109-127.

Miele, M., and Evans, A. (2010). When foods become animals, Ruminations on ethics and responsibility in care-full practices of consumption (in press).

Murphy, B.K. (2000). International NGOs and the challenge of modernity, Development in Practice, 10: 3/4, 330-347.

Newel, P. (2000). Globalisation and the new politics of sustainable development: In: Bendell, J. (ed.). In Terms for Endearment, Business, NGOs and Sustainable Development. Greenleaf Publishing, UK.

Nicoll, A., Thayaparan, B., Newell, M. and Rundall, P. (2002). Breast Feeding Policy, Promotion and practice in Europe. Results of a survey with Non-governmental organisations, Journal of Nutritional & Environmental Medicine 12: 3, 255-264.

Pyakuryal, B. (1989). Non-governmental organisations: their roles in family planning program in Nepal, Econ J Nepal 12: 4, 10-30.

Skobly, S.I. (1996). The role of international NGOs in Promoting nutrition rights, Food Policy 21:1,111-121.

SN Supermarket News. (2009). Available at: http://subscribers.supermarketnews.com/mag/sns_top_worldwide_3/index.htm.

Wade, J.A. (2001). Stakeholders, Ethics and Social Responsibility in food supply Chain. In: Estham, J F. Liz Sharpless, L. and Ball, S D. (eds.). Food supply chain management issues for the hospitality and retail sector.

Wakeman, T. (2005). East Anglia Food Link: an NGO working on sustainable food, Grassroots Initiatives for Sustainable Development 10 June 2005. Available at: http://www.eafl.org.uk.

Wilkins, J.F. (2005). Eating right here: Moving from consumer to food citizen, Agriculture and Human Values 22: 269-273.

Yakovleva, N. (2009). Sustainable Benchmarking of Food Supply Chains, GPMI Working Papers No. 2009-02, April 2009.

Zompetti, J. (2006). The role of advocacy in civil society, Augumentation 20:167-183.

Governance addressing societal values

V. Almaas and M. Kaiser
University of Bergen, Centre for the Study of the Sciences and the Humanities, Allegt. 34, 5020 Bergen, Norway; vibeke.almaas@svt.uib.no

Abstract

The question of good governance has become a central issue for European food policies, as it is for questions of science and technology in general. Good governance is assumedly relating to central values, among them ethical values, that are common in society and that are believed to influence consumer attitudes and choices. Addressing these values more directly has become a concern of politics, in the food area as well as in science and technology in general. After all, values are believed to be central to questions of social identity. Philosophers, sociologists and social psychologists have pointed out that our self-conceptions are strongly influenced by what we believe to be values worth striving for. Politicians started to worry about the 'European identity' for real especially after the rejection of the European Constitutional Treaty by two member states in 2005. They demanded more focus on what the coined the 'European values'. The EU project 'Value Isobars' attempts to probe some of the implicit assumptions behind this assumption. The following intuitions are participants of this project (in alphabetic order): Centro de Estudos sobre a Mudança Socioeconómica, DINAMIA (Portugal), dialog<>gentechnik (Austria), Eberhard Karls Universität Tübingen (Germany), London School of Economics and Political Science (UK), Universiteit Maastricht (Netherlands) and University of Bergen (Norway; coordinator). The landscape of European values is assumedly complex. It is proposed to introduce the concept of ethical value 'isobars', a term borrowed from meteorology. Isobars in their common weather forecast meaning are line drawn between points of common atmospheric pressures. How much do we really know about the nature of values? How could we actually chart existing values in different societies and compare them? And how could we get into a constructive dialogue with societal groups about their values and be part of normative dynamics? And what are the legal or other instruments available to policy makers? Is soft law a good approach? These are the main questions that the project Value Isobars attempts to answer. This project is based on a specific conception of what it means to discuss values in relation to science and technology. In our presentation we attempt to point out the specific relevance for governance in the food sector, since it is obvious to us that this area is particularly prone to be influenced by value-based attitudes. Food security and the question of equity is one such example, but also questions of animal welfare and sustainability. Many have realized that any merely top-down policies will not work here. Engaging the public and addressing their values is supposedly crucial for robust long-term policies. However, for this to work in actual practice – so we argue – we need better insights into the nature, characteristics and dynamics of societal values.

Keywords: governance, values, food, consumers, attitudes, surveys

Introduction

During the last years food industries have experienced a number of crises. Outbreaks of bovine spongiform encephalopathy (BSE) or mad-cow disease in red meat markets, avian flu in poultry, foot and mouth disease in cattle and sheep in addition to *Salmonella* in eggs, and dioxins and other pollutants in some of the largest food and drink brand names have all generated distrust and reduced confidence in food safety. The crises have resulted in a focus on food preferences by the consumers: health benefits, animal welfare and environmental aspects (Kjærnes, 2010; Lien and Nerlich, 2004). Further, more consumers pay attention to their own consumption of food and are aware what they are buying. Important factors and attributes for the consumers' habits can be: fair trade, social responsibility within and outside the food sector, food safety, risk assessment, GM food, organic food, ecological food, local food, food miles, animal welfare and environmental impacts such as protection of overexploited fish stocks for instance.

There appears to be a more conscious movement among the consumers which reflects in general a more active and critical consumer influencing markets in the form of political consumerism. Extended effects of the crises have emerged as an issue on the public, partly also of the political agenda in many European countries. For instance, in recent years the market for organic food has changed considerably. This underlines that the role of citizens as consumers of food has changed significantly; from being protected by a welfare state, people appear to be accorded more freedom and responsibility as individuals. The notion of 'consumer choice' redirects attention from the state towards market relations (Kjærnes, 2010). It seems that consumerism has become more political. Studies of political and ethical consumerism indicate that in many cases there is a link between individualised 'mobilisation' in the market in the form of political or ethical consumption, collective mobilisation and public attention (Micheletti *et al.*, 2006).

Markets have responded to these concerns through an expanding array of certification and labelling schemes operated on a transnational basis, but often with different standards and sometimes conflicting interests. The political responses and frameworks have become increasingly transnational. In light of the previously mentioned crises this has resulted in certifications, ethical guidelines, stronger labelling, but also the establishment of consumer organisations and governmental oversight bodies, as for instance EFSA (The European Food Safety Authority (EFSA) is the keystone of European Union (EU) risk assessment regarding food and feed safety, http://www.efsa.europa.eu/). These events have combined to create a renewed interest in provenance of food products with transparent and verifiable systems of food traceability along the chain. Consumers' reflections and demands underline that governance is value-based and therefore addressing consumers' values more directly seems essential for a more effective food dialogue. Schemes of citizens' involvement and consultation are seen as promising forms of governance.

Terms and definitions

What does governance and participation actually mean? We see governance as not merely a structure of leadership or a system of government, but also as a way to lead long-term processes which are harmonious with the stakeholders, i.e. among others the involved industries and the consumers. Participation allows the public to voice their opinion and ideally contribute to a decision making process. While from the administrative point of view participation might be seen as a process to build public support for specific projects, citizens might be motivated by the perspective to influence decisions. Participation is seen as a way to clarify different, often opposite, views and interests regarding a specific problem, and a way to improve the quality of the decision making by ensuring relevant information and fair and democratic processes (Bulkeley and Mol, 2003).

As the project Value Isobars is still running, it is not our intention in this paper to provide a final definition of how to conceive of a value. Indeed the improved clarification of this concept is one of our main goals. However, the notion of 'value' can broadly be characterized as 'an ambiguous concept that governs human behaviour. The concept of values has its origins in the economic sphere and a set of values may be placed into the notion of a value system. Values are considered subjective and vary across people and cultures. Types of values include ethical/moral values, doctrinal/ideological (political, religious) values, social values, and aesthetic values (www.wikipedia.org, accessed 14 March 2010). In this project, however, we use 'social values' more as an umbrella term for all of the above mentioned types, acknowledging the heterogeneous sources and contexts of what values are conceived of. At the same time, for a systematic analysis, we shall address this problem in more detail in our project (cf. Description of work, www.value-isobars.no). Even if the term value is a bit vague we believe values provide common identity and visions, they stimulate positive or negative attitudes, and their conflict is experienced as a major obstacle to action.

European values are also an important term in this project, which refers to common cultural traditions and common historical and social reference points. The focus at values is very clear and visible in EU in a general way. Such values may indeed constitute a European identity and their function is important for the appreciation of our selves. In 2009, the President of the European Commission Manuel Jose Barroso stated that the Commission:

> '...has faced some of the most challenging times the European Union has ever experienced, at a time of huge economic and social change. Our record shows that we have held fast to the core values and core goals that have made the Union so successful, and will leave a powerful contribution to the future development of the European project. I am particularly proud of our response to the economic and financial crisis, our ambitious agenda to fight climate change and the creation of a real European energy policy. The European Union is now taking the lead in shaping globalisation with European values and in promoting the European interest worldwide' (European Commission, 2009a).

A report by the Science Policy Research Group (2004) concluded that social values can be seen as central drivers for peoples' attitudes to science and technology, but that one is still largely unable to understand and penetrate basic concepts to acquire good and reliable data on European social values, to perform an intelligent reading of existing data, and to adjust policy adequately to empirical finding. A mapping of social values is needed if the various platforms of debate on science and technology in Europe are to lead to sustainable innovation. Values are one of the key and unexplored parameters responsible for changing attitudes on science and technology.

The basic questions of Value Isobars

The overall aim of the project is a more flexible and value informed science and technology governance. The basic questions are:

1. What are values? What is a value, and what is a social value – these are still vague and undefined concepts in spite of the many uses to which they are put. However, specific for our project is the focus on values with regard to science, technology and its policies. There is little doubt that technology developments and innovations must be compatible with values of those concerned in order to be accepted. Recognizing the problems in the introduction, many people realise the values of the different stakeholders. Values need to be more addressed move explicitly because values may reveal underlying causes for attitudes. That means we need to pay more attention to consumers' values. Since we see values as important drivers that create identity and social cohesion, it is necessary that we pay attention to them.
2. How to measure and detect values? How do we discover important social values and how do we detect and measure them? Social values cannot be ignored in the process of technological innovation, in particular not in questions of food. Science and technology make possible a number of different opportunities and social futures; those that are consistent with peoples' values are likely to be supported, while those that clash with peoples' values are likely to create controversy and resistance. However, in actual practice we see that surveys (e.g. Eurobarometers) that pretend to map citizen values seldom provide useful or reliable information. It is astonishing that a whole scholarly community working with such surveys has paid so little attention to the question how we best can measure and detect values.
3. How to have dialogues about value issues? There are several ways to see how values are addressed and how these may change. There are a variety of methods for assessing technology, ranging from participatory methods such as citizens' panels, consensus conferences, scenario workshops and open hearings, to interdisciplinary working groups on the expert level. Common for all these methods are

participation from citizen, consumers and stakeholders. The 'voice of the people' has always been important in the political decision-making process in a democratic society. Freedom of speech is one of the basic elements of a democratic society and laypeople or non-experts are important components of the collective representations. Expert knowledge is often strictly limited to one specific subject and ordinary citizens have a different approach to technology because they tend to see it from the perspective of their own life: 'how could this possibly affect my work situation, my health and the life of my family?'(Andersen and Jæger, 1999).

4. Tools related to value issues. What legal instruments relate to value issues; regulation, law, institutional efforts like ethics committees, guidelines or what? How can EU take values into consideration while dealing with food policy or are this really left to the market? Instruments are needed to allow policy makers to map social values, to detect significant changes in them and to relate these to current and emerging attitudes to science and technology. Legal, regulatory and other mechanisms, including soft-law (like guidelines), are to reviewed and critically discussed in regard to their potential to stimulate value-based and value-informed flexible governance and to respond to existing value isobars in European publics. For instance, the debates over GM food have among others brought about a wealth of laws and regulations, both on a national and a supra- national level. Basically these regulations aim at dealing with the risks of the products concerning public health and the environment. To some extent, some regulations also cover the processes involved in a product, like labelling of GM food products within the EU (Kaiser and Lien, 2006). Soft law refers to quasi-legal instruments that do not have the binding force of laws. Typically one finds soft law in the international arena, such as in various declarations in international bodies, like the UN. For instance, the EU communication on the 'Precaution principle' can be seen as one such example of soft-law (Kaiser, 2003).

Food sector

Food raises major issues of the distribution of welfare and risks of social concerns, of economic power, justice, cultural values, animal welfare and environmental issues. Buying and eating food are the objects of regulation. Recently people in their capacity as 'consumer' have become active agents in this politics of food consumption. In that way, one might say that 'consumer choice' has become a mode of regulation. The shifts are characterised not only by specific types of state engagement, but of characteristic relationships between states, markets and households. There are indications that ethical consumerism is a growing trend and moral responsibility is a relevant buying motivation among various consumers groups (Carrigan *et al.*, 2004). Consumers, particularly in industrialised countries, are concerned about the application and genetic engineering on food production, and for instance GM food is a growing concern for consumers. Politics of food is no longer confined to policy-making within the national state, but more closely connected to innovations and discourses that take place on the transnational arenas of science and technology (Lien and Nerlich, 2004).

Survey

To identify peoples' attitudes, trends and opinions an important tool is the Euro barometer. Eurobarometers have usually been a tool to cover and compare peoples' attitudes and opinions on different topics in Europe and cover a series of surveys regularly performed on behalf of the European Commission since 1973. It produces reports of public opinion on certain issues relating to the European Union across the member states. The Eurobarometer is a way to track and analyze public opinion in all European Member States and to improve the information and communication policy of European decision-makers (www.wikipedia.org, accessed 14 March 2010).

However, attitudes thus captured are a spontaneous reaction and a snapshot which gives an expression of the present situation, while we see values as a more stable, fundamental, and reflective factor. Attitudes can quickly change and are more receptive for inputs from external circumstances. Therefore, attitudes can be influenced and changed by information, context or, as in a survey; by the way the questions are asked or formulated and the context in which they appear. This mean that surveys which deliver results based on attitudes are usually not very useful in regard to long-term perspectives. We see values as more basic and therefore not so easily subjected to change compared to attitudes, trends and opinions. People do not change values as often as they change attitudes. In this perspective; a survey depicting peoples' values would make easier to indicate a long-term perspective and provide more information to predict consumer behaviour and inform long term governance and policy. In the spite of the broad attention to social values we still do not have the conceptual and methodological means to understand what these values actually are or how strongly they are embedded within social identities. Values are recognized as drivers of important attitudes that one need to address in European policy and that should therefore be integrated into a deliberative democratic culture.

Improve the mechanisms

What we receive in surveys is a wealth of data, i.e. information that is largely a collection of individual measurement points relating to attitudes. Only when surveys are carried out in regular intervals one can hope to receive some information about trends and developments. Yet even those are beset with large difficulties and uncertainties, especially in regard to values as drivers of attitudes. Sustainable innovation is value-based innovation (Eurobarometer, 2005) and technological progress can never be free of ethical implications. So how do we discover and identify consumers' values? Our principal answer in the project is that we need to mimic the theoretical efforts that are well-known from other parts of science, i.e. we need to start out with clearly defined concepts (e.g. values), need to construct out of empirical data and conceptual analysis our basic phenomena (e.g. value change or value-driven attitudes), and then within a theoretical framework we need to identify potential causal pathways. This is what turned meteorology into a successful and predictive enterprise. Vilhelm Bjerknes was the main figure to encourage the move from isolated data collections to synchronise and systematic large scale observations. Isobars could be drawn on weather maps and thus allowed the reading of movements and trends. Isobars in their common weather forecast meaning are lines drawn between points of common atmospheric pressures. It is this effort, metaphorically speaking, which we should aim at for value-based governance.

Conclusions

The project Value Isobars aims at clarifying the underlying issues and theoretical frameworks that may lead to the early identification of ethical issues and a greater effectiveness of European science and technology policy.

Instruments are needed that allow policy makers to map social values, to detect significant changes in them, and to relate these to current and emerging attitudes to science and technology, and in particular in the food sector. We need more than merely 'numbers' or attitudes; we need reliable information about consumers' values. These need then to be transformed into a map with causal dynamics, which is what good governance would need. Currently we have the 'numbers', but we need to transform those into corresponding isobars.

Acknowledgements

We acknowledge the funding of the European Commission for this work (contract number 230557). We furthermore acknowledge the provision of important discussion points from our partners, in particular (alphabetically) from: Ines Gameiro, Maria Eduarda Goncalves, Stefan Grünert, Brigitte Gschmeidler, Nicole Kronberger, Laurens Landeweerd, Simon Meisch, Thomas Potthast, Stavroula Tsirogianni, David Townend. However, the full responsibility for the content, style, errors and inaccuracies is ours, the authors, of this paper.

References

Andersen, I-E. and Jæger, B. (1999). Danish participatory models. Science and Public Policy. October.

Bulkeley, H. and Mol, A. (2003). Participation and environmental governance: consensus, ambivalence and debate. Environmental Values 12. White Horse Press.

Carrigan, M., Marinova, S. and Szmigin, I. (2004). Ethics and international marketing: research background and challenges. International Marketing Review, 22(5), 481-493.

Eurobarometer (2005). Social values, Science and technology. June 2005 Brussels. Available at http://ec.europa/public opinion/archives/ebs/ebs 225 report en.pdf.

European Commission (2009). Delivering real benefits for citizens across Europe: Commission achievements 2004-2009, 28th May 2009. Online available: http://ec.europa.eu/commission_barroso/president/priorities/bilan/index_en.htm.

Henseleit, M. and Kubitzki, S. (2008). GMO-Free' Labels – Enhancing Transparency or Deceiving Consumers? Institute of Agricultural Policy and Market Research, Justus-Liebig University Giessen, Germany.

Kaiser, M. and Lien, M.E. (eds.) (2006): Ethics and the Politics of Food. Wageningen Academic Publishers, Wageningen, the Netherlands.

Keeney, R.L. and Raiffa, H. (1976). Decisions with Multiple Objectives: Preferences and Value Tradeoffs. New York: John Wiley & Sons.

Kjærnes, U. (2010). Sustainable Food Consumption. Some contemporary European issues. National Institute for consumer research, Norway.

Lien, M. and Nerlich, B. (2004). The politics of food. Oxford, Berg Publishers.

Micheletti, M., Follesdal, A. and Stolle, D. (2006). Politics, Products, and Markets, Exploring Political Consumerism Past and Present. Transaction Publishers.

Reducing meat consumption: ethical arguments in the context of today's consumer society

E. de Bakker and H. Dagevos
LEI, Wageningen UR (Agricultural Economics Research Institute), P.O. Box 29703, 2502 LS The Hague, the Netherlands; erik.debakker@wur.nl

Abstract

If Western diets containing a lot of animal proteins spread over the globe during the next forty years, the consequences will be catastrophic, particularly for poorer countries. The ethical and political discussions about (reducing) meat consumption are characterised by a paradox: contrary to the importance of moral responsibility is the widespread assumption that modern consumers are not equipped to take such responsibility. This results in a pessimistic view on the possibilities for change on the demand side that will increase sustainable consumption patterns. A broader perspective can be opened by assuming that no clear boundaries exist between 'citizens' and 'consumers' as can be learned from consumption sociological studies. This could encourage practical policy strategies on the demand side for reducing our meat consumption that escapes the rigid contrast of responsibility versus irresponsibility. We conclude with three routes of transition as an example of how modern consumers can be approached as allies instead of opponents. In fact, these routes embody different ambitions and expectations about consumers, taking into account the surrounding socio-economic, political and cultural environment. The perspective we unfold seems to be aligned with pragmatic ethics in which virtues are seen in a wider context.

Keywords: sustainability, transition, alliances, deontology, utilitarianism, virtues

Introduction

> 'Concede your main point about Western cultural arrogance, I do think it is appropriate that those who pioneered the industrialization of animal lives and the commodification of animal flesh should be at the forefront of trying to atone for it' (Coetzee, 2003: 107).

When Elizabeth Costello, the creation of the famous novelist J.M. Coetzee, is caught up in a debate at an American college about how we treat animals in modern societies, she admits that the western countries which laid the foundations of an industrial system that excessively exploits the lives of animals are also primarily responsible for redeeming the consequences of this industrialisation. From her point of view, this means that we should stop eating animals and assume a vegetarian diet. Provocatively comparing modern slaughterhouses with the Holocaust of the Jews during the Second World War, she states that we should not close our eyes to the way we instrumentally use and mistreat other living beings. Her fictional character didn't prevent Costello's moral arguments becoming a touchstone for more academic reflection on our moral responsibilities towards animals (Cavell *et al.*, 2008). However, one could also enter the debate on eating animals from an environmental perspective. In recent years, it has become increasingly clear that the increasing consumption of meat and dairy products will create huge environmental problems (greenhouse gas emissions, land degradation and water shortage) (Audsley *et al.*, 2009; FAO, 2010; GEO 4, 2007; UNEP, 2007; Keyzer *et al.*, 2005; Pimental and Pimental, 2003, 2008; Steinfeld *et al.*, 2006). If Western diets containing a lot of animal proteins spread over the globe during the next forty years, the consequences will be catastrophic, particularly for poorer countries. Reducing our meat consumption in the near future seems inevitable if we want to take a responsible approach. In this context, we want discuss the ethical arguments concerning our meat consumption and moral responsibilities and how these arguments relate to the reality of today's consumer society. We will try to demonstrate the following propositions:

- There is a paradox between ethical reasons that refer to the importance of moral responsibility and the widespread assumption that modern consumers, apart from a very small minority, are not inclined to become ethical consumers.
- This paradox can be solved by assuming that no clear boundaries exist between 'citizens' and 'consumers' and by practical policy visions that are more explicit about the alliances they wish to enter into with consumers.

Different ethical arguments

According to the fictitious novel character Elizabeth Costello, we should ban meat from our plate. Although Costello seriously doubts the role of reason with respect to moral behaviour, it seems nevertheless appropriate to render her moral position as deolontogical ethics: we do not have the right to treat animals as factory products and that is why we should stop eating meat produced by such systems. A more moderate and less rigid position in the same line of reasoning, albeit with a very different dietary outcome, is that we should only buy meat from production chains applying high standards of animal welfare. Although this last view will inevitably question whether the chosen standards are high enough to guarantee 'just' killing of animals for human needs, the common factor to both ethical positions is that their main focus is our relationship with (non-human) animals.

However, reducing our meat consumption could also be related to our responsibilities with respect to other human beings or our natural environment. As already indicated in the introduction, due to its environmental impact our western 'meat diet' has become subject of discussion. The ethical positions that take into account the environmental impact of our livestock and meat and dairy consumption on a global level often combine both deontological and utilitarian reasons that revolve around the concept of distributive justice. Firstly, it is argued that we are obliged to strive for fair distribution and that we should show solidarity with populations or regions that are in need because their subsistence is threatened. Secondly, we should strive towards a situation in which an optimal degree of happiness and welfare for everyone is achieved, and this also means that we should find sustainable solutions to avoid the deterioration of our natural environment. However, the sense of urgency to take action to undo the unjust distribution of joys and burdens need not be based on ethical considerations. This can also be based on the idea of 'enlightened self-interest', resulting from the belief that we will finally go down altogether if we do not find sustainable solutions.

Responsible citizen, irresponsible consumer

Only a small reduction of meat consumption could result in a substantial environmental impact (Stehfest *et al.*, 2008). If 'rich' consumers were to eat 10 kg/capita plant proteins per year to replace their animal proteins, the reduction in global emissions would be 4% for NH_3, 0.2% for CH_4 and 3.7% for N_2O (Zhu *et al.*, 2006). Not eating meat and dairy on one day a week would probably exceed the positive environmental effects that could be achieved by a similar reduction of our motorised transport. The bottom line is that relatively small changes in our lifestyle could make a major difference in favour of a more sustainable environment. Taking responsibility for the environmental and social consequences of our meat consumption could start with fairly modest changes. However, where sustainable policies are recommended, these recommendations focus mainly on the supply side. The keywords here are resource-use efficiency, technological innovations, environmental policy and production structure changes (Steinfeld *et al.*, 2006: 276; GEO 4, 2007; Zu *et al.*, 2007). One might find a general suggestion to support the development of ecological labels, but generally consumers are not seen as an important factor. Brief and to the point, in many publications discussing the consequences of our meat consumption, the modern consumer appears as an obstacle to sustainability.

Individual obstacles: Homo economicus

The reasons for having little or no trust in the consumer as an actor who will adopt a more sustainable lifestyle can be framed more individually or institutionally. Taking a more individual perspective, the line of reasoning often hangs on a yawning gap between citizens and consumers. As a citizen, we might want to have a sustainable environment and society but as a consumer, these wishes and intentions are quickly forgotten and we buy the products that are cheapest without caring much about our environment or other issues. The consumer is primarily seen as a *Homo economicus* driven by self-interest who individually weighs the costs and benefits of the products offered to him. And in a situation when a consumer might be willing to consider a sustainable purchase, he/she gets caught in the prisoner's dilemma – the problem that there will always be 'free riders' who will not change their behaviour and that it does not seem rational to pay more while other consumers who are not prepared do the same will benefit from this altruistic behaviour. Only a central government has the power to break out of this dilemma (this view of a social dilemma is described in reports by the Netherlands Environmental Assessment Agency).

Institutional obstacles: the passive and powerless consumer

Besides this perspective based on the assumption of individual rationalism and the importance of self-interest, it is also possible to point at institutional conditions that block moral behaviour. These could be described as structural and societal features that work as obstacles to sustainable food choices that would serve the common good. The economic interests of big food companies, marketing strategies of large retailers and dominant cultural norms can all be regarded as examples of institutional conditions that prevent consumers becoming more aware of their eating habits and encourage them to change their lifestyle. 'Ethical consumers' who feel responsible about the consequences of their purchase behaviour are seen as a marginal group that will never become mainstream and that will always be confined to a small minority of often well-educated people with sufficient means. But in most cases, comes the reply, people are strongly influenced by institutional conditions that shape their environment of consumption. The market situation is not a place for moral considerations and one can only hope that the domain of politics and public opinion has the preconditions for freely discussing societal issues and reaching fair decisions and policies (this view is, more or less, represented by Dutch NGOs dealing with the environmental and global effects of meat consumption).

Although both lines of reasoning differ substantially, the outcome is more or less the same with regard to ethical arguments resulting in a certain moral responsibility that one has as a human being, whether this individual responsibility is based on deontological or utilitarian reasons or even on enlightened self-interest. The crucial point is that such responsibility can only be applied to moral citizens but not to (mostly) 'non-ethical' consumers. And when it comes down to practical strategies for sustainable development, it does not really matter whether the 'lack of consumer morals' is caused by individual rationality or by institutional conditions. The bottom line is that one cannot expect consumers to be responsible actors who really get things moving. Certainly, the most pessimistic view of western consumers arises when the image of a *Homo economicus* is combined with the belief in structural features that obstruct further considerations about food production methods. In this case, one sees on the one hand an active consumer insofar as he/she rationally pursues his economic interests, and on the other hand a passive and rather powerless consumer insofar as he/she is influenced by institutional circumstances that hinder moral considerations.

In search of the virtuous consumer

The policy debate about responsible solutions for sustainable livestock production and the growing demand for meat and dairy is to a certain extent characterised by the idea that there is an insolvable contrast between responsible citizens and irresponsible consumers. This debate mainly concentrates on improving efficiency or changing supply side structures while the consumption patterns are given little attention. Without underestimating the importance of supply side policy and more efficiency, we believe that the demand side deserves more attention in the context of sustainable consumption. We will argue that the contrast between consumers and citizens rests on one-sided assumptions and that there is a more subtle approach to this matter. For a start, we can agree with Jessica Nihlén Fahlquist (2009) who does not frame the problem on the basis of a contrast that seems far too rigid. Inspired by a practical Aristotelian approach, she argues that the responsibility of individuals is always interdependent on an 'enabling environment' and cannot be judged by looking at consumption patterns as a purely individual activity. In this context, she points to the responsibility of governments and corporations to 'create systems to make it easier for individuals to respond to the emerging norm that we ought to act in environmentally friendly ways' (Nihlén Fahlquist, 2009: 120). But we want to take this approach one step further and connect it to more recent insights found in consumption sociology, in particular the concept of modern consumer profiles that distance themselves from the distinction between citizen and consumer and which focus on the social and ethical motivations that people take into account in their decisions and choices. (Harrison *et al.*, 2005; Jackson, 2006; Micheletti, 2003; Shah *et al.*, 2007). A similar orientation directly links consumer wishes to citizen virtues like solidarity, justice, consideration, etc. The archetypal boundary between the definitions of 'citizen' and 'consumer' becomes blurred, although it does not disappear, if it is deemed possible that justice, sustainability or animal welfare count in people's considerations in their capacity as consumer. In this respect, food consumers can be called food citizens.

When the distinction between citizen and consumer becomes the subject of debate, it becomes possible to touch on another artificiality: the assumed primate of self-interest as a core assumption of the rational choice model. Consumer behaviour is not then excluded from ethical or emotional influences or pro-social or long(er) term action perspectives. Citizen values penetrate, as it were, into consumer wishes and needs. High ideals blend into individual advantage; a pragmatic disposition is combined with principles. Following this line of argument, we should search for the 'virtuous consumer' who is driven by various different individual motives related to a surrounding socio-economic, political and cultural environment.

Alliances with consumers: three routes of transition

The direct environment in which consumers make their purchases should become a civic domain in which private and public interests are inextricably intertwined. Taking a broader approach where no clear boundaries exist between 'citizens' and 'consumers', it also becomes possible to develop practical policy visions based on a wider variety of alliances with consumers. This practical perspective seems more aligned with Aristotelian ethics in which virtues are seen in a wider context. We want to conclude with three routes of transition change for reducing meat consumption: 'passive sustainability', 'active change', and 'cultural revolution'. In fact, these assumptions embody different ambitions and expectations about the alliances that one wishes to enter into with consumers.

Passive sustainability

In this route, food consumers are mainly defined as passive end users. People are fairly passive when it comes to their food. They do not generally concern themselves with difficult questions about food and feel no real need to do so. They will accept sustainable food innovations, particularly if they are not very noticeable. By incorporating plant semi-manufactures in sausages, hamburgers or mince products, a hybrid product assortment is created whereby eating sustainable products gradually becomes more accessible. The alliance entered into with consumers here is characterised by passive partners on the consumer side.

Active change

This change route assumes that consumers as active and engaged citizens are part of a civil society, also where food consumption is concerned. In this route, the idea is that social debate and taking small practical steps that appear to be very feasible, will make our consumption more sustainable. Practical action perspectives can be found in moderating meat consumption through smaller portions or regularly incorporating a meat-free day. In relation to the government, the consumer is a discussion partner with whom a moderate dialogue is entered into.

Cultural revolution

This route is an extension of the active change route, but in this case it is about policy and lifestyle alternatives that form a clear break with current practice. Followers of this route are citizen-consumers who eat little or no meat and whose food choices take into account production method, animal welfare or the environment. These socially and environmentally aware consumers could be part of critical alliances for the government where it concerns the actual achieving of certain ambitions with regard to sustainability and socialisation of food and agriculture.

These different routes of transition serve as an example of how one could shape alliances with modern consumers without getting caught in rigid views or fruitless discussions on moral responsibility versus irresponsible behaviour. We emphasise the fact that the three routes are not mutually exclusive. The question that should constantly be asked in policy is whether and to what extent a certain change route is effective for promoting sustainable consumption. It is mainly about looking for combinations bearing in mind the different motives of consumers and the circumstances in which consumers make their choices.

References

Audsley, E., Brander, M., Chatterton, J., Murphy-Bokern, D., Webster, C. and Williams, A. (2009). How low can we go? An assessment of greenhouse gas emissions from the UK food system and the scope to reduce them by 2050. FCRN-WWF-UK.

Cavell, S., Diamond, C., McDowell, J., Hacking, I. and Wolfe, C. (2008). Philosophy and animal life. Columbia University Press, New York.

Coetzee, J.M. (2003). Elizabeth Costello. Secker & Warburg, London, 233 pp.

Food and Agriculture Organization of the United Nations (FAO). (2010). The state of food and agriculture 2009: livestock in the balance. FAO, Rome.

Global Environment Outlook 4 (GEO 4). (2007). Global Environment Outlook 4: Environment for development. United Nations Environment Program (UNEP, Progess Press, Valeletta, Malta.

Harrison, R., Newholm, T. and Shaw, D. (eds.) (2005). The ethical consumer. Sage, London.

Jackson, T. (ed.) (2006). The Earthscan reader in sustainable consumption. Earthscan, London.

Keyzer, M.A., Merbis, M.D., Pavel, I.F.P.W., and Van Wesenbeeck, C.F.A. (2005). Diet shifts towards meat and the effects on cereal use: can we feed the animals in 2030? Ecological Economics 55: 187-202.

Micheletti, M. (2003). Political virtue and shopping: individuals, consumerism and collective action. Palgrave Macmillan, New York.

Nihlén Fahlquist, J. (2009). Moral responsibility for environmental problems – individual or institutional? Journal of Agricultural and Environmental Ethics 22, pp. 109-124.

Pimentel, D. and Pimentel, M.H. (2003). Sustainability of meat-based and plant-based diets and the environment. American Journal of Clinical Nutrition 78, pp. 660S-663S

Pimentel, D. and Pimentel, M. (2008). Food, energy and society. [3rd edn.] CRC Press, Boca Raton.

Shah, D.V., McLeod, D.M., Friedland, L. and Nelson, M.R. (2007). The politics of consumption/The consumption of politics. In: D.V. Shah *et al.* (eds.) The annals of the American academy of political and social science. Sage, Los Angeles, pp. 6-13.

Steinfeld, H., Gerber, P., Wassenaar, T., Gastel, V., Rosales, M. and De Haan, C. (2006). Livestock's Long shadow: environmental issues and options. Food and Agricultural Organization, Rome.

United Nations Environment Programme (UNEP). (2007). Global Environmental Outlook 4: Environment for development. Progress Press, Valletta, Malta.

Zu, X., Van Wesenbeeck, L. and Van Ierland, E.C. (2006). Impacts of Novel Protein Foods on sustainable production and consumption: lifestyle change and environmental policy. Environmental & Resource Economics 35, pp. 59-87.

Making sense of demands for the integration of ethics into the regulation of GMOs

F. Lord[1] *and L. Létourneau*[2]
[1]*Faculty of Law, Université Laval, 1030 Avenue des Sciences Humaines, Québec, Canada;* [2]*Animal Sciences Department, Université Laval, 1030 Avenue des Sciences Humaines, Québec, Canada; lyne.letourneau@fsaa.ulaval.ca*

Abstract

Many promote the potential of ethics for sound and legitimate biotechnology policy-making. One can appreciate these efforts within the regulation of GMOs. Whereas some jurisdictions have granted ethics a formal influence in this area others have strongly resisted these attempts. However, there is still much unresolved confusion and difficulties associated with the proposition to integrate ethics into the regulation of GMOs. Authors have put forward varied conceptions of ethics and of its appropriate role within policy-making. As a result, the 'integration' of ethics into policy-making ends up with fluctuating, if sometimes contradictory, meanings. We need to make sense of the demands for the integration of ethics, as well as of its underlying implications, in order to meet with success the regulatory challenge techno-scientific developments represent. Our paper aims to clarify the goals and strategies of proponents of the integration of ethics. The objectives pursued can be summarized as follows: (1) to exert control over professional science and its products; and (2) to increase the 'legitimacies' of regulation and policy-making. Strategies proposed to achieve these goals will be explored, along with regulatory systems reputedly integrating ethics.

Keywords: regulation of GMOs, ethics, policy making, legitimacies

Introduction

Ethics is often promoted as an essential tool to achieve sound biotechnology policy-making. With regard to the regulation of genetically modified organisms (GMOs), a number of commentators have emphasised ethical reflection and analysis as a tool to determine whether GMOs should be introduced into the market and the environment. While some countries have formalized this recommendation within regulatory processes, many countries still govern the release of GMOs by resorting predominantly or exclusively to technical- or science-based considerations. This 'science-based regulation' model requires applicants to scientifically demonstrate the safety of GMOs if they wish to commercialize or otherwise distribute these products on a given territory, with no or little consideration to ethical concerns.

To date, no consensus has been reached over what should be the role of ethics within regulation, whether in theoretical or practical terms. Commentators have put forward various proposals for the integration of ethics within regulation. However, they do not always make clear how and why this should be the case. Even what constitutes 'ethics' varies from one commentator to another: sometimes ethics refers to a set of substantial principles; at other times it encompasses procedural elements. As a result, the proposal to 'integrate' ethics within regulation remains somewhat fluctuating and confused. In this paper, we aim to shed light on the substance of this recommendation. To do so, we will discuss the means (1) and the ends (2) commonly associated with the proposal to integrate ethics.

The means to integrate ethics

Many jurisdictions have already put in place independent and multidisciplinary advisory committees to address ethical concerns related to biotechnology – or similar ones. The remit of these committees often include at least one of the following items: researching ethical concerns, engaging public participation, and recommending legislative and policy initiatives. Yet, these committees usually play a distant role in regulation. In contrast, more direct means have been advocated in order to secure for ethics a role more closely associated with the regulation of GMOs. Two of them are outlined below: (1) the resolution of value judgments occurring at the time of risk assessment, and (2) the enforcement of ethical standards in addition to scientific ones.

The resolution of value judgments in risk assessment

According to science and technology studies, implicit value judgments occur routinely as an integral part of science-based risk assessment (Brunk *et al.*, 1991). One example is the allocation of the burden of proof: should the applicant or the regulator have the responsibility to demonstrate the safety (or lack thereof) of a GMO? The party on which this obligation rests will be required to produce the most convincing evidence, while the other will provide contradictory evidence, which is a lesser burden. The allocation of the burden of proof is a normative (or ethical) issue because it involves values (about the social role of the regulator) rather than scientific expertise. Very often, such value judgments go unnoticed because an assessor just rules conflicting evidence in favour of one party without referring to any notion of 'burden of proof', and motivates her decision solely in scientific terms.

Authors have disputed the practice of entrusting the resolution of ethical issues to civil servants instructed to resolve technical ones (Haller and Gerrie, 2007; Waring and Lemmens, 2004). According to them, the solution is not to purge risk assessment from value judgments, but to make sure that its inherent normative content is dealt with appropriately. In order to achieve this, value judgments could be resolved prior to risk assessment (Millstone, 2007). It is proposed that normative issues be resolved beforehand by qualified and legitimate means. The problem, however, is that not all value judgments can be resolved this way. Some of them only present themselves in the assessment of specific products (Brunk *et al.*, 1991). Therefore, it is virtually impossible to anticipate all normative issues involved in the risk assessment of products yet to be regulated, or even designed. Some value judgments will (and maybe should) rest in assessors' hands. Notwithstanding, which issues should be left to the discretionary power of government scientists, and which should not, is controversial (Jasanoff, 1994).

An investigation of relevant legislative and administrative documents reveals that many normative issues are indeed resolved prior to risk assessment. For example, in relation to the allocation of the burden of proof, for instance, the European Union's regulation states that applicants to the approval of GM food products have the duty to prove their safety (paragraph 4(3) Regulation (EC) No 1829/2003). Similarly, Canada's Pest Control Products Act (PCPA) states that the promoter of a new pesticide has the burden to prove its safety (para. 7(6)a) PCPA). In contrast, with respect to novel food regulation, the same requirement is merely hinted at within the Canadian Food and Drug Regulations and relevant administrative guidelines (section B.28.002; Health Canada, 2006). It should nonetheless be mentioned that the current regulatory situation in Canada is not one where anything goes.

The enforcement of ethical standards

While the potential impacts of GMOs are diversified, many commentators deplore the fact that science-based regulation is only concerned with a narrow set of issues. Its technical focus, indeed, restricts regulators to the monitoring of the safety and effectiveness of GMOs (André, 2002). Under this view, regulators should thus go beyond regular assessments and take into account relevant socioeconomic and ethical considerations, for these are essential to determine how benefits and risks of new technologies are distributed, among other things (Eduljee, 2000; Webster, 2004). It is worth mentioning that, for some commentators, the incorporation of ethical and socioeconomic concerns within regulatory decision-making would not be satisfactory unless these apprehensions could lead to the restriction or prohibition of specific GMOs (Busch *et al.*, 1991).

The adoption and enforcement of 'regulatory' ethical standards is put forward to make sure that GMOs released into the market and the environment are ethically acceptable. In addition to the safety and efficacy requirements, GMOs would therefore need to meet some set of 'ethical standards' defined in the regulatory framework (Lacy, 2000). This proposition, however, is rarely tackled with in terms of its practical implications at any level. For instance, deciding on any such ethical standard is likely to involve a lengthy and difficult political, legislative, and administrative process aimed at determining which concern is sufficiently relevant and important to direct the introduction of new technologies. Moreover, its enforcement would no doubt require an accurate and accountable method to establish whether a specific product meets the standard or not. The appropriate enforcement of an ethical standard could also require the extension of a regulatory agency's human resources, knowledge base, regulatory practices, and therefore budget.

Some jurisdictions have adopted regulatory ethical standards. Within the EU, Directive 2001/18/EC states that member countries 'may take into consideration ethical aspects when GMOs are deliberately released or placed on the market as or in products' (Preamble, 9th). Regulation (EC) No 1829/2003 allows EU members to resort to 'other legitimate factors' when science proves to be insufficient (Preamble, 32nd). The European regulation does not identify specifically which ethical aspects should be taken into consideration; neither does it detail what their regulatory role should be. This lack of precision may explain why regulators prefer to resort to the more established scientific considerations (Sand, 2001). More definite ethical standard exists. The Norwegian Gene Technology Act notably requires GMOs to present tangible social benefits beyond their direct use (NBAB, 2009). In Canada, the PCPA allows regulators to take into account social and economic considerations to assess the 'value' of new pesticides, in addition to their safety (paragraphs 2(2), 4(2)d) and 8(1) PCPA). However, such initiatives could still remain unsatisfactory to the proponents of the integration of ethics if they do not appear to pursue their preferred ends.

The ends of the integration of ethics

Propositions to integrate ethics within the regulation of GMOs seem to spring from diverse objectives. One of them feeds on the critique of science-based regulation and relates to conceptions of legitimate regulatory decision-making. Another end is to control techno-scientific development through a different regulatory model. To do so, however, commentators first seek to reduce the authority of scientific expertise within regulatory decision-making.

The promotion of different conceptions of legitimate regulatory decision-making

The means to integrate ethics discussed in Part 1 both carry the suggestion that GMOs should be regulated on a more comprehensive and thorough basis - one that science alone cannot achieve. Proponents of the integration of ethics, indeed, seek to invalidate science-based regulation as a dominant model. To be sure, if they demonstrate that science alone fails to provide legitimate regulatory decisions, then the case for the integration of ethics (or other previously excluded considerations) becomes stronger. Therefore, commentators aim to show that regulation without ethics is intellectually unsound. This critique becomes political when science-based regulation is accused of failing to meet valuable public goals such as balancing different perspectives about GMOs, earning public trust, or engaging public participation. To meet those shortcomings, ethics is advocated as a necessary component for legitimate regulatory decision-making.

One such strategy to promote the integration of ethics is to reject the positivist ideology predominant in regulatory settings, where scientific expertise is represented as unflinchingly objective, neutral and rational (Shrader-Frechette, 1985). As mentioned above, by exposing positivism, proponents of the integration of ethics reinforce their case (Amy, 1984). Science and technology studies have long sought (and managed) to show that positivism is a fiction, because it simply blurs the distinction between scientific knowledge and values (Haller and Gerrie, 2007). Instead of justifying science-based regulation with a mistaken assumption about scientific expertise, they propose to legitimize regulatory decision-making with a model that explicitly includes ethics.

The issues of public trust and public participation also serve as incentives to promote the integration of ethics within regulation. It is alleged that ethical sensibility could generate public trust towards regulation (Thompson, 2000). Under this view, more science wouldn't increase public trust (Jensen and Sandøe, 2003), but ethics might. Opening up regulation to ethical considerations would also, according to some, facilitate public participation. Ethics is indeed considered to be closer to the interests and knowledge of laypersons, at least when compared to the expertise required for scientific knowledge and discussion (Abergel and Barrett, 2002; Hagendijk and Irwin, 2006). Proponents of the integration of ethics are putting forward a conception of democratic governance opposed to technical democracy: one where decision-makers are not accountable for the technical expertise available to them, but for inclusive and transparent ethical analysis.

The control of techno-scientific development

In many respects, the integration of ethics aims for a heightened control over techno-scientific development. This attempt at control first presents itself as a power struggle over the normative control of regulation. What should motivate regulatory decisions is here at stake, down to the very value judgments of the experts performing risk assessment. At the heart of the matter thus lies the distribution of authority over disputed issues. For when one issue ceases to be considered solely as a scientific question and transforms into an ethical one, it cannot be resolved anymore as it used to be: previous experts lose their influence and others increase theirs. Such power contests are not new: this is a common tactic within regulatory processes for scientists and policy-makers alike to dispute the nature of issues before them so they can put them under their administrative jurisdiction (Jasanoff, 1994). The proposal to integrate ethics within GMO regulation is reminiscent of these conflicts.

Another implicit stance associated with the integration of ethics is linked to the moral evaluation of law. As Létourneau explains in the context of animal biotechnology: 'Moral neutrality does not exist in the ... law on animal biotechnology. No matter what system of regulation is put into place, the regulatory framework will reflect a particular stance on the double ethical dilemma raised by the genetic engineering of animals. For, either explicitly or implicitly, it will reflect both a conception of the relationship between humans and animals, as well as a moral position with respect to the acceptability of genetically modifying animals' (Létourneau, 2005). That being the case, some proponents of the integration of ethics appear to be seeking to prevent the deployment of technologies and products that they deem unethical, and encourage ones that best answer their moral concerns. Regulation is indeed a socially powerful relay of ethics. Yet whether this is the purpose of regulation to do so does not seem to be preoccupying them.

Conclusion

The first section of this paper reveals that some jurisdictions have attempted to integrate ethics - although some may have done so unknowingly or half-heartedly. Nothing suggests that ethics is intrinsically relevant or irrelevant to regulation. In some respect, ethics is inextricably linked with GMO regulation for it reflects implicit ethical choices. But this is true in other area of law. On the other hand, how knowledge is generated and used is generally recognized as important, as this is an important part of what makes governmental intervention legitimate. How ethics will be affected and perceived as a professional, intellectual and philosophical undertaking within regulation is not discussed at all in the surveyed literature. While criticizing the regulatory use of scientific expertise, commentators should prepare for a similar disapproval of ethics in regulatory settings. Ethical analysis also rests on value judgments, and critics will comment its performance as a regulatory tool. If proponents of the integration of ethics believe that science is not a reliable basis for regulation, they should start thinking about how ethics can do better.

Our discussion about the ends of the integration of ethics suggests that its proponents might pursue undisclosed agendas. Two hypotheses can be formulated. First, proponents of the integration of ethics might individually seek to crystallise their own values within regulation. If this were the case, then they would only be satisfied when regulation reflects their ethical preferences. Or, they might collectively seek to increase the social and political influence of their expertise within policy-making. However, what should not be ignored is that if ethical considerations become 'integrated' within regulation in the form of legal norms, the real experts become lawyers. Unless ethics is somehow epistemologically insulated from the rest of regulation, the integration of ethics could lead to securing the authority of legal expertise in regulatory decision-making and the transposition of legal experts' values in the interpretation of 'ethical yet regulatory' norms. In addition to make clearer demands, proponents of the integration of ethics need to think these issues thoroughly, and in very practical terms.

References

Abergel, E. and K. Barrett (2002). Putting the cart before the horse: A review of biotechnology policy in Canada. Journal Of Canadian Studies 37: 135-161.

Amy, D. (1984). Why policy analysis and ethics are incompatible. Journal of Policy Analysis and Management 3: 573-591.

Andrée, P. (2002). Biopolitics of GMO in Canada. Journal of Canadian Studies 37: 162-191.

Brunk, C., L. Haworth and B. Lee (1991). Value assumptions in risk assessment: A case study of the Alachlor controversy. Wilfrid Laurier University Press. Waterloo. Canada. 161 pp.

Busch, L., W.B. Lacy, J. Burkhardt and L.R. Lacy (1991). Plants, power, and profit: Social, economic, and ethical consequences of the new biotechnologies. Blackwell. Oxford. U-K. 275 pp.

Eduljee, G.H. (2000). Trends in risk assessment and risk management. The Science of the Total Environment 249: 13-23.

Haller, S.F., and J. Gerrie (2007). The role of science in public policy: Higher reason, or reason for hire? Journal of Agricultural and Environmental Ethics 20: 139-165.

Health Canada (2006). Guidelines for the safety assessment of novel foods. Health Canada. Ottawa. Canada. 115 pp.

Jasanoff, S. (1994). The fifth branch: Science advisers as policymakers. Harvard University Press. Cambridge. USA. 302 pp.

Jensen, K.K. and P. Sandøe (2002). Food safety and ethics: The interplay between science and values. Journal of Agriculture and Environmental Ethics 15: 245-253.

Lacy, W.B. (2000). Agricultural biotechnology, socioeconomic issues, and the fourth criterion. In T.H. Murray and M.J. Mehlman (eds). Encyclopedia of ethical, legal and policy issues in biotechnology. John Wiley & Sons. New York. USA. 1,250 pp.

Létourneau, L. (2005). The regulation of animal biotechnology: At the crossroads of law and ethics. In: E. F. Einsiedel et F. Timmermans (eds). Crossing over – genomics in the public arena. University of Calgary Press. Calgary. Canada. pp. 173-192.

Millstone, E. (2007). Can food safety policy-making be both scientifically and democratically legitimated? If so, how? Journal of Agricultural and Environmental Ethics 20: 483-508.

NBAB (Norwegian Biotechnology Advisory Board) (2009). Sustainability, benefit to the community and ethics in the assessment of genetically modified organisms. Avalaible at: http://www.bion.no/publikasjoner/Diskusjonsnotat%20b%E6rekraft%20engelsk%202009_11_18.pdf.

Sand, I.-J. (2001). The legal regulation of the environment and new technologies: In view of changing relations between law, politics and science – the case of applied genetic technology. Zeitschrift für Reschtssoziologie 22: 169-206.

Shrader-Frechette, K.S. (1985). Science policy, ethics, and economic methodology: Some problems of technology assessment and environmental-impact analysis. Kluwer Academic Publishers. Boston. USA. 321 pp.

Thompson, P. (2000). Intégration de facteurs d'éthique à la biotechnologie alimentaire et agricole. Industry Canada. Ottawa. Canada. 45 pp.

Waring, D.R., and T. Lemmens (2004). Integrating values in risk analysis of biomedical research: The case for regulatory and law reform. The University of Toronto Law Journal 54: 249-290.

Webster, A. (2004). State of the art: Risk, science and policy – researching the social management of uncertainty. Policy Studies 25: 5-18.

Sociable swine: engaging society's values into a sustainable pig breeding research programme that aims for balanced improvement in productivity and welfare

M. Benard and T. de Cock Buning
Athena Institute for Research on Innovation and Communication in Health and Life Sciences, Faculty of Earth and Life Sciences, Vrije Universiteit Amsterdam, De Boelelaan 1085, 1081 HV Amsterdam, the Netherlands; marianne.benard@falw.vu.nl

Abstract

We present the results of the exploratory phase of a project, which aims to delineate the conflicts between value frames of various stakeholders (pig industry; animal welfare organisations; consumers; scientists) involved around a new selection strategy in breeding. By adding a social behaviour value to the commercial breeding programme, undesired aggression of pigs might be replaced by positive social interactions: a balanced improvement of welfare and production might be achieved. By means of an interdisciplinary approach that combines the expertise of ethology, physiology, quantitative genetics, and ethics/technology assessment, the opportunities and (dis)advantages of this strategy will be explored. A constructive dialogue regarding the upcoming laboratory results will run parallel with the technical research programme to decide on future directions. Exploration through in-depth interviews and focus groups showed, among others, non-scientists appeared to be concerned that economic motives are the main drivers behind this research strategy, instead of the expressed motives of improving animal welfare. Moreover, the approach to include social interactions in breeding was perceived as an end-of-pipe solution and an implicit acceptation of the intensive pig husbandry. These results might indicate irresolvable value conflicts between the stakeholders and scientists involved. This paper ends with a discussion of options to proceed constructively.

Keywords: animal welfare, pig husbandry, sustainable breeding, value frames, science-society dialogue

Introduction

Pigs live a complex social life, and their social interactions are known to profoundly affect their wellbeing. Negative interactions such as excessive aggression and oral manipulation of pen-mates can be harmful, whereas affiliative relationships can be beneficial by offering 'social support' (Keeling and Gonyou, 2001). Until now, social interactions of cattle and poultry have been ignored in breeding programs (Rodenburg *et al.*, 2010). This might have led to unintended co-breeding of negative social behavioural traits like competition and aggression between animals (Craig and Muir, 1996). As a result of the Dutch political debate on animal welfare and agro ethics, in the near future a ban can be expected regarding tail-docking and castration, as well as the obligation of group housing of sows (LNV, 2007). However, this combination will put pressure on the social interactions between pigs and will increase the risks of negative interactions between pen-mates. These negative side effects for animal welfare might be solvable by changing the current approach of the commercial breeding system, i.e. new breeding strategies that include social interactions of pigs. Recently, a new selection strategy has been designed that adds a 'social breeding value[1]' to the genetic breeding programme that awards the heritable social effect on (dis)advantages group members (Bijma *et al.*, 2007a,b). As this strategy explicitly selects on social

[1] Breeding value is an index indicating an estimation of the genetical base of pigs for one or more phenotypical characters relative to the phenotypical specification of his particular breed or breed lines.

interactions, e.g. to stimulate the co-operative behaviour in piglets, it might result in a win-win situation for animal welfare and production (Rodenburg *et al.,* 2010).

By means of an interdisciplinary approach which combines the expertise of ethology, physiology, quantitative genetics, and ethics/technology assessment, the opportunities and the (dis)advantages of the new breeding strategy will be explored, during its development. Early introduction of participatory processes in science and technology development is acknowledged to be important for a better societal embedding (Schot and Rip, 1997). Guston and Sarewitz (2002) emphasize the need for approaches, like real time assessment, that integrate society and science at an early stage. For the new breeding strategy this means that parallel with the technical research programme, a constructive dialogue regarding the upcoming laboratory results will be set up to decide on future directions. This dialogue will be organized between the scientists directly involved and various relevant stakeholders (e.g. pig industry; animal welfare organisations; branch organisations). In addition, the opinions of society at large, such as consumers, will be inventoried and these will be compared with the input of the stakeholders. Although Dutch consumers are a differentiated group, striking similarities in perceptions on animal welfare issues stand out (Te Velde *et al.*, 2003). Most consumers have the conviction – similar to farmers – that it is acceptable to eat meat. However, consumers have – opposed to farmers – in general a negative perception of the welfare of farm animals. Consumers worry about the environment of the pig and the pig's ability to live an acceptable live. The values of the consumer are not only related to the health of the animal, but add the need for space to move freely and to fulfil natural behaviour (Aarts and van Woerkum, 2005).

If these contrasts are, to some extend, due to different 'understandings' and 'moral valuations' of breeding programmes and animal welfare, a mutual learning process might be beneficial for all parties, including the piglets. Successive dialogue sessions, being tailor-made designed, can contribute to a constructive mutual learning process which explores win-win options in research and breeding strategies. However, controversial issues – such as animal welfare – have the tendency to run into a deadlock as result of sensitive contrasts or repeating arguments (Beekman and Van der Weele, 2004). This is frequently the result of implicit value conflicts. Such a conflict does not arise on the question if single values are wrong or right but on the order in which a set of values is prioritized (Kupper *et al.,* 2007). The whole set of values together with norms, convictions, and interests constitute a person's frame of reference. This frame influences the way a person perceives and judges a certain problem/issue and how he responds to that (Kupper *et al.*, 2007; Schön and Rein, 1994).

In this paper we (1) analyse the results of a qualitative probe to describe the issues at stake, (2) discuss possible value conflicts around the breeding strategy, and (3) identify appropriate options to deal with these value conflicts.

Methodology

To identify the societal issues at stake, nine relevant stakeholders (two animal welfare organisations, meat industry, two branch organisations, branch media, the Dutch ministry of agriculture, pigs genetic institute, animal welfare science) and six scientists who are directly involved in the programme were interviewed by means of semi-structured interviews. Selection of the informants was based on a literature inventory of the relevant stakeholder groups. The inclusion of the opinion of the Dutch consumer was done by the use of focus groups. Although most people are not aware of using values, a focus group is a suitable method to provide insight into the deeper motivations that lie behind opinion and ideas (Kupper, 2007). We performed three focus group sessions with each nine consumers (N=27), whom were recruited by a commercial recruitment company. The consumers were mixed in gender, age and education, and not selected/invited on the animal welfare issue.

At the beginning of the interviews and focus groups, we solicited for first associations regarding the specific aim of the new breeding strategy, but also regarding more contextual aspects such as the current situation for pig husbandry, their attitude towards the pigs, and ideas on animal welfare. By asking for first associations and letting people declare his or her associations, we derived insights in the person's moral intuitions. By interfering as less as possible (except for W-questions), the issues brought forward were own choices, enabling us to assess the primary negative or positive attitude (and arguments behind) towards the new breeding strategy.

Results

To structure the results, the data was grouped according to the stakes of the different stakeholder groups. The *Scientists* included the scientists directly involved in the project. The *Social Actors* included the informants of the two animal welfare organisations and consumers and the *Intermediaries* included the remaining stakeholders.

The interviews and focus groups provided insights into various issues that the stakeholders came up with related to the new breeding strategy. These issues were grouped into four clusters (1) Support for the *Aim of Research* to increase welfare by breeding more sociable pigs, (2) Breeding as a *Suitable Solution* for welfare problems, (3) Comments regarding *Breeding as Method,* and (4) *Respecting the animal* (Figure 1).

The first cluster, '*Aim of Research*', shows a steep divide in perspective between the involved scientists on the one side, and the social actors on the other side regarding the contribution of the breeding strategy towards animal welfare. Scientists mentioned an improvement of the welfare of pigs as significant or primary aim of the strategy (1). In contrast, the social actors believed the primary aim was economical (2), and animal welfare at best at a second place. The social actors shared the idea that as long as economical aims have priority in the current husbandry and research strategies, animal welfare cannot truly be improved. The consumers perceived the aim of the strategy as a combination of economical aims fed by consumer greed (3). These reasons were, according to the consumers, unacceptable to adapt the animal.

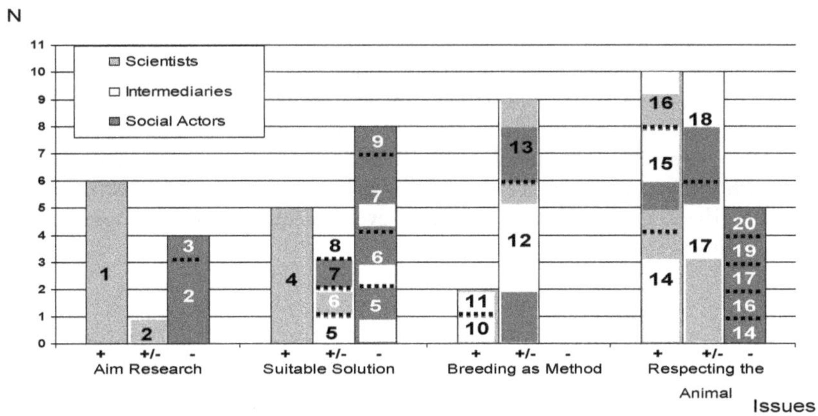

Figure 1. Responses to the new breeding strategy.
Number of times issues mentioned, grouped per thematic cluster and per stakeholder groups. The Y-axis represents how often an issue was mentioned. The numbers in the bars refer to the specific issues and the shades of gray in the bars indicate by which groups these issues were stated (in a positive=+, negative= -, or neutral=+/- sense). For details see text.

The cluster '*Suitable Solution*' shows the issues involved the improvement of animal welfare by the use of breeding. According to the scientists, breeding was a suitable method (4). They emphasized that the pig husbandry is a complex system with different interests involved, including the environment and the 'farmer as entrepreneur'. Other options that improve animal welfare – such as more space or the adding of straw in sties – would might have a better result for the animal but excluded other interests with the consequence that they were unrealistic (on short term) or undesirable. Both intermediaries and consumers had moral problems with the strategy because they saw it as adapting the animal to its environment (small spaces) – instead of the environment to the animal (more space) (5). In addition, some of the intermediaries and social groups believed that this strategy would not increase animal welfare, since the bad circumstances would still induce the adapted pig to be unhappy. In a similar line of thought, one scientist agreed that it is questionable whether welfare can be improved when the physical environment is left untouched (6). Other intermediaries and social actors perceived the breeding strategy as a plausible strategy, but perceived it mainly as an end-of-pipe solution. Aggressive behaviour or unhappiness of pigs was seen as the result of the bare environment and the management by the farmers. Therefore, the strategy could help to improve animal welfare but was not accepted as the right solution (7). For one social actor, the best solution was to stop eating meat (9). One intermediary had no problems with adapting the animal to the environment but questioned if it is acceptable to sort out needs of animals (8).

'*Breeding*' in general was not valued as positive or negative, but it did raise concerns of the intermediaries and social actors. Pig breeding, because of the probalistic character of statistics, one should be careful. As in every generation, breeding produces only a proportion of the desired phenotypes, which implies that also a proportion of undesired phenotypes is produced. This might decrease the welfare of those animals (13). Breeding could also harm the integrity of the animal as a result of inbred (12). Poultry and dogs were frequently used as extreme examples of harming the naturalness or integrity of the animal. Positive valued by an intermediary was the changed attitude of breeders who tried more to breed in a responsible manner (10), and the potential of the method (11).

The fourth cluster '*Respecting the Animal*' contains the comments with respect to the relation between our normative position towards animals and the new breeding strategy. Diverse stakeholders, primary scientists and intermediaries, believed that the naturalness (14), integrity (16), and the animal itself (15) would not directly be harmed by the new research strategy. However, the animal, its naturalness or its integrity could be harmed when the concerns on breeding – discussed under number 12 and 13 in text – would not be taken into account or when the strategy would make pigs indifferent for their environment (17). In addition, one should be alert not to use this strategy as argument to restrict the available space of the animal (18). The social actors shared the same concerns but were more pessimistic in comparison with the intermediaries because the new strategy emphasized the instrumental position of the animal (19). The strategy could also harm the development of the species: aggression might be necessary for surviving of the species (20).

Values

Besides the new breeding strategy, we also asked during the focus groups and interviews for the personal attitudes towards the pig. Eight of the nine stakeholder groups, except for the animal welfare organisations, saw the pig as a production animal, as a means to an end (food). Nevertheless, the use of the same issues as positive, neutral and negative comments by different stakeholder groups (Figure 1: 1,5,6,7,14,17), shows that each informant had its own particular consideration to draw the line regarding the acceptability of this project. Where some regarded it as acceptable to adapt the animal as long as the animal was not harmed (consequentialism), others regarded the action in itself as unacceptable (deontology). Although all the scientists, intermediaries and consumers positioned the pig as a production animal, they differed in their valuation of the human-animal relationship. For example, many of the intermediaries stated the necessity to keep an emotional distance of the pig. Especially the farm representatives rejected the projection of human norms to the pig. In contrast, the consumers did assign human norms to the pig when discussing the current pig husbandry and the new breeding strategy. This brings them more in line with the representatives of the animal welfare organisations: they regarded the pig as a subject, a respectful end in its self. This distinction might explain the different valuing of the intensive pig husbandry. The scientists and the intermediaries accepted that the current commercial pig husbandry is economically driven and commented from that perspective, while the frame of reference of consumers and animal welfare organisations is a husbandry that will satisfy the natural needs of the animal.

Conclusion and discussion

This explorative study provided insights into the various issues related to the new breeding strategy. Especially the social actors were primary negative towards the aim and the strategy; while the intermediaries expressed more nuanced concerns; and the scientists were primary positive. It also appeared that within the groups different considerations were made when and how the intrinsic value of the pig was harmed. We suggest that these differences between and within groups are correlated with the positioning of the pig against other values that influence the acceptability of the breeding program. If this is true, is there a way out to escape from an endless and frustrating deadlocks and repeating arguments for deaf ears (Beekman and Van der Weele, 2004)? An option to deal from a governance perspective with irresolvable fact-value conflicts is provided by Hisschemöller's and Hoppe's (1995-1996) *'general policy theory'.* They identified four different societal problem types based on consensus and dissensus about values and facts. As for the new breeding strategy no agreement among scientists and social actors exists on the weighting of animal and market values and uncertainty exists on the available knowledge, the breeding programme can therefore be perceived as an 'unstructured problem'. The general policy theory claims that each of the four types of fact-value problems has a related policy strategy. Applying the right strategy to the wrong problem will lead to predictable failures. The strategy for an unstructured problem focuses on the problem itself instead on the solution: acknowledge that stakeholders do possess different perspectives on facts and values at stake. The usual solution, assigning only the scientists as the expert regarding facts will lead to failure, hence a persistent problem. Instead the strategy should focus on a shared vision on the problem; only then solutions can be created. In other words, a learning process among the stakeholders should be facilitated.

Next steps in this project will be devoted to experiments in joint learning. Conditions will be set in which the stakeholders will be invited to explore together win/win options that respect the various value positions and result in enhanced animal well being. Two to three times a year methodological workshops will organized 2010-2012 with the researchers and other stakeholders to discuss the data, the interpretation and the follow up decisions. However, this approach is not without risks. Conflicts might reach the media and consequently damage the program and the private sphere of the researchers. It is important that for such kinds of sensitive learning processes a commitment is asked by the participants, in the line of Habermas rules 'Machtsfreie' Dialogue (Habermas, 1981).

References

Aarts, N. and Van Woerkum, C. (2005). Perceptions of farm animal welfare: the issue of ambivalence. In: De Jonge, F. and Van den Bos, R. The human-animal relationship. Royal Van Gorcum, Assen, The Netherlands, pp. 140-151.

Beekman, V. and Van der Weele, C. (2004). Naar een gereedschapskist voor constructieve ethiek. Report 7.04.10, LEI, Den Haag, ISBN-90-55242-932-4.

Bijma, P., Muir, W.M. and Van Arendonk, J.A.M. (2007a). Multilevel selection 1: Quantitative genetics of inheritance and response to selection. Genetics, 175, 277-288.

Bijma, P., Muir, W.M., Ellen, E.D., Wolf, J.B. and Van Arendonk, J.A.M. (2007b). Multilevel Selection 2: Estimating the Genetic Parameters Determining Inheritance and Response to Selection. Genetics, 175: 289-299.

Craig, JV., and Muir, WM. (1996). Group selection for adaptation to multiple-hen cages: beak-related mortality, feathering, and body weight responses. Poultry Science 75: 294-302.

Guston, D.H., and Sarewitz, D. (2002). Real-time technology assessment. Technology in Society: an International Journal, 24(1): 93-110.

Habermas, J., (1981).The Theory of Communicative Action. translated by Thomas McCarthy (1984), Cambridge, Polity.

Hisschemöller, M. and Hoppe, R. (1995). Coping with intractable controversies: The case for problem structuring in policy design and analysis. Knowledge and Policy: The international Journal of Knowledge Transfer and Utilization 8 winter 1995-1996: 40-60.

Keeling, L. and Gonyou, H. (2001). Social Behaviour in Farm Animals. CAB International, UK.

Kupper, F., Krijgsman, L., Bout, H. and De Cock Buning, T. (2007). The value lab: exploring moral frameworks in the deliberation of values in the animal biotechnology debate. Science and Public Policy, 34, November 2007, 657-670.

Ministerie van Landbouw, Natuur en Voedselkwaliteit (2002). Nota Dierenwelzijn. Available at: http://www.minlnv.nl/portal/page?_pageid=116,1640803&_dad=portal&_schema=PORTAL&p_news_item_id=18778.

Rodenburg, T.B., Bijma, P., Ellen, E.D., Bergsma, R., De Vries, S., Bolhuis, J.E. Kemp, B. and Van Arendonk, J.A.M. (2010). Breeding amiable animals? Improving farm animal welfare by including social effects in breeding programmes. Aminal Welfare 2010, 19 (in press).

Schön, D.A., and Rein, M. (1994) Frame reflection: Toward the resolution of intractable policy controversies. New York: Basic Books.

Schot, J. and Rip, A. (1997). The past and future of constructive technology assessment. Technological Forecasting and Social Change, 54(2-3), 251-268.

Te Velde, H., Aarts N. and Van Woerkum, C. (2002) Dealing with ambivalence: farmers' and consumers' perceptions of animal welfare in livestock breeding. Journal of Agricultural and Environmental Ethics 15(2), 203-219.

Part 8. Ethics of bioenergy production

Bioenergy and agricultural land use: examining the framing of the ethical debate

C. Gamborg[1], *K. Millar*[2] *and P. Sandøe*[1]
[1]*Danish Centre for Bioethics and Risk Assessment, University of Copenhagen, Rolighedsvej 25, 1958 Frederiksberg C, Denmark;* [2]*Centre for Applied Bioethics, University of Nottingham, Sutton Bonington Campus, Loughborough, Leics, LE12 5RD, United Kingdom; chg@life.ku.dk*

Abstract

Bioenergy research and technology development is attracting increasing attention, with bioenergy being promoted as an essential component of sustainable energy strategies. The increasing global production of first generation bioenergy and significant investment in research has resulted in notable public scrutiny and debate. As a result a number of ethical issues have been identified, including the impact of increased bioenergy production and use on food prices, regional development, poverty alleviation policies, fair trade, biodiversity, greenhouse gas emission and notably, on land use. A prominent and recurrent dimension of the ethical debate focuses on the impact of producing bioenergy from viable agricultural land (i.e. land that is suitable for food production). In particular, concerns are raised as to whether it is 'unfair' to use land for bioenergy at a time of food insecurity and significant food poverty. Moreover, as the amount of land used for bioenergy production steadily increases, the ethical issue of land displacement becomes more prominent; which in a broader context involves questions about what constitutes sustainable land use. Focusing on the land use issue, this paper will examine the nature of the arguments that assert: (1) it is ethically unacceptable to use agricultural land for bioenergy (particularly liquid biofuel production) at a time of significant global hunger and food insecurity; and (2) second generation biofuels are more acceptable than first generation. It will be argued that at least when applying a consequentialist ethical approach, it may be difficult to uphold an ethical distinction between first and second generation biofuels.

Keywords: biofuel, bioenergy, ethics, food, second generation

Introduction: framing of bioenergy

Bioenergy is framed, as well as promoted, as an essential component of sustainable energy strategies. Here, drawing on the FAO definition, *bioenergy* refers to renewable energy from biological sources (wood energy and agro-energy) which can be used for heat, electricity and transportation fuel, and their co-products. Although there is notable inconsistency in the use of terminology, in this paper the term 'biofuel' refers to liquid fuels produced from the aforementioned biological sources. Currently, the production of biofuels appears to be one of the major controversial issues within the agricultural and environmental sectors, somewhat akin to previous discussions of the industrialisation of agriculture (Thompson, 2008) or the GMO debate of the late 1990s (Mol, 2007). In particular two general arguments appear to recur within both the academic literature and the reported public/stakeholder debates: the claim that (1) first generation bioenergy crop production for biofuels (i.e. fuels that are derived from basic feedstock, such as seeds or grains, that could in theory enter the animal or human food chain) is ethically problematic and; (2) second generation biofuels are far more acceptable since the application of new biomass utilisation technologies allow the use of 'waste' (lignocellulosic biomass) materials in the production system.

An example of the first argument can be found in the address made by UN Special Rapporteur on the Right to Food (2000-2008), Jean Zeigler in October 2007, 'It is a crime against humanity to divert arable land to the production of crops which are then burned for fuel.'[1] And in a similar vein, referring to liquid biofuels, it has been argued that if 'you start to fuel cars with crops ... you are instantly putting the world's one billion starving people in competition with the world's one billion motorists.'[2] Runge and Senauer (2007) conducted a food and fuel balance calculation, stating that 'filling the 25-gallon tank of an SUV with pure ethanol requires over 450 pounds of corn – which contains enough calories to feed one person for a year.' Statements and calculations such as these have sparked fierce debate. The type of argument presented by Zeigler and others have been countered by commentators such as Matthews who point to the need for economic development in many countries, arguing that 'the real crime against humanity is to block Africa's potential, by blocking its exports and stunting its growth.' (Mathews, 2008: 99). In essence it is claimed that developing countries should be given the economic opportunity to 'grow not only food but also cash crops, and that the cash crop in greatest demand right now is biofuels' (*ibid.*). However, according to the FAO (2008), it is exactly this driver that has contributed to higher food prices, posing an immediate threat to food security impacting on poor food buyers in urban and rural areas.

Examining the nature of the second argument, it has been claimed that 'many of the problems associated with 1st-generation biofuels can be addressed by the production of biofuels manufactured from agricultural and forest residues and from non-food crop feedstocks.' (IEA, 2008: 6). This implies that current concerns regarding the contribution of first generation biofuels to higher food prices, due to competition with food crops and accelerating deforestation (and other potentially indirect land use effects), can be countered if second generation biofuel production is used instead. This view is echoed by Doornsbosch and Steenblick (2007: 4) arguing that these 'so-called second-generation technologies could, in theory, make it possible to avoid competing land use claims by growing biomass feedstocks on marginal and degraded land and using residual biomass materials.'

The aim of this paper is therefore to examine the arguments present against growing crops for biofuels and arguments for the use of second generation biofuels, pointing out some of the key premises which require further discussion. Initially the arguments are presented in greater detail and this is then followed by a review of the ethical framing of the bioenergy and agricultural land use debate. The main ethical concerns raised by potential agricultural land use changes that may occur as a result of increased bioenergy (particular biofuel) production are listed and discussed. As part of this initial analysis some of the broader concerns will also be examined, for example regarding the ethical implications of using energy crops grown in developing countries to produce transport biofuels.

Two main arguments in the debate

First argument: It is ethically unacceptable to use arable land for biofuel production when food insecurity exists in many regions of the world. In an OECD report on the sustainability of biofuels – 'Biofuels: is the cure worse than the disease?' – the authors stated that the 'rush to energy crops threatens to cause food shortages and damage to biodiversity with limited benefits.' (Doornsbosch and Steenblick, 2007: 4). They argued that theoretically speaking enough land might be available to feed an ever increasing world population – and produce sufficient biomass feedstock at the same time. However, it is more likely that land-use constraints will limit the amount of new land that can be brought into production, and, according to OECD, leading to a notable food-versus-fuel conflict (*ibid.*). It is precisely this 'food-versus-fuel' conflict which has prompted some observers to deem the growth in biofuels production as unethical: 'Ten years from now the rapid expansion of biofuel production may look foolish, or worse – unethical, if it leads to environmental degradation, high food prices, and increases

[1] http://news.bbc.co.uk/2/hi/americas/7065061.stm) accessed 20 March 2010.

[2] http://news.bbc.co.uk/2/hi/business/7026105.stm) accessed 20 March 2010.

the number of undernourished people.' (Cassman and Liska, 2007: 22). Recently, Rice (2010) tried to estimate the impact of industrial biofuels on food poverty and came to the conclusion that global biofuel consumption is likely to increase from about 70 billion litres in 2008 to 250 billion litres in 2020 – noting that for the EU, the increase will be steeper – from 13 billion litres to about 55 billion litres. As a result, in order to reach the EU 10% target on transportation fuels alone (Rice, 2010) the total land area directly required to grow industrial biofuels in developing countries could reach '17.5 million hectares [of developing country land], well over half the size of Italy' (*ibid.*). This would unavoidably displace food and animal feed crops to new areas of so-called marginal or idle land, potentially leading to ecological impacts such as loss of biodiversity, although the extent of this has not been estimated.

According to Runge and Senauer (2007) up to 80% of the total household income of the poorest (such as landless labourers or rural subsistence farmers) is spent on food, highlighting the risk presented by increases in the price of staple foods which may quickly lead to malnutrition, hunger and inevitable increases in mortality rates due to increases in malnutrition-related diseases. These statements have been countered by Braun and Pachauri (2006) who claim that biofuel-targeted crop production does not necessarily compete with (or even drive out) food production. They suggest that food production can be combined with energy production, and moreover, energy crops could be targeted to more marginal lands. In addition, they claim that food insecurity is a result not simply of a lack of food availability, but poverty, which is dependent on policies related to access to a number of basic requirements, such as technological resources, improved infrastructure and open markets.

Thus this first argument seems to rely on two controversial assumptions, firstly that the main limiting factor in being able to address global malnutrition is the availability of agricultural land, and secondly that abstaining from growing bioenergy crops will allow for increased food production that will alleviate food poverty. Regarding the first assumption some have questioned whether land use changes to increase biofuel production in developed countries will significant impact on food available in developing regions (Pimentel *et al.,* 2009). Diversion of land to biofuel production within Europe and North America may provide markets for farmers outside of Europe to produce stable crops at better global market prices. However significant challenges may relate to equitable market access for farmers in developing regions who wish to produce food and/or biofuel crops. Therefore the potential for bioenergy to impact on market prices may once again highlight the inequalities of the global food system which applies to all forms of cash crop production. Regarding the second assumption it is important to note some of the possible inefficiencies and inequalities of current food production systems. Livestock production is a very inefficient way to produce food, using grain as animal feed that can be directly consumed by humans. Hence in terms of tackling food insecurity, it could be argued that reducing levels of animal production may be a better option than totally abstaining from growing crops in order to produce first generation biofuels.

Second argument: Second generation biofuel is ethically more acceptable than first generation. The idea of using inedible 'waste' from food crops (e.g. straw) or domestic waste as feedstock in the production of biofuels has been discussed for several years (e.g. Pimentel *et al.,* 1981) but only in the past few years has it been a much targeted research topic. The technology is, however, not currently sufficiently developed to render second generation biofuels production commercially viable (Connor and Hernandez, 2009). Whether or not a technological breakthrough in this area is eminent depends on a number of factors. Over the coming decade, it is possible that the technology will be greatly refined, but not economically viable (Antizar-Ladislao and Turrion-Gomez, 2008; Howarth and Bringezu, 2008). However, the precisely 'true' production cost is also seen as one of the potential selling points of second generation fuels. Thus it is claimed by some that first generation biofuels are expensive to produce when total production costs are taken into account (e.g. government grants and subsidies are excluded) and they provide only moderate greenhouse gas reduction benefits (Yuan *et al.,* 2008). In fact, some argue that food crop-based biofuels, through the conversion of peatland or grasslands (e.g. Southeast Asia or United States), result in a net carbon *debt* (Fargione *et al.,* 2008; Searchinger *et al.,* 2008).

Although second generation biofuels rely on the use of 'waste', so-called indirect land use may be problematic. According to Howarth and Bringezu (2008: 68) 'indirect land use will be responsible for substantially more carbon loss (up to twice as much) than direct land use; however, because of predicted increases in fertilizer use, nitrous oxide emissions will be more important than carbon losses themselves in terms of warming potential'. If correct, the use of some second generation biofuels may lead to higher GHG emissions than their fossil fuel counterparts, and according to some estimates this will result in substantially more carbon loss (up to twice as much) than direct land use (Bindraban *et al.,* 2009).

Moreover, it seems that two further controversial assumptions underlie this argument and the overall debate about the desirability of second generation versus first generation biofuels in terms of feedstock use. The first seems be the assumption that straw is 'waste'. The second assumption concerns whether straw or other forms of 'waste' from crop production is freely assessable and unlimited resources – in terms of costs. Thus, these assumptions may affect the feasibility of using residue 'waste' material as feedstock on a large scale. In terms of resource availability, straw and other forms of so-called waste may turn out to be a very limited feedstock; and as highlighted above it is debatable to which extent this material is a 'waste'. It is important to note that straw is still part of the crop, i.e. what is grown, therefore there is a 'natural' limit to how much is available. The amount of straw available is dependent on factors such as crop rotation, field management practices and climate (Pahkala and Kontiri, 2009). More importantly, straw is useful as a farm resource in number of ways, for example used as bedding for animals. In addition, it may be used (when 'put back' into the soil) to improve soil fertility, control erosion, and maintain soil carbon (Gomiero *et al.*, 2009). This discussion therefore highlights that the specific 'details' of production, the specific differences, will affect any sustainability assessment of both first and second generation biofuels and in turn, if applying a consequential approach, it will affect the assessment of ethical acceptability. For example, diverting land currently used to produce animal feed for highly (energy) inefficient meat production systems, to first generation biofuel production may be a more acceptable and sustainable practice.

Going beyond the argument presented for the increased investment in second generation biofuels, as discussed above, it is important to note an extension of the debate that requires a reframing of a consequentialist analysis. A key aspect would therefore be the need to consider another 'detail' which concerns identifying which *form* of bioenergy will be the most efficient, social acceptable and environmentally friendly. This is likely to depend on the outcome of a full Life Cycle Analysis (LCA), which includes a social impact assessment.

Hence it may be argued that other forms of bioenergy are more efficient than the production of biofuels, either first or second generation. For example, the growing of crops such as *Miscanthus* or willow in short rotation coppice (SRC) may make good use of the land, potentially being very efficient if these crops are used in an effective way in stationary facilities such as combined heat and power plants, rather than use to produce biofuels. These crops are high yielding, yet low-input perennial crops with little or no fertiliser requirements so they may also provide a number of other environmental benefits (Rawlings, 2007). Thus it is claimed by some that biomass may be more efficiently used in combined heat and power plants instead of being turned into transportation fuels for vehicles (Campbell *et al.,* 2009). Again, a number of these arguments need to be further analysed and substantiated.

Conclusion

Global bioenergy production offers opportunities as well as potentially posing a food, energy and environment trilemma (Tilman *et al.* 2009). After initially examining two of the prominent claims – it is ethically unacceptable to use arable land for biofuel production when food insecurity exists in many regions of the world, and the claim that second generation biofuel is ethically more acceptable than first generation – it appears that rather than defining the ethical acceptability of bioenergy by the type of biomass feedstock used (e.g. grain or straw), the critical issue for any analysis is the outcome of a detailed sustainability assessment which takes into account not only energy balances but also wider societal needs and impacts. The initial analysis presented here has not set out to refute some of the 'prominent' arguments against particular bioenergy production systems. However this paper has attempted to demonstrate the *complexity* of the current debate. From a consequentialist view, the issue is not about an 'either, or' approach, in terms of first or second generation bioenergy, but about identifying the best possible options for biomass use. In order to assess what constitutes the 'best' options, more extensive analysis and wider ethical reflection is needed.

References

Antizar-Ladislao, B. and Turrion-Gomez, J.L. (2008). Second-generation biofuels and local bioenergy systems. Biofuels, Bioproducts & Biorefining 2: 455-469.

Bindraban, P.S., Bulte, E.H., and Conijn, S.G. (2009). Can large-scale biofuels production be sustainable by 2020? Agricultural Systems, 101, 197-199.

Campbell, J.E., Lobell, D.B. and Field, C.B. (2009). Greater transportation energy and GHG offsets from bioelectricity than ethanol. Science 324: 1055-1057.

Cassman, K.G. and Liska, A.J. (2007). Food and fuel for all: realistic or foolish? Biofuels, Bioprod. Bioref. 1:18-23.

Connor, D.J. and C.G. Hernandez (2009). Crops for biofuels: current status and prospects for the future. In: R.W. Howarth and S. Bringezu (eds.) Biofuels: Environmental consequences and interactions with changing land use, pp. 65-80. Available at http://cip.cornell.edu/biofuels/. Accessed 2 March 2010.

Doornsbosch, R. and Steenblick, R. (2007). Biofuels: Is the cure worse than the disease? OECD Roundtable on Sustainable Development. Available at http://media.ft.com/cms/fb8b5078-5fdb-11dc-b0fe-0000779fd2ac.pdf. Accessed 2 March 2010.

FAO (2008). Bioenergy, food security and sustainability – towards an international framework. FAO. Available at: http://www.fao.org/fileadmin/user_upload/foodclimate/HLCdocs/HLC08-inf-3-E.pdf. Accessed 2 March 2010.

Fargione, J., Hill, J., Tilman, D., Polasky, S. and Hawthorne, S. (2008). Land clearing and the biofuel carbon debt. Science 319: 1235-1238.

Giampietro, M., and Mayumi, K. (2009). The biofuel delusion: The fallacy of large scale agro-biofuels production. Earthscan, London.

Gomiero, T., Paoletti, M.G. and Pimentel, D. (2009). Biofuels: Efficiency, ethics, and limits to human appropriation of ecosystem services. Journal of Agricultural and Environmental Ethics. DOI 10.1007/s10806-009-9218-x.

Howarth, R.W. and Bringezu, S. (eds.) (2008). Biofuels: Environmental consequences and interactions with changing land use. Available at http://cip.cornell.edu/biofuels/. Accessed 2 March 2010.

IEA (2008). From 1st- to 2nd-generation biofuels technologies. http://www.iea.org/papers/2008/2nd_Biofuel_Gen.pdf. Accessed 20 March 2010.

Mathews, J.A. (2008). Opinion: is growing biofuel crops a crime against humanity? Biofuels, Bioprod. Bioref. 2: 97-99.

Mol, A.P.J. (2007). Boundless biofuels? Between environmental sustainability and vulnerability. Sociologica Ruralis 47(4): 297-315.

OECD/IEA (2010). Sustainable production of second-generation biofuels. Availble at http://www.iea.org/papers/2010/second_generation_biofuels.pdf. Accessed 2 March 2010.

Pahkala, K. and Kontturi, M. (2009). Potential of straw as bio-energy raw material in northern European countries. NJF Report 5(3): 56.

Pimentel, D., Marklein, A., Toth, M.A., Karpoff, M.N., Paul, G.S., McCormack, R., Kyriazis, J. and Krueger, T. (2009). Food versus biofuels: Environmental and economic costs. Human Ecology 37: 1-12.

Pimentel, D., Moran, M.A., Fast, S., Weber, G., Bukantis, R., Balliett, L., Boveng, P., Cleveland, C., Hindman, S. and Young, M. (1981). Biomass energy from crop and forest residues. Science 212: 1110-1115.

Rawlings, A. (2007). An expanding liquid biofuel market: Investigating the likely impacts on welfare and environment. Environmental Ethics 1: 26-37.

Rice, T. (2010). Meals per gallon. The impact of industrial biofuels on people and global hunger. London: ActionAid UK.Available at http://www.europolitics.info/pdf/gratuit_en/266253-en.pdf. Accessed 19 March 2010.

Runge, F. and B. Senauer (2007) How biofuels could starve the poor. Foreign Affairs May/June 2007: 41–53.

Searchinger,T., Heimlich,R., Houghton, R.A., Dong,F., Elobeid, A., Fabiosa, J. , Tokgoz, S., Hayes, D. and Yu, T. (2008). Use of U.S. croplands for biofuels increases greenhouse gases through emissions from land-use change. Science 319: 1238-1240.

Tilman, T., Socolow, R., Foley, J. A., Hill, J., Larson, E., Lynd, L. (2009). Beneficial biofuels – the food, energy, and environment trilemma. Science, 325, 270-271.

Thompson, P. A. (2008). The agricultural ethics of biofuels: A first look. Journal of Agricultural and Environmental Ethics 21(2): 183-198.

Von Braun, J. and R.K. Pachauri (2006). The promises and challenges of biofuels for the poor in developing countries. International Food Policy Research Institute , Washington, DC.

Yuan, J.S., Tiller, K.H., Al-Ahmad, H., Stewart, N.R. and Stewart, C.N. (2008). Plants to power: bioenergy to fuel the future. Trends in Plant Science 13(8): 421-429.

Searching words for sustainable biofuels

L. Landeweerd, M. van der Veen and P. Osseweijer
Delft University of Technology, Kluyver Centre for Genomics of Industrial Fermentation, Julianalaan 67, 2628 BC Delft, the Netherlands; l.landeweerd@tudelft.nl

Abstract

Agriculture has always been used to produce food, feed, fiber, fuel and feedstock. The balance between these uses has changed over the centuries. There is now pressure on land and water resources to produce enough food and other products for an increasing population and to improve our carbon footprint by increased forestation. Modern technology makes it possible to use agricultural products much more effectively for fuel or feedstock. To be able to meet the demands for energy, it has been suggested that crops for biomass production should be grown in the developing countries, where more land is available than in developed countries, specifcially Europe. Furthermore, the production potential of tropical developing countries is much higher than temperate countries. Growing biomass and processing it in developing countries may aid the economies involved, increasing local welfare. But due to an increase in the demand for biomass, specifically in the developed countries, pressure will increase on the use of land of emerging economies. NGOs and others have expressed serious concerns with regard to the protection of rainforests and other ecosystems but also to the rights of the populations of the developing countries. About 900 million people suffer from malnourishment globally, only a marginal percentage lives in the developed countries. If the land that would be used for bioenergy would replace the food production this problem may increase. In that case it would not be defendable to produce the raw material for biofuels and feedstock for the developed countries in developing countries. On the other hand a large percentage of the energy use in Africa presently comes from locally collected wood, a very unsustainable use of biomass. Experts therefore suggest that integration of agriculture for food and energy would actually enhance the local quality of life. This however requires a mutual vision and coerced strategy for Sub-Saharan Africa. Second generation biofuels using modified yeasts may enhance this solution, but it would also demand adjustments of the local infrastructures. The concept sustainability that is currently used functions too much as a 'thick concept' to analyse the ethical issues involved. Distributive justice cannot be applied in this context of a global justice as there is too much asymmetry in the issues involved and in the power between the different stakeholders at hand. In this paper, we will use the concept of human dignity to 'refine' the analysis of the current debate on biofuel production.

Keywords: biofuels, moral vocabularies, Rawls, sustainability, capabilities approach

Introduction

Until now, fossil fuels have played an important role in our society. Gasoline, diesel and LPG cover the need for transportation fuels whilst coal and oil are also used for the generation of electricity in power plants. Due to some important drawbacks of these fuels this is changing. This change is driven by several factors. Fossil fuel resources will be depleted at some time in the future. A possibly more pressing matter is that usage of fuels derived from fossil oils or gasses pollutes the environment, and is the main responsible for the greenhouse effect: in 1996, transportation was responsible for some 28 per cent of the gasses that cause the greenhouse effect in the EU (European Commission, 2000) and if our economy develops as expected, CO_2 emissions are expected to have increased with 50 per cent between 1990 and 2010. It was suggested that the invention and further development of biofuels could well serve as a basis for a more sustainable and less polluting economy, although many recent reports tend to be critical (Ros, 2010; Wiskerke and Winckers, 2010).

The advantage of biofuels over fossil fuels is relatively easy to explain: only CO_2 bound by the growth of the biomass from which it is derived, rather than CO_2 that has been bound for millions of years (oil or gas), is released when using biofuels. Furthermore, in theory, biofuels can be produced infinitely, due to their renewable nature. Biofuels have the added advantage that they do not need to be produced in politically unstable regions. Therefore, there would be a more a secure access to these products. In practice, these advantages do not necessarily work out.

The EU and its Member States committed to the reduction of their CO_2 emission to counter the greenhouse effect and ratifying the Kyoto Protocol in May 2002. The protocol acknowledges the importance of the problem of climate change (Malça, 2006) and requires that industrialised countries reduce the emissions of six greenhouse gases (CO_2, which is the most important one, methane, nitrous oxide, hydrofluorocarbons, perfluorocarbons and sulphur hexafluoride) to an amount that is 5.2% lower than the 1990 level. This aim is to be reached during the first commitment period, which lasts from 2008 to 2012, but the EU-member states agreed to reduce the greenhouse effect from 2008 to 2012 by more than 8 percent under the 1990 levels (United Nations Framework Convention on Climate Change, 1998). Several governments drew up new policy lines to give incentives for the implementation of biofuels. The idea is to introduce biofuels to the market on a short term by providing subsidies and facilitating bank loans, changing the current infrastructure and researching new production methods. More and more biofuels are therefore used as an alternative.

Europe has less land available for growing biomass for biofuels than any other continent in the world. In the United States, about 40% of the land that can be used for agriculture is used for agriculture; in Africa, this is 10%, in Europe and Asia 70% (House of Lords, 1998). Biofuels is a product that is presently very agriculture-intensive (Food and Agricultural Organisation, 2006). If Europe wants to keep providing in its own food and feed production and also wishes to replace a substantial amount of their transport fuels by biofuels, the production of biomass for European biofuels will have to take place elsewhere. Next to importing biomass or biofuels from possible exporters such as Brazil, using lands in Africa seems to be the logical next step. Biofuel production might go at a cost of food and feed production (Wakker 2005), since currently the production of biofuels is still highly land intensive (Food and Agricultural Organisation 2006). Although the ethical discussion in Europe is also concerned with this competition, the consequences for non-industrialized countries will be much more diverse. Recently a European group of experts calculated that 40 M hectares could be made available in Europe for producing biofuels in a sustainable way and without decreasing the capacity for food production. This would supply up to 30% of the biofuel requirements could be supplied from the Ukraine. Even if their land could produce the biomass at competitive costs, it still may provide incentives for additional import from elsewhere. Therefore, regions that already have to cope with food and feed shortage might be inclined to produce crops for biomass. As recognised by the African experts themselves, this necessitates a specific approach to the production of biofuels (African Convention of the Global Sustainable Bioenergy Project, 2010).

Sustainability

The term sustainability is omnipresent in current debates on biofuel production. It is a standard concept for the ethical assessment of biofuel production. The problem area described by it is not new. Already in 1798 Thomas Malthus remarked: '...the power of population is infinitely greater than the power in the earth to produce subsistence for man.' (Malthus, 1798). His perspective on things was perceived of as overly pessimistic, until the growing problem of poverty and population growth, combined with the depletion of natural resources began to force us to adopt a more self-critical stance. In the past ten years, the depletion of our natural resources has not been out of the media for longer than a day. The depletion of fossilised oil is not the only problem, it's apparent devastating effect on the climate is no longer contested, in spite of recent criticism on the IPCC. It appears mankind is not very efficient in his dealings with his environment.

The concept of sustainability is rather young. It was first coined in a report by the commission Brundtland of the United Nations entitled 'Our Common Future' (1987). In this report, economic growth, environmental issues and poverty issues in the developing countries were linked for the first time. Sustainability links issues of economic, ecologic and poverty related issues, by aiming at a more efficient use of our natural resources so future generations will also be able to reach an acceptable welfare level, without destroying the environment. At the 'Convention on Biological Diversity' in Rio de Janeiro (1992) it was agreed to aim for sustainable economic development. The Kyoto protocol (1997), de World Convention for Sustainable Development in Johannesburg (2002) and the United Nations Conference on Climate Change in Copenhagen (2009) were further milestones. The development of biofuels is considered to be a key to a more sustainable energy source.

In the many reports and documents on biofuel production, sustainability has slowly developed as a master value, a value to which all other values are reduced, whilst itself not being reducible. The demands put on the concept are ever increasing, incorporating domestic as well as foreign social demands, both labour conditions and welfare, both efficiency and environmental impact, both ethical and economic factors. It encompasses both measurable dimensions and immeasurable political ideals. The disadvantage of this semantic outcrop is that the term is starting to refer to every possible dimension, and therefore is increasingly void of meaning and applicability. It no longer creates the possibility of a perspective of agency. It rather frustrates the development of such perspectives. The controversial competition between the production of food and fuel can therefore not be resolved with the abstract term of sustainability. An alternative might be found in current theories of justice. The theory that stands out in this respect is the influential theory of justice that was proposed by the philosopher John Rawls.

Rawls' distributive justice

Rawls developed two principles of justice on the basis of rational agreement, and the concept of an original position to ensure the justness of this rationality. Rawls stated:

> '[the principles of justice] are the principles that free and rational persons concerned to further their own interests would accept in an initial position of equality as defining the fundamental terms of their association. These principles are to regulate all further agreements; they specify the kinds of social cooperation that can be entered into and the forms of government that can be established' (Rawls, 1992 (1971): 11).

In Rawlsian philosophy the basic principles of justice should balance individual liberty with an equal distribution of liberty. This should be combined with a provision of the greatest benefit for the least advantaged:

> 'Social and economic inequalities are to be arranged so that they are (a) to the greatest benefit of the least advantaged, consistent with the just savings principle, and (b) attached to offices and positions open to all under conditions of fair equality of opportunity' (Rawls, 1992 (1971): 302).

People should come to this agreement through rational reflection, envisaging themselves not to have a specific place or background in society; one should regard oneself as not knowing what position one would have in society, whilst one is envisaging what society would be a just one. In Rawls' vision, 'those who are at the same level of talent and ability, and have the same willingness to use them, should have the same prospects of success regardless of their initial place in the social system' (Rawls, 1992 (1971): 43-44). Social class, gender or any other contingency should not have a negative influence on the liberty someone is to have, to pursue his goals in life. With this construction, Rawls tried to find a tool that could balance freedom, or liberty, and equality and omit any prejudices stemming from one's class or one's ethnic, linguistic, religious or cultural background. However, social and economic inequalities should be distributed in such a way that they can reasonably be expected to be advantageous to all those who are the worst off in the first place. Elaborating on his principles of distributive justice in his *Political Liberalism* (1993) Rawls poses the question how one can create a stable society if within this society different groups and cultures with different value systems can coexist. To this aim, he introduces the concept of overlapping consensus. This concept, as elaboration of his principles of justice, is also used to analyse the issues at play in the debate on sustainable biofuels (Soltau, 2009; Meckled-Garcia, 2009; Bührs, 2009). There are however some disadvantages.

In Rawls' thought, as in most liberalist theories of justice, the combination of egoism and freedom of choice are supposed to eventually lead to a choice for solidarity. But the problem with his approach to justice is that if there is too large an asymmetry between different parties, the combination between a natural egotism on the one hand and freedom of choice on the other is no longer a guarantee for a basis for solidarity through a social contract. The existing liberalist framework for a distributive justice is therefore indirectly advantageous for western liberal democracies. The problem with the implementation and introduction of biofuels is that the ethically relevant issues these carry along do not merely concern individuals and states, but rather communities and multinationals. Furthermore, they concern cross-national relationships, and can therefore not be resolved with a system originally designed to delimitate governmental power through delimitating the power of one individual over the other.

Liberalist theories of justice assess tend to propose negative approaches to ethics, in the sense of a 'do not meddle, except'. This works fine in theory for a meritocracy in which all individuals are small entrepreneurs. But such a 'greengrocers' society' does not exist in reality. Specifically with regard to the moral assessment of the introduction of biofuels, this moral vocabulary needs radical revision. When we do not, we may be able to hush our consciences with regard to the environment and the pending but everdoubted climate change, but at the expense of local communities in other parts of the world.

In search of an alternative

The developed states are increasingly sensitive to the necessity of a cleaner (and often more expensive) basis for energy production and consumption, but they cannot expect the rest of the world to adjust to such needs without proper incentives. Bioenergy production could both decrease and increase welfare in Africa. It would decrease when it would replace food production, it would increase if it would intensify agriculture. A transition to bioenergy production has to be directed by non-commercial approaches that are aimed at communities rather than individual interests. A sustainable world needs a sustainable Africa, but one cannot expect Africa to pay for this. The problem of food security is due to not merely poverty and food scarcity but also access of food. It is also related to increased use of energy and of meat consumption when a community becomes richer. These put much more severe burdens on agriculture.

There is a need for an alternative moral vocabulary to map the different ethical dimensions of the implementation of biofuels. Revitalising the old moral concept of human dignity might offer a solution. Martha Nussbaum and Amaryta Sen (Sen, 1985; Nussbaum and Sen, 1993; Nussbaum, 2000, 2006) developed a way to operationalise this concept with their *capabilities approach*. At the time, it was presented as an alternative for the more quantitative model of measuring welfare levels by looking at the gross national product. (Nussbaum, 2000). An operationalisation of this type is argued by De Vries en Petersen (2008). The capability approach aims to connect the 'subjective experience of a good quality of life with the objective resource-oriented aspects (cf. Costanza *et al.*, 2007; cf. Petersen and De Vries, 2008: 1008). A similar elaboration was proposed by this author in a recent publication (Landeweerd, 2010). Well-being is not measured, nor reduced to the simple and more neoliberal economic notion of freedom of choice and freedom of opportunity. The turn from a 'resource-based' view on ethics to a 'capabilities' view on ethics renders it possible to take a personal, as well as a collective (environmental, social, institutional) perspective. As Petersen and de Vries state: 'An unstable climate, polluted drinking water or a corrupt police force are examples of limitations to capabilities and, hence, to functionings in the environmental and institutional (or governance) domain' (Petersen and de Vries, 2008: 1008). With this approach, Nussbaum and Sen were able to revitalise the concept of human dignity. It offers an alternative for the rather abstract sustainability criteria that have been defined until now (Lausanne Round Table, 2008; Cramer Commission, 2006). They conceive of it as a good alternative of the over-economical nature of sustainability and the over-procedural nature of 'justice'.

Conclusion

The dilemma of the competition between food and fuel production is much more complex and interrelated than it is now presented to be. It cannot be resolved with the current moral vocabulary that builds on the concept of sustainability. Traditional approaches in liberal philosophy do not seem to offer the proper tool either. They even seem to blind us for the most important normative aspects of the introduction of biofuels; the effects for local communities on an international scale and the responsibilities of multinationals. To bring about the transition of our economy from fossil fuel to biofuels in a socially responsible fashion, we will need to look at other aspects of our current concept of justice. With the help of Sen and Nussbaum's capabilities approach, the concept of human dignity can be revitalised in an operationable fashion, and provide for a better alternative to assess the moral acceptability of different biofuels, and different strategies to produce them and the biomass that they are produced from.

Acknowledgements

I would like to thank Jan Willem Ebbinge for his useful comments.

References

African Convention of the Global Sustainable Bioenergy Project (2010). http://academic.sun.ac.za/sa/biofuels/Convention/Draftresolution.html. Accessed 28 March 2010.

Bührs, T. (2009). Environmental Space as a Basis for Legitimating Global Governance of Environmental Limits Global Environmental Politics – 9:4, pp. 111-135.

Brundtland Commission (1987). Report of the World Commission on Environment and Development: Our Common Future. Transmitted to the General Assembly as an Annex to document A/42/427 – Development and International Co-operation: Environment. http://www.un-documents.net/wced-ocf.htm. Laatst bekeken: 30 december 2009.

Costanza, R., B. Fisher, S. Ali, C. Beer, L. Bond, R. Boumans, N.L. Danigelis, J. Dickinson, C. Elliott, J. Farley, D. Elliott Gayer, L. MacDonald Glenn, T. Hudspeth, D. Mahoney, L. McCahill, B. McIntosh, B. Reed, S. Abu Turab Rizvi, D.M. Rizzo, T. Simpatico and R. Snapp (2007). Quality of life: an approach integrating opportunities, human needs, and subjective well-being. Ecol. Econ. 61, 267-276.

Cramer Commissie (2006). Testing framework for sustainable biomass. http://www.lowcvp.org.uk/assets/reports/070427-Cramer-FinalReport_EN.pdf. Accessed 20 January 2010.

Food and Agricultural Organisation (2006). 'The state of food security in the world 2006: Taking stock 10 years after the World Food Summit'. Rome, Food and Agricultural Organisation of the United Nations.

House of Lords (1998). Select Committee on European Communities – Second Report 15 December 1998. http://www.publications.parliament.uk/pa/ld199899/ldselect/ldeucom/11/8121501.htm. Accessed 20 January 2010.

Kyoto Protocol (1997). http://unfccc.int/resource/docs/convkp/kpeng.html. Accessed: 20 January 2010.

Landeweerd, L. (2010) Duurzame Biobrandstof, een Beperkt Vocalulaire'. In Huijer, M. Smits, M. (eds.). Moralicide. Uitgeverij Klement, Kampen, the Netherlands, pp. 194-208.

Malthus, T.R. (2004 [1798]). 'An essay on the principle of population' hoofdstuk 1, pag. 13. Oxford World's Classics (herdruk). Oxford: Oxford University Press.

Meckled-Garcia, S. (2009). 'Do Transnational Economic Effects Violate Human Rights?', Journal of Ethics and Global Politics 2(3), pp. 259-276.

Nussbaum, M. and Sen, A. (eds.) (1993). The Quality of Life Oxford: Clarendon Press.

Nussbaum, M.C. (2000) Women and Human Development: The Capabilities Approach. Cambridge: Cambridge University Press.

Nussbaum, M. (2006) Frontiers of Justice. Disability, Nationality, Species Membership. Cambridge, Massachusetts: Harvard University Press.

Rawls, J. (1992 [1971]). A Theory of Justice. Cambridge, Massachusetts: Harvard University Press.

Rawls, J. (1993). Political Liberalism. Columbia: Columbia University Press.

Round Table for Sustainable Biofuels (2008). http://cgse.epfl.ch/webdav/site/cgse/shared/Biofuels/VersionZero/Version%20Zero_RSB_Std_en.pdf. Accessed 20 January 2010.

Ros, J.P.M., K.P. Overmars, E. Stehfest, A.G. Prins, J. Notenboom and M. van Oorschot (2010) Identifying the indirect effects of bio-energy production. Netherlands Environmental Assessment Agency (PBL), Bilthoven.

Sen, A. (1985). Commodities and Capabilities. Oxford: Oxford University Press.

Soltau, F. (2009). Fairness in International Climate Change Law and Policy. Cambridge: Cambridge University Press.

United Nations Framework Convention on Climate Change (1998). http://unfccc.int/resource/docs/cop4/16a01.pdf. Accessed 20 January 2010.

Wakker, E. (2005). Greasy palms: The social and ecological impacts of large-scale oil palm plantation development in South East Asia. Friends of the Earth. http://www.foe.co.uk/resource/reports/greasy_palms_impacts.pdf. Accessed 20 January 2010.

Wiskerke, W. and Winckers, J. (2010) Het verborgen klimaat effect van biobrandstoffen. Stichting Natuur en Milieu, Utrecht.

Bioenergies: unveiling the ethos of the agrarian research

M. de Lattre-Gasquet[1], D. Vermersch[2], M. Bursztyn[3] and P.-H. Duée[4]
[1]Cirad, 42 rue Scheffer, 75116 Paris, France; [2]Agrocampus Ouest, Département Economie Rurale et Gestion, Rennes, France; [3]University of Brasilia, Centre for sustainable development, Brasilia, Brazil; [4]Inra, rue de Universite, 75007 Paris, France; marie.de_lattre-gasquet@cirad.fr

Abstract

At present, biofuels arouse both hopes and fears, although agriculture has always produced, transformed and consumed various forms and quantities of energy. There would be no major ethical question if agricultural production was sufficient to feed the world, if biodiversity, culture and social rights were respected, and if pressure on cultivated land did not lead to ecosystems degradation. However, more than one billion people are malnourished and agriculture must simultaneously also respond to demands for non-food products. Agricultural research institutions (and researchers individually) are increasingly constrained by ethical issues related to biofuels. In France, the Joint Ethics Committee of Inra (Institut national de la recherche agronomique) and Cirad (Centre de coopération internationale en recherche agronomique pour le développement) has recently worked on this question. Biofuels appear as a revealing (in the chemical sense) of a crack in the *ethos* of those institutions. Cirad and Inra both carry out mission-oriented research. For years, researchers have worked to increase productivity and agricultural revenues, improve food security and market stabilization, reduce inequalities, etc. This mission-oriented research was a moral bail to agricultural productivity. Environmental damages provoked by agriculture and development policies leading to a wide range of problems are undermining the *ethos* of the two research institutions. Biofuels research, in large terms, is revealing difficulties as different research issues are competing among themselves. The Committee recommended revising the meaning of 'mission-oriented research' and made specific recommendations on how to deal with some ethical questions related to biofuels research.

Keywords: ethics, biofuels, ethos, agricultural research

Biofuels: hopes and fears

Beginning 2008, the chairmen of the Board of two French agricultural research organisations, the *Institut national de la recherche agronomique* (Inra) and the *Centre de coopération international en recherche agronomique pour le développement* (Cirad) have asked their joint ethics committee *(comité consultatif commun d'éthique pour la recherche agronomique)* to reflect upon the question of biofuels and the non-food valorisation of agricultural production. Cirad and Inra are increasingly asked by their partners, whether public or private, to improve traditional bioenergy and accelerate the transition towards more sustainable means of energy production. The institutions and the researchers individually are increasingly constrained by these solicitations which lead to ethical questions related to the principles guiding choices of research themes and priorities, as well as conditions of partnerships, especially in developing countries. The ethical committee analysed the long and hectic relationship between agriculture and energy and the current controversies. It looked at the ethical impact of biofuels and made recommendations.

As it produces biomass from solar energy, agriculture produces, transforms and consumes various forms and quantities of energy whose variation depends on the latitude, the technical level of the production systems, the extent of product transformation and the animal's contribution to the final production.... Until recently, agriculture fulfilled its own energy needs, the archetype being animal traction used on close to 20% of European farmlands. But the energy thus produced was for internal consumption. The link between agricultural markets and energy, on the output side, grew weaker during the 20th

century as the transport sector was increasingly dominated by fossil fuels. As agriculture became more dependent on motorisation and chemical fertilisers derived from fossil fuels, energy grew in importance as an input. This was also borne out through energy-intensive activities such as produce warehousing, transformation and distribution.

As a result, autonomous agriculture that produced its own inputs became increasingly heteronomous in the sense used by I. Illich (1973) in its production modes, which are dependent on agriculture both technically and economically, both upstream and downstream. Heteronomy resulted from the adoption of technological innovations, on the one hand and, on the other, the logic of division of labour, production specialisation and, hence, the opportunity to apply economies of scale thanks to the size of the production units.

Since biofuels (firewood, charcoal, animal excrements, etc.) generate less than 2% of the world's current transport fuels, during the last few years international discussions on sustainable development have focused sharply on biofuels as the only substitute available in the short term to replace fossil fuels used in transport. Doubts are being raised about their potential impact on food security and environmental safety, although by the year 2030 estimates indicate that less than 4% of the arable lands will be used for the production of liquid biofuels (OECD/IEA, 2006). Biofuels are complex, since they involve very different local situations, production conditions and scales, and timeframes. Their competitiveness depends on the relative prices of fossil fuels and agricultural commodities, hence crop types, farming modes and production sites. Excluding exceptional cases such Brazilian ethanol made from sugar cane, subsidies and tax benefits are needed to guarantee competitiveness even when oil prices are high. But incentive policies do not favour the developing countries which have comparative advantages in the production side, but lack adequate processing and export capacity. Yet, the production of agricultural commodities as feedstock for biofuels could provide an opportunity for agricultural growth and rural development in these countries (FAO, 2008).

Further, the impact of biofuels has not been clearly established. Assessments, which vary according to site and process, remain partial as a result of goals multiplicity and lack of goal ranking. Further, how greatly biofuels contribute to reducing greenhouse gas emissions is still a controversial issue.

Biofuels: unveiling the ethos of the French agricultural research

The ethics of biofuels is not limited to arbitration between developing the value of food and non-food agricultural products (Thomson, 2008a,b). This question resonates deeply, especially upstream, when effective mastery of finalised research depends on goal ranking, in other words, on the ethical objectives of which the public research institution is both the custodian and the guarantor. This is what we call here the *ethos* of agronomic research, in other words a set of action-oriented principles stemming from a widely shared ethical target. The etymology of the word *ethos* indicates that the term is polysemic but refers to a characteristic of the person. Immerged in a changing world, human beings feel the need for a residence and stable rules of existence. This has led to the idea that to 'reside' under morally just conditions requires a 'residence' or 'habitat'. By extension, this means an organic set of shared ethical arrangements, convictions and ways of living that make life together sustainable.

At both Inra and Cirad, the biofuels issue has brought up many questions of ethics. Their ethos, today, seems overstretc.hed on two scores: rationality dictated by technological and economic goals that impose form on innovations, and rationality dictated by sharply rising present-day values such as global ecological awareness, biodiversity, food sovereignty, agricultural and rural multifunctionality, etc. Referring to the distinction suggested by Max Weber (1971) between instrumental rationality and axiological rationality (rationality relative to values) sometimes could lead to the minimalisation of the antithesis between what is called the ethics of responsibility and the ethics of conviction. This said, ethical questions on biofuels crystallise this exaggeration, with the risk of corrupting the founding ethical principles of agricultural research, through, for instance, two-sided innovative research in which biofuels could, in the long run, lead to higher food prices.

Illustration: the Inra case

The institutionalisation of French agricultural research, at least as far as Inra is concerned, was influenced by an original ethos that, in the main, fit in with the original objectives of the Common Agricultural Policy (CAP) introduced in the early 1960s, namely, to increase productivity, ensure a fair standard of living for the agricultural community, stabilise markets, secure availability of supplies, and provide consumers with food at reasonable prices. The moral principles of this ethos, thus, are stated as economic necessities or as a series of interconnected ultimate goals that guarantee the ethics of agricultural productivism, thus referred to as 'moral productivism'. This ethos was broadly shared by the scientific community and the agricultural community as a whole, as shown in its solidarity-based development vision, a vision adopted when Cirad was created, e.g. in its policy towards the hot regions.

By the end of the 1970s in France, this agricultural productivism had gradually lost its moral aura after production system intensification proved patently harmful to the environment and as the gap in farm revenues grew. At that time, the scientific community started focusing on consumer interests and more explicitly on consumer health. Environmental issues were considered later and became topical towards the end of the 1980s, which coincided with the first CAP reform, stoked by the budgetary cost of surpluses and then by environmental and territorial constraints that, from then on, would dictate public aid to production.

Quite clearly, these lapidary historical comments call for explanation and qualification, but the successive CAP reforms have diluted the Policy into a pale depiction of the original statement of ethos that bolstered the first steps in agricultural research.

Illustration: Cirad and oil palm research

The institutes created during the colonial period were grouped under Cirad, whose mission is to conduct research to improve production and develop outlets. To cope with the shortage of petroleum products in Europe during the second world war, the Cirad tropical forests technical section created a forest fuels division for heating and carbonisation, while the *Institut de recherche pour les huiles et oléagineux* (IRHO) grew increasingly interested in fuel oils and worked on catalytic cracking of plant oils. Neither of the two sought large-scale development of energy crops; both were anxious to respond to needs for greater exploitation of available local resources. The drop in oil prices brought these production projects to a halt ... until the first energy crisis in the 1970s. By 1985, anthropic pressure in the dry tropics, especially in Africa, had grown so intense that research started focusing on technological processes, analytical tools for industries *(filière)* and their organisational methods. In 2007 Cirad decided that the second of its six priority lines of research would be: 'Studying how we can ensure that the emergence and development of bioenergies favour people in developing countries' in order to provide political and private sector decision-makers with rational, reliable information. Further, in response to the growing demand for palm oil, Cirad, together with Sofiproteol created PalmElit SAS, a subsidiary for the production and distribution of oil palm seed. The combination of the new strategic line of research and the creation of PalmElit has caused some tension within and outside of Cirad.

Palm oil is traditionally used in the agro-food industry (approx. 80% of production) and in oleo-chemistry (approx. 20%), has an exceptional yield close to four tons of oil per ha (palm oil and palm kernel oil), in other words ten times more than soybean oil and four times more than rapeseed oil. Production costs are the lowest among the vegetable oils. It is considered to have high potential for biodiesel production. In less than 30 years, the land area given to oil palms has increased threefold to reach close to 14 million hectares. The rapid development of these plantations, especially in Indonesia but also in Malaysia and several countries in Africa has triggered extremely sharp controversy. Palm tree plantations have unquestionably been responsible for vast deforestation of primary and secondary forestlands, with very negative consequences leading to habitat conversion and loss of biodiversity. Although this phenomenon accounts for less than 3% of total deforestation in Indonesia, where the polemic is the most bitter, certain companies receive concessions ostensibly to create palm trees plantations. They then fell the trees and disappear without planting a single palm tree. Further, palm tree cultivation causes pollution: herbicides, pesticides, other inputs, and factory effluents. Depending on the production techniques, palm trees have a negative or positive effect on climate change. The effects are negative when land is cleared by burning and when methane is released from the waste settlement ponds without anything being done to contain it. Positive effects: carbon storage (if palm trees are planted in forest-free zones, the savannah and previously deforested areas), and savings in fossil fuels (if the methane from effluent treatments is captured and used to economise on fossil energy). The palm tree is at the heart of numerous social conflicts with the local communities that feel despoiled by companies that obtain concessionary rights from the authorities, without prior clarification and settlement of their customary rights. Deforestation can jeopardise the survival of the most impoverished rural population. The power play between the village planters and the factory representatives who process the planters' yields is often unequal. And the labour law is not systematically applied in certain industrial plantations. Some plantations are being accused of connections with the Mafia, armed rebels, etc. Last, in 2007, when the price of palm oil soared (+70%), certain countries started to export rather than serve domestic markets, causing local shortages that have serious effects on the diet of poor consumers.

When palm oil was not used for biofuels, Cirad and other national research institutions conducted research without problems. The production of biofuels and the increase of cropping zones with the above-mentioned problems leads to greater demand for research. It is now that Cirad's research is confronted with goal-oriented (technological and economic) rationalities and value-oriented rationalities.

Science and the market: chimerical coherency

Whether they like it or not, public research institutions are being propelled to the crossroads of science and markets as a result of unprecedented hybridisation of scientific and economic rationality which leads stalwartly into a knowledge economy. Henceforth efforts to develop knowledge become the focus of economic interests. In a globalised world where most barriers to trade are gradually removed, secure income is often temporary, as are various forms of division of labour. Ultimately importance is only given to the innovation race, which is both the cause and effect of technological autonomisation often judged inevitable.

Public research institutions do not only think that they must follow this so called liberal turn, but also lend it theory and justification, even take it as an example. In time this could transform public research into an industrial sector or a service sector. Without querying the need for partnerships between research and industry in the biofuels sector, we need to recognise the risk of institutional and mimetic isomorphism between public and private research. Isomorphism can be traced to the present-day context, ridden with uncertainties. It is depicted especially through the rhetoric on competitiveness, which invades the space and discourse of scientific policy.

Revisiting the ethos of agricultural research and development-oriented research

Inra and Cirad are developing mission-oriented research whose starting point includes demands of a political and socioeconomic nature entailing choices made difficult by the stakes involved, in other words, choices that always affect both the stakeholders and the scientists. The new context for innovations in agriculture is marked by the growing problem of scarcity (of soils, water, fossil energy, food production, etc.) that requiring a new construct for questions calling for answers. This includes the downstream conversion of problems into research issues, with the participation of all parties concerned. It is related to the 'philosophy of agriculture' (Thompson, 2008).

To cope with this situation, the ethics committee suggested refocusing on the ethos of agricultural research and development-oriented research (including sustainable development) and then reformulating the objectives. With this in mind, it is important to distinguish between instrumental rationality and axiological rationality on the one hand, and between the various axiological rationalities on the other. The first difference dwells on the distinction between effectiveness (instrumental rationality) and the horizon of values from which we assess actions (axiological rationality). The importance of effectiveness must not allow us to forget the importance of values. The linkage between these two rationalities is too sensitive to be governed solely by market dynamics. We need to think explicitly about the horizon of value from which we assess choices and actions. The second difference concerns the axiological evaluation of value, which can either be an assessment of the action's goals or a reference to values defined independent of the means and the ends, i.e. the intrinsic values. With regard to agriculture, the final goal plays an important role in the common opinion and brings out representations that suggest the existence of a hierarchy of purposes. This may lead, *inter alia,* to the assertion that in the agricultural sector priority should be given to food supply thus relegating biofuels to the position of product, co-product or by-product. For agricultural activities, this assessment of the final goal includes a question on the relationship between these activities and the natural environment, and, more broadly with nature, which is thus and again solicited in two ways, as a technical system of references and as an ethical system of references.

The intrinsic values are composed of people's rights and freedoms: the difference between the price and dignity (of the person) is based on these rights and freedoms composed of fundamental freedoms (of action, speech and conscience) as well as basic rights (basic needs, which, if not fulfilled, prevent the attainment of freedoms). The question of justice can be broached on the basis of these two types of values.

For the proponents of a more ecocentred ethical approach, or even moral naturalism, the brief is to seek the *'conditions of equilibrium'*, or homeostasis research in which physical principles inevitably upstage moral principles. For the others, the brief is to seek *'universalisable reciprocity'* in order to transcend partial collusion of interests. This requires rationalisation and hence universalisation of values even if it means suggesting new counterfactual ideals (for example, biodiversity, sustainability, multifunctionality) capable of inspiring the research finalisation mode. But, on the one hand, the effort to universalise values clashes with legitimate cultural diversity; on the other hand, it has been surpassed by another universalism, economic universalism that underpins trade globalisation.

The ethics committee then also proposed a grid of principles and values that juxtaposed or even compounded physical principles (limited nature of resources, systemic approach, hardiness and adaptability, etc.) and moral principles (human dignity, freedom, equality, etc.). Nature, thus, in a double role of technical and ethical system of references, has not been rejected as a possible benchmark on the road to the universalisation of values.

The Committee also recommended that Inra and Cirad continue studying the definition of mission-oriented research, research, which explores various potential paths to problem solving, is based on transdisciplinary approaches and involves a large variety of stakeholders. Since the term 'sustainable development' is commonly used by both organisations, the Committee also recommended studying the present-day significance of sustainable management of development on the one hand and, on the other, the effects of the stakes of sustainable development on the way to approach and conduct research, especially research on biofuels.

Last, the Committee suggested revisiting the question of the methodology of transdisciplinarity, which uses a different method in the organisation of questions on scientists' competency and in the management of new controversial issues.

References

Food and Agricultural Organisation (2008). La situation mondiale de l'alimentation et de l'agriculture. Les biocarburants: perspectives, risques et opportunités. Rome: FAO, décembre 2008. 156 pages.

Illich, I. (1973). La Convivialité. Paris: ed. Seuil, coll. Points. (Eng. Title 'Tools for Conviviality).

Organisation for Economic Cooperation and Development (OECD)/International Energy Agency (IEA) (2006). World Energy Outlook, 2006.

Thompson, P.B. (2008a). The agricultural ethics of biofuels: a first look. Journal of Agricultural and Environmental Ethics, 21:183-198.

Thompson, P.B. (2008b). Agricultural Biofuels: Two Ethical Issues. In: NABC Report 20: Reshaping American Agriculture to Meet its Biofuel and Biopolymer Roles.

Weber, M. (1971), Économie et société, Paris, Plon.

Biomass production for energy usage in Bavaria: an ethical assessment using a case study comparison of rapeseed and sorghum

C. Dürnberger[1], B. Formowitz[2], H. Grimm[1], A. Uhl[2] and M. Zichy[1]
[1]Institut TTN (Institute Technology, Theology and Natural Sciences), 80335 Munich, Germany; [2]Technologie- und Förderzentrum Straubing (Technology and Support Centre in the Centre of Excellence for Renewable Resources [TFZ]), 94315 Straubing, Germany; christian.duernberger@elkb.de; anne.uhl@tfz.bayern.de

Abstract

The Institute Technology, Theology and Science (TTN) and the Technology and Support Centre (TFZ) have set up a joint research project to explore whether the production of biomass for energy use in Bavaria can be ethically sound. In a first step, a three-phase 'Ethical Delphi' was carried out. In a second step, the results of the Ethical Delphi were correlated with paradigmatic and normative concepts that influence the debate in order to create a variation of the 'Ethical Matrix' tailored to energy from biomass. The so-created ethical tool was then applied to two different scenarios which were rapeseed for small-scaled oil production and sorghum for biogas production. In this paper, this newly developed ethical tool will be introduced and first interim results of the case studies presented.

Keywords: energy from biomass, sustainability, ethical matrix

Introduction

Energy shortage and climate change have made the search for alternative energy sources one of the key issues of our time. Energy production from biomass is considered to be one possible alternative or at least supplement to the use of fossil fuels. Contrary to other energy alternatives, biomass-to-energy has given rise to a controversial ethical debate revolving around agriculture's social responsibility to feed people and the need for ethically sound ways to produce bioenergy (Mitchell, 2008; MNP, 2008; Streffer *et al.* 2005; Vogt, 2002; Karafyllis, 2000).

The Institute Technology, Theology and Science (TTN) and the Technology and Support Centre (TFZ), both situated in Bavaria in Southern Germany, have set up a joint research project to explore whether the production of biomass for energy use in Bavaria can be ethically sound. In a first step, a three-phase 'Ethical Delphi' was carried out that allowed controversial issues to be structured according to expert opinion. A body of experts, including natural scientists, representatives of the humanities, agricultural practitioners, etc., were invited to share their views on the issue and to investigate decision-making processes. First results of the Ethical Delphi were presented at the EurSafe Congress 2009 in Nottingham (Dürnberger *et al.*, 2008). In a second step, the results of the Ethical Delphi were correlated with paradigmatic and normative concepts that influence the debate in order to create a variation of the 'Ethical Matrix' (Mepham *et al.*, 2006) tailored to production of biomass for energy use. The so-created ethical tool was then applied to two different scenarios which were rapeseed for small-scaled oil production, which is used as pure plant oil and sorghum for biogas production in Bavaria. The biogas is used either for electricity and heat or for directly deploying biogas as natural gas. In a third step, the results of step one and two will be validated and discussed in a 'Citizen Conference' in 2011.

In this paper, the newly developed ethical tool, consisting of two different matrices, including the classical Ethical Matrix, will be introduced and interim results of the case studies presented.

Methodology

The overall objective of the project is to explore whether the biomass production for energy use in Bavaria can be ethically sound. Due to the fact that the simple question 'Is biomass production for energy use in Bavaria ethically sound?' cannot be answered in a general way because there are too many contingent factors involved, the scope of the project has been limited by the following premises:

- The primary user of the ethical evaluation is the farmer, as a person who decides what to grow and who has to stand up to public ethical concerns.
- The ethical evaluation should be done on two case studies: rapeseed for small-scaled oil production and sorghum for biogas production. Narrowing down the question to these two scenarios helps decisively in making the ethical analysis and judgement substantial and reliable. Besides that, these two particular scenarios have been chosen because they are realistic options for Bavaria.
- Beyond the ethical evaluation of these two scenarios the project should deliver an ethical tool that generally helps to make ethical judgements on biomass production. In particular, it should allow not only farmers, but also politicians and other interested persons, to gain an overview of possible areas of tension and future conflicts, and of what to take into account when deciding on what to grow. Moreover, the tool should help to improve the communication between stakeholders, scientists and consumers. This meant designing an ethical tool which is easy to handle, not too time-consuming to use and not too complicated.

Regardless of whether a narrow scenario or relevant to biomass production as a whole is to be judged, the ethical evaluation of biomass production has to be done in three different dimensions: environmentally, societal and culturally. As a consequence, for the ethical assessment a four-step ethical tool has been developed, consisting of the following parts:

- Step 1: ethical assessment of the possible environmental impacts.
- Step 2: ethical assessment of the possible societal impacts.
- Step 3: ethical assessment of the involved or affected cultural aspects.
- Step 4: Final ethical evaluation and judgement.

In the following outline, the methodology of the three first steps and their application to the two scenarios will be described. By this, the developed ethical tool will be presented at here with some interim results of the case studies.

Environmental dimension

Methodological remarks

There is a broad public, political and expert consensus that analysing biomass production from an ethical point of view clearly involves evaluating its possible impacts on the environment.

Being a highly complex task, a detailed assessment of the environmental impact is beyond farmers' responsibilities and competencies, however, a number of institutions are engaged in sustainability assessments. Moreover, the nature of the environmental impact is extremely dependent on the specific circumstances (weather, soil conditions, etc.) that are hard to record. Therefore, generalised statements can barely be given. Finally, Germany has implemented relatively high legal standards of environmental sustainability for growing biomass.

As a result of these reasons, the environmental dimension is dealt with in this assessment in the following rather simple way. For both processes, biomass production and the conversion pathway, the main impact factors and possible options for enhancement are listed in two different matrices with regard to the five basic principles of environmental protection: (1) soil, (2) water, (3) air, (4) climate and (5) biodiversity protection. The so-gained matrices (the matrix for biomass production is presented in Table 1) give the practitioner an informative yet not overburdensome view on the environmental side of this practice. This allows individuals to check which measures have already taken and which could still applied. The matrix for processing will be described in the final project report. The possible differences between centralized and decentralized processing and use will be considered.

Regarding the ethical assessment of the environmental impact, the matrices allow a initial rough appraisal of the ethical relevant differences between practices or cultivation scenarios.

Table 1. The environmental-ethical matrix for biomass production.

Protection of	Some factors of influence	Good professional practice (Gute Fachliche Praxis) and some possibilities for optimisation
Soil	Erosion Cultivation Soil compaction Farm Machine time Fertilization Crop Rotations	Protective measures against erosion (contour ploughing, soil cover over the whole year, planting windbreaks, etc.) Optimizing cultivation techniques, only driving on dry soil, reducing/ adapting tire pressure, etc. Reduction of machine running time Fertilization according to the regulation for fertilization, reduction of fertilization to a minimum, consideration of soil nutrient concentrations for calculation of fertilizers requirements, etc. Well balanced crop rotations with humus increasing and decreasing crops; planting crops with different root systems (deep-rooting and shallow-rooting plants), etc.
Water	Emissions of fertilizer Emissions of pesticides Soil type and its cultivation Weather	Additional measures for water protection (according to the water framework directive) Optimal fertilization (according to regulation for fertilization) Optimal usage of pesticides (according to the plant protection legislation)
Air	Odour emissions Greenhouse gas emissions NH_3-emissions Dust emissions Emissions of pesticides/ herbicides Emissions of fertilizer	Optimal usage of fertilizer (according to regulation for fertilization) Optimal usage of pesticides (according to the plant protection legislation) Installation of filters (particulate filter, dust filter, etc.) Minimizing greenhouse gas emissions
Climate	Greenhouse gases: Fertilizer fabrication Machinery production Emissions emerging through cultivation and transport By- and co-products	Sustainability directive, etc. Optimal cultivation practice Conservation of energy
Biodiversity	Changes in land use Alteration of agricultural practices (towards monocultures) Pesticides / herbicides	No usage of land with a high floral and faunal diversity; compensation measures/ compensation area Wider crop rotations, permanent cropping (permanent grassland, etc.), using different species or varieties, perm culture, etc. Minimization of pesticide and herbicide use

Case study results

In general, rapeseed and sorghum do not notably diverge in terms of the issues raised in an environmental-ethical reflection. Their cultivation and environmental impact depend mainly on location factors (climate, soil, etc.). Furthermore the cultivation techniques, e.g. organic and conventional farming systems, etc., and the machinery used are important when considering the effect on the environment. Therefore it is difficult to compare the cultivation and cultivars of different enterprises, if various conditions prevail. Additionally, the conversion pathways, e.g. decentralized oil production or biogas production, vary immensely in terms of technical equipment and levels of efficiency.

However, some differences can be identified. In Germany, mainly winter rapeseed is cultivated, while sorghum is grown as a summer crop. Nevertheless, rapeseed generates a moderate surface cover of the soil till the beginning of winter and thus soil erosion might occur. For sorghum it depends on the crop rotation and which catch crop (before the main crop) is cultivated. It is vital to consider if the intermediate crop builds soil coverage comparable to rapeseed, or if a cultivar is taken, such as perennial ryegrass, which forms an extensive coverage thus protecting soil against erosion.

In contrast to sorghum, rapeseed is known as an intensive crop, although the primary cultivation and weed control is similar, except the use of herbicides. These differences of active substances might have different impacts on the environment. Depending on the soil fertility and the disease and pest pressure, generally more plant protection and fertilization steps are needed. More working steps associated with longer machine running time increase the labour input and exhaust emissions. With an appropriate application of plant protection products or fertilizer no further water and air contaminations should arise. On the other hand rapeseed has the benefit to attract insects in its flowering state, such as bees and bumble-bees.

In oil mills dust emissions occur during storage and rearrangement of rapeseeds, while gas emissions can emerge in biogas plants, without appropriate coverage of the digesters or final depots. There is a big difference in the amount of water used during the conversion. Biogas plants run exclusively with biomass from renewable resources, without the input of slurry, have to dump water to initialize fermentation. If no technical preparation of the biogas digestate takes place, such as separation into solid and fluid fractions, the water stays in the digestates and no residual water arises. Thus, biogas plants can be seen as a closed system.

Societal dimension

Methodological remarks

Besides an environmental assessment growing and processing biomass for energy use also has to be evaluated with regard to its possible societal impacts. For that, Mepham's classical 'Ethical Matrix' is used. Based on an inquiry in academic literature and on the results of the Ethical Delphi mentioned above, the relevant stakeholders have been identified (see the societal-ethical matrix, Table 2). As ethical criteria serve the three well-established, both theoretically and practically justified, ethical principles which also can be found in Mepham's matrix: respect for wellbeing, respect for autonomy and respect for justice.

All in all, the societal ethical matrix offers a relatively comprehensive view on possible societal impacts and fields of tension and allows for the identifyication of critical points quickly. Additionally, it will help to improve the communication between stakeholders, especially between farmers, consumers, scientists and other affected individuals.

Table 2. The societal-ethical matrix for biomass production.

Stakeholder	Wellbeing	Autonomy / freedom	Fairness
Farmers	Economic wellbeing	Freedom of choice between economic alternatives	Competitiveness
	Working conditions		
National energy consumers	Secure and economical energy supply	Freedom of choice between different energy sources	
		Reduction of dependence from single energy sources	
National and international food consumers	Secure and affordable high value food products	Autonomy (depends on food security)	International justice
Fodder consumers	Secure and affordable high value feeding stuff	Freedom of choice (greater assortment of fodder products, e.g. Rapeseed provides press cake; Sorghum can be taken as fodder, but has no fodder value in Germany)	
Regional Communities	Maintenance of the natural basis of life (refers to environmental protection)	Participation (right to have their opinion, see cultural dimension)	
	Strengthening the economics of a region		
	No essential reduction in quality of life (through noise, odour, architecture, etc.)		
Fellow human (international)	Secure and affordable high value food	Autonomy and freedom (depending on food security and preservation of natural basis of life)	International justice
	Preservation of the natural basis of life		
Future Generations	Preservation of the natural basis of life	Autonomy and freedom (understood as chances of life, depending on natural basis of life)	Intergenerational justice
Others e.g. bee keepers	Wellbeing of bees		

Case study results

When examined more closely, there are some differences between rapeseed and sorghum cultivation concerning their societal impacts. With regard to the working conditions, rapeseed is known as an intensive crop, although the primary cultivation practice is similar to sorghum.

With respect to the fodder industry, one co-product of rapeseed oil processing is press cake which amounts to two thirds of the input raw material. Press cake is known to be a high value fodder, mainly consisting of protein. In contrary, in the process of biogas production out of sorghum no co-product which can be used as fodder is produced. Ruminants in certain regions, e.g. tropical countries, feed on sorghum, but in Germany this system is not adapted. On the other hand biogas digestates, remaining after biogas production, can be used as a high value organic fertilizer. Since rapeseed has the benefit of attract bees when in a flowering state, beekeepers are another groups of stakeholders preferring this form of cultivation.

So, the differences between rapeseed and sorghum are rather marginal. As a consequence, these case studies cannot reveal the strengths of the matrix above. However, if applied to 'bigger' questions, e.g. energy from biomass vs. wind energy vs. hydropower vs. solar energy, the matrix will serve as a powerful tool for an ethical discussion.

Cultural dimension

The controversy about bioenergy not only is a controversy about risks and benefits for nature and fellow humans but also about deeply rooted cultural prejudices. An ethical assessment only focusing on the environmental and the societal dimension is insufficient because of fundamental, culturally embedded views about agriculture and the feeding purpose of plants, which stand against energy from biomass. Although these views are – strictly speaking – barely of ethical relevance and may therefore be easily overridden, they should nevertheless be taken into account even from an ethical point of view. There are two reasons for that, firstly these views are the legitimate expressions of an individual's autonomy and must therefore be respected. Secondly, these views might be a serious obstacle to the energy production from biomass. Dealing with these views in a responsible way is both strategically and ethically indispensable.

Concerning the debate on energy from biomass, there are at least three influential cultural and historical ideas that must be taken into consideration:
- The symbolic value of plants normally used for nutrition. This is especially true for corn which is the basis for bread, a ubiquitous foodstuff.
- The social role of agriculture as a traditional supplier of food.
- The public image of agriculture as idyll.

All these concepts are powerful cognitive frames in the perception of bioenergy. As a result, any successful communication strategy has to take into account these attitudes and believes. The impact for rapeseed and sorghum has not yet been analysed in detail. However, the fact that rapeseed is traditionally grown in Germany for edible oil production while sorghum is not commonly grown will certainly play a role in this context. Furthermore the initial situation might be important. If a crop is already cultivated in a large scale and its production area is increased, it might occur that only that crop is planted in this area. If an unpopular crop is introduced into the crop rotations, if at all, it might be noticed as an additional cultivar.

Conclusion and outlook

What is clear from this work is that *general* ethical valuations about biomass production for energy usage are impossible to make. There are too many factors varying significantly according to engaged technology, time, place and other circumstances (e.g. weather, soil condition, rain, political decisions, market developments, etc.). A substantial ethical evaluation is only feasible when focused on specific cases. However even at this level, clear statements are difficult. As to the comparison between the production of rapeseed and sorghum in Bavaria for energy production, only minor ethical differences have been identified. If at all, rapeseed performed slightly better in the ethical matrix.

Besides that, the developed ethical tool delivers a methodological framework which can guide in-depth ethical analyses as well as quick ethical surveys of biomass-to-energy. It especially seems suitable to compare the ethical aspects of different alternative energy technologies, e.g. energy plants, solar energy, wind energy, etc. Further data from this study will be published in 2011.

References

Dürnberger, Ch., Formowitz, B., Grimm, H. and Uhl, A. (2009). Technological innovation and social responsibility: challenges to Bavarian Agriculture and the provision of bioenergy. In: Ethical futures: bioscience and food horizons. Millar, K., Hobson West, P., Nerlich, B. (eds.). Wageningen Academic Publishers, Wageningen, 99-105.

Karafyllis, N.C. (2000). Nachwachsende Rohstoffe. Technikbewertung zwischen den Leitbildern Wachstum und Nachhaltigkeit. Reihe Soziologie und Ökologie. Band. 5. Leske und Budrich, Obladen.

Mepham, B., Millar, K., Tomkins, S., Thorstensen, E. and Kaiser, M. (2006). Ethical Delphi. Manual. LEI, Den Haag.

Mitchell, D. (2008). A note on rising food prices. Policy research working paper of the World Bank/Development Prospects Group. Available on http://www-wds.worldbank.org/external/default/WDSContentServer/IW3P/IB/2008/07/28/000020439_20080728103002/Rendered/PDF/WP4682.pdf [Date of consultation 26/03/10].

MNP (Milieu en Naturr Planbureau) (2008). Local and global consequences of the EU renewable directive for biofuels – Testing the sustainability criteria. Available on http://www.mnp.nl/bibliotheek/rapporten/500143001.pdf [Date of consultation 26/03/10].

Streffer, Ch., Gethmann, C.F., Heinloth, K., Rumpff, K. and Witt, A. (2005). Ethische Probleme einer langfristig globalen Energieversorgung. Walter de Gruyter, Berlin.

Vogt, M. (2002). Ethische Aspekte nachwachsender Rohstoffe. In: Nachwachsende Rohstoffe für Baden-Württemberg. Tagungsband eines Symposiums vom 21. 10. 2002. Ministerium für Ernährung und Ländlichen Raum (eds.) Karlsruhe, 1-12.

Part 9. Animal welfare

Designing a computer game for pigs, to explore and redefine the interface between animal welfare, science and ethics

C.P.G. Driessen[1], M.B.M. Bracke[2] and M. Copier[3]
[1]Applied Philosophy Group, Wageningen University, P.O. Box 8130, 6700 EW Wageningen, the Netherlands; [2]Wageningen UR, Livestock Research, P.O. Box 65, 8200 AB Lelystad, the Netherlands; [3]Design for Playful Impact, Utrecht School of the Arts, P.O. Box 2471, 1200 CL Hilversum, the Netherlands; clemens.driessen@wur.nl

Abstract

The regulatory requirement to provide enrichment materials to farmed pigs has been translated into a project to design a computer game for pigs. A team of researchers of animals, ethics and game design have gathered to develop an enriched environment meant to challenge the cognitive abilities of domestic pigs, and to offer a site for interaction between humans and animals in a playful and meaningful way. With designing an interactive device, scientific questions on subjective experience and cognition can be connected in a tangible and practical way with philosophical and ethical questions regarding suffering, capabilities, and enrichment. This generates an opportunity to experience and explore these 'heavy' questions in new ways by researchers and the public. This paper identifies philosophical issues relevant to the design process and to our understanding of the nature of animal cognition, subjective experience and ethical relations between humans and animals. Several ethological experiments and enrichment devices are discussed to illustrate their particular implications for understanding relevant philosophical and ethical questions. The paper ends by exploring possible meanings of cognition, naturalness, and the implications for pig housing design and for moral views on the practice of pig farming.

Keywords: pig housing enrichment, play design, animal cognition, capabilities approach, post-humanist ethics

Introduction: challenging pigs

Pigs have a reputation for being intelligent animals, but ethical implications of their supposed capacities do not seem to have materialised yet. Farmed pigs in the European Union are required to have access to 'enrichment materials', in order to allow them to perform their (natural) behaviour, to reduce boredom and tail biting, and, hence, to reduce the need for tail docking. According to directive 2001/93/EC, 'pigs must have permanent access to a sufficient quantity of material to enable proper investigation and manipulation activities'. This requirement has led farmers to provide materials such as a plastic ball or a metal chain with some plastic piping. However, these 'toys' for the most part neither appeal to the cognitive abilities of the pigs nor do they resolve societal concerns related to pig welfare (e.g. barren housing and injuries due to tail biting) and reduced animal integrity (tail docking; modifying the animal to 'fit' the system). To allow for a more interesting development of behaviour, and in the process of creating an opportunity to integrate ethical reflection and animal research, a team of game designers and researchers in animal welfare and ethics have joined to design a computer game for pigs. In this way we intend to explore possible meanings of animal welfare and shed new light on the ethical dimensions at the interface between animals (pigs), humans (farmers, consumers, scientists and game-designers) and technology (computer game, pig housing). Our working hypothesis is that the meaning of moral concepts co-evolves with technological developments, and, hence, that ethical issues can be studied as part of design processes in which designers, researchers, farmers and animals have an active role. The development of a computer game will trigger cognitive and behavioural processes as well as moral deliberations, which can then be studied in real time as they unfold.

The aim of the project is not only to create a device that helps farmers meet the Directive or to help reduce boredom in pigs. It also aims to show that different design choices may propose different understandings of animal behaviour and their capabilities, generating particular implications on what is adequate pig housing and how we ought to treat pigs. Depending on the environments we provide for them, we allow pigs to emerge in specific ways, to reveal themselves with certain qualities and behavioural repertoires, and as such to appeal to our moral commitments. As part of the design process we want to create a web-based interface through which pigs and humans could interact in a playful manner. The public will be invited to engage with existing farm practices, e.g. interacting with a pig in a competitive or collaborative multi player computer game. In this way a new 'hybrid' animal practice could emerge, mixing features from human and animal behaviour in practices of farm animals, performance animals, ethological research and pets. This new practice can offer a fresh angle to debate our ambivalent behaviour towards animals involving both tendencies for exploitation (as consumer) and concern (as citizen). Such a design activity, therefore, is likely to generate reflexive experiences of our societal dealings with intensively farmed pigs in a creative setting, less 'heavy', more indirectly and implicitly.

Our moral views about farm animals are as much a function of the relations between animals and particular humans (farmers, consumers and citizens) and evolved meanings in existing technological environments (free market system; barns with slatted floors), as of the rights they are granted in ethical theory. In this project we propose to explore ethics as a practical and shared mode of inquiry, in addition to being a matter of explicit conceptual analysis and syllogistic reasoning. In this way ethics can be practiced as an integral part of reflexive animal welfare research and critical technological design, and be a process of tinkering and learning (cf. Korthals, 2004; Mol and Law, 2004). The aim is then to create circumstances in which new connections, between philosophy and science and between humans and animals, can emerge.

This paper offers a brief overview of the philosophical part of the design process, as it involves a practical and tangible way of relating scientific, philosophical, technical, and ethical questions. First, we will describe a few research projects into animal cognition and enrichment devices. For each intervention it can be argued that a particular understanding of cognition is implied, as well as a particular mode of ethical thinking. Different modes of ethical thinking relate differently to cognition. For example: utilitarians relate it to animal sentience and suffering, while deontologists focus on what animal beliefs, desires and emotions imply for making them 'subjects-of-a-life'. After discussing these examples and the specific ways cognition and related ethical aspects emerge, we will explore the implications of these interwoven questions for designing a computer game for playful pigs and humans.

Material environments, cognition and ethics

Pigs are invariably called intelligent in popular culture, but despite their large numbers we have not yet managed to allow pigs to reveal their cognitive potential. The barren housing conditions of intensive farming facilities in which most pigs live offer little room to pursue the elaborate behavioural repertoires and strategies pigs, by nature, are capable of. The lack of cognitive challenges is likely to produce frustration and boredom (Wemelsfelder, 1994). This generates welfare problems: increased aggression and abnormal behaviours such as tail-, ear- and flank biting in pigs.

Domestic pigs are still able to live natural lives

From the 80s a group of ethologists led by David Wood-Gush designed research environment called 'Edinburgh Pig Park'. By placing production animals in a more natural environment, their behaviour could be compared to that in commercial housing conditions (e.g. Stolba and Wood-Gush, 1989). The land race pig, bred for centuries into the production animal we know now, showed the whole range of behaviours found in its wild ancestor, the wild boar. Pigs still operate behavioural rules that derive from their natural environment, e.g. they still have a win-shift feeding strategy (Held *et al.,* 2009). Instead of searching for food in the place where it was last found (the feeding trough), pigs prefer to search for food in novel places as if their environment still resembles the natural characteristics of food depletion after foraging. Thus, pigs still have many properties and capacities that derive from their evolutionary history in their original habitats, and this largely determines what pigs need in order to be happy. This may include using their cognitive abilities, as cognition is an integral part of pigs' welfare needs (Bracke and Hopster, 2006). The Pig Park experiment refuted the (often implicit) argument that domestic pigs no longer have the behavioural needs associated with its ancestral environment; that they have become adapted to living in a barn. In fact, modern domesticated pigs still have 'original' morphological characteristics such as rooting disks which they no longer need in captivity, and they are still very well able to survive in the wild (as feral pigs in many parts of the world do). The pig housing design paradigm developed upon these findings (the Family pen system for pigs) provides an example of a system that goes beyond the objectives of a narrowly conceived utilitarianism that focuses on the absence of animal suffering and pain. It appears to incorporate a more comprehensive notion of well-being related to animal flourishing, or even *telos.*

Pigs can use mirrors, but did not recognize themselves

Recently an animal experiment at the University of Cambridge has generated evidence of pigs' ability to use mirrors. After five hours of experience, seven out of eight pigs were able to adequately use a mirror to find food. Instead of looking behind the mirror, the pigs immediately went to look for the food in the right place. Non-mirror-acquainted pigs did first look behind the mirrors (Broom *et al.,* 2009). The pigs failed to pass the classic self-recognition test, i.e. they did not respond to dye on their faces that had been applied while they were unconscious (asleep), but perhaps this was due to the fact that pigs are often 'dirty', so they are used to having 'spots' on their faces (Milius, 2009).

This cognition experiment has a more artificial set-up than the Edinburgh Pig Park. Compared to the binary pass/fail set-up of the tool-use and self-recognition experiment, the Pig Park allowed for the expression of animal agency in a much more multi-faceted and open setting. The mirror experiment appears to assume a more disembodied understanding of cognition, as if a Cartesian mind is making inner projections of the outside world, solving analytical problems based on primarily visual information. In public media interviews Broom expressed his hope that with our increasing knowledge of the abilities of pigs to use tools or complex devices, their moral status would follow suit and their conditions of living be improved: 'if an animal is known to be clever it is less likely to be treated as a food-producing machine and more like a sentient being. Perhaps the conditions in which pigs are raised ... may be improved as a result of the study'(Milius, 2009). This hope seems based on the assumption that moral concern is supervenient on intrinsic characteristics, that we attribute moral respect based on features of resemblance to ourselves and our understanding of intelligence, and that using a mirror indicates stepping up the ladder towards human capabilities and a legitimate claim to more humane treatment.

A mandrill playing an electronic game with zoo visitors

Other species with even more cognitive potential, as well as with welfare concerns related to confinement, are apes and monkeys. Zoo enrichment mostly aims to simulate –at least in the eye of visitors- the animals' natural habitats (Shepherdson, 1998), but in some cases the interventions are more artificial and playful. In the 1970s a male mandrill in a zoo in Washington state who behaved too aggressively towards the other mandrills in the enclosure was given an electronic device to play with. It was a simple three button reaction-time game, that offered the opportunity to win a piece of food either by competing with visitors – who had access to a similar device – or a computer that had a fixed time interval. After a few months of practice the mandrill managed to respond quicker than visitors most of the time, and he preferred to play human contestants over playing against the computer, presumably as he had figured out that this offered a bigger chance of winning. The device led to a reduction in aggression towards his pen mates (Markowitz *et al.,* 1982).

In this example the reduction in aggression is taken as a sign of improved well-being of the mandrill, and aggression was thereby revealed as related to an apparent lack of cognitive challenges. Thus, here affect and cognition are found to be linked. A utilitarian argument can be associated with this set-up: preventing the suffering of penmates by distracting the aggressor. Alternatively, it may be associated with a capabilities approach aiming at animal flourishing (Nussbaum, 2004) focusing on how the device allows the mandrill to develop and use its cognitive capabilities.

An orang-utan on facebook

Recently the orang-utan Nonja, living in a Viennese zoo, has gone online. She has started uploading images she makes with her compact camera to her own facebook profile. There, Nonja has already gathered over 80,000 friends. Alas, most of the shots are blurry and without an eye for composition: the animal seems not much interested in the device apart from receiving the raisins produced with each click of the camera that she manages to generate. The zoo management thought this could be a way to entertain the apes in their confinement, and to generate interest from the public (Nonja), stressing the vulnerability of the apes and their endangered natural habitats. Apart from the at first sight anthropomorphic character of an ape making pictures and being on facebook, on closer inspection the animal does not truly interact and is not in a situation in which new cognitive abilities can be revealed, besides the ability to push a button to receive a raisin. Instead of providing the opportunity for human animal interaction, the set-up creates a divide between her and us; under the banner of friendship a rather one-sided mode of communication is produced. Instead of offering a cognitive challenge, the device mostly helps to express problems associated with confinement. The ethical approach that is implied here seems to be an ethics of care, as the focus is on (though one-sided and superficial) personal relations between an object of care and caregivers. On a positive note this example shows how one device (camera) can have two interfaces with entirely different meanings (Nonja working for raisins; public being informed about apes on facebook).

Designing a computer game for pigs and humans

The two pig research projects illustrated that the cognitive abilities of pigs are not catered for in current pig farming facilities, and that particular material environments may imply specific understandings of cognition and ethics. In the remainder of this paper we will propose to create a cognitively challenging 'game' environment for pigs and a human public that would allow further exploration of our understanding of animal cognition and that could foster new modes of public engagement with ethical questions regarding pig farming.

Ethical approaches to be explored and accompanying understandings of cognition

An interactive device would implicitly allow the public to immerse themselves in the experience of questions such as 'do they suffer', 'is their life worth living', 'do we respect them as subjects'. Rather than attempting to separate the philosophical, scientific and ethical questions regarding cognition, a game could foster the active participation of pigs, scientists and a lay audience in deliberating on these issues in a more integrated and participatory way. Besides the classic approaches to ethics (consequentialism and a deontological rights-view of animal ethics), two forms of understanding ethics particularly seem promising to explore with the game design: the 'capabilities approach', and an approach that could be labelled 'post-humanist ethics'.

The capabilities approach takes our ethical duties not to be aimed simply at meeting needs, but as allowing animals to flourish by developing their basic capabilities. Animals, instead of mere passive recipients of pleasure and pain (as in utilitarianism), figure in a more active role as subjects realizing their own development. Not all desires of animals should be granted though: their subjectivity and claim to autonomy are understood as limited, analogous to the way we assess the preferences and choices of children or mentally disabled. It is up to us, in a process of 'intelligently respectful paternalism' (Nussbaum, 2004), to determine what capabilities are to be considered important and valuable, and to decide on dilemmas that arise when interests collide. Confined animals that are under the control of humans are considered to be alienated, further compromising the autonomy of their desires.

Another approach to granting agency and subjectivity to animals and attending to their cognitive needs can be put under the banner of 'post-humanist ethics' (e.g. Haraway, 2008; Whatmore, 2002). This mode of thinking further develops the value of human-animal relations, and focuses in a more symmetrical way on the processes in which both humans and animals are granted particular forms of cognition and subjectivity. Rather than starting in an essentialist and perfectionist way from basic capabilities as in Nussbaum's capabilities approach, the post-humanist approach leaves the abilities and qualities of animals more open and considers these as emergent, changing, and dynamic features of relations with humans and a specific environment. Our duties towards animals then are best formulated by thoughtful and possibly intensive interaction and continuous tinkering with the complexities and paradoxes that are prevalent in human-animal relations.

Designing a computer game for pigs and humans to playfully interact

The game aims at multi-level human-animal interaction, and its mode of operation is not specified at the outset. Hence, the post-humanist mode of ethical thought seems appropriate to inform the design process. The challenge then is to create opportunities for open behavioural patterns as in the Pig Park, but without necessary commitment to the ideal of naturalness and ancestral habitats (cf. Segerdahl, 2007). As in the mirror test we would be offering an artificial device, but in our case we do not provide a binary test, and aim for the emergence of a more embodied type of cognition. With offering a game, cognition is not operationalized primarily as tool use or solving complex problems, but can also start off from what motivates and maintains elaborate play behaviours (in both pigs and people). As in the example of the orang-utan on facebook, we hope to generate wider public involvement and new modes for sharing experiences of the worlds of animals, but in addition we also aim to reduce boredom and offer a more interactive, cognitively challenging and playful mode of engagement, as was apparently better achieved in the case of the Mandrill playing with its visitors.

With cognition regarded as a relational and emergent characteristic, rather than as a property inherent in the species in its natural habitat, potential cognitive abilities may be revealed that are not part of known behavioural repertoires (as in the case of using a mirror to find food). And when cognition is not just seen as a quality of the minds animals that may be tested in a laboratory by being isolated from their bodies and interaction with others, but as a quality of pigs that may partly also be 'discovered' in the development and use of the game, cognition can be viewed as a dynamic quality of the animal and its surroundings. As the development and expression of human cognition can be understood also to emerge after increasingly complex interactions with social and material environments, it is due time to offer the animals under our care a fitting material culture, to together with us explore their species specific capabilities.

The project to design a computer game in which pigs and humans are to interact might be considered an all too playful way to approach the issue of welfare on pig farms, especially when compared to the common solemnity expressed in the genre of animal ethics. Nevertheless, playing together is regarded as the precondition for developing a shared culture (Huizinga, 1949). Generating such a (partly material) culture of shared experiences of play can provide a fresh angle into the ways that understandings of cognitive abilities, ethical claims, and technical challenges could be related. Thereby with this design project, adding to the Kantian question: 'Are they rational?' And Bentham's question: 'Can they suffer?' We propose to put to the test the question: 'Can we play together?'

References

Bracke, M.B.M. and Hopster, H. (2006). Assessing the importance of natural behavior for animal welfare, Journal of Agricultural and Environmental Ethics 19: 77-89.

Broom, D.M., Sena, H., Moynihan, K.L. (2009). Pigs learn what a mirror image represents and use it to obtain information, Animal Behaviour, 78(5): 1037-1041.

Haraway, D. (2008) When species meet, Minnesota University Press, Minneapolis, USA, 423 pp.

Held, S., Cooper, J.J., Mendl, M.T. (2009). Advances in the study of cognition, behavioural priorities and emotion. In: Marchant-Forde, J.N., The welfare of pigs. Springer, Dordrecht, the Netherlands, pp. 47-94.

Huizinga, J. (1949 (1938)). *Homo ludens*, a study of the play element in culture, Routledge, London, UK, 220 pp.

Korthals, M. (2004). Before dinner: philosophy and ethics of food, Springer, Dordrecht, the Netherlands, 256 pp.

Markowitz, H., Stevens, V. Mellen, J. and Barrow, B. (1982). Performance of a mandrill (*Mandrillus sphinx*) in competition with zoo visitors and computer on a reaction-time game. Acta Zoologica et Pathologica Antverpiensia 76: 169-180.

Milius, S. (2009). Pigs use mirrors; smart swine can figure out how to find a bowl of food, Sciencenews Oct 7 http://www.sciencenews.org/view/generic/id/48133/title/Pigs_use_mirrors. Accessed February 18, 2010.

Mol, A. and Law, J. (2004). Embodied Action, Enacted Bodies. The Example of Hypoglycaemia, Body and Society 10: 43-62.

Nonja (2010). http://www.facebook.com/pages/Nonja/190010092116. Accessed January 10, 2010.

Nussbaum, M.C. (2004) Beyond compassion and humanity: Justice for nonhuman animals. In: C.R. Sunstein and M.C. Nussbaum (eds.), Animal rights. Current debates and new directions, Oxford University Press, Oxford, UK, pp. 299-320.

Segerdahl, P. (2007) Can natural behaviour be cultivated? The farm as a local human/animal culture. Journal of Agricultural and Environmental Ethics 20: 167-193.

Shepherdson, D.J. Mellen, J.D., Hutchins, M. (1998). Second nature, environmental enrichment for captive animals, the Smithsonian Institution, Washington, USA, 350 pp.

Stolba, A., D. G. M. Wood-Gush (1989), The Behaviour of Pigs in a Semi-natural Environment, Animal Production 48: 419-425.

Wemelsfelder, F. (1993) Animal Boredom: Towards an Empirical Approach of Animal Subjectivity. Doctoral Thesis, Rijksuniversiteit Leiden.

Whatmore, S. (2002) Hybrid Geographies: Natures, Cultures, Spaces. Sage, London, UK, 225pp.

Specifying animal welfare standards: ethical issues

P.B. Thompson
Michigan State University, USA

Abstract

Ethical debates over animal welfare in agricultural production systems have thus far focused either on the general rationale for intervention on behalf of animals, or on tensions and potential contradictions that arise between alternative ways of conceptualizing what matters about animal welfare. Appleby and Sandøe, for example, offered a compelling analysis of these tensions in a 2002 paper. This paper examines how these philosophical tensions emerge at the practical level of specifying standards for animal production systems. A theoretical framework developed by Lawrence Busch in a 2000 paper is used to characterize alternative approaches to the specification of food and agricultural standards. In the case of animal welfare, Busch's framework shows how three distinct domains of specification become embroiled in the moral tensions identified by Appleby and Sandøe. First, there is a tension between engineering and performance approaches to developing standards. Second, the possibility of writing standards as what Busch calls divisions, Olympic, ranks and filters allows for multiple ways of accommodating the factors identified by Appleby and Sandøe. Finally, given this multiplicity, there are ethical political questions that arise in connection to the procedure for establishing standards, exemplified especially in the tension between optimizing strategies (as envisioned by the European Welfare Quality Project) and more chaotic, market based approaches that rely heavily on consumer demand and the power of retailers. The paper argues that in either case the turn to procedural ethics does not necessarily serve animal interests well.

Keywords: livestock production, regulation, perfectionism, welfarism, standard setting

Introduction

As discussed here, an animal welfare standard is an explicit set of requirements for a food animal production system that have been developed to improve, assure or ameliorate the quality of life for animals being kept in that system. Much of the literature on standards emphasizes the role of standards in coordination schemes. Technical standards such as PAL (phase alternating line) and ATSC (Advanced Television Systems Committee) specify instructions for encoding electronic pulses so that they may be reproduced on a television set. Equipment designed for PAL is not compatible with equipment designed for ATSC, so these standards can be understood as devices that allow different manufacturers of electronic equipment to coordinate their activities. Many standards are developed solely for the purpose of coordinating business activities in an efficient and reliable manner. In contrast, regulatory standards such as the allowable levels for various contaminants in or processed foods are established for reasons that are extrinsic to coordination and trade, though once established they play an important coordinating role.

Standards for food safety determine access to markets for producers, for example. Such standards may be developed either by government agencies or by private firms such as food retailers, but it is useful from a coordination perspective for food safety standards to be uniform. David Singh Grewal has analyzed how once a standard starts to become widely used in an industry, the standard takes on a characteristic that Grewal calls 'network power'. Irrespective of intrinsic goods secured by compliance to the standard, utilization of the standard becomes mandatory if one is to have access to the networks of suppliers, manufacturers, retailers and users that are bound together by the standard, (Grewal 2008). Thus, though it is perfectly legal to sell television sets built to the PAL standard in the United States, they are very difficult to find in virtue of the network power associated with the widespread use of the ATSC standard. A manufacturer who wishes to supply televisions to the U.S. market has no choice but to adopt the ATSC standard. Something similar happens in connection with food safety. Although products containing GMOs may be legally sold in Europe, European retailers have adopted standards discouraging the use of GMOs that have achieved significant network power.

Animal welfare standards are more like food safety standards than technical standards for television sets, but they, too, can be expected to exhibit network power once they become widely adopted. Efforts such as the European Animal Welfare Quality Project are attempting to develop explicit requirements for food animal production that can be expected to have ramifications well beyond Europe. The explicitness of standards distinguishes them from animal welfare ideals, which express ethical conceptions of what a good quality of life would be and may stipulate responsibilities for humans who keep or use animals. As will be discussed below, ideals that appear to be action guiding from a philosophical perspective may be vague, highly general and even inapplicable in a given setting. In contrast, a standard implies the possibility of a test, that is, a set of measures or procedures that can be applied to operationally defined indicators to yield an unambiguous result as to whether the standard has or has not been met. Testable indicators are taken to constitute acceptable evidence of compliance with the standard in question.

Thus, for example, an animal standard might be 'animals should able to fully extend their limbs'. An indicator of this standard might then be whatever spatial dimensions of the enclosure in which animals are kept that would be sufficient for animals of the relevant species to fully extend their limbs. A test for this standard would consist in measuring the spatial dimensions of actual production enclosures. Note that none of these concepts refer directly to the conceptualizations of welfare that would articulate why being able to fully extend one's limbs would be thought to contribute to the quality of life for animals. Philosophical conceptions of animal welfare often combine standard veterinary health measures such as morbidity and mortality with ideas that stress the animal's cognitive well-being (e.g. the creature's subjective state of relative satisfaction or distress) and the ability to perform species typical behaviors. Extending one's limbs might bear on either cognitive well-being, if frustration of this behavior was a source of dissatisfaction and discomfort, or on natural behavior, if full extension of limbs is a behavior typical of individuals in that species. 'Welfarism' and 'perfectionism' represent competing philosophical ideals that place stress on cognitive vs. species-typicality interpretations of why behaviors matter, respectively.

Busch provides a typology for standards. *Filters* are standards that establish minimums that must be met by all actors participating in a domain. With respect to animal welfare, a filter distinguishes between acceptable and unacceptable levels of well-being for animals in farm production systems. *Olympics* in contrast are standards that represent 'the highest' or 'the best'. An Olympic standard for animal welfare would represent the maximal state of well-being achievable for farm animal production. *Ranks* partition goods ordinally, while *divisions* partition goods as kinds with no implication that one is better than another. There have been some attempts to introduce ranks into animal welfare standards among European retailers, so that consumers may coordinate their feelings for animals with their willingness-to-pay. In contrast, standards for alternative animal production systems are represented as divisions in the United States, though of course consumers may impute rankings to divisional standards according to their own preferences. Busch's typology is represented by the easy to remember acronym FORD, (Busch, 2000).

Animal welfare standards may arise in the public sector as government regulations, or they may emerge in the private sector different several different configurations. As government regulations, welfare standards have functioned primarily as filters that determine which animal products will be permitted for sale in public markets. In the private sector, firms have attempted to specify Olympic standards that will place them in a position to reap price premiums. Niman Ranch, for example, is a US company that represents their standards as reflecting the highest levels of animal welfare. Yet because Niman Ranch purchases from a number of standards, the internal effect of standards among Niman Ranch suppliers (such as their prohibition of genetically modified feeds) is to serve as filters that stipulate what suppliers must do in order to have access to Niman's final customers. Private standards for distinct production systems are often represented as divisions rather than ranks. Thus the United Egg Producers have distinct standards for caged and non-caged production systems, but denies any putative ranking of the two systems with respect to animal welfare.

The potential for gaps between welfare ideals and welfare standards, or between standards and their indicators or tests means that is entirely possible for an animal welfare standard to fail in its quest to improve the quality of life for farm animals. In particular, five modes of failure can be observed in current approaches to animal welfare standards. Thus far, mainstream animal ethics has been somewhat inattentive to these potential failures, in part because of philosophers' tendency to overlook the practical circumstances in which action actually occurs, but also because philosophers and animal advocates alike have neglected the nature of standard setting and monitoring itself.

Failure mode 1: the theory/practice gap

To start with the obvious, animal welfare standards have not escaped the longstanding problem of gaps between theory and practice. Attempts to specify idealized conceptions of animal welfare as explicit production standards have led to failures and fiascos. Animals themselves have often been the primary victims. Moral theories purport to provide a basis for decision-making and action; hence one would expect a moral theory to provide a basis for standard setting. In some cases, however, moral theories are inapplicable to standard setting because following their dictates would preclude ever arriving at the need to develop standards for farm animal production. For example, Tom Regan embeds his claim that animals are the subject of a life within a larger ethical theory that he refers to as 'the rights view'. Under any obvious interpretation the rights view forbids the use of animals in food production altogether.

Peter Singer's approach to animal ethics provides a somewhat more interesting case in that his call for balancing of welfare interests among all affected parties might be thought to provide some basis for evaluating standards. Singer stresses animal suffering, and since animal welfare standards are intended to mitigate suffering in production systems it would seem as if Singer's theory might be useful. In fact, however, Singer's development of the theory is pitched toward the perspective of a consumer considering whether it is morally acceptable to consume animal products than it is to the perspective of a producer considering conditions under which it would be morally acceptable to produce them. In broad terms, Singer's theory leaves open the possibility that the pleasure and other benefits enjoyed by humans who eat animal products would outweigh any suffering endured by animals, especially if this suffering could be sufficiently reduced from what Singer evidently takes to be unacceptable levels associated with modern industrial animal farming. Since animal welfare standards are intended to reduce animal pain and distress, it is possible that they may be indicated by Singer's theory.

There are, however, numerous theory-practice gaps that remain. First, Singer does not specify any but the broadest means to quantify animal welfare. At a minimum, Singer's approach needs to be augmented with a more detailed account of the dimensions of animal welfare, such as those developed by Fraser, Weary and Pajor or Appleby. Second, in focusing on the consumer perspective Singer provides information on current alternatives to eating meat, milk and eggs, but he does not provide comparable detail about how the status quo is to be regarded from a producer's perspective. Most importantly he does not address how far a producer is required to go in pursuit of improved animal welfare given the economic environment in which producers currently operate. Given competitive markets for animal products, it may not be feasible to implement animal welfare standards except as an industry-wide initiative or as a government regulation. But seeking improved welfare in either theater implies a much more complex decision situation than that of individual consumer choice. Factors that must be faced in seeking co-operation among producers or achieving and implementing a political decision are simply not addressed in Singer's writings.

Indeed, from a practical standpoint any attempt to instate animal welfare standards in an industrialized country today would presuppose a number of constraints that are unlikely to be reflected in any philosophical theory of animal ethics. First, animal products will continue to be produced and distributed at a scale that is not significantly smaller than that of the status quo. While there are many good reasons for reducing the global trajectory of growth in the production and consumption of agricultural animals, global demand can be expected to offer attractive economic opportunities for animal producers. Second and closely related, while it is reasonable to expect price increases to follow implementation of animal welfare standards, rapid and dramatic increases in the price of food are met with vociferous and sometimes violent protest. There is no obvious reason to assume that increase in the price of animal foods are an exception to this general rule, so there will be significant pressure to insure that price adjustments are moderate and gradual.

There are also ethical arguments to constrain rapid rises in the price of animal products, especially in economically challenged populations. Preparing and eating meals with meat, milk and eggs is culturally appropriate for many if not most of world's poorest people: they know how to prepare these foods and want to eat them. While economically better off consumers may be eating too much animal protein, increased consumption levels among those below the poverty line and the extreme poverty line would very likely translate into improved health and quality of life. Poor people living in developed countries may lack skills for preparing healthy vegetarian meals and may be especially vulnerable to diets overloaded with calorie dense sugars and fats when the prices of meat, milk and eggs rise. Clearly, it is theoretically possible to provide healthy alternatives even for the poorest people, but the means and political will to undertake such efforts are among the factors that are most likely to fall into a theory-practice gap. One can say, with Kant, that plans which fail in practice actually don't work in theory, either, but philosophical theories of animal ethics have yet to articulate principles that would be readily implementable to animal welfare standards development.

Failure mode 2: unintended consequences and the human dimension

The potential of unanticipated outcomes is almost as obvious as the theory-practice distinction. Unwanted outcomes can occur for many reasons, but one can trace the failure to anticipate outcomes to a philosophical mistake when a problem has been characterized too narrowly for the purpose of standards development. When European countries first started to implement regulatory bans on caged layer production, it was as if eliminating cages would automatically result in improved welfare. However, anecdotal evidence from egg producers suggests that the immediate results were often quite adverse for birds. Tracing these unintended outcomes reveals that a simple 'no cages' standard is insufficiently attentive to the multiple dimensions of animal welfare.

Generally, producers shifted to systems with relatively large flocks in large open buildings. These systems had several problems. Most fundamentally, in large groups of unconfined birds, more aggressive hens are persistently stimulated to engage in dominance behaviors that may be a component of developing both an internal pecking order in smaller flocks, and in protecting group boundaries from penetration by birds from other groups in the wild. In large flocks, aggressive birds may be experiencing almost constant stress that results in feather pecking and bullying of less dominant birds. Meanwhile less dominant birds are subjected to constant pecking and are subject to both injury and even death as a result.

In addition to this basic behavioral issue, which continues to be a potential problem in non-cage systems, some early non cage systems used methods for managing manure and distributing feeds that cause both stress and disease in flocks. Exposure to fecal bacteria was an issue in some cases. More frequent interactions with human caregivers had the potential to elicit fear responses in some houses. Many of these problems arose simply because producers and their employees lacked the husbandry skills needed to manage birds in non-cage environments. Within a few years, husbandry improved and new systems were designed that provided weaker birds with avenues of escape from pecking behaviors.

Thus while it would be erroneous to conclude that cage bans resulted in long term decline in hen welfare, it would be equally mistaken to think that implementation of cage bans led to unambiguous improvements in welfare. In hindsight, one can assert that producers should have used the phase in time that was allowed before cage bans took place to experiment with improved methods and husbandry. Yet while regulatory filters mandated the elimination of cages, requirements for experimentation and husbandry were not explicitly address by these new standards. Thus one might also see the failure here as one that occurred at the policy level. yet the development of animal welfare standards has not escaped the most familiar kind of error that arises when one moves to a new standard entirely too quickly.

Failure mode 3: indicators and misplaced concreteness

Although other models are being developed, most animal welfare standards are in fact filters that articulate explicit minimum criteria that must be met if the production practices are to be deemed morally and commercially acceptable. These filters can focus on specific production practices, such as feed restriction or gestation crates, or they can be written to require performance minimums for various aspects of animal welfare. In the latter case, the move from an animal welfare ideal to a standard looks forward to the need for indicators and tests. The temptation to carry language such as that used in the Brambell Report's Five Freedoms directly into standards is tempered by the recognition that it may be very difficult to specify indicators and tests for a standard such as 'freedom from pain and injury'. One response may specify the standard as 'no detectable marks or evidence of injury,' such as bruising, cuts or scars. In some cases, this effectively makes the testable indicator itself serve as the technical standard for animal welfare.

However, studies have shown that swine kept in group housing are much more likely to exhibit bruises, cuts and scars. An American Veterinary Medical Association Committee reviewing welfare and swine housing systems took this to be an indicator of pig welfare that offset advantages associated with greater freedom of movement and ability to exhibit natural behavior in group housing systems. This judgment contributed to the committee's overall reluctance to conclude that welfare in group housing systems is superior to welfare in confined systems. Arguably, however, minor bruises, cuts and scars arising from contact with other pigs do not represent serious compromises to the Brambell Committee's 'freedom from pain and injury' criteria. Here it seems that the indicator was taken to be equivalent to the underlying welfare standard in a manner that can be challenged.

An even more serious gap between indicators and welfare can arise when indicators are derived from sources that do not reflect animal needs. Bryan Norton argues that shifting the discussion from underlying standards to indicators can be a way circumvent tedious and contentious debates: people who cannot agree on the standard may indeed agree on a particular indicator. In such cases, as above, the indicator effectively *becomes* the standard, and agreement on the indicator allows one to avoid needless philosophical debates about the standard itself. But studies that have derived indicators of animal well-being from surveys and focus groups have the potential to enshrine indicators that may be quite misleading and erroneous. Studies of the public show that people assume that farm animals are much like people are, wanting to enjoy time outdoors in the fresh air and sunshine. But animals may shun open spaces provided for them because innate fear of predators conditions them to avoid open spaces. An indicator of well-being derived in the manor suggested by Norton may thus be quite misleading when applied to a given production system. Studies to derive indicators of animal welfare from public opinion have in fact been conducted.

Failure mode 4: coordination breakdown

Appleby and Sandøe (2002) describe 'perfectionist' and 'welfarist' philosophies of animal welfare. Succinctly, the perfectionist takes species typical behaviors to be constitutive of animal welfare, while the welfarist sees the ethical significance of any behavior as depending on veterinary health and cognitive experiences of satisfaction, pain or distress. That is, frustration or prevention of species-typical behaviors in a given production environment is not in itself problematic from the welfarist position, though the suffering, disease or mental anguish that accompanies the inability to engage in species-typical behavior *is* ethically significant. For the perfectionist, on the other hand, animals that *do not* experience anguish when species-typical behavior is stifled are, in a sense, doubly harmed: first by the fact that they cannot engage in a natural behavior and second by behavioral or genetic modification that has robbed them of the *desire* to engage in said behavior.

In the terminology used above, perfectionism and welfarism represent ideals for animal welfare. Here, the philosophical tension between these ideals can be replicated in the form of standards. The problems with indicators noted above can both conceal and exacerbate this tension. If perfectionists and welfarists both agree that 'ability to fully extend one's limbs and turn around' is an indicator of welfare, whether one takes the perfectionist or the welfarist view of this behavior, there will likely be little conflict between perfectionists and welfarists in writing a standard for production that requires enough space in the production environment to engage in this behavior. In this case, the standard could be said to resolve the conflict between a perfectionist and a welfarist, but it could just as easily be said to conceal it. It is reasonable to think that the conflict has been concealed when new standards surface the conflict once again. There is currently a discussion underway as to whether colony housing systems, or 'furnished cages', should acceptable for egg production. These systems accommodate many natural behaviors, but they are, at the end of the day, cages, and cages are not likely to win support from perfectionists, who may see freedom of movement to be a strong indicator of natural welfare. Because colony housing involves smaller group size, there is speculation that it may be associated with reduced cognitive stress and confusion because hens are able to establish a stable pecking order, something that they cannot do when their movement is unrestricted in large group housing situations. There is, in the case, simply a transfer of the well-understood philosophical tension between perfectionism and welfarism to the level of a production standard: what seems good to the perfectionist seems bad to the welfarist, and vice versa.

But a second layer of failure can occur in virtue of the role that animal welfare standards play in satisfying the expectations of people who wish to coordinate their behavior. Recent research on both sides of the Atlantic suggests that the public itself may be split along the lines of the perfectionist/welfarist divide. If so, it will be challenging to develop a set of standards that address consumer expectations about animal welfare. What one group will see as a meaningful standard will be seen as misleading and deceitful by the other group.

Failure mode 5: agency risk

The principal/agent model applies well to animal welfare. Animals are principals whose welfare is sought and promoted by agents, including individual stock handlers, owners and even animal protection activists. In all principal/agent situations, there is the potential for agency risk: the situation in which interests or concerns of importance to the agent interfere with or even override those of the principal.

Indeed, the divide between the interests of producers and those of farm animals is without question a key source of ethical tension within farm animal production itself.

A more interesting form of agency risk occurs in connection with animal welfare standards. Whether acting alone or through organizations, animal protectionists have sought tougher animal welfare standards as a way to promote the interests of farm animals. But an ironic tension can arise between the protectionist's long term goal of improving the quality of life for farm animals and the immediate interests of animals actually being kept in production settings. This tension is well-illustrated by the U.S. controversy over voluntary standards for caged layer production introduced by the United Egg Producers in about 2000. UEP proposed a voluntary standard for stocking density of approximately 450 cm. sq. minimum. This filter was to replace an existing industry average stocking density of 325 cm^2 (an improvement of approximately 40%). The UEP standard had been fully implemented by about 80% of U.S. producers who applied the standard to 100% of the flock after a two year phase in period.

Although one can question whether 450 cm^2 is adequate space for a U.S. laying hen (U.S. hens are smaller than European hens), there is no question that this change made a significant improvement in the welfare of laying hens, an improvement measurable in terms of reduced mortality and morbidity (Webster, 2007). However, animal protectionists have been sharply critical of this action by UEP. They have castigated UEP for their approach and argued that European law provides greater protection for laying hens (Braunschweig-Norris, 2005), and have campaigned for laws that would replicate the European animal environment in U.S. states (Erbe, 2008). While the eventual effect of such laws might indeed produce improvements in animal welfare, they have lengthy phase in periods and would, if implement in the U.S. on a state-by-state basis, lead producers to relocate in unregulated areas (including Mexico). The laws permit housing at the 325 cm^2 average during this phase in period, as well as for out-of-state producers supplying states where strong regulatory filters have been instituted.

Whether the animal protectionists' or the UEP's approach is better for the birds is at least a debatable point. On one hand, the voluntary and incremental improvements reflected in UEP standards accept pragmatic constraints that would be rejected by a more ambitious approach. The original UEP standards, for example, were configured so that UEP members would be able to adopt them without needing to make large investments in new physical infrastructure. On the other hand, this approach was able to improve quality of life for a large number of animals in a relatively short period of time, especially when compared to more ambitious regulatory reform efforts. Standards sought by animal protectionists would more fully realize philosophical ideals of animal welfare, but protectionists themselves are split over the ideals they hope to realize. Those motivated by the vegan ideal of animal rights may see standards as a means to make animal production more expensive, inconvenient and unpopular. They may find more success in a measure that affect a comparatively small number of animals, especially when doing so drives a wedge between animal producers and the broader public (see HSUS, 2009). There is thus room to explore various pros and cons that are associated with each of these competing approaches to animal welfare standards.

Conclusion

A close examination of animal welfare standards reveals a number of opportunities for philosophical analysis and debate. More classically philosophical approaches to animal ethics have not thus far engaged the questions that arise in connection with the development of standards. To some extent, the general orientation of philosophical theories of animal ethics limits their applicability to standard setting exercises, yet these theories do articulate moral ideals that inform current standards, and that might do so more explicitly and thoughtfully if philosophers were to develop a more complete understanding of the role, function, and possible strengths and weaknesses of various standard setting approaches.

References

Appleby, M.C. and P.T. Sandøe (2002). Philosophical Debate on the Nature of Well-being: Implications for Animal Welfare. Animal Welfare 11: 283-294.

Braunschweig-Norris, J. (2005). The U.S. Egg Industry – Not All It's Cracked up to Be for the Welfare of the Laying Hen: A Comparative Look at United States and European Union Welfare Laws. Drake Journal of Agricultural Law 10: 511-539.

Busch, L. (2000). The moral economy of grades and standards. Journal of Rural Studies 16: 273-283.

Erbe, B. (2008). California Animal Cruelty Could be Cured by Proposition 2. U.S. News and World Report.

Grewal, D. S. (2008). Network Power: The Social Dynamics of Globalization. New Haven, CT, Yale University Press.

HSUS (The Humane Society of the U.S.) (2009). Wendy's Gets Applause for Its New Policy on Cage-Free Eggs. Retrieved March 30, 2010, from http://www.humanesociety.org/news/news/2009/05/wendys_cage_free_052209.html.

Webster, A.B. (2007). Animal Care Guidelines and Future Directions. Poultry Science 86(6): 1253-1259.

Animal welfare: between governmental and individual responsibility

L. Heerwagen
Institute of Food and Resource Economics, University of Copenhagen, Rolighedsvej 25, 1958 Frederiksberg C, Denmark; lh@life.ku.dk

Abstract

The improvement of animal welfare is major goal of the European Union and its member states. The significance of this task is emphasized not only by its mention in the Treaty of Amsterdam, but also by the policy initiatives launched since then. Thus, it might seem that the task of improving animal welfare is a strictly governmental responsibility at national and supranational levels, but this is not the case. An important vision in the European Commission's action plan on animal welfare is a market approach, wherein consumers, informed by improved food labelling, drive a virtuous cycle, to the benefit of producers as well as farm animals. But this approach also gives individual consumers a responsibility, since they are now expected to decide individually whether or not animal welfare should be improved. This paper undertakes a discussion of the justifiability of such transfers of responsibility from an institutional to an individual level. It is proposed that political decision-makers might share responsibility for animal welfare improvements with consumers, if they are better able to handle such a task. However, when ability and resources are the principles for a division of labour, it is also implied that political decision-makers should provide appropriate circumstances for consumer responsibility-taking. This leads to a discussion of specific ways in which such conditions could be provided, in the form of food labelling and narratives that relate consumption to political action.

Keywords: ethical consumption, EU, food labelling, informed consumers

Introduction

The improvement of the welfare of farm animals is major goal of the European Union. With the Treaty of Amsterdam of 1997, it was agreed that the European Union and its member states 'shall pay full regard to the welfare requirements of animals'. In accordance with the Treaty, in 2006 the European Commission initiated an action plan to improve animal welfare. European consumers of animal products have a significant part to play in this plan. Hence, the action plan includes the objective of having the general public and other stakeholders 'fully appreciate their role in promoting animal protection and welfare' (European Commission, 2006: 11). This also means that Europeans, in their role as consumers, purchase so-called 'animal welfare friendly' products for a price premium.

There are already a number of such products on the European markets, but the Commission expects that these markets can be increased by providing consumers with improved information about animal welfare – particularly in the form of food labelling. According to the Commission, this approach could lead to improved welfare for farm animals, driven by consumer demand:

> Improved information to consumers offers the prospect of a virtuous cycle where consumers create a demand for food products sourced in a more animal welfare friendly manner which is transmitted through the supply chain back to the primary producer, who may be able to receive a premium price for their product and thus recoup a portion of any associated higher production costs. (European Commission, 2009: 2)

The Commission suggests that animal welfare standards can be improved by strengthening consumer demand for animal welfare friendly food. The development of new labelling programmes is not only about making some consumers better able to satisfy personal preferences through well-informed choices; rather, it is anticipated that they will 'appreciate their role' and purchase certain food products, in order to improve animal welfare. Thus, the question of whether the welfare of sentient beings should be improved is to some degree determined by the choices of individual consumers. Therefore, consumers can be said to assume responsibility for these improvements. Furthermore, this responsibility is the result of a transfer, since the question of improving animal welfare is shifted from an institutional to an individual level.

The justifiability of this responsibility transfer is unclear. According to Tim Lang, an informed consumer approach, as stressed by the Commission, not only places responsibility for animal welfare on consumers; it also absolves policy-makers of responsibility, if consumers fail in this task (Lang, 2006: 261). Thus, a discussion of the justifiability of transferring responsibility for animal welfare improvements from political institutions to individuals seems relevant; this is the purpose of this paper. First, the paper proposes that the public's ability to influence a societal task can be seen as a way of justifying the shift in responsibility from a governmental to an individual level. Second, the paper discusses the sharing of responsibility between consumers and political decision-makers that is actualized in the case of the Commission's animal welfare initiative. Third, the paper suggests food labelling and relevant narratives as two ways of providing appropriate circumstances for responsibility-taking.

Responsibility-fleeing and responsibility-sharing

Responsibility is traditionally regarded as a construct that refers to the praise- and blameworthiness of past acts. This is not the kind of responsibility which is under discussion in the Commission's policy initiative. In this case, the notion of responsibility is about considering animal welfare as a societal task, which should be promoted at an institutional as well an individual level. Such a 'forward-looking' responsibility, assumed from different social roles, could be characterized in this way:

> We also say, however, that people have certain responsibilities by virtue of their social roles or positions, as when we say that a teacher has specific responsibilities, or when we appeal to our responsibilities as citizens. In this meaning, finding an agent responsible does not imply finding the agent at fault or liable for a past wrong, but rather refers to agents' carrying out activities in a morally appropriate way and aiming for certain outcomes (Young, 2006: 119).

In this way, social roles, functions, and positions in which we are engaged, assign to us the responsibility for certain tasks. Such responsibilities may be rather concrete and familiar, as is the case when teachers are held responsible for the teaching of students and their learning, and are assessed accordingly. But responsibilities can also appear in less familiar forms, as is the case when consumers must assume responsibility for improvements to animal welfare standards.

Societal tasks can be distributed and shifted, and the same may therefore be said of the responsibilities that accompany these tasks. The instance of political institutions preparing individuals, in their role as consumers, to determine levels of animal welfare, is a case of responsibility-shifting (Kjærnes, 2008: 4). However, it may seem unclear how this affects governmental responsibility for animal welfare improvements. One option is to view transfers of responsibility for societal tasks from a governmental to an individual level as an expression of governmental responsibility-fleeing, since the handling of societal tasks is essentially the responsibility of governmental institutions (Sørensen, 2005: 71-74). However, another view emphasizes that responsibility should be shared according to capacity, power, and resources:

> There is a common intuition in contemporary societies that if an agent, whether individual or institutional, has the capacity, power, and resources to contribute to solving a social problem, they have a responsibility to do so, i.e. that power and capacity entail responsibility (Fahlquist, 2009: 118).

Arguably, such a view would allow a transfer of governmental responsibility to individual consumers. This would be the case, if individuals in their role as consumers were able to address a given societal task, while political institutions were not. It has been asserted in a number of cases that national governments are incapable of addressing problems that have a somewhat global character. In such cases, the consumers' ability to change the direction of production has been proposed as one way of filling the 'responsibility vacuum' caused by the inability of political decision-makers to do so (Micheletti and Stolle, 2007: 160).

Responsibility-sharing for animal welfare improvements

Initially, the question of whether a transfer of responsibility to consumers is justifiable seems to depend on the inability of political institutions to address the need for improved animal welfare standards, and the corresponding ability of consumers to do precisely this. Both these requirements can be found in the Commission's action plan on animal welfare. The Commission refers to the rules of the World Trade Organization (WTO), which presumably does not recognize animal welfare as a reason for impeding trade. This significantly complicates improvements to animal welfare by legislative means. Consumer demand is therefore seen as a way in which production can be pushed in a more animal welfare friendly direction, without harming economic interests (European Commission, 2006).

When the ability to influence a societal task is proposed as a principle for the sharing of responsibility, it may obligate consumers to do whatever is in their power to improve animal welfare standards. On the other hand, the same principle also obligates other agents act according to their capabilities. With regard to environmental issues, it has been suggested that one governmental responsibility might consist of providing individuals with opportunities that would make it easier for them to assume responsibility:

> They have it in their power to make it easier and less expensive for individuals to choose the environmentally friendly option and they can provide information that is easily accessible and as straightforward as possible. In essence, they are responsible because they have the power to create opportunities for individuals to do what is right. Another way of phrasing it is to say that institutions can make it easier for individuals to assume forward-looking responsibility. (Fahlquist, 2009: 119)

It seems reasonable to believe that the same approach to governmental responsibility is valid for the improvement of animal welfare standards. Under given circumstances, it may be impossible for governments to improve animal welfare regulations, and they may therefore be justified in sharing responsibility with consumers. However, these institutions should still be able to provide consumers with appropriate circumstances for taking responsibility, and in such cases, they are also responsible for doing so. This follows from the very same principle that justified the sharing of responsibility in the first place. Accordingly, the discussion of how political institutions may, in a justifiable manner, share responsibility for animal welfare improvements, should also imply a discussion of the opportunities provided by these institutions to consumers.

Improving conditions for consumer responsibility-taking

The assertion that policy-makers should provide consumers with appropriate circumstances for assuming responsibility with regard to animal welfare is quite open-ended. As mentioned, the Commission regards labelling as a primary obstacle for consumers who wish to purchase animal welfare friendly food. Animal welfare is a so-called 'credence attribute', which the consumers are not able to verify when animal welfare friendly food is purchased. Animal welfare labelling and the credibility of such labelling are therefore important, when consumers are to identify foods that have been produced with extraordinary considerations for animal welfare (European Commission, 2009: 5). However, a 2006 survey revealed that many European consumers were dissatisfied with animal welfare labelling. Hence, 55% of the respondents disagreed to some extent with the claim that it is easy for consumers to find information about products sourced from animal welfare friendly production systems. Furthermore, 56% of the respondents denied to some degree that the current labelling programs allowed them to identify animal welfare friendly products (European Commission 2007: 41-46).

There seems to be support for the belief that appropriate circumstances for consumer responsibility-taking should include appropriate food labelling. The European labelling of shell eggs is a well-known example of how the right kind of labelling can help push production in a more animal welfare friendly direction. Although they are official, mandatory marketing standards, the categories, 'eggs from caged hens', 'barn eggs', 'free-range eggs', and 'organic eggs' are generally perceived by consumers as indicators of animal welfare standards. Since the introduction of these standards, the market for animal welfare friendly eggs, that is, barn, free-range, and organic eggs, has increased significantly; in some Northern European countries, close to 50% of shell eggs sold at retail level are animal welfare friendly (European Commission, 2009: 15).

There may also be another way in which the consumption of animal welfare friendly eggs makes a case for appropriate circumstances for consumer responsibility-taking. Owing to a 1999 directive on the welfare of laying hens, the use of traditional battery cages is to be phased out by 2012, although so-called 'enriched cages' will still be allowed (Appleby 2003: 159). To some extent, this development is also the result of consumer efforts. Thus, it was recently stressed by Markos Kyprianou, Commissioner for Health, in a comment on the approaching curtailment, that 'the Commission listened to the demands of EU consumers and has taken concrete action to improve the welfare of laying hens' (Anonymous, 2008: 1).

This move from consumer behaviour to political action may be important for the provision of appropriate circumstances for consumer responsibility-taking in general. The feeling of powerlessness and the lack of belief in personal responsibility-taking among consumers are phenomena which have been identified in a number of studies. A 2001 study emphasized that:

> Many consumers believe that they are powerless, in terms of their purchasing behaviour, to affect animal welfare standards. This is contradictory to all current marketing strategies, where consumer preferences are viewed as market drivers. This issue may be compounded again by consumer reluctance to accept responsibility for animal welfare, through their demand for animal-based food products. In this way, consumers adopt a fatalistic approach to food choice. Policies to address this should encourage consumers to acknowledge their role in current animal welfare standards and ways in which they can assist in improving standards (Harper and Henson, 2001: 10-11).

The study underlines the contrast between the power which consumers are claimed to possess, and the sense of powerlessness perceived by consumers themselves. It is therefore also suggested that policies should convince consumers that they actually have a serious, important part to play, if animal welfare standards are to be improved. The suggestion also seems important in a discussion of how policy initiatives can be justified, by providing consumers with appropriate circumstances for taking responsibility. If consumers actually perceive a tremendous distance between the single acts of individuals and some aggregate result, they need to be assured that they are nevertheless part of an overall political endeavour. The development of the market for animal welfare friendly eggs demonstrates that 'piecemeal change is worthwhile in itself, and finally leads to wholesale change' (Appleby, 2003: 171). In this way, consumption as a policy driver becomes not only a question of capturing lesser or larger market shares for animal welfare friendly food, but also of preparing political action.

Conclusion

The European Commission lays emphasis on the crucial role consumers have to play, if welfare standards for farm animals in the EU are to be improved. An important part of the Commission's action plan is therefore a market approach to animal welfare, where well-informed consumers push production in a more animal welfare friendly direction by purchasing animal welfare friendly food. It could be claimed, however, that this transfer of responsibility from political institutions to individual consumers demands justification. This justification is provided by reference to a principle that allows a distribution of responsibility, according to the involved agents' opportunities to influence the task at hand. Such a principle is double-edged. It may justify regarding consumers as responsible and involved in the solution of societal problems and tasks, but it can also lead to new responsibilities. Hence, it seems reasonable to believe that the very same institutions that hold consumers accountable for outcomes should also provide consumers with good circumstances for taking on these responsibilities. The development of improved animal welfare labelling, as proposed by the Commission, is arguably an important part of providing such appropriate circumstances. However, narratives that relate consumption to a shared political endeavour may also be significant.

References

Anonymous (2008). Animal Welfare: Commission report confirms the potential benefits of banning conventional battery cages for laying hens, Press release. Available at: http://europa.eu/rapid/pressReleasesAction.do?reference=IP/08/19. Accessed 15 march 2010.

Appleby, M. (2003). The EU Ban on Battery Cages: History and Prospects. In: Rowan, A.N., Salem, D.J. (eds.). The State of the Animals II: 2003, Humane Society Press, Washington D.C., USA, pp. 159-174.

European Commission (2006). On a Community Action Plan on the Protection and Welfare of Animals 2006-2010. Available at: http://ec.europa.eu/food/animal/welfare/work_doc_strategic_basis230106_en.pdf. Accessed 15 march 2010.

European Commission (2007). Attitudes of EU Citizens towards Animal Welfare. Special Eurobarometer 270. Available at: http://ec.europa.eu/public_opinion/archives/ebs/ebs_270_en.pdf. Accessed 15 march 2010.

European Commission (2009). Options for animal welfare labelling and the establishment of a European Network of Reference Centres for the protection and welfare of animals. Available at: http://ec.europa.eu/food/animal/welfare/farm/docs/options_animal_welfare_labelling_summary_en.pdf. Accessed 15 march 2010.

Fahlquist, J.N. (2009). Moral Responsibility for Environmental Problems – Individual or Institutional? Journal of Agricultural and Environmental Ethics 22: 109-124.

Harper, G. and Henson, S. (2001). Consumer Concern about Animal Welfare and the Impact on Food Choice – Final Report. Available at: http://ec.europa.eu/food/animal/welfare/eu_fair_project_en.pdf. Accessed 15 march 2010.

Kjærnes, U., Jacobsen, E. and Dulsrud, A. (2008). Consumer choice as a mode of governance. The case of farm animal welfare in Europe. Available at: www.suscons.ulg.ac.be/IMG/ArlonPapers/Jacobsen%20consumerchoice%20as%20a%20mode%20of%20governance.doc. Accessed 15 march 2010.

Lang, T. (2006). Achieving Access to Ethical Food: Animal and Human Health Come Together. In: Turner, J., d'Silva, J. (eds). Animals, Ethics and Trade: The Challenge of Animal Sentience. Earthscan, London, United Kingdom, pp. 261-272.

Micheletti, M., Stolle, D. (2007). Mobilizing Consumers to Take Responsibility for Global Social Justice. The Annals of the American Academy of Political and Social Science 611: 157-175.

Nordic Council of Ministers (2004). Ethical Labelling of Food. Available at: http://www.norden.org/fi/julkaisut/julkaisut/2004-741/at_download/publicationfile. Accessed 15 march 2010.

Sørensen, M.P. (2005). Liberal Democratic Society and Political Consumerism – A Problematic Combination. In: Boström, M., Føllesdal, A., Klintman, M., Micheletti, M., Sørensen, M.P. (eds.). Political Consumerism: Its motivations, power, and conditions in the Nordic countries and elsewhere. Proceeding from the 2nd International Seminar on Political Consumerism, Oslo August 27-29 2004, pp. 59-76.

Young, I.M. (2006): Responsibility and Global Justice: A Social Connection Model. Social Philosophy and Policy 23: 102-130.

The ethical basis of using animals in agricultural production: consumer understandings

S. Kupsala, P. Jokinen and M. Vinnari
University of Eastern Finland, Department of Geographical and Historical Studies, P.O. Box 111, 80101 Joensuu, Finland; saara.kupsala@ymparisto.fi

Abstract

Consumers' ability to gain information about current animal farming practices is weakening because of the concentration of animal farming and urbanisation. However, at the same time farm animal issues have received increasingly attention in the media and in the policy arena. The growing distance to animal farming among consumers and the politicisation of animal issues raise questions how consumers perceive the ethical basis of using animals in agricultural production. In this paper we approach this question by drawing from four group interviews with citizen-consumers (organic consumers, vegetarians, gastronomes and a random group of ordinary food shoppers) carried out in Finland in 2009. The examination of transcribed discussions shows that consumers feel much moral ambiguity on meat consumption since they have become increasingly disconnected from livestock production and since they encounter highly contradictory information on animal farming. Consumers have responded in diverse, often conflicting ways to these moral ambiguities. Ordinary shoppers maintained a vague, habit-based trust that 'things are fine' in animal production and they also aspired to sustain a certain kind of emotional distance to livestock production and animal slaughter. In contrast, for organic consumers and gastronomes, the re-gaining of connection with food production worked as a resolution for their moral ambiguities. Acknowledging the animal origin of food and knowing personally how animals are treated worked as a form of legitimation to consume animal products. This paper concludes that consumers can take various citizen-consumer positions and moral strategies to influence animal welfare in everyday contexts. Due to the growing gap between food production and consumption, novel social institutions to facilitate welfare-friendly choices may be required.

Keywords: farm animal welfare; consumer understandings; moral standing of animals; politics of consumption

Introduction

In late modern societies, there is critical reflexivity about food production methods, and food issues have become increasingly politicised and contested. In particular, the utilisation of animals in agricultural production poses complex moral dilemmas for consumers (e.g. Fraser, 2008; Knight and Barnett, 2008). Moral questions have become increasingly demanding for consumers since they have become ever more distanced from animal farming and slaughtering due to urbanisation and the concentration of the livestock industry: nowadays few have personal knowledge how animals (or the particular animal to be consumed) have been reared and slaughtered (e.g. Hobson, 2008). Similarly, only few is still under the influence of the traditional farming ethics of agricultural communities that would give guidance how one should relate with farm animals and how animals should be utilised and treated. Furthermore, consumers have increasingly encountered a diversity of competing discourses concerning the moral value of animals as the societal discussion on animal farming has increased.

The industrialisation of livestock production has posed even more moral dilemmas for consumers: consumers tend to associate animal welfare with the 'natural' life of animals, with the ideas of 'traditional' and small-scale farming and with the individual care of animals – all of which divert clearly from the industrial mode of livestock production (e.g. Miele and Evans, 2006; Skarstad *et al.,* 2007). Consequently, consumer studies indicate that consumers tend to view the level of animal welfare in intensive forms of animal farming highly negatively (e.g. Phan-Huy and Fawaz, 2003; Verbeke and Viaene, 2000). However, it seems that only a fraction of consumers are using consumption as a site for political engagement, adopting a more welfare-friendly consumption practices, while for the majority a central resolution for their moral ambiguities appears to be the diversity of dissociation and avoidance practices prevalent in meat consumption (e.g. Holm and Møhl, 2000; Schröder and McEachern, 2004).

In this paper, we explore these issues further, analysing what kinds of citizen-consumer positions and moral strategies consumers can adopt in farm animal welfare issues. Based on group interviews among Finnish consumers, we focus especially on the question of the distance to animal production and what kind of role it plays in consumers' moral reflections on animal farming.

Methods

This paper draws from four group interviews with citizen-consumers carried out in Finland in autumn 2009. The focus group study is a part of a wider research project *Politicised animals – The consumer and farm animals* at the University of Eastern Finland, in which we study consumption politics and cultural meanings associated to animal agriculture with both qualitative and quantitative methods (www.joensuu.fi/polle). Focus groups were used to provide in-depth qualitative data on consumers' perceptions and experiences of farm animal welfare issues. This method allows studying the collective construction of social meanings as a part of group interaction (e.g. Krueger and Casey, 2000).

The group interviews included three groups that have an acknowledged special interest in animal welfare issues: vegans/vegetarians, organic consumers and gastronomes. To complement these special interest groups, we also carried out a group with randomly selected consumers to broadly reflect on the cultural argumentation concerning food consumption. Each group included 8-9 participants, 33 in total. The groups of vegetarians, organic consumers and random consumers included mainly women, while the gastronome group consisted mainly of male participants. The vegetarian and random consumer groups consisted of relatively young participants (with the average ages of 29 and 31 years respectively), while the organic group and particularly the gastronome group included much older participants (with the average ages of 41 years and 61 years respectively). All group discussions lasted for two hours. All interviews were recorded and transcribed verbatim, and the data was analysed by coding it with the help of the qualitative analysis software NVivo.

Findings

Distance to animal production and consumers' moral ambiguities

When we started unfolding consumers' moral reflections on animal farming, a central issue for the interviewees was the question of distance to modern livestock production. In their discussions, participants produced familiar notions of having a contact only with 'traditional' or 'idyllic' forms of animal farming, which was clearly contrasted with the modern industrial livestock production. Participants had often had rather extensive childhood experiences in their relatives' small-scale farms, in which animals were looked after as individuals and they were valued in a fundamentally different way than in current large-scale farms.

> My imagery [of animal farming] originates from certain kinds of idyllic experiences. My cousin's family had a farm and livestock. Animals had room to move around there and they moved freely in the cowshed. It was a bit different [than nowadays]. Then it created just a positive feeling (Vegetarians, F).

Consequently, as participants had little connection with modern farming, they showed much uncertainty how animals are faring in current farms. It was noted that media circulates highly contradictory images on animal farming. The food industry distributes positive images of content animals in idyllic conditions, while the undercover material released by animal advocacy movements paints a much gloomier picture of current livestock production. Both actors seem to disseminate only extreme images to promote their own agendas. Since consumers had to rely on mediated information, they showed much uncertainty which sources to trust when forming opinions on animal farming.

> I recall a video clip from a hennery shot by the Rights for Animals shown in a current affairs programme. When I was watching the programme, I started to think who is speaking truth here. People in the Rights for Animals want to shoot only the filthiest and most horrible things and show that this is the truth about livestock production, while the adversary wants to show only the marketing material. I think that the ordinary consumer has no possibilities to really know what the truth is. It must be somewhere between these two [portrayals] (Organic consumers, F).

The growing distance to mainstream animal farming and the existence of contradictory information implied that consumers felt much moral ambiguity in their consumption of animal products. Consumers expressed diverse moral strategies to deal with these ambiguities.

Habitual trust and dissociation practices among ordinary shoppers

For the participants in the random consumer group, the notion of habitual trust played an important role when dealing with their ambivalent feelings. These consumers maintained a vague trust that 'things are fine' in animal production. There was an aspiration to sustain a belief that the animal protection legislation and governmental inspection function well in Finland. This trust was not a reflexive trust, based on active reflection and conscious choices, but rather, it was a habitual trust – it was something taken for granted and its basis could not be articulated and explained (c.f. Bildtgård, 2008). It was just somehow in 'the back of the head', as one participant put it. In international comparisons, Finnish citizens tend to express very high levels of trust in the conventional livestock production and in the public authorities that regulate it (e.g. Adam, 2008; EC, 2005).

> Moderator: When you are thinking about the conditions of farm animals in current farms in Finland, what kinds of opinions do you have on it?
>
> F1: You just want to think that they are having good conditions, but when you don't know the real situation...
>
> F2: But somehow I have a trusting feeling that things are fine there.

For ordinary shoppers, another way to deal with their moral ambiguities included dissociation and avoidance practices – a typical sarcophage orientation identified in numerous other studies on meat consumption (see Buller and Cesar, 2007). Participants often noted how slaughtering and the cutting of meat was 'hidden inside four walls' and how they preferred to consume meat in a form that does not remind of the animal it comes from. As one interviewee put it, 'It's easier to buy a sausage or a meatball in a clean package from a super market than to be really involved in the manufacturing process.' The living animal the person had encountered was never associated to the product as if there would not be any kind of connection between them, 'You may touch a cow and you do not then think that some day it can produce milk to your hot chocolate or end up to your plate.' The farm animal that was to be consumed should not appear as too individualised or human-like subjective agent, but it should stay as an anonymous and abstract idea.

> Animals are marketed to children as domestic animals or a farm's animals, but not as production animals. In the day-care centre we read to children about a funny Mimmi-cow who wanted to have a swing. They are awfully human, and Putte-pig's name day and whatever. I don't want either to read

to children that here Mimmi-cow wants to have a swing, but then it becomes mince meat of which this meatball is made (Ordinary shoppers, F).

Likewise, the distinction between a 'pet', subjected to food taboos, and 'livestock', defined as edible, appeared as an important categorisation among ordinary shoppers (comp. Eder, 1996: 86-88; Fiddes, 1991: 132-135). It was important to maintain a certain kind of emotional distance to farm animals. Interviewees noted how there should not be any kind of personal relationship to the animal that was to be consumed. If an animal transgressed somehow the pet-livestock distinction, it was deemed as inedible.

My mum had some hens, they were so called pet hens, so that they were living quite freely. And it was not even conceived that when a hen dies it would be eaten. They were buried, and a cross was put to their graves (Ordinary shoppers, F).

Moral engagement and seeking connection

Organic consumers and gastronomes had a highly different orientation from the ordinary shoppers. Their meat consumption ethics was based on a zoophage orientation, instead of the sarcophage orientation, identified in the random consumer group. Zoophages have no problems in acknowledging the animality in food products and they do not avoid products that remind of the animal the products come from (see Buller and Cesar, 2007). Seeking connection to animal farming was an important moral orientation among organic consumers. Knowing about the life of animals worked as a form of legitimation to consume them. If one is able to watch an animal 'from eye to eye' and to acknowledge what kind of life it has lived and accept it, then s/he is entitled to consume products from that animal.

[When I lived in a biodynamic farm] I ate also meat because ... you could watch that animal from eye to eye and then you could watch it in your plate. ... [People say] that they cannot eat certain meat because they know how that cow had taught them to milk cows. But in my head it turns the other way round so that in that case I can eat meat ... So from my ethical perspective then I can allow it to myself (Organic shoppers, F).

We have had sheep and hens... Of course it was hard to eat them for the first time, but somehow you thought that if you cannot eat this meat, then you cannot eat any meat, as you had raised that sheep with your own hands (Organic shoppers, F).

Organic consumers aspired to counteract the tendency of increasingly lengthy commodity chains and to regain the connection with food production. They liked buying animal products directly from producers or through organic food clubs.

Among gastronomes, animality in meat was strongly associated to its quality. Purifying animality from meat implied that its quality declined: nutriments and aromas were cleansed off at the same time as fats, bones and fascia were removed from the product. Hence, gastronomes viewed the dissociation tendencies in wider population in highly negative terms. As the participants were older than in other groups, they could notice how the availability of different animal products had decreased in last decades: nowadays it was difficult to get certain parts of carcasses, offal, bones or ox tales and even 'fish is sold as cleaned without bones, without head'.

Due to their relatively old age, gastronomes had a clearly stronger contact to animal farming and slaughtering than participants in the other groups – often having witnessed animal slaughter by themselves. They had continued to maintain their contacts with farmers. Hence, for gastronomes the moral question of utilising animals was unambiguous. As they had seen with their own eyes that animals had had a good life, there was 'nothing extraordinary' in utilising animals in agricultural production.

> I don't think there is anything extraordinary [in animal production]. Of course you always think that you have a background, that you have seen how those animals have grown, and how they have been looked after well and seen how they have been killed (Gastronomes, M).

Conclusion

Consumption has been traditionally associated with highly individualised and non-political activities; it has been likened with the pursuit of personal pleasures rather than communal care and public engagement (Soper and Trentmann, 2008). However, consumption has become increasingly a strengthening public arena of activism as well as an expression of citizenship (e.g. Clarke *et al.,* 2007; Seyfang, 2005). In farm animal welfare issues, citizen-consumers can take various positions and moral strategies when influencing in farm animal welfare in their everyday contexts.

In the politics of consumption, a central question has been how consumers can care at a distance and take moral responsibility for their actions as the commodity chains have lengthened (e.g. Clarke *et al.,* 2007). This is a particularly heightened question in food issues as consumers have become increasingly disconnected from food production. In our group interviews, consumers developed a diversity of moral strategies to deal with their ambiguities as they had to increasingly rely on mediated, often contradictory information on animal farming. Ordinary shoppers maintained a vague, habit-based trust that 'things are fine' in animal production and they also expressed desires to keep the animal origin of food as hidden. In contrast, organic consumers and gastronomes produced a highly morally engaging citizen-consumer subjectivity that aspired to cherish the contact with food production. For organic consumers knowing about the life of animals gave a moral legitimacy to consume them, and gastronomes even embraced the animality of meat products on food ethics, quality, taste and nutrition grounds.

In this paper, we have been able to touch only shortly upon the diversity of citizen-consumer positions and moral strategies involved in the consumption of livestock products. There is a need to investigate further how different social institutions encourage the development of various citizen-consumer positions. In particular, due to the growing gap between food production and consumption, novel social institutions to facilitate welfare-friendly choices may be required.

References

Adam, F. (2008). Mapping social capital across Europe: Findings, trends and methodological shortcomings of cross-national surveys. Social Science Information 47: 159-186.

Bildtgård, T. (2008). Trust in food in modern and late-modern societies. Social Science Information 47: 99-128.

Buller, H. and Cesar, C. (2007). Eating well, eating fare: Farm animal welfare in France. International Journal of Sociology of Food and Agriculture 15: 45-58.

Clarke, N., Barnett, C., Cloke, P. and Malpass, A. (2007). Globalising the consumer: Doing politics in an ethical register. Political Geography 26: 231-249.

EC (2005). Special Eurobarometer 229: Attitudes of consumers towards the welfare of farmed animals. European Commission, Brussels.

Eder, K. (1996). The social construction of nature: A sociology of ecological enlightenment. Sage, London.

Fiddes, N. (1991). Meat: A natural symbol. Routledge, London.

Fraser, D. (2008). Understanding animal welfare: The science in its cultural context. Blackwell, Oxford.

Hobson, P. (2008). Beasts and boundaries: An introduction to animals in sociology, science and society. Qualitative Sociology Review 3: 23-41.

Holm, L. and Møhl, M. (2000). The role of meat in everyday food culture: An analysis of an interview study in Copenhagen. Appetite 34: 277-283.

Knight, S. and Barnett, L. (2008). Justifying attitudes toward animal use: A qualitative study of people's views and beliefs. Anthrozoös 21: 31-42.
Krueger, R.A. and Casey, M.A. (2000). Focus groups: A practical guide for applied research. Sage, London.
Miele, M. and Evans, A. (2006) Negotiating signs of pleasure and pain: Towards a democratic-deliberative model of animal welfare monitoring. In: Kaiser, M. and Lien, M. (eds.) Ethics and the politics of food. Wageningen Academic Publishers, Wageningen, the Netherlands, pp. 190-196.
Phan-Huy, S.A. and Fawaz, R.B. (2003). Swiss market for meat from animal-friendly production: Responses of public and private actors in Switzerland. Journal of Agricultural and Environmental Ethics 16: 119-136.
Schröder, M.J.A. and McEachern, M.G. (2004). Consumer value conflicts surrounding ethical food purchase decisions: A focus on animal welfare. International Journal of Consumer Studies 28: 168-177.
Seyfang, G. (2005). Shopping for sustainability: Can sustainable consumption promote ecological citizenship? Environmental Politics 14: 290-306.
Skarstad, G.Å., Terragni, L. and Torjusen, H. (2007). Animal welfare according to Norwegian consumers and producers: Definitions and implications. International Journal of Sociology of Food and Agriculture 15: 74-90.
Soper, K. and Trentmann, F. (eds.) (2008). Citizenship and consumption. Palgrave Macmillan, Houndmills.
Verbeke, W.A.J. and Viaene, J. (2000). Ethical challenges for livestock production: Meeting consumer concerns about meat safety and animal welfare. Journal of Agricultural and Environmental Ethics 12: 141-151.

Assessment of farm animals and intersubjectivity

F. Karlsson
Uppsala University, Department of Theology, Box 511, 75120 Uppsala, Sweden; fredrik.karlsson@teol.uu.se

Abstract

The purpose of this paper is to outline an approach, called the textual approach, to theoretically confirm the intersubjectivity of assessments of animal welfare. The field of medical hermeneutics is used as a source for inspiration of how to correlate results of disparate empirical methods in order to better understand the health status of another being. The textual approach suggests that farm animals should be assessed as physiological objects as well as experiencing subjects. Stress-measurements add information about the animal as an object, while the adjective-based method adds information about the animal as a subject. The textual approach also implies, morally, that the assessed animal should be treated as an intrinsically valued being. The concern for intersubjectivity is thus not only based on scientific concerns, but also concerns for the animals in themselves.

Keywords: animal welfare, Wemelsfelder, medical hermeneutics, animal ethics

Introduction

The purpose of this paper is to outline a theoretical approach to confirm intersubjectivity of assessments of animal welfare. Analogies from the field of medical hermeneutics will be drawn in order to suggest how results of disparate empirical methods to assess animal welfare may be cross-examined for intersubjectivity.

Intersubjectivity can be understood as a sub-criterion of objectivity (Galtung, 1967: 28). Alternatively, intersubjectivity is understood as a kind of objectivity (Kaplan, 1964: 128). An observation is intersubjective when it has been confirmed by several observing subjects. An intersubjective observation is, or can reasonably be assumed, to be objectively true.

A criterion of intersubjectivity may be understood as assuming that the involved subjects make the observations with, for example, identical methods, values, and competence (Israel, 1970). This understanding of intersubjectivity may be helpful for the work of particular research groups, working closely together. It is, however, not helpful as a criterion for science in general.

The criterion can also be understood to be satisfied when intersubjectivity of the results is achieved (Bergström, 1972: 108). This is the understanding of intersubjectivity that is relevant here. The results of various manners to assess animal welfare should be intersubjectively confirmable or challengeable.

By intersubjectivity, it is assumed that subjects interpret percepts of an object with the particular belief system of each subject. It is also assumed that there actually is an external object, for example an animal, to perceive, along which similarities in the percepts of each subject order themselves, or at least tend to do so. The challenge of intersubjectivity in animal welfare-assessment is to identify those similarities among percepts or observations of the assessors. Once those similarities have been identified, it may also be concluded along what, if anything, the similarities have arranged themselves. That 'what' is the observed object.

Medical hermeneutics and animals

Medical hermeneutics is a phenomenological ethics for clinical situations. Hermeneutical procedures are generally characterized by attempts to interpret perceived phenomena in an intersubjective manner. A common point for medical hermeneutics is to consider patient care as the starting point of any interpretation (cf. Agich, 2005). Another common claim is that approaching the patient hermeneutically includes more aspects of the patient's condition than can be described by scientific instruments (Daniel, 1990; Leder, 1990; cf. Svenaeus, 2000).

A physician attempting to understand the condition of the patient is doing something that seems similar to a veterinarian attempting to understand the condition of a farm animal. The veterinarian may be as concerned about the health of the animals, as the physician is concerned about the health of the patient. The power relation between veterinarian and farm animal is at least as asymmetrical as the power relation between physician and patient. The setting in a modern farm is as artificial as the clinical setting. The veterinarian and the animal may be strangers to each other, just as the physician and the patient may be. Insights within the field of medical hermeneutics may, thus, very well also be useful for understanding perceptions of animals.

Still, the ultimate purpose of treating a farm animal is very different from the purpose of treating a human patient. Also, the experiential background, knowledge and terminological habits of a human patient may be quite distinct from the physician's, but the distance is most definitely farther between veterinarians and animals.

A suggested approach to acquire intersubjectivity within medical hermeneutics is to use text as a metaphor for the human patient (Daniel, 1986, 1990; Leder, 1990). I will call this approach 'the textual approach'.

The textual approach considers the patient as a primary text. In order to understand this text, a number of secondary texts must be construed. Intersubjectivity is considered to be acquired by correlating a number of secondary texts.

In the human case, the secondary texts have been suggested to be the physical, instrumental, experiential text, and the narrative text. Intersubjectivity is suggested to be acquired by correlating the texts to each other according to coherence between the texts, degree of mutual collaboration between physician and patient, and possibility of successful treatment (Leder, 1990). Due to lack of space, I will only discuss how the secondary texts may be understood in the animal case, while the criteria of correlation cannot be further discussed here.

The physical text is told by the patient's body to the physician, either by direct sensation (palpation) or by tools that augment direct sensations (Leder, 1990). This notion of a text can be extended directly to the animal case. The human patient, as well as the animal, is treated as a palpable object. No claims about the subjectivity of the patient, nor the animal, are necessary.

The instrumental text is the most artificial text, consisting of translations of the patient's body, as an object, to intellectual signs – typically graphs and numbers (Leder, 1990). This is probably the text most familiar, or at least most accessible, to the modern farmer and veterinarian. Again, the animal and the human patient alike are assessed as objects. This makes the application of the text to the animal a case of direct rather than analogical extension.

The experiential text refers to the patient's experiences of the illness prior to the clinical situation (Leder, 1990). The extension to the animal case would be by analogy because it would include references to the subjective perspective of the animal. An analogy to the animal case would assume that it is possible to know sufficiently about the animal's experiences prior to the assessment of its welfare. Such experiences may be reasonably deduced by a person, for example the farmer, which has followed the animal somewhat closely.

The narrative text is constituted by the bodily condition, the story told by the patient about the condition, and the physician who interprets the story into medical terminology (Leder, 1990). In the animal case, the condition is told by the animal by way of species-specific communication. This communication can be improved by improving the human counterpart's knowledge and/or experiences about the particular species-specific communication and goods for animals. It can also be improved by improving the opportunities of the animal to communicate with human counterparts.

The textual approach and methods to assess animal welfare

The secondary texts are here intended as a kind of analytical categories. The relation to particular methods to assess animal welfare may differ. Certain methodologies can fit in one single category, meaning it adds content to one kind of text only. The common methods of assessing animal welfare by way of measuring physiological stress by heart rate or neuro-hormonal markers (Von Borell *et al.*, 2007; Mormède *et al.*, 2007), for example, adds content only to the instrumental text. Similarly, palpation only adds to the physical text.

Methods to assess fear, based on physiological markers as well as behavioralist-psychological methods (Forkman *et al.*, 2007; Boissy *et al.*, 2007), however, transgress several of these texts. The methods treat the animal as an object, resulting in intellectual signs as well as behaviourist descriptions, while also assuming that a subjective state like fear is assessed.

The textual approach does not suggest that a method is better because it adds content to more kinds of text. The approach, however, suggests that methods that do not add significant content to all texts should be used together with other methods, completing the assessment.

The experiential and narrative text assumes a subjectivity of the assessed animal. The experiential text is heavily dependent on the attention of the farmer or veterinarian to the animals throughout their lives. A formal method may be possible, but the key issue would still be that the assessing person knows the individual animal as an individual with its own life and history.

The narrative text inherently includes the perspective of both the animal and the human perceiver. The issue of perceiving animals as communicative subjects thus gets even more significant than in the case of the experiential text. I will outline a method to add content to the narrative text a bit more carefully than the more established methods mentioned above.

Methodology commonly used in studies of animal personality has been helpful in developing methods to assess animal welfare that adds to the narrative text. Within this methodology, descriptions of animals by adjectives assess individual distinctiveness (Gosling, 1999; Stevenson-Hinde *et al.,* 1980; Svartberg and Forkman, 2002). It has recently been demonstrated that using adjectives to study animal behaviour have results that converge with the results of conventional, behaviourist methods. The usage of predefined lists of adjectives, however, creates some divergence between the results of the methods (Uher and Asendorpf, 2008).

Françoise Wemelsfelder and others have used descriptions of animals by adjectives to assess the welfare of perceived farm animals. The test subjects are asked to spontaneously find suitable adjectives, rather than select them from predefined lists. An underlying assumption is that the adjectives refer to percepts of the whole animal. The notion of animal as subject is clearly present, considering that the resulting adjectives have been 'playful' and 'irritable', and that the method stems from personality studies (Wemelsfelder *et al.,* 2000, 2001).

The resulting variations of particular terms and score scales are standardized by the statistical method Generalized Procustes Analysis (GPA). The method attempts to identify intersubjective content among the adjectives. GPA may be used to suggest that two observers probably have a similar understanding of the behaviour, albeit they use different, but similar, terms to describe the behaviour. The method does this not by using the semantical content of the assertions, but by analyzing the practical function of the assertions (Wemelsfelder *et al.,* 2000).

Results have been obtained and confirmed for several species of farm animals saying that there is a highly significant correlation between how different human observers describe the moods of the animals (Minero *et al.,* 2009; Napolitano *et al.,* 2008; Rousing and Wemelsfelder, 2006; Wemelsfelder, 2007). The effect of professional background of the test subjects, and the effect of background environment of the perceived animals have been shown not to significantly confound the results (Wemelsfelder, 2007; Wemelsfelder *et al.,* 2009). Also, there are some results, unfortunately still unpublished, showing that there is reason to believe that the described mood also had physiological correlations in the observed animals (Wemelsfelder, personal communication). Furthermore, the results have been shown to be helpful for veterinary inspectors (Wemelsfelder, 2005; Wemelsfelder and Lawrence, 2001).

This adjective-based approach to assess animal welfare adds content to the narrative text. The textual approach suggests that the experiential, narrative, instrumental, and physical texts have the status of corroborating or contradicting each other. Within the suggested approach, the adjective-based approach assuming the animal as a subject has similar status to the more common, reductionist and instrumental methods, but a different role (namely, adding content to another text).

Moral implications

To consider methods to assess animal welfare that assumes an animal subjectivity may imply that certain non-human animals carry intrinsic moral value, although this is not a necessary implication. It is, by now, a well-established fact that many animals can suffer, implying that they have, at the very least, subjectivity sufficient to suffer. This means that it is necessary to take the subjective perspective of the animal into account when assessing animal welfare as long as 'animal' refers to the whole animal being.

If animal welfare is instead understood as probability of animal survival, then animal welfare can be addressed by treating the animal as a physiological object only. If animals are viewed as physiological production units, and no more, then the suggested approach of cross-checking methodologies is overly elaborated. It is then instead sufficient to correlate the interpretations of the physical and instrumental texts with each other.

The reduction of the animal to a mere object could be made out of ignorance of the established fact that many non-human animals do have some kind of subjective perspective. It could, however, also be made because that fact is ignored out of moral reasons. Animals may be acknowledged as having subjectivity, but still not be considered to be worthy to be treated according to their whole nature. They may, for example, be argued not to have the right kind of subjectivity.

To include the subjectivity of the animal as a factor when assessing animal welfare could, similarily, be both because of non-moral and moral reasons. The inclusion could be because of a wish to more completely assess animal welfare. It could be an expression for a wish to use and acquire the right facts while assessing animal welfare.

Also, it could, in addition, be because of a notion about the dignity, intrinsic value or other inherent moral quality of the animal, making it a receptor of moral concern. This latter reason inherently underlies the suggested textual approach.

In the human case, the textual approach considers the good for the human patient as basic for the assessment. In the animal case, the good for the animal is basic. Because of the nature of the activity of animal husbandry and animal breeding, the animal is, as opposed to the human patient, eventually killed. It is not necessary that killing the animal always is a violation of the moral concern that is implied when animal welfare is assessed in the suggested manner, neither is there any space here to further elaborate on that problem. The point here is that the textual approach not only is an attempt to treat the assessed animal as a whole in accordance of known facts about the animal, but also that this is done out of moral reasons.

Conclusion

The benefit of the suggested approach is that it largely correlates with the already established habit of confirming and re-confirming empirical results by using different methods. It challenges, however, the notion that treating animals as subjects would be bad science (a Cartesian notion that surprisingly is still encountered). The results of Wemelsfelder *et al.* show that some methods treating the assessed animals as subjects have intersubjective qualities. The intellectual signs resulting from the conventional stress-measurements – treating the assessed animals as objects – are easily checked by various human subjects, possibly confirming the results intersubjectively.

The textual approach suggests a further control for intersubjectivity by correlating the different texts, once they have been filled with content by disparate methods. The approach thus recommends a suspicious attitude towards particular empirical methods, while it corroborates empirical methodology as such. The intersubjectivity of each particular empirical method is viewed as too provincial to be the finally validating kind of intersubjectivity. Instead, the aspect of assessments of animal welfare that bridges disparate methods should be identified and considered to hold a more stable kind of intersubjectivity.

The suggested approach to combine disparate methodologies to assess animal welfare implies a number of consequences for anyone that claim to assess animal welfare. Skills, knowledge and experience are needed to do this that goes beyond technical, reductionist methods. Species-specific communication is already taught at universities, and should also be taught as a practical skill as necessary as being able to take biopsies or draw blood in order to perform well as a veterinarian.

Also, to spend time in order to know individual animals would be a vital aspect of assessing animal welfare. This entrusts the farmer an active role in the assessment of animal welfare. It, however, also turns a critical edge towards large-scale farms, at least if they are sparsely staffed. Farm animals must be looked at more often in order to assess how they fare. This is true to the extent that farm animals are viewed as both objects and subjects; and there are good reasons to think they should be viewed as such.

References

Agich, G.J. (2005). What Kind of Doing is Clinical Ethics? Theoretical Medicine 26: 7-24.

Bergström, L. (1972). Objektivitet. Prisma, Stockholm, 129 pp.

Boissy, A., Manteuffel, G., Bak Jensen, M., Oppermann Moe, R., Spruijt, B., Keeling, L.J., Winckler, C., Forkman, B., Dimitrov, I., Langbein, J., Bakken, M., Veissier, I. and Aubert, A. (2007). Assessment of positive emotions in animals to improve their welfare. Physiology & Behavior 92: 375-97.

Daniel, S.L. (1986). The Patient as Text: A Model of Clinical Hermeneutics. Theoretical Medicine. 7: 195-210.

Daniel, S.L. (1990) Interpretation in Medicine: An Introduction. Theoretical Medicine, 11: 5-8.

Forkman, B., Boissy, A., Meunier-Salaün, M.-C., Canali, E. and Jones, R.B. (2007). A critical review of fear tests used on cattle, pigs, sheep, poultry and horses. Physiology & Behavior 92: 340-374.

Galtung, J. (1967). Theory and Methods of Social Research. Columbia University Press, New York, 534 pp.

Gosling, S.D. (1999). Personality Dimensions in Nonhuman Animals: A Cross-Species Review. Current Directions in Psychological Science 8: 69-75.

Israel, J. (1970). Objektivitet och värderingsfrihet inom samhällsvetenskapen: sociologisk diskurs eller häxjakt? Sociologisk forskning 7: 65-73.

Kaplan, A. (1964). The Conduct of Inquiry. Chandler, San Francisco, 428 pp.

Leder, D. (1990). Clinical Interpretation: The Hermeneutics of Medicine. Theoretical Medicine 11: 9-24.

Minero, M., Tosi, M.V., Canali, E. and Wemelsfelder, F. (2009). Quantitative and Qualitative Assessment of the Response of Foals to the Presence of an Unfamiliar Human. Applied Animal Behaviour Science 116: 74-81.

Mormède, P., Andanson, S., Aupérin, B., Beerda, B., Guémené, D., Malmkvist, J., Manteca, X., Manteuffel, G., Prunet, P., van Reenen, C.G., Richard, S., Veissier, I. (2007). Exploration of the hypothalamic–pituitary–adrenal function as a tool to evaluate animal welfare. Physiology & Behavior 92: 317-339.

Napolitano, F., De Rosa, G., Braghieri, A., Grasso, F., Bordi, A., Wemelsfelder, F. (2008). The Qualitative Assessment of Responsiveness to Environmental Challenge in Horses and Ponies. Applied Animal Behaviour Science 109: 342-54.

Rousing, T. and Wemelsfelder, F. (2006). Qualitative Assessment of Social Behaviour of Dairy Cows Housed in Loose Housing Systems. Applied Animal Behaviour Science 101: 40-53.

Svartberg, K. and Forkman, B. (2002). Personality Traits in the Domestic Dog (*Canis familiaris*). Applied Animal Behaviour Science 79: 133-55.

Stevenson-Hinde, J., Stillwell-Barnes, R. and Zunz, M. (1980). Subjective Assessment of Rhesus Monkeys over Four Successive Years. Primates 21: 66-82.

Svenaeus, F. (2000). Hermeneutics of Clinical Practice: The Question of Textuality. Theoretical Medicine and Bioethics 21: 171-89.

Uher, J. and Asendorpf, J.B. (2008). Personality assessment in the Great Apes: Comparing Ecologically Valid Behavior Measures, Behavior Ratings, and Adjective Ratings. Journal of Research in Personality 42: 821-38.

Von Borell, E., Langbein, J., Després, G., Hansen, S., Leterrier, C., Marchant-Forde, J., Marchant-Forde, R., Minero, M., Mohr, E., Prunier, A., Valance, D. and Veissier, I. (2007). Heart rate variability as a measure of autonomic regulation of cardiac activity for assessing stress and welfare in farm animals: A review. Physiology & Behavior, 92: 293-316.

Wemelsfelder, F. (2005). The Reliability of Qualitative Behaviour Assessment in On-Farm Conditions (Abstract). Third International Workshop on the Assessment of Animal Welfare at Farm and Group Level. University of Natural Resources and Life Sciences, Vienna, Austria.

Wemelsfelder, F. (2007). How Animals Communicate Quality of Life: The Qualitative Assessment of Animal Behaviour. Animal Welfare 16 (S): 25-31.

Wemelsfelder, F. and Lawrence, A.B. (2001). Qualitative Assessment of Animal Behaviour as an On-Farm Monitoring Tool. Acta Agriculturae Scandinavica, Section A – Animal Sciences 51 (S): 21-25.

Wemelsfelder, F., Hunter, T.E.A., Mendl, M.T. and Lawrence, A.B. (2000) The Spontaneous Qualitative Assessment of Behavioural Expression in Pigs: First Explorations of a Novel Methodology for Integrative Animal Welfare Measurement. Applied Animal Behaviour Science 67: 193-215.

Wemelsfelder, F., Hunter, T.E.A., Mendl, M.T. and Lawrence, A.B. (2001). Assessing the 'Whole-Animal': A Free-Choice Profiling Approach. Animal Behaviour 62: 209-20.

Wemelsfelder, F., Nevison, I. and Lawrence, A.B. (2009). The Effect of Perceived Environmental Background on Qualitative Assessments of Pig Behaviour. Animal Behaviour 78: 477-84.

Animal welfare regulations in the EU: a matter of ethics or a matter of human economic interest?[1]

M. Landera-Luri
University of the Basque Country, Faculty of Law, Barrio Sarriena s/n, 48940 Lejona, Spain; mariamercedes.landera@ehu.es

Abstract

This contribution dealt with the more relevant European norms concerning the breeding conditions and care of farm animals. The application of these norms takes place in the European Union built, fundamentally, upon a market forces based approach. This implies their understanding as an instrument of homogenized practices in the sector which seek to prevent inequalities of conditions in competition. Beyond the critical evaluation, which, from an ethical point of view, is derived from subordinate animal protection to the structural and organizational necessities of the market, this fact conditions, notably, the practical reach of these provisions.

Keywords: farm animals, European legislation, unnecessary suffering, free movement of goods, fair competition

Introduction

As its original name, the European Economic Community, suggests the EU was formed primarily to engender peace in Europe by bringing the member states together through economic means. The principles of animal welfare were included much later as a response to concerns over animal welfare and were not originally included in the Treaty of Rome. As economics, not morality, is the cornerstone of the European Union, the issue of animal welfare has always sat a little uncomfortably within the original economic objectives of the Treaty.

The Treaty of Rome of 1957 sets out a common commercial policy to progressively abolish restrictions on communitarian trade as one of its basic objectives. The free movement of goods (Article 28) is, in this context, a basic prerequisite for the efficient functioning of the European market, one of the world´s largest single markets. Animals are included in this market as another form of merchandise, more precisely, as agricultural products (Article 38.1). Specifically, animals reared for human consumption fall into this category and the term 'farm animals' is applied to them.

At the same time, the EU has approved a series of norms establishing minimum requirements for the breeding and care of farm animals. These norms are part of the provisions known as Animal Welfare Regulations. The key text in this body of norms is the European Convention for the protection of animals kept for farming purposes which was adopted in 1976. This Convention is applicable to animals in modern intensive stock-farming systems. The animals dealt with in the Convention are animals bred or kept for the production of food, wool, skin or fur or for other farming purposes.

It is said that the objective of these norms is twofold: the protection of farm animals and the reducing of inequalities in conditions of competition. In this way, clearly, in Case C-128/94, Hans Hönig v Stadt Stockach of 19 October 1995, Reference for a preliminary ruling in, ECR, 1995, I-03389.

[1] To Titín: to the person who decided that a cat`s life isn´t worth enough to authorize the medication needed to fight against feline AIDS.

The objective pursued by the Community legislature is twofold. It is on the one hand to protect the animals farmed, and on the other hand to reduce inequalities in conditions of competition as between Member States in this particular sector (the market in eggs and poultry), which are attributable to the different national rules in this area. In order to strike the *right balance between the conflicting interests of the economic necessity of reducing production costs, and the moral and ethical necessity of introducing rules protecting animals* and to reconcile them, the Community legislature sought to lay down, as a first step, common minimum standards.

It is my hypothesis that these two objectives possess different status. The avoiding of inequalities in competition would be the main objective. Therefore, the main objective is to protect human economical interests. The protection of animals would be a finality of second category.

This hierarchy becomes clear in the cases in which it is not possible to fulfil both objectives at the same time. The cases in which these objectives come into conflict.

Hypothesis

My hypothesis is that the resolution of the conflicts of interest previously referred to, usually results in the downgrading or weakening of the principles of Animal Welfare in favour of those of the organization and functioning of the market.

Development of the hypothesis

In this section we will see if this hypothesis can be sustained in the light of the political statements which explain the reasons for the approval of the laws concerned, the configuration of the norms themselves, and the practical of the European Court of Justice.

The objective of the legislator in passing the law, ratio legis

The motive of market harmonization which guides all of these dispositions is clearly reflected in all of the corresponding explanations of motives. To begin with the basic norm in this area, the Council Decision of 19 June 1978 concerning the conclusion of the European Convention for the Protection of Animals Kept for Farming Purposes (78/923/EEC), already shows the context in which Animal Welfare provisions come to have meaning...

Whereas *the protection of animals is not in itself one of the objectives of the Community:* whereas, however, there are disparities between existing national laws on the protection of animals kept for farming purposes which may give rise to unequal conditions of competition and which may consequently have an indirect effect on the proper functioning of the common market;

Likewise, the Preamble of the Council Directive 98/58 of 20 July 1998.

Whereas differences which may distort conditions of competition interfere with the smooth running of the organisation of the market in animals.

Whereas there is therefore a need to establish common minimum standards for the protection of animals kept for farming purposes *in order to ensure rational development of production and to facilitate the organisation of the market in animals.*

Whereas a comparative examination of animal welfare provisions applicable in the Community and in certain non-member countries together with an appraisal thereof should be undertaken with a view to determining the nature Community initiatives *aimed at eliminating distortions of competition.*

These considerations are reproduced, to a greater or lesser length, in all the norms derived from the foregoing one. Among them are the development directives of the principles of Animal Welfare for application to the different species of farm animals.

Economic references in the configuration of the norms

In this section, concepts contained in the structure of the norms that make reference, directly or arrived at by way of interpretation, to economic questions, will be examined.

Economical considerations constitute a criterion for the interpretation of the *Animal Welfare principles (Article 2 of the European Convention).* These principles are, predominantly, formulated in an open way. As the Advocate General Léger explains in his opinion, delivered on 15 July 1997, about The Queen v Minister of Agriculture, Fisheries and Food, ex parte Compassion in World Farming Ltd. Reference for a preliminary ruling: High Court of Justice, Queen´s Bench Division, United Kingdom, European Court reports [1998] Page I-01251.

Although the imperative form is used in the provisions of the Convention, with the word `shall´ being systematically used to lay down the principles contained in Chapter I, those provisions are for the most part characterized by imprecision ... the conditional nature of this rules [he is referring to Recommendation of the Standing Committee of1998 concerning Cattle] precludes them from being recognized as having the slightest binding force and that where, on the other hand, a standard is laid down in mandatory terms, its imprecision renders it unenforceable (paragraph 129, 139 and 146).

The Convention does not define the content or the limits of the general principles set out within it. In its place, the concrete content of these proclamations, the categorical *Five Freedoms,* ends up being determined by content criteria, the economic one among them.

Established experience and scientific knowledge

All the Animal Welfare requirements: the housing and provision with food, water and care, the freedom of movement, the space, the lighting, temperature, humidity, air circulation, ventilation and other environmental conditions such as gas concentration or noise, are to be put into practice on the basis of their conformity with *the physiological and ethological needs in accordance with established experience and scientific knowledge.* The question that must be answered is, therefore, what is understood by *established experience* and *scientific knowledge*. For present purposes, the meaning of the first expression is of particular importance.

Established experience. This interpretative criterion reinforced the economic perspective which guides these dispositions when it was defined by the Standing Committee of the European Convention for the Protection of Animals kept for Farming Purposes as knowledge that is put beyond doubt and that is gained through *direct participation in animal husbandry* or clinical veterinary practice. In this way, the livestock breeders are installed as sources of interpretation of the principles of Animal welfare.

Revealingly, the Spanish State in *Real Decreto 348/2000* of the 10th of March, which incorporated into legislation the protection of farm animals set out in Directive 98/58/CE, specifies that within the interpretive criterion of established experience and in order to put into practice the concrete demands of the general principles, preference will be given to the *experience of production.*

Unnecessary pain, suffering or injury

The desired categoricalness of the so-called '*Five freedoms*' is placed in doubt when it comes to ascribing concrete content to each of them. The objective of the protection of animals is placed under question by the recurrence of such dubious concepts as *unnecessary* suffering or injury.

The Explanatory Report on the European Convention on the protection of animals kept for farming purposes lays down that the underlying idea is to avoid any unnecessary suffering or injury and to secure conditions that shall be in conformity with physiological and ethological needs of the individual animals. These provisions were conceived as guiding principles for the subsequent elaboration of detailed measures.

The meaning of this term, which is of great relevance, has not been defined by the EU. Basing myself on commentators on this regulation, I consider revealing the definition proposed by some commentator of unnecessary, as referring to injury or suffering that is neither morally nor economically justified.

In the jurisprudence of the European Court of Justice, Advocate General Léger has pointed out the idea of proportionality in the judging of the unnecessity of a practice. In his opinion, the harm to the health or life of animals resulting from the rearing method in question is established by objective scientific evidence and the measures adopted are proportionate to the objective pursued (Case C-1/96 The Queen v MAFF ex parte Compassion in World Farming, Preliminary Ruling, paragraph 151 (1)).

This idea of proportionality between the measure questioned according to the principles of Animal Welfare and the objective being pursued by that measure, seems to be present also in this sentence handed down by the court of a member state, the *Court of Appeal of the United Kingdom.* The case concerns the possible violations of Community norms, among them the Treaty of Rome, in so far as it concerns the free movement of goods, by the 2004 Hunting Act, that prohibits the hunting with dogs of foxes, deer, hare and mink. The Court held that in the Case of Regina (Countryside Alliance and others) v. Attorney General and another, Regina (Darwin and others) v. Same, Regina (Friend and another) v. Same.

...unnecessary suffering is not exclusively a question of scientific fact. Judgment and evaluation are required. Evaluation requires consideration of the need for suffering or the asserted need, a moral or ethical judgment and the striking of a balance between competing factors such as the degree of and necessity of suffering. One factor Parliament can legitimately take into account is the reason why the activity is engaged in, for example sport. For that reason there is no moral equivalence between the lengthy pursuit of a mammal before it is killed and the possible accidental failure to kill a mammal cleanly by shooting it for pest control purposes.

Arising from this approach, suffering may be considered *unnecessary* when it is not essential to achieve the proposed objective or when there is a less burdensome way of achieving it available. The objectives can be numerous. Fair competition and free movement of goods objectives, included. In general, human economical objectives. The production of *foie gras*, whose production requires the force feeding of geese and ducks, is probably one of the cases that best illustrates the prioritizing of human economical interests over those of the minimal legal demands of Animal Welfare. The *Report of the Scientific Committee on Animal Health and Animal Welfare on Welfare Aspects of the production of foie gras in ducks and geese* adopted the 16 December 1998 cites the many negative effects on animals of force feeding and concludes that force feeding, as currently practised, is detrimental to the welfare of the birds. However, its recommendations propose the continuation of the production of *foie gras*, with the implementation of Animal Welfare measures. Economic arguments are decisive in the decision to reject the prohibition of force feeding, a ban on force feeding. The ban is likely to cause a considerable reaction from those involved in the foie gras industry, especially among the *farmers and processors*, as well as the *public in general*. The irritation of the 300,000 *people directly concerned with the production...* The French industry employs about 30,000 *people involved in all aspects of production*. In this context, the connection between Animal Welfare and competition, in this case, competition with third countries, is patent; 'if the third country producers improved animal welfare and hence the public image of their product, before E.U. producers did so, the E.U. producers lose much of their market'.

Arguments from jurisprudence

On more than one occasion the European Court has had to decide whether the protection of the life and health of animals might prevail over the economic requirements of the free movement of goods or fair competition. In other words, which is the principal objective in these twofold norms. As a starting point, The Court of Justice of the European Community contextualizes the problems arising from the application of Animal Welfare in terms of the general rules on the free movement of goods and on the common organizations of the markets in the agricultural sector (Joined Cases 141 to 143/81, Gerrit Holdijk and others *[1982] ECR 01299*).

In this context, and referring to free movement principle, from a general perspective, the Court maintains that:

> 'The question of whether or not national measures on animal health may include an import licensing system without infringing Article 36 of the Treaty cannot be answered uniformly in every case. In fact, the reply to that question depends upon the relationship in an individual case, between, on the one hand, the inconvenience caused by the administrative and financial burdens imposed under such a system and, on the other hand, the dangers and risks for animal health resulting from the imports in question' (Case 74/82, Commission of the European Communities v Ireland, paragraph 51, [1984] ECR 00317).

But this general consideration has, in the practical, a highly relevant restriction consolidated in jurisprudential European practice, *that the protection of life or health of animals is no longer possible where Community directives provide for harmonization of the measures necessary to achieve the specific objective which would be furthered by reliance upon this provision* (*Case C-5/94 The Queen v Ministry of Agriculture, Fisheries and Food, ex parte: Hedley Lomas (Ireland) Ltd* [1996] ECR I-02553 and *Case C-1/96 The Queen v MAFF ex parte Compassion in World Farming [1998] ECR I-01251, paragraph 18 and 47, respectively).*

The aforementioned twofold objective in the norms of Animal Welfare becomes patently clear when the conflict arose between the structural principle of fair competition between producers from different member states and the protection for animals kept for farming purposes. In *Case C-128/94, Hans Hönig v Stadt Stockach* the Court sentences that:

> ...the Directive seeks first to provide protection for animals kept for farming purposes and secondly to reduce the disparities in conditions of competition as between Member States on the market in eggs and poultry (paragraph 14[1995] ECR I-03389).

The court decided, differently, in *Case 131/86, United Kingdom of Great Britain and Northern Ireland v Council of the European Communities [1988] ECR 00905.* In this case, the Court must decide on the primary objective of Directive 86/113 which lays down minimum standards for the protection of laying hens in battery.

However, the set of preparatory acts relied upon the parties follows that the harmonization of the minimum requirements for animals kept for farming was decided essentially to eliminate the unequal conditions of competition in this area (paragraph 26).

And to cite the Council and the Commission:

> The Council and the Commission also argues that the directive in question pursues the objectives of Article 39 of the Treaty and, above all, to prevent competition in the market for poultry and eggs due to operating conditions different in the United States (paragraph 10).

Conclusion

The balance between human economic interests, broadly understood of those involved in the animal production market, and the interest of farm animals, incorporated in the idea of welfare, is certainly precarious. The intended double objective of the norms of Animal Welfare is placed in doubt in case of conflict. This is due to the distinct status attributed to each of these objectives. Instrumental *Status* for the idea of protection of animals, as a way of harmonizing animal management practices in a predominantly homogenous market.

Acknowledgements

This paper has been supported by the Spanish Government, under project DER 2009-08167 JURI.

References

Brooman, S. and Legge, D. (1997). Law Relating to animals. Cavendish Publishing Limited, London-Sydney, p. 189-190.
Poblador. (2008). Resumen comentado de la normativa de Bienestar Animal, ITG, Instituto Técnico Ganadero, p. 2.

Effects of feed supplements on meat quality: an ethical analysis that includes the welfare and health of production animals and human consumers

F.L.B. Meijboom[1], M.B.M. Bracke[2] and F.J. van der Staay[3]
[1]Utrecht University, Faculty of Veterinary Medicine, Department of Animals in Science and Society, P.O. Box 80.166, 3508 TD Utrecht, the Netherlands; [2]Wageningen University and Research Center, Livestock Research, Edelhertweg 15, 8219 PH, Lelystad, the Netherlands; [3]Utrecht University, Faculty of Veterinary Medicine, Department of Farm Animal Health, P.O. Box 80151, 3508 TD Utrecht, the Netherlands; f.l.b.meijboom@uu.nl

Abstract

It has been recognized that the feed of animals and the resulting food for the human end-consumer are intimately linked. A few years ago, Bügel (2003) posed the question whether the human micronutrient status of food can be improved by supplementing production animals with components that are believed to improve human health. The basic idea is feeding these putative health-promoting nutrients to domestic animals and to let them accumulate in the animal's tissue in order to make them available to the consumer in the resulting meat and dairy product(s). This may result in programs for developing 'added value' or functional foods derived from livestock products. We present this development and define its ethical implications in terms of human and animal health and welfare, human autonomy and respect for animals. By defining these questions, we aim to show that an ethical evaluation of enriched meat products should not be postponed to the implementation of this development, but has to be taken into account in the current process of research and development.

Keywords: functional foods, animal welfare, food safety, respect

Increased interest in healthy food

The impact of food on human well-being, health status and healthy aging arouses public interest, especially in industrialized countries (Green *et al.*, 2001). However, it is of equivalent interest for the health status in developing countries (Bender, 1992). In particular, the quality of food attracts growing interest, both on the part of the consumer and the producer. There is a shift from 'health risk' concerns due to, for example, contamination of food products with dioxins or residues of antibiotics as growth promoters, to the potential beneficial effects of food and food composition (e.g. MacRea *et al.*, 2005). Especially the role of food constituents in the prevention of diseases through their ability to modulate physiological systems is gaining growing attention.

As a result new products are developed and introduced at the interface of food and health. In this process it is interesting to note that the food industry is moving towards the pharmaceutical industry (Brännback *et al.*, 2002), whereas the pharmaceutical industry shows growing interest in health-promoting constituents in nutrition. As a consequence, the distinction between functional food and pharmaceuticals is rather vague (Hasler, 2000), as the indications for applying them may overlap considerably.

Animals, feed and healthy food

In the context of increased attention to healthy food animal products play a double role. On the one hand, there is serious discussion about their healthiness, for instance in terms of an over-consumption of proteins. At the same time, a discussion has started questioning whether the human micronutrient status of food can be improved by using feed supplements for production animals (Bügel, 2003). The basic idea is that feeding putative health-promoting nutrients to domestic animals will lead to accumulation in the animal's tissue in order to make enhance availability to the consumer in the resulting meat product(s). This may lead to 'added value' or functional foods derived from livestock products, including those that meet specific nutritional requirements of human genetic conditions. Although this is still in the process of development, a number of patents on this technique have already been filed.

This development does not only provide new opportunities in preventive medicine or marketing, but also raises ethical questions, especially about human health and safety. However, we argue that questions also arise with respect to animal health and welfare.

Ethical questions related to human health, safety, and autonomy

Any program to develop healthier meat through supplementing the feed of production animals with bioactive components should act along the line of basic ethical principles such as the prevention of harm, fairness and respect for autonomy of the human end-consumer (Mepham, 1996; Schroeder, 2007). In other words, these products should be safe, include a valid claim and should be fair for all involved parties.

With respect to safety, these new products raise interesting questions. On the one hand, they require technical competence. These products and applications are relatively new and the common-sense views on safety and risks often do not suffice in this context. For instance, a considerable number of end-consumers often favours the use of bioactive compounds from natural sources above synthetic compounds (pharmaceuticals). This appears to be based on the beliefs that (a) these compounds, being natural, are implicitly safe; (b) they do not have side-effects or (c) are a universal remedy (Chang, 2000). However, such beliefs, partly exaggerations and fallacies, do not only result out of technical ignorance, but (also) highlight an ethical dimension. Although, safety is paramount in every introduction of a product in both the food sector and health sector, the specific interpretation of what safety means in relation to pharmaceutical products may be/is different from food products (Meijboom, 2007). This variation is not only the result of differences in risk-benefit analyses, but is also rooted in differences in the moral evaluation of the product's aim. In practice this implies that the issue of the safety of nutriceuticals does not only require a risk-benefit analysis, but also a moral evaluation of the value of the aim of the product: is the aim feeding, or preventing or even curing diseases?

Also at the level of consumer autonomy questions arise. Even though, most people will agree that it does not make sense to start enriching meat without scientific underpinning of the expected health benefits, there is still discussion on the soundness of claims of many functional foods. This equally holds for the use of bioactive components in meat products. Especially the dosing may be problematic, because of the large variation in plants used in animal feed. Plants are complex mixtures of active compounds. They may amplify, reduce or neutralize the effects of other nutritional constituents (Essers *et al.*, 1998) both with respect to desirable and undesirable (even toxic) effects. On top of this, the nutritional state of the human end-consumer with respect to these supplements may be extremely different between developing countries and western, industrialized countries (e.g. Murphy and Allen, 2007), but may also differ between males and females, and between individuals [e.g. due to different levels of appetite and intake]. This complicates the interpretation of the principle of due respect for the autonomy, e.g. it becomes difficult to inform consumers on the claims and effects of these products in a trustworthy way.

Animal health and well-being

Furthermore, most papers addressing these putative health-promoting effects of bioactive food supplements are concerned with effects on humans. Only a few papers address effects on health and well-being of production animals. This implies a further ethical dimension. On the one hand, issues of health and well-being of animals are likely to arise. Although there appears to be growing consent between farmers, veterinarians and consumers that animal health and animal well-being is and should be a major concern in livestock production, this new development may raise three types of ethical questions related to the animals.

- First, feed supplements explicate that – despite apparent agreement on the moral relevance of animal welfare – we are far from agreeing what exactly constitutes health and well-being (Sundrum *et al.*, 2006), i.e. what the morally relevant features for health and welfare are.
- Secondly, it raises questions on how animal health and welfare should be weighted if it is in conflict with other values. Given the common sense view that animal welfare matters morally, but that the same holds for human health, it is in practice not self-evident how potential conflicts between both values should be addressed. In fact, the aspect of uncertainty complicates the process of weighing. In some cases the impact of enriched meat products on animal health or welfare may be relatively clear, e.g. high doses of bioactive compounds may compromise the animals' welfare, and in the worst case, their health. However, in the case of feed supplements such as herbs this picture is complicate by the unpredictable (adverse) effects and interactions with some of the commonly used (veterinary) medicines.
- Finally, there certainly is no agreement among consumers or stakeholders in the production chain on the ethical assessment of animal use, e.g. some consumers are vegetarians, while others would like to see cheaper animal products and don't seem to care much about animal welfare.

All these questions ask for ethical reflection rather than exclusive focus on further scientific research. On the other hand, the production of enriched meat products also touches, beyond health and welfare, upon the problem of respect for animals. The animals are not treated for their own sake (i.e. for improving their health), but to improve meat quality and in order to supply the human end-consumer with bioactive compounds. This is in conflict with animal-rights views, which attribute inherent value to animals (e.g. Regan, 1983).

Conclusion

Improving meat quality and human health through providing by supplements in the feed of production animals is a groundbreaking idea that deserves further consideration. However, this concept is still in its fledgling stages. Crucial studies are still missing that have put this idea to the test. This holds for the scientific questions as well as for its ethical dimensions. Our introduction to the variety of ethical questions related to the development of enriched meat products illustrates that the ethical evaluation should not be postponed to the later stages of implementation. Instead, these issues ought to be taken into account in the current process of research and development.

References

Bender, A. (1992). Meat and meat products in human nutrition in devloping countries, FAO, Food and Nutrition paper 53, http://www.fao.org/docrep/T0562E/T0562E00.htm.

Brännback, M., de Heer, A.J., and Wiklund, P. (2002). The convergence of the pharmaceutical and the food industry through functional food: strategic change and business opportunity or an illusion? Pharmaceuticals Policy and Law 5: 63-78.

Bügel, S. (2003). Can human micronutrient status be improved by supplementing domestic animals? Proceedings of the Nutrition Society 62: 399-402.
Chang, J. (2000). Medicinal herbs: drugs or dietary supplements? Biochemical Pharmacology 59: 211-219.
Essers, A.J.A., Alink, G.M., Speijers, G.J.A., Alexander, J., Bouwmeister, P.-J., van den Brandt, P.A., Ciere, S., Gry, J., Herrman, J., Kuiper, H.A., Mortby, E., Renwick, A.G., Shrimpton, D.H., Vainio, H., Vittozzi, L., and Koeman, J.H. (1998). Food plant toxicants and safety: risk assessment and regulation of inherent toxicants in platn foods. Environmental Toxicology and Pharmacology 5: 155-172.
Greene, H.L., Prior, T. and Frier, H.I. (2001). Foods, health claims, and the law: comparisons of the United States and Europe. Obesity Research 9: 276S-283S.
Hasler, C.M. (2000). The changing face of functional foods. Journal of the American College of Nutrition 19: 499S-506S.
MacRae, J., O'Reilly, L. and Morgan, P. (2005). Desirable characteristics of animal products from a human health perspective. Livestock Production Science 94: 95-103.
Meijboom, F.L.B. (2007). Trust, Food, and Health. Questions of Trust at the Interface between Food and Health, Journal of Agricultural and Environmental Ethics, 20/3, pp. 231-245.
Mepham, B. (1996). Food Ethics, Routledge: London.
Murphy, S.P. and Allen, L.H. (2007). Nutritional importance of animal source foods. The Journal of Nutrition 133: 3933S-3935S.
Regan, T. (1983). The Case for Animal Rights, University of California Press.
Schroeder, D. (2007). 'Public Health, Ethics, and Functional Foods', Journal of Agricultural and Environmental Ethics, 20/3, pp. 247-259.
Sundrum, A., Padel, S., Arsenos, G., Kuzniar, A., Hendriksen, B.I.F., Walkenhorst, M. and Vaarst, M. (2006). Current and proposed EU legislation on organic livestock production, with a focus on animal health, welfare and food safety: a review. In: Rymer, C. *et al.* (eds.), Future perspective of animal health on organic farms: main findings, conclusions and recommendations from SAFO network. Proceedings of the 5th SAFO workshop. Odense, Denmark: SAFO.

Part 10. Animal biotechnology

Animal ethics committees, democracy and decision-making

H. Röcklinsberg
Swedish Univeristy of Agricultural Sciences, SLU, Dept of Animal Environment and Health, Box 7068, 750 07 Uppsala, Sweden; helena.rocklinsberg@hmh.slu.se

Abstract

The general task of an animal ethics committee is to balance potential human benefit with e.g. pain and stress inflicted on animals. Above legislation the composition of the committee and it's routines are vital to fulfill this task. Some countries (e.g. Sweden and Australia) have laymen representation in the ethics boards in order to keep citizens' insight and trust in research goals and methods. Researcher's task is to guarantee a scientific judgment of the applied research. One frequently used category of animals, genetically modified (GM) animals, are designed to e.g. develop human diseases, or to produce certain substances. Assessment of suffering due to the genetic modification is to be kept apart from pain inflicted in the experiment in the committee's evaluation of an application. However, studies of ethical committees in Sweden show decisions are sometimes taken in spite of lacking information about the scientific necessity to use GM animals, etc. Hence there is a risk of less strict scrutinizing of applications than an ordinary citizen may have reason to expect. If so, and if this becomes known to the public, a severe risk of democratic deficit and lacking trust emerges, not only in animal ethics committees but possibly also more general in research. Hence there is a challenge for a democratic society to handle decision-making in interdisciplinary and hierarchical contexts where expert knowledge is needed. After a first section on relating pros and cons regarding animal experimentation and public control, a second on ideal control and assessment in reality, I elaborate on models for democratic decisions, and finally suggest a hermeneutic approach for interdisciplinary settings that aims at strengthening assessment and hence democracy through respect, autonomy and equal opportunities to expression.

Keywords: animal experiment, democratic decision-making, hermeneutics

Relating arguments pro and con animal experimentation with arguments pro and con public control

Animal experimentation is a practice that evokes strong feelings among both proponents and critics, and has a long history of being debated. Some very early documented experiments on animals were performed in Alexandria in the third century B.C. by physicians interested in anatomy. What reactions this evoked is not known. It is however interesting to note that during the 17th century both an Irish and a British physiologist Robert Boyle and Robert Hooke 'showed genuine concern for the welfare of some of their experimental subjects' and refrained from repeating some experiments that were described as too cruel (Monamy, 2000: 16). Nevertheless potentially increased knowledge and probable benefits for humanity did not stop them from performing other animal experiments. Exactly this line of thought evoked criticism – *doubt about the relevance of using animals* – perceived as fundamentally different from humans in anatomy and spirit – for human benefit. During the 18th century in e.g. England and Italy public awareness and reaction slowly grew into anti-vivisectionism now combining the criticism of extrapolating results from animals to humans with empathy with the animals, and further rejecting Descartes description of animals as 'automata', and instead arguing for animals' having moral status and humans a responsibility towards them (Monamy, 2000; Preece 2002). In England and Wales restrictions of painful experiments were put into force through the 1876 Cruelty to Animals Act, which was stricter than Martins Act (from 1822). Among other things it claimed that large animals (horses, cats and dogs, donkey and mules) should not be used if smaller animals could be used, demand for anesthetics and animals should be killed after the experiment. This Act inspired other countries, e.g. Finland, and Germany (Alexius Borgström, 2009).

In Sweden criminalization of cruelty towards animals was debated in government from the early 1840ies, and inspired by the British also discussed vivisection. But a debate is no decision, and a series of proposals for restriction of animal experimentation were turned down with either of three arguments: (a) cruelty towards 'creatures' was criminalized since 1857, which prohibited 'obvious cruelty towards creatures' and was regarded strict enough and argued to cover also animals used for experiments, (b) it is a Christian moral duty to treat animals well and better to educate the public in respect for animals than formulate a law, and (c) it would be difficult to draw a line between 'unnecessary' suffering compared with benefits for humanity (Alexius Borgström, 2009: chapter 3, (my extrapolation)). The third line of thought is still used in Swedish Animal welfare regulation and animal ethics committees. Hence this utilitarian thought model of balancing between potential benefits in human medicine on the one hand and animal suffering on the other has been used in Sweden since the 19th century. Although the level of punishment for cruelty towards animals was changed from paying a fine to prison in 1900, and wild and laboratory animals were included in the 1857 act, there were no court cases as regards laboratory animals (Alexius Borgström, 2009). Sweden became its first Animal Protection Act as late as 1944.

Interesting for this study of ethics and democracy in animal ethics committees is another concern in the vivisectionist debate, namely whether there is or is not a *need of public control* of science in general and animal experimentation in specific. Through Cruelty to Animals Act a license became necessary for experiments in education, although this control mechanism seems not to have restricted the research very much due to researchers in favour of experiments being in charge of the decision (Mendelsohn, 1987 cited in Alexius Borgstrom, 2009). The idea of public control was explicit also in the Swedish debate between mid 19th century and beginning of 20th century, but didn't find its legislative form until the 1988 Animal welfare Act. According to Alexius Borgström there are two clusters of ideas prevalent in the early law discussion both departing from human having a moral superiority: on the one hand in terms of our right or even duty to use animals at any level of suffering for the sake of human benefit and scientific gains, and hence regulations and public insight or control would risk limiting or prohibiting experiments, and on the other hand the view that the general morality in society should be overarching all human research enterprises, not ensuring science a moral chart blanche, and that it is every Christian's duty to treat humans and animals well, combined with the view that comparison between humans and animals is heresy (Alexius Borgström, 2009). I.e. there was a claim that researchers should adhere to a certain moral setting or interpretation of Christianity, and implicitly that all animal experiments should be ethically justifiable for ordinary citizens. However, the reaction from Sweden's regulative body was that scientists have the right to assess the importance of research tasks, and as shown above, if potential human medical benefits are expected animal suffering was no reason to restrict an experiment.

It is possible to conclude that the presented early arguments against animal experimentation differ from today's by drawing not primarily on empathy and sentience, but on doubts about the scientific relevance, as well as considering it a heresy to compare homo sapiens with any other species and hence to extrapolate knowledge about anatomy and physiology from animals to humans. Today's theological arguments take the opposite turn, stressing the common origin in creation and similarities in physiology (e.g. Linzay, 1994). Early arguments pro animal experimentation on the other hand are very similar to today's debate. In both of the contexts the scientific validity is seldom questioned, the scientific interpretation of an experiment and the assessment of relevance are claimed to be correct, and the experiment is argued to be of special importance. Thus a humanistic or theological interpretation of human options, duties and rights subordinated compared to a 'scientific'. It has further been shown that arguments contra public control followed the same pattern of argumentation as contra animal experimentation. No such clear parallelism occurs for early arguments pro public control, as this view rested on an idea of Christian duty to consider animal suffering as morally relevant whereas arguments contra animal experimentation at this time built on the idea of comparison being a heresy as well as questioning of the scientific validity.

Today public control in terms of animal ethics committees is a precondition for performing experiments in many countries, often underpinned with the argument that this ensures that only research methods and objectives acceptable by the public will be performed. Having sketc.hed core issues evoked by animal experimentation, the following section deals with the assignment of AEC which is scrutinized by examples from studies of how AEC work in Sweden.

Ideal and reality: discrepancy between regulation of animal ethics committees and actual application (example of Sweden)

According to current Swedish law the animal ethics committees should contain 14 members in an equal number of researchers or animal technicians and laypersons (including a representative from an animal welfare organization), and two judges as head and vice head of the committee. The task of the committee members is to make a cost-benefit analysis by evaluating the potential benefits for humans with the potential suffering for animals used in a particular research protocol, stating that human interests may triumph animal suffering if the goal is justifiable and animals are not caused undue suffering, and if not an unnecessary repetition of research (SFS, 1988: 534). Thus the ethical assessment is supposed to be performed within a teleological framework, maximizing the human benefit and minimizing animal suffering. Such balancing is not an easy task, and requires skills in recent (medical) research, animal science and in ethics, and is rather time consuming. As a consequence it is difficult to find new board members, and they often stay at their board for more than one election period of four years (some even more than fifteen years) which of course gives them valuable committee experience, but also might involve a risk of developing committee habituations and traditions with regard to how the ethical discussions are proceeding.

Lately several studies on how Swedish AEC work have shown both disparate perceptions of 'ethics' (Ideland, 2009) and very limited discussions on ethical issues in animal ethics committees in Sweden (Alexius Bergström, 2009; Holmberg and Ideland, 2009; Ideland, 2009; Nordgren and Röcklinsberg, 2005; see also Schuppli *et al.*, 2004, 2005 on the limited awareness and use of 3R in Canadian animal ethics committees). Explanations given in these studies point at strong focus on 'details' such as small changes in number of animals used, on much time spent on clarifications of the aim of the study through medical discussions and on wording of the protocol (that should be, but often is not, understandable to laymen and researchers from other fields) in many committees. It is further shown that the cost-benefit analysis is understood in a narrow sense, and is setting the boundaries of the ethical reflection together with the 3Rs Reduction (Holmberg and Ideland, 2009). The Swedish Animal Welfare Agency asked Kristina Dahlborn to investigate the committee meetings' working practice, and found a large variety of discussion and decision methods as well as differences in assessment procedure (Dahlborn, 2006). Further a study on AEC's ethical assessment of use and production discussion of GM-animals a number of ethically relevant questions e.g. methods of research, choice of kind of animals (GM animals or not), method of producing GM animals, their genotype and phenotype, and assessment of suffering are seldom discussed although these aspects have direct ethical relevance for assessment of animal suffering (Nordgren and Röcklinsberg, 2005). One is tempted to conclude that the committees' balancing process seems to build on more thorough information on the side of human interests or even benefit than on the animal's situation or suffering.

Even if this last conclusion would proof to be too far reaching, the picture painted by the above mentioned studies of often and continuously limited or lacking ethical dimensions in the committees' discussions lead to least three problems. They have special relevance for the issue of ethics, decision-making and democracy. *Firstly*, limited and not nuanced information about the animals and their situation make balancing according to the task given in regulation impossible. *Secondly*, due to large variety in aims of and methods for committee discussions, standards for assessment and levels of suffering accepted set in each AEC a certain application might become not only different assessment but also meet different claims depending on committee. Since researchers are to apply to the AEC in their geographical region such differences might cause practical difficulties when researchers from different regions are involved in the same project. Further it might have negative impact on researchers' trust in the assessment process and the AEC validity as interpreter of the Animal welfare regulation. From the perspective of these two first issues a *third*, and democratically most relevant problem arises, namely public trust in AEC and their assessment process and result. According to Swedish regulation, AEC are constructed to mirror interests of civil society, and decisions are to be taken on a democratic basis (SFS, 1988: 534).

In the following section I will discuss some models of decision-making which could be suitable for AEC not only as models of thought, but also to handle the difficulties mentioned above. The hypothesis is that theories on democratic decision-making substantially could contribute to a model of decision-making in the AEC that improves the ethical balancing process in assessing applications on animal experimentation.

Models of democratic decision-making in interdisciplinary committees

There are at least two dimensions of the legislator's demand that AEC should be democratic. The first relates to AEC being an agency in a democratic state, structured to manifest some level of representativeness as regards society's interests, and the second relates to the actual decision-making within each committee. In this paper the first claim is taken as given, stated by the legislation and not in focus of debate. Since it is a prerequisite for the form of AEC that at the moment is not under reconstruction, I regard it more fruitful to consider *how* this democratic structure can contribute in the assessment process, i.e. the second dimension of formulating democratic decisions. Further, such a focus opens for elaborating on the question of whether the assessment process could be improved through higher democratic awareness or standards.

In political science three frequently used overarching models of how to establish a democratic decision are (a) by reference to a majority vote, (b) through a participatory process and (c) deliberative democracy (Gilljam and Hermansson, 2003). Translated to the much smaller setting of AEC the model of majority based decisions is of less interest than the other two. It might be relevant to solve a long debate with seemingly no end, but as a first approach of agreement it is too insensitive to ethical elaboration. Given the common situation that researchers in general vote for acceptance (possibly after agreed modifications) any other delegates potentially against may not miss out since the committees have a very small number of delegates. Further, as shown above the main problem is the too limited ethical evaluation and hence a decision model of majority vote would rather justify a quick process without much discussion than inspire to ethical elaboration. The participatory process model is formulated in order to actively include citizens and alternative perspectives on political issues, and have been used for debate on e.g. GM-crops or research on stemcells (Schicktanz, 2003). In principle Swedish AEC could perform public meetings on a scene inviting everybody to listen, but not formally participate, and there seems to be practical obstacles of such a solution. A better point of departure with regard to AEC is the deliberative democracy that could contribute substantially as regards both form and content for the ethical discussion (Gutmann and Thompson, 1996). 'The form of deliberations we defend is constrained by both the content and the condition principles, which are each independently valuable' (Gutmann and Thompson, 1999). In the case of AEC it is inevitable to include some content, refraining from the strict formal rules of communication (Habermas 1996), not least due to regulation stating at least two values: human benefit and animal welfare.

A similar point is made by Hedlund who, in scrutinizing the Swedish legislative process and debate on gene technology from the perspective of political science, formulates a 'democratic norm' with value elements relevant for both participatory and deliberative democracy (Hedlund 2007). This norm comes close to what has above been shown missing in the AEC discussions – listening to non-researcher's views, openness to alternative methods and approaches to an issue, and discussion on basic ethical issues – and is hence interesting to consider. Hedlund proposes 'equality' and 'autonomy' as central values and further stresses the importance of citizen perspective and an open reasoning process for a democratic decision-making. In her study citizen perspective is shown to be crucial for making 'a variety of standpoints become visible and open to debate – a basic prerequisite for democratic accountability' (Hedlund, 2007: 258). She further suggest this could counter balance focus on experts' perspectives, and thereby, to some extent improve the 'opportunities of expression' and call for decision-makers' 'responsivity', two further concepts Hedlund derives from the democratic norm. Regarding the role of experts in decision-making she concludes that, in case of gene technology, they 'have shared a two-pronged advantage as agenda setters and problem definers.'(Hedlund, 2007: 260). It is further shown that experts' privilege to define the problem influences what issues are debated and to what extent arguments are considered in the process as a whole. In sum, I'd argue these findings show an undemocratic decision structure which is parallel with the situation in many AEC, and that her democratic norm is useful in relation to the structure of decisions in many AEC. The next task is to find a decision-model that does justice to individuals and ethical aspects involved as well as to legislation on AEC and Animal welfare.

Fusion of horizons in interdisciplinary hierarchical settings

According to Alexius Bergström the Swedish legislative process and law on AEC express a vision of a 'corporativist model' for decision-making: decisions through consensus after deliberation in a non-hierarchical atmosphere ensuring arguments and not social status to be decisive (Alexius Bergström, 2009). Although reality deviates from this ideal, some core elements in the vision are still valuable, and elaborated in leading theories on democratic discourse (Habermas, 1996). Inspired by, consensus ethics, I have elsewhere argued that five core values are relevant for decision-making in a heterogeneous group: respect for each discussion partner, context sensitivity, respect for arguments including emotions, a shared picture of the situation, and finally relating theory and practice (Röcklinsberg, 2006). Building on this, I suggest a hermeneutic approach could meet also the 'democratic norm' (respect and autonomy) in combination with the corporativist ideal of the law to create a non-hierarchical democratic setting. Following Gadamer's version of respect – remaining open to the other's 'alterity' (Andersheit), and oneself prepared to learn something from the other – helps to overcome hierarchical and interdisciplinary structures. Further a genuine dialogue is an open-minded non-aggressive meeting between listening parts, 'a mutual learning process in the course of which partners gain a better understanding of both the subject matter and themselves' (Dallmayr, 2009). Fusion of horizons is then not giving up oneself, but rather creating a new common horizon and temporary point of departure for decision. A committee member should ideally see other member's interpretation of a research application, questions and final weighing of animal welfare and human benefit in a certain way as an insight in an equally valid perspective, or in Hedlund's terms equal opportunity of expression. This does not mean total conformity with this other view, but rather preparation to seriously relate it to the own situation and perspective that then will be altered. When this hermeneutical approach is applied in real life a praxis emerges that is an interpretation of a situation that relates back to praxis and a new interpretation, etc. Gadamer's ideal could also be described in terms of virtues or dispositions that should be practiced and promoted among AEC members. An AEC that followed the democratic norm, practiced Gadamer's respect and ensured equal opportunity of expression would most probably strengthen democracy through a more elaborate ethical assessment of applications than today, and hereby deserve public trust.

References

Alexius Borgström, K. (2009). Djuren, läkarna och lagen – en rättslig studie om djurförsöksetik, Iustus förlag, pp. 259. (Animals, medical doctors and the law – a judicial study on ethics in animal experimentation, with a summary in English).

Dahlborn, K. (2006). PM om de djurförsöksetiska nämndernas arbetssätt, Djurskyddsmyndigheten. (PM about animal ethics committees' work).

Dallmayr, F. (2009). Hermeneutics and inter-cultural dialogue: linking theory and practice. Ethics and Global Politics, Vol 2, No 1: 23-39.

Gilljam och Hermansson (2003). Demokratins mekanismer, Liber, Malmö, pp. 297.

Gutmann and Thompson (1999) in Macedo, S. Deliberative politics, Oxford University Press, New York.

Habermas, J. (1996). Between Facts and Norms: Contributions to a Discourse Theory of Law and Democracy. Transl. William Rehg. London. Polity.

Hedlund, M. (2007). Demokratiska genvägar. Expertinflytande i den svenska lagstiftningsprocessen om medicinsk genteknik. Doctoral thesis, Lund Political Studies, Lund, 297 pp. (Democratic Shortcuts: Expert Influence in the legislative process of medical Gene Technology in Sweden. With a summary in English).

Holmberg T. and Ideland M. (2009). Transgenic silences? The rhetoric of comparison and the transgenic mice as 'ordinary treasures', Biosocieties, 4 (3-4).

Ideland M. (2009) Different views on ethics: how animal ethics is situated in a committee culture. J Med Ethics 2009:35, 258-261.

Linzey, A. (1994). Animal Theology. SCM Press, London, 214 pp.

Monamy, V. (2000). Animal Experimentation. A Guide to the issues. Cambridge University Press, Cambridge, 109 pp.

Nordgren, A. and Röcklinsberg, H. (2005). Genetically modified animals in research: an analysis of applications submitted to ethics committees on animal experimention in Sweden. Animal Welfare 14: 239-248.

Preece, R. (2002). Awe for the Tiger, Love for the Lamb. A Chronicle of Sensibility to Animals. UBC Press, Vancouver, 413 pp.

Röcklinsberg, H. (2006). Consent and Consensus in Policies Related to Food – Five Core Values. Journal of Agricultural and Environmental Ethics, 19:285-299.

Schicktanz (2003). Schicktanz, Silke und Naumann, Jörg (Hrsg) Bürgerkonferenz: Steritfall Gendiagnostik. Ein Modellprojekt der Bürgerbeteiligung am bioethischen Diskurs. Leske + Budrich, Opladen 2003

Schuppli, C.A., Fraser D., McDonald M. (2004). Expanding the Three Rs to Meet New Challenges in Humane Animal Experimentation. ATLA 32, 525-532.

Schuppli C.A. and Fraser, D. (2005). The Interpretation and Application of the Three Rs by Animal Ethics Committee Members, ATLA, 487-500.

SFS (1988) 534/Swedish legislation on animal experimentation and animal ethics committees.

Is there more to life than welfare? How the concept of animal integrity can contribute to discussions of animal ethics

M. Gjerris[1] and C. Gamborg[2]
[1]Danish Centre for Bioethics and Risk Assessment, Institute of Food and Resource Economics, The University of Copenhagen, Rolighedsvej 25, 1958 Frederiksberg C, Denmark; [2]Danish Centre for Bioethics and Risk Assessment, Forest & Landscape, The University of Copenhagen, Rolighedsvej 23, 1958 Frederiksberg C, Denmark; mgj@life.ku.dk

Abstract

Using animals for agricultural or medical purposes evidently raises a range of ethical questions. Often these are divided into three categories: (1) risks to human health or the environment, (2) risks to animal welfare and (3) other moral concerns. Concerns discussed under (3) are typically concerns about the social and economic consequences of the case in hand, whether or not the proposed use of the animals can be said to be unnatural in any way and if the case in hand in some way violates the integrity of the animals. However, whereas the other concepts seem more well-defined and closer to distinct meanings, integrity still remains a somewhat nebulous notion. In the paper, we propose that the concept of integrity can be viewed from a mainly biological as well as a predominantly philosophical perspective (sometimes mixed up and obscured) and within each perspective on an individual and a species level. The paper clarifies these four understandings, and discusses how they may be utilized in discussions of the use of animals (in a broad sense) to make clear why some still express hesitation of the use of animals although risk managers, human health experts and animal welfare scientists have said that there from their point of view are no or only few ethical concerns. We propose that especially from a philosophical perspective on an individual level the notion of animal integrity and the ethical concerns that follows, may be used to explain why some think that there is more to life than welfare and that this is an important aspect of any discussion of the ethics of using animals.

Keywords: biology, philosophy, welfare, unnatural, scepticism

Introduction

Concepts are more than exclamations referring to entities in the world outside of language. Concepts are also creative and innovative and in some ways a catalyst for new insights into the world we live in (Løgstrup, 1995). At the same time concepts can be used as rhetorical devices, twisted and turned to fit the interests of the speaker. Concepts such as *sustainability*, (Gamborg and Sandøe, 2005) *public dialogue* (Gjerris, 2008) and *animal welfare* (Gjerris *et al.*, 2009) are examples of this. They all seek to describe new insights into the human-nature relationship and at the same time they are all used in a plurality of meanings – more or less consciously shaped to fit the interests of certain participants in the debate. The result of this is that the concept often looses its initial relation to new insights and in a worst case scenario is reduced to a mere rhetorical device.

In this article the focus is on the concept of *animal integrity*. Exhaustive overviews of the different ways the concept has been used are becoming plentiful, as are the criticisms of very special positions. The aim here is to interpret the concept as an expression of important insights into the human-animal relationship – as triggered by practices such as genetic modification of animals evoking strong public concerns. This will be done by describing the possible content of the concept from a biological as well as philosophical perspective at the species level and the individual level. The aim is to elucidate how the concept of *animal integrity* can make a productive contribution to discussions of animal ethics that otherwise tend to focus on animal welfare issues.

Discussions about the use of animals

Human use of animals has since the early 1970s been under increased ethical scrutiny (Sandøe and Christiansen, 2008). A prominent example of this is the discussions related the use of modern biotechnologies like cloning and genetic modification on animals. Ever since the presentation of the cloned sheep Dolly in 1997 the ethical limits of animal biotechnology such as cloning have been fervently debated (Gamborg *et al.*, 2009). Attempts to generalize the discussions obviously run the risk of finding what you are looking for, but none the less we will claim that in general the issues discussed have been (1) risks to humans and the environment (and thereby humans), (2) risks to animal welfare – often focusing on rather narrow interpretations of the concept of animal welfare and (3) other moral concerns (Lassen *et al.*, 2006). In general we find it fair to say that the focus has been on the first two groups of concerns (EGE, 2008). These are also the groups of concern that most easily can be discussed on a background of science, thus seemingly ensuring less subjective discussions. The basic problem with this approach is twofold. (1) It overlooks the multitude of value-based decision that underlies the first two groups of concerns (Meyer *et al.*, 2005, Verhoog, 2000, Sandøe *et al.*, 2006) and (2) it does not reflect the importance of concerns as they are expressed in surveys of public attitudes to biotechnology where issues as integrity, naturalness, dignity play a larger role than typically expressed in ethical debates (Lassen *et al.*, 2006).

Looking at the literature on animal ethics the concept of integrity plays a small yet increasingly persistent role, essentially adding that there *are* other things than animal welfare that could be important within the field of animal ethics. The notable exception to the silence around the concept of animal integrity is within the discussion of the ethics of animal biotechnology (see for instance Verhoog, 1992, 2000; Sandøe *et al.*, 1996, Rollin, 1996; Vorstenbosch *et al.*, 1999, Bovenkerk *et al.*, 2002; De Vries, 2006; Hauskeller, 2007). Drawing on these sources in our own interpretation of the concept of animal integrity it is the hope that it can contribute to bridging the gap between the public skepticism towards animal biotechnology and the widespread notion within the scientific community that the skepticism is basically an expression of irrationality, fear and biological illiteracy (Meyer, 2006) as well as addressing the apparent fright in the debate about human use of animals of dealing with a concept as far from science as animal integrity.

The concept of animal integrity

There are several definitions of the concept of animal integrity available, yet it seems that the notion is still clouded by some form of intangibleness, eluding clear understanding. The word is derived from the Latin *integer* meaning *wholeness, completed, untouched, unharmed.* In the literature, Rutger and Heegers' (1999) definition is often seen as a point of departure, seeing integrity as the: 'wholeness and intactness of the animal and its species-specific balance, as well as the capacity to sustain itself in an environment suitable to the species'. De Vries (2006) interprets this definition as mainly aiming towards physical, biological or genetic wholeness. To a large extent violations of animal integrity will therefore also violate the welfare of the animal (if animal welfare is understood as the mental experiences of the animal) (Vorstenbosch, 1993). But also other practices can be seen as violation of integrity including breeding animals that fits the production systems better rather than developing housing systems that satisfy the behavioral needs of the animals (Shriver, 2009) or the insertion of genes from other species into pigs that that enables them to mimic human diseases as e.g. Alzheimer's (Prather *et al.*, 2008). Integrity has also been as a way to express apprehension or precaution in relation to various uses of animals, thus interpreting the idea of animal integrity essentially as a precautionary attitude to (more) radical animal modification and manipulation as e.g. expressed with germline transgensis in certain monkeys, so-called marmosets (Olsson and Sandøe, 2010).

Other important notions brought into play in trying to elucidate the wider discussion integrity include *telos* (Rollin, 1995) and a notion of inherent worth (e.g. Hauskeller 2007). *The* discussions about animal integrity in the literature to a large degree end up evaluating whether the close link to biological and genetic processes that most definitions of animal integrity rely on makes sense (see also Vorstenbosch, 1993; Rollin, 1995; Verhoog, 2000). Thus De Vries (2006) find the definition of Rutger and Heeger too broad and seek to limit violations of animal integrity to situations where 'the physical and mental ability of the animal to live a good life and flourish' is affected. He suggests developing other concepts for moral experiences that moves beyond these situations or quite simply rejecting them as irrelevant. Along the same line Bovenkerk *et al.* (2002) find the concept of integrity too unclear in this form to be helpful, since it does not 'refer to a state of affair that can be assessed empirically'; an issue discussed especially in relation to applied animal behavioral sciences (Würbel, 2009). They do however find it more useful than De Vries as it fill 'a gap between moral theory and moral experience'. Here we follow the line of Bovenkerk *et al.*, but suggest a way to fill up the gap more than just accepting it. By looking at the concept of animal integrity from two different perspectives on two different levels we hope to make clear how the 'scientification' of the concept present in much of the discussion *prevents* the concept from being seen as full of content as the concerns regarding risks to humans and the environment and risks to animal welfare.

The species level from a biological perspective

When examining whether the concept of integrity makes sense, we have decided to look at the question from two perspectives: a biological one and a philosophical one. These terms should be understood rather broadly. *Biological* means that what is in focus is empirically discernable changes in the animal and/or changes that affect the welfare of the animal. Welfare should here be seen as comprising both the mental experiences of the animal and the possibility that the animal can perform its species-specific behavior. *Philosophical* means that what is in focus is the pre-scientific experience of the animal and the human-animal relationship.

Usually the claim that there is integrity at the species level signifies that it is a violation of this integrity to change the species through genetic modification. However, as also Sandøe and Holtug (1996) points out this implies that also more traditional ways of changing the genome of a species (such as selective breeding) would be wrong. Furthermore the argument would need to discern between changes that where done to benefit humans e.g. for higher productivity and changes done to benefit the animals e.g. for disease resistance (which could obviously also be done for the sake of productivity) (Verhoog, 1992). Finally it would also be a claim that the extinction of species is also a violation of their integrity since this wipes out the genome of the species. A problem – which we will not dwell on here – is that the concept of *species* itself is rather fuzzy, including something between 9 and 22 concepts of species (Robert and Bayliss, 2003). It makes the notion of violation of integrity meaningless as it is so closely tied to empirically discernable changes.

The individual level from a biological perspective

If the concept of integrity were linked closely to biological of physiological empirically discernable phenomena, it would seem that some moral experiences of violation of integrity become troublesome. The frequently used example is the breeding of animals either through genetic modification or more traditional technologies to produce animals that have lesser mental capabilities for suffering, thus being more able to cope with the production pressure in intensive production systems (Rollin, 1995; Shriver,

2009). Claiming that the loss of species specific behavior or the ability to cope with the environment is causing the violation seems to beg the question as the animals have been changed. As expressed in Bovenkerk *et al.* (2002): `though we have a prima facie duty not to interfere with an animal's interests, there is nothing wrong with *changing* these interests`. However, this seems to be one of the situations where there is a gap between our moral theory and our moral experience (Bovenkerk *et al.,* 2002), suggesting a philosophical rather than a biological perspective is more fitting.

The species level from a philosophical perspective

The basic problem with tying the concept of integrity to a more or less empirical and biological framework is that it loses some of its ability to express moral experiences about human use of animals. On the species level the concept becomes vulnerable to the fuzziness of the concept of biological species; whereas the notion that something could be problematic although not inflicting on the animal welfare or possibility to perform species-specific behavior becomes vulnerable on the individual level. To Bovenkerk *et al.* (2002) the problem is that the more the concept of integrity is removed from biology the less empirical it becomes, but at the same time it becomes better at containing the moral experiences that initially gave rise to it – `it fills the gap`.

If *integrity* is taken to mean *whole, untouched* and *complete* it could on the species level be seen as expressing the moral experience of the importance of the possibility of species to adapt through evolution to the universe of contingencies that they find themselves in. Violation of integrity would then be situations where humans would interfere in this process, whether it would be for the benefit of the animals or themselves. The ideal is non-interference in the development of living beings and the communities they live in and to the extent it is done anyway, whatever the purpose, integrity is being violated. Obviously good reasons can be given to interfere. But this will need to be done on a case by case basis and with the knowledge that integrity is an *absolute concept* in the sense that it cannot be said to be dependent on the reasons for our actions or something that can be more or less violated. Interfering is always violating integrity – but sometimes it can be necessary to protect other morally relevant qualities. The integrity of a species is thus basically an expression of its freedom from human intervention in its development.

However, it is important to note, that integrity does not disappear when violated. It is not so that species that have been domesticated or otherwise manipulated lack integrity. It just presents itself in a different way than wild, untouched animals. Moreover, this understanding of what could be meant with integrity on a species level is far from unproblematic. A problem is seeing humans as inherently `unnatural`. Obviously all species are constantly being shaped by the communities they are part of and the environment they unfold in. Evolution is in this sense marked by *inter*dependence. Why is it then that especially human participation in this mutual influence is ethically problematic? The short answer to that objection is that humans have the possibility to choose how much of an influence they wish to have.

The individual level from a philosophical perspective

The real advantage of cutting the ties to the mainly biological understandings of the concepts of integrity seems to be on the individual level, though. And it is also here most moral experiences that lead to the notion that there is more to life than welfare take place. It is exactly here that biologically founded notions of integrity in our view fail, since they cannot incorporate the wide held moral experiences of integrity being violated because the state of affairs as experienced by the animal does not change (Sandøe and Holtug, 1996; De Vries, 2006). Nevertheless there are people who would find it to be a violation of the integrity of an animal, it if had been bred to be less capable of experience suffering so as to better fit the production systems of modern agriculture (Olsson *et al.,* 2006) or if it had been genetically modified to produce proteins normally only found in other species.

Here De Vries (2006) suggests either rejecting the experiences or using other terms. Bovenkerk *et al.* (2002) sees integrity as a something that fills the gap, but remain silent on what it could be that is experienced as wrong. Our suggestion is to remain in a pre-scientific understanding of the animal, not reducing it to an object for examination, but relating to it as another being-in-flesh as ourselves. And at this level the 'completeness' of the animal that is present to out immediate experience before we show up with our abilities to change the animals into something that can satisfy our needs comes into play. Evidently, we can kill an animal, but we cannot breathe life into it. In essence *integrity* is the experience that the animal is whole, complete, full, finished, when we meet it. We cannot add to it, only take away (Løgstrup, 1995; Abram, 1996).

If the concept of integrity is developed along these lines we find it likely that it will be able to encompass many of the moral experiences that often are handled (and even dismissed) under the heading of `other moral concerns`. By interpreting integrity as a basic 'zone of untouchableness' given by the sheer existence of the animal, the moral concerns about interfering in the lives of animals can be seen to make sense, even if they go beyond the welfare interests of the animal. It may thus be used to explain why some think that there is more to life than welfare and that this is an important aspect of any discussion of the ethics of using animals. This does not mean that integrity should always be protected, but that possible violations should always be justified.

References

Abram, D. (1996). The Spell of the Sensuous. Vintage Books.

Bovenkerk, B., Brom, F.W.A. and Van Den Bergh, B.J. (2002). Brave New Birds: The Use of 'Animal Integrity' in Animal Ethics. The Hastings Center Report, Vol. 32.

De Vries, R. (2006). Genetic Engineering and the Integrity of Animals. Journal of Agricultural and Environmental Ethics, 19:5, pp. 469-493.

European Group on Ethics in Science and New Technologies to the European Commission (2008): Ethical aspects of animal cloning for food supply. Opinion No 23.

Gamborg, C. and Sandøe, P. (2005). Applying the notion of sustainability – dilemmas and the need for dialogue. In: Holm, S. and Gunning, J. (eds.) Ethics, Law & Society. Hants: Ashgate, pp. 123-130.

Gamborg, C., Gjerris, M. and Sandøe, P. (2009). The ethics of animal cloning. In: Gunning, J., Holm, S. and Kenway, I. (eds.) Ethics, Law & Society. Vol. IV. Hants: Ashgate, pp. 43-54.

Gjerris M (2008): The Three Teachings of Biotechnology, in David K. & Thompson P (eds.): What can Nano learn from Bio? Elsevier pp. 91-106.

Gjerris, M., Olsson, I.A.S., Lassen, J. and Sandøe, P. (2009). Ethical perspectives on animal biotechnology, In: Atkinson, P., Glasner, P. and Lock, M. (eds.): Handbook of Genetics and Society:Mapping the new genomic era. Genetics and Society Book Series. Routledge, pp. 382-398.

Hauskeller, M. (2007). Biotechnology and the integrity of life. Taking public fears seriously. Ashgate Publishing Limited

Lassen, J., Gjerris, M. and Sandøe, P. (2006). After Dolly – Ethical limits to the use of biotechnology on animals. Theriogenology 65, pp. 992-1004.

Løgstrup, K.E. (1995). Vidde og prægnans. Sprogfilosofiske betragtninger. Metafysik I. 2. udg., Gyldendal, København.

Meyer, G., Folker, A.P., Jørgensen, R.B., Von Krauss, M.K., Sandøe, P. and Tveit, G. (2005). The factualization of uncertainty. Risk, politics and genetically modified crops – a case of rape. Agriculture and Human Values, 22 (2): 235-242.

Meyer, G. (2006). Why clone farm animals? Goals, motives, assumptions, values and concerns among European scientists working with cloning of farm animals. Danish Centre for bioethics and Risk Assessment.

Olsson, A., Gamborg, C. and Sandøe, P. (2006). Taking ethics into account in farm animal breeding: what can the breeding companies achieve? Journal of Agricultural and Environmental Ethics 19(1): 37-46.

Olsson, A. and Sandøe, P. (2010). What's wrong with my monkey? Ethical perspectives on germline transgenesis in marmosets. Transgenic Research 19: 181-186.

Prather, R.S., Shen, M. and Dai, Y. (2008). Genetically Modified Pigs for Medicine and Agriculture. Biotechnology and Genetic Engineering Reviews 25, 245-266.

Robert, J.S. and Baylis, F. (2003). Crossing Species Boundaries. The American Journal of Bioethics Vol. 3:3. Taylor & Francis Group, pp 1-13.

Rollin, B.E. (1995). The Frankenstein Syndrome. Ethical and social issues in the genetic engineering of animals. Cambridge University Press, Cambridge.

Rutgers, B. and Heeger, R. (1999). Inherent Worth and Respect for Animal Integrity, In: Dol, M., Van Vlissingen, M.F., Kasanmoentalib, S., Visser, T. and Zwart, H. (eds.): Recognizing the Intrinsic Value of Nature. Assen: Van Gorcum, pp. 41-53.

Sandøe, P., Holtug, N., Simonsen, H.B. (1996). Ethical Limits to Domestication. Journal of Agricultural and Environmental Ethics, 19:5, pp. 469-493.

Sandøe, P., Christiansen, S.B. and Forkman, B. (2006). Animal Welfare: What is the Role of Science? In: Turner, J. and D'Silva, J. (eds.): Animals, Ethics and Trade: The Challenge of Animal Sentience, London & Sterling VA: Earthscan, 41-5.

Shriver, A. (2009). Knocking Out Pain in Livestock: Can Technology Succeed Where Morality has Stalled? Neuroethics, Vol 2:3 115-24.

Verhoog, H. (1992). The concept of intrinsic value and transgenic animals. Journal of Agricultural and Environmental Ethics, 5/2: 147-160.

Verhoog, H. (2000). Defining positive welfare and animal integrity. 2nd NAHWOA Workshop. Proceedings of NAHWOA – Network for Animal Health and Welfare in Organic Agriculture, Córdoba, Spain, 8-11 January, 2000.

Vorstenbosch, J. (1993): The concept of integrity. Its significance for the ethical discussion on biotechnology and animals. Livestock Production Science, 36:109-112.

Würbel, H. (2009). Ethology applied to animal ethics. Applied Animal Behaviour Science 118: 118-127.

Some legal and ethical aspects of xenotransplantation in light of the proposed revision of the Directive on the protection of animals used for scientific purposes

M. Jorqui Azofra
Inter-University Chair BBVA Foundation – Provincial Government of Biscay in Law and the Human Genome University of Deusto and University of the Basque Country, Spain; mjorqui@genomelaw.deusto.es

Abstract

Unlike what happened in other times, society in general, and especially the scientific community has become aware that animals share our sensitivity to pain and the capacity to suffer. In this regard, it is generally accepted that animals must be protected from all types of abuses. In fact, it is unavoidable today that animals used in scientific experiments enjoy the maximum degree of protection and well being. This is a view based on an ecocentric notion of living matter opposed to the traditional anthropocentric approach, as it has become evident that ethics should not be limited to those belonging to the same species. Likewise, there is a broad consensus – with the exception of members of certain animal protection groups – regarding the need to experiment with animals, when no alternative methods (AM) are available, given that the current state of scientific knowledge still does not allow for this type of experimentation to be entirely abolished. The proposal for a Directive on the protection of Animals used for scientific purposes (PD) – European Commission 'Proposal for a directive of the European Parliament and of the Council on the protection of animals used for scientific purposes', 5 November 2008, Brussels – tries to strike a compromise between ensuring that research can continue in the European Union (EU) and improving animal welfare. From an animal welfare perspective it is a step forward. In this article we will review some of the main ethical criteria that serve as a basis for balancing the scale; in other words, to guide and legalise animal experimentation in the field of xenotransplantation (XT). To that end, we will take into account some aspects of the PD, in order to reflect, in turn, on the following issue: Where is European institutional ethics headed on this issue?

Keywords: animal welfare, '3R's' principle, alternative methods, xenotransplantation, primates

Introduction

EC Directive 86/609/CEE was adopted to harmonise animal experimentation practices in the EU. Nevertheless, this objective has been frustrated due to a series of deficiencies in said Directive. Some Member States have established considerably more advanced national measures, while others only apply the minimum rules. As a result, there are different levels of protection in Europe. Specifically, among the priority objectives of the proposal for a Directive on the protection of Animals used for scientific purposes (PD) – European Commission 'Proposal for a directive of the European Parliament and of the Council on the protection of animals used for scientific purposes', 5 November 2008, Brussels – , is to guarantee equal conditions throughout the EU and to establish stricter and more transparent measures in the area of animal experimentation. Through the principles that we will examine below, we will see that in light of what is proposed in the PD, the following objectives are expected to be achieved in this area and therefore in the specific field of xenotransplantation (XT), where procedures are carried out using animals for experimentation: intensify promotion of the development, validation, acceptance and application of altenative methods (AM), and that the number of animals used in the respective XT procedures be reduced to the minimum; get Member States to improve breeding, housing and care, as well as the methods used in each XT procedure to eliminate or reduce to the minimum any possible pain, suffering, anxiety or lasting injury caused to the animals in said procedures. All of which contributes to the full application of the principle of replacement, reduction and refinement of animal experimentation – known as the '3R's' principle.

Ethical-legal principles that will serve as a guide for animal experimentation in the XT field in light of the PD

Reducing animal experimentation by introducing AM

The basic objective of this proposal is *to achieve full replacement of procedures using live animals for scientific purposes as soon as scientifically possible.* Technical and scientific advances in the XT field can be rapid, as can improvement in knowledge of the factors that influence animal welfare. In this sense, the new PD includes periodic review of the Directive. As a priority these reviews must examine the possible replacement of animal use, specifically the use of primates, in cases in which it could be possible taking into account scientific advances. Nevertheless, with the scientific knowledge available today, it is still necessary to use primates in XT procedures. However, given their genetic similarity to human beings and their highly developed social skills, using them in such procedures poses important ethical questions, considering that their behavioural, environmental and social needs are often frustrated within a laboratory setting. The general criteria followed by the PD refers that primates will not be used in procedures given their particularly high level of neurophysiologic, sensory and cognitive development, unless their use is limited only to those essential biomedical areas in benefit to human beings, where there is still no other replacement alternative method, and only in cases in which those procedures are undertaken with a view to the avoidance, prevention, diagnosis or treatment of clinical conditions in humans or these species that are life-threatening, debilitating or endanger the survival of the species. In this sense, it is observed that there are different human purposes that are prioritised today over animal welfare. However specific provisions are included in the PD to reduce primate use to an absolute minimum. Certain requirements regarding animal origin are also included, and tracking mechanisms are proposed to guarantee the efficiency of the proposed measures that seek to achieve full replacement of primates in scientific procedures. It is therefore necessary to provide a scientific and ethical justification that the purpose of the procedure cannot be achieved using a species other than primates. To ensure that primates are only used under exceptional or very special circumstances, any project involving these species would require the highest level of justification and review.

Ethical and scientific evaluation as part of the authorization process

Therefore, in the PD it is necessary to *guarantee based on science and ethics that each use of animals is subject to a detailed assessment regarding the scientific validity, the usefulness and fitness for that particular purpose.* Respect for the animals as living, feeling beings requires taking into consideration a series of criteria to evaluate the ethical and scientific justification of each XT procedure. In this sense, it is unavoidable, as reflected in the PD, that there be a transparent system of authorisations to legitimise said procedures. This system must take into account the following aspects, among others: the goals and possible benefits of the project, as well as the probability of achieving those goals; the number and species of animals to be used; the effects that the project will have on the animals used; what consideration is given to the principle of the 3Rs; what problems could application of the 3Rs imply; what measures would contribute to overcoming them; and for what reasons is the possibility of applying AM rejected. Therefore, justification of a XT procedure shall only be carried out after weighing the scientific value of the project against its possible effects on the welfare of the animals. Therefore, legalising it should take into account all aspects that may have an adverse impact on animals. In this regard, it must also be kept in mind that respect for animal welfare standards in the XT field, together with good laboratory practices, depend largely on the quality and professional competency of the personnel supervising scientific procedures that use animals, as well as those who carry them out.

Quality of the establishments and professional competency as a guarantee of animal welfare

To guarantee a high degree of competency among the people who use animals and of the procedures themselves, these activities must be carried out only in establishments and by people authorised by the competent authorities. Therefore, the *principle of responsibility* is also encouraged, whereby researchers who use animals in XT procedures are responsible for everything surrounding the animal welfare. To take on these responsibilities, it is essential that researchers be informed of all the factors associated with each XT procedure that may affect the animals' welfare, of the mechanisms that may minimise the adverse effects in this regard, as well as measures that should be taken in the event of such adverse effects. The objective set by the PD regarding obtaining and maintaining an adequate degree of competency is therefore plausible, and must be demonstrated before authorisation is granted or renewed to those persons. Therefore, it is established that authorisation by a competent authority and certification of having successfully passed the relevant training courses must be the object of mutual recognition of all the Member States.

Minimum housing and care requirements

It should be noted that animal care and housing requirements also vary between Member States. On the other hand, some of these requirements no longer reflect the most recent knowledge regarding the effects of housing and care on the welfare of the animals, and on the respective scientific results of each XT procedure. Therefore, the PD includes the *need to establish minimum requirements for housing and care, which will always be subject to advances based on new scientific evidence.* In light of these, all animals used in XT procedures will be provided with adequate care in regards to their health and welfare, satisfying their different ethological and physical needs.

Effective promotion and application of the 3Rs

The PD expressly includes *the need to apply the 3Rs principle to the care and use of live animals for scientific purposes.* This internationally recognised principle is a key factor to promoting animal welfare and the quality of the scientific results. As we have already pointed out, one of the fundamental aspects to legitimise the use of animals in XT procedures is based on the need to achieve the research objectives through other procedures that do not use live animals (*principle of replacement*). Currently there are numerous problems to covering the gap between the demand for transplants and the supply of donor organs. Before using animals for this purpose, AM should first be verified, and if these are scientifically satisfactory for achieving the desired results. In the field of XT, said methods refer, among others, to reducing the demand for human organs and tissues by introducing public health methods to prevent conditions that currently require treatment by transplant; applying measures to increase the supply of human organs; developing transplant material from stem cells; and developing artificial organs and tissues.

In turn, the type of barriers that could be posed by applying these and other AM in this field would have to be verified. There may be a series of factors, scientific as well as economic, that will stand in the way of replacing animals. To advance in the development and application of the replacement principle, Member States – as envisioned in the PD – will contribute financially and in other ways to the development and scientific validation of AM intended to provide a comparable level of information, the same or greater than that obtained in procedures with animals, but which do not use, or use fewer animals or entail less painful procedures.

Similarly, the Member States must take the necessary steps to promote research in this field. In regards to this principle, it is anticipated that the scope of responsibility of the European Centre for the Validation of Alternative Methods (ECVAM) will be expanded to include the coordination and promotion of development and use of alternatives to animal experiments. As has been suggested by the Expert Group of European scientists coordinated by the European Science Foundation (ESF) and the European Medical Research Councils (EMRC), the role of this centre shall be strengthened so that it includes the co-ordination and promotion on the validation and use of alternatives to animal procedures including applied and basic biomedical and veterinary research and regulatory testing. Furthermore, ECVAM shall communicate information on the availability and application of alternative methods to the relevant authorities of the Member States.

On the other hand, it is crucial that the *principle of reduction* be applied to every XT procedure. This refers to the need to reduce the number of animals used in procedures to the necessary minimum without compromising the quality of the research, and the validity of the scientific results, nor animal welfare. To improve the application of the principle of reduction in the field of XT, the choice of methods and the species to be used must be assessed, as this choice has a direct impact on the number of animals used as well as their welfare. In this sense, the methods applied must use the least number of animals with which reliable results can be obtained, and choose the species with the lowest degree of neurophysiologic sensitivity. Specifically, in the XT field, while primates were an attractive alternative for use as a source of organs for human transplant, ethical and practical issues have driven the need to focus attention on pigs as possible organ donors.

It is also worth noting the following, among other, criteria for improving the application of the cited principle: the choice of appropriate research techniques; best statistical analyses; prevent duplication of procedures and non-productive studies with animals; reuse of animals, which would have to be analysed on a case by case basis and limited exclusively to procedures in which the scientific objective and welfare of the animal are not compromised.

The PD, in turn, provides for the use of animal tissue and organs in the development of *in vitro* methods. Therefore, to apply the principle of reduction, it is necessary for Member States to establish programs for sharing animal organs and tissues. It also includes the need to create veterinary bio-banks on a large scale, to support the 3Rs, using leftover tissue extracted in the framework of clinical procedures. In XT procedures it will be crucial that animals be raised specifically for this purpose, so that people who carry out the respective procedures will know their genetic, biological and behavioural background. Thanks to that knowledge, the scientific quality and reliability of the results will increase, and the variability will decrease, which will ultimately lead to a reduction in procedures and the number of animals used.

Therefore, the *refinement principle* consists of using methods that guarantee the reduction to the minimum of any type of pain and suffering to the animals, and which improves the care, treatment and living conditions of each animal to improve its welfare, taking into account the experience over its entire life. This principle has great transcendence from the ethical point of view – as it is directly related to the reduction of animal pain and suffering-as well as from a scientific point of view-given that the physical and psychological welfare of the animals subject to experimentation can have a significant effect on the scientific validity of the results themselves. The possibilities for applying this principle are fundamentally centred on the following areas: housing, breeding and care, experimental procedures, evaluation of pain and appropriate methods for killing an animal.

Among the steps to follow to eliminate or reduce the degree and duration of animal suffering in a XT procedure and to comply with the principles of reduction and refinement principles, the following are worth noting: choice of animal, housing, breeding and care conditions, as well as acclimation, in accordance with said choice with the scientific objective; choice of appropriate techniques; appropriate use of sedatives, tranquilizers, analgesics and anaesthetic; choice of the appropriate measures to assess animal pain and suffering; the need for training and authorisation to kill an animal using humane methods considered appropriate to the species; appropriate supervision of the animals, etc. Given that the PD gives maximum priority to the considerations on matters of animal welfare, each establishment must have a permanent ethical control entity with the primary function of fostering an adequate environment of care and to provide the necessary instruments for the effective application of scientific advances and recent techniques in regards to the principle of the 3Rs, in order to improve the lifetime experience of the animals. It is anticipated that, among the members there will be an expert in applying the 3Rs principle.

Assessment of the criteria put forth

The criteria set forth show that the PD is headed in the direction of animal welfare, structuring a series of principles to serve as the philosophical basis for the ethical assessment of actions that are injurious to the animals when their interests conflict with human interests, as happens in the field of XT. We think that, effectively, the PD tries to strike the necessary balance; on the one hand, it promotes competitiveness and research in Europe and, on the other hand, it takes full account of the welfare requirements of animals, attending to their behavioural, environmental and social needs; although this requires passing through the sieve of human morality, making protection of animal interests into a human interest. This focus means that animal rights will be recognised when they are understood to be a responsibility that falls on the human species. It thus seems that these rights in reality refer to the 'duties' of the human species. This means, for example, that not just anything can be done to them, and that actions regarding the same have a moral significance, above all if it has a positive or negative effect on their needs, interests or welfare. In this sense, we think that the PD is taking positive steps toward a less anthropocentric ethic, trying to reconcile modern philosophical anthropocentrism – which inspires most European legislation – with new moral demands in favour of animal welfare.

In this sense, it seems positive to us that the PD includes periodic reviews, which take into account the most recent scientific knowledge and the evolution of public opinion on these matters, which is increasingly showing a preoccupation with the need to reduce and eliminate procedures causing pain, suffering, anxiety, or lasting injury to the animals. We believe, at the same time, that the capacity of animals to experience pain prevents us, as we have seen, from justifying the negative of extending the circle of compassion and moral consideration to animals. Therefore, financing research in cognitive ethology also seems positive to us, as it will help us to answer those questions regarding the mental nature of the animals. This, in turn, has great transcendence in order to go into the field of animal ethics in depth and to therefore try to understand the moral statute of animals.

Conclusions

According to what was set forth, the need can be appreciated to carry out an assessment of the damages and benefits of each XT procedure to determine if the injury to animals (suffering, pain and anxiety) and, should it come to it, the environment, are ethically justified by anticipated scientific advances that, in this field, could benefit human beings. Certainly, the final objective of XT is to save human lives in the future. However, despite the undeniable possibilities that this therapeutic option entails, there is a preoccupation on an international scale with many aspects related to research underway, and more concretely, with a hypothetical clinical phase of the same. It is therefore inescapable, to cover the more important problems for the patient and/or public health before proceeding to the clinical phase, and to grant the maximum priority to animal welfare in the context of maintenance, breeding and use of animals in the different projects that are carried out in the pre-clinical phase. The permanent ethical control entity of each establishment will have a key role in guaranteeing the adjustment of the respective methodological, ethical and legal aspects of XT. We think that, in effect, the need to guarantee the efficacy of measures seeking to promote animal welfare and, in particular, effective application of the

3Rs is a basic pillar in this field. However, it would have to be reinforced with national inspections that would act as an essential guarantee of compliance with the cited principles, which seem to be oriented toward the idea of conceiving our treatment of animals as a matter of justice.

Acknowledgements

This work has been funded by a Grant to Research Groups by the Basque University System, ref. IT-360-07, Department of Education, Universities and Research of the Basque Government.

References

Commission of the European Communities. (2008). Brussels, 5.11.2008 COM 543 final; 2008/0211(COD). Proposal for a Directive of the European Parliament and of the Council on the protection of animals used for scientific purposes.

Daar, A.S. (2003). Xenotransplantation: recent scientific developments and continuing ethical discourse. Transplantation Procedings, 35: 7, 2821-2822.

ESF-EMRC (2009). Position on the Proposal for a Directive on the Protection of Animals used for Scientific Purposes, 2nd Edition March 2009: http://www.esf.org/index.php?eID=tx_nawsecuredl&u=0&file=fileadmin/be_user/publications/AnimalsProtection_2nded.pdf&t=1269509005&hash=a1476dd62c21384e340d3c7ed7e764fa. Accessed on: 25 March 2010.

Festing, M.F.W. (2000). Reduction, model development and efficient experimental design, in: Progress in Reduction, Refinement and Replacement of Animal Experimentation, Balls M, van Zeller AM and Halder ME (eds), Elsevier Science, Amsterdam, 721-727.

Francescotti, R. (2007). 'Animal mind and animal ethics: an introduction', The Journal of Ethics 11: 239-252.

NHMRC (National Health and Medical Research Council) (2003). Policy on the care and use of non-human primates for scientific purposes. NHMRC, Canberra, 2003. http://www.nhmrc.gov.au/_files_nhmrc/file/publications/synopses/nonhum2.pdf. Accessed on: 14 Nov 2009.

Nuffield Council on Bioethics (2005). The Ethics of Research Involving Animals. Nuffield Council on Bioethics, London, Chapters 11 and 12, 2005: http://www.nuffieldbioethics.org.

Romeo Casabona, C.M. and Urruela Mora, A. (2008). 'New legal developments in xenotransplantation: the Spanish approach', Revista de Derecho y Genoma Humano, 29: 111-129.

Russell, W.M.S. and Burch, R.L. (1959). The Principles of Humane Experimental Technique, Methuen & Co. Ltd., London.

The Scientific Steering Committee (2002). The need for non-human primates in biomedical research, statement adopted 4-5 April 2002. http://ec.europa.eu/food/fs/sc/ssc/out253_en.pdf: Accessed on: 10 Nov 2009.

Transgenic animals: Spanish legislation

A. Emaldi Cirión
Lecturer in Law, Bioethics and Biotechnology. University of Deusto. Researcher at the Inter-university Chair BBVA Foundation, Provincial Government of Biscay Chair in Law and the Human Genome, University of the Basque Country, University of Deusto, Spain; aemaldi@der.deusto.es

Abstract

An area that is attracting a lot interest of today in the legal, scientific and ethical aspects within the field of biotechnology is the creation of genetically modified animals. A genetically modified organism is an organism that carries a gene in each of its cells, of which the sequence does not belong to that organism. In animals this entails introducing the transgene during the first stages of the fertilization process; and reconstructing that animal from cells that carry the transgene through the creation of chimeras or nuclear transfer. The uses for these transgenic animals are currently broad and beneficial since they are used in the field of human health, as well as in production. Indeed, there are several research projects underway examining the use of transgenic animals in the field of biomedicine. The purpose of the research is to expand our knowledge about genes potentially involved in human pathologies, define their functions and metabolic pathways of application, and determine possible therapeutic targets. In addition, they are used to evaluate the safety and effectiveness of possible therapeutic compounds, in order to obtain reliable and safe organs for allotransplant, etc. They are also utilised in industries, such as the pharmaceutical industry, which is attempting to obtain therapeutic compounds from transgenic animals, or the agricultural industry, which is trying to improve the quality and quantity of fish and stock breeding species, or to obtain derived products, with bigger or better food characteristics. In conclusion, all these industries have committed themselves to this technology, expecting that it will bring about an improvement in their products and future benefits. However, we must be aware that this process must respect legal regulations.

Keywords: transgenic, animals, animal protection, Spain

Introduction

An area that is attracting a lot interest today in the legal, scientific and ethical aspects in the field of biotechnology is the creation of genetically modified animals. A genetically modified organism is one that carries a gene in each of its cells, of which the sequence does not belong to that organism. In animals this entails introducing the transgene during the first stages of the fertilization process; and reconstructing that animal from cells that carry the transgene through the creation of chimeras or nuclear transfer.

The uses for these transgenic animals are currently broad and beneficial since they are used in the field of human health, as well as in production. Indeed, there are several research projects underway examining the use of transgenic animals in the field of biomedicine to expand our knowledge about the genes potentially involved in human pathologies, to define their functions and metabolic pathways of application, and to determine possible therapeutic targets. In addition, they are used to evaluate the safety and effectiveness of possible therapeutic compounds, in order to obtain reliable and safe organs for allotransplants, etc. They are also utilised in industries, such as the pharmaceutical industry, which is attempting to obtain therapeutic compounds from transgenic animals, or the agricultural industry, which is trying to improve the quality and quantity of fish and stock breeding species, or to obtain derived products, with bigger or better food characteristics. Finally, all these industries have committed themselves to this technology, expecting that it will bring about an improvement in their products and future benefits.

Therefore, as happens in other applications of genetics, our society is confronted with two *a priori* conflicting visions. On the one hand, there has never been such ethical sensitivity regarding the respect for animals, even when their use may result in benefits for humanity. On the other hand, the applications of this technology reaches fields that have an enormous social and economic interest. Consequently, there has been a need to legislate protection for animals used in research, experimentation and commercialisation.

Legal protection for genetically modified animals

There is an increasing interest in protecting the animals. In fact, in the 19th century the regulation that offered the right to animals was as an object of ownership, of appropriation, etc. Starting in the 20th century, there is a concern for their protection from the actions of man, and a need to specifically legislate legal protection that should be granted to animals, in general, and genetically modified in particular. The first international declarations and accords appear, regulating the issue and serving as a support to countries to write their own regulations. Among the most significant we may highlight:

- The Universal Declaration of Animal Rights (15 October 1978), which primarily includes: the prohibition of cruelty, humane sacrifice, research when necessary and there are no other alternatives, research without suffering and protection guaranteed by law and by the States.
- The European Convention for the protection of vertebrate animals used for experimental and other scientific purposes of the Council of Europe (18 March 1986).
- Directive 86/609/CEE of the Council of European Communities (24 November 1986) regarding the protection of animals used for experimentation and other scientific purposes, which establishes common standards that incorporate the main principles, objectives and provisions of said Convention. This Directive has been amended by Directive 2003/65/CE of the European Parlament and of the Council of 22 July 2003. Spain incorporated in Royal Decree 223/1988, of 14 March this Directive.
- In the European Union, following the lines of the general biotechnology policy, specific measures have been provided for genetically modified animals. Therefore, in this regard in 1992, the European Union created Directive 90/219/CEE of the Council on 23 April 1990, regarding the confined use of genetically modified micro-organisms later replaced by Directive CE/81/98. It applies to the confined use of any genetically modified organism, including not only micro-organisms, but also plants and animals. In these types of instruments, the confinement measures are adjusted to each of the activities to be performed (from the creation of, up to the later use of a genetically modified animal, including confined transport) and they depend on the level of risk associated with the transgenic animal in question. Similarly, measures applicable to the intentional release of genetically modified animals have been implemented, for experimental purposes as well as commercial purposes, in the latter case included in Directive 18/2001/CE Cross border movements, in turn, are the subject of discussion in the Cartagena Protocol on Biosafety for countries that are members of the same, measures that in turn have been implemented in the European Union in the Regulation on the Cross Border Movement of GMOs.

In light of this regulation, Spain has a general standard on animal protection in criminal law and in civil law, and a specific regulation that protects transgenic animals:

In Spanish Civil Law, the animal is an object, a moveable thing, and therefore, it cannot be a holder of rights (art. 333). In effect, under Civil Law, only people have rights. Identity is the quality that allows one to be considered subject to rights and obligations, in other words, holder of legal relationships (arts. 29 and 32). Therefore, even when the legal system does not allow another holder of rights, however, it considers animals as an object of protection and therefore, there are several precepts of the civil code that reference them. Specifically, said articles refer to the offspring of these, which are considered fruits (art. 355), to their possession (art. 465), appropriation (art. 610), classification (art. 465), as an object of sale (art. 1494), responsibility for damages caused by the animals (art. 1905 and 1906), among others.

On the other hand, in Spanish Criminal Law, the animals is considered both object of property – thing – and therefore, must be protected from abduction, damages, etc. The last Criminal Code reform (Organic Law 15/2003, of November 25, which modifies Organic Law 10/1995, of November 23 of the Criminal Code) includes, nonetheless, some new items, such as: the crime of art. 337 of abuse with cruelty of domestic animals provoking their death or severe injury; and the crime of art. 631.2 of abandonment of domestic animals, endangering their lives or integrity, which joins those already existing in art. 631 (now art. 631.1, of caretakers who do not restrain ferocious animals, which is obviously not directed toward animal protection) and of art. 632 (cruel mistreatment of domestic and unauthorised show animals).

In regards to specific regulation for transgenic animals' protection, we firstly find Royal Decree 223/1998 of 14 March, about protection of animals used for experimental and other scientific purposes, currently modified by Royal Decree 1201/2005, of 10 October, regardless the protection of animals used for experimental and other scientific purposes. The purpose of this Royal Decree is to ensure animal protection, in other words, that they are granted the appropriate care, that they are not caused unnecessary pain, suffering, stress or prolonged injury, and that the number of animals used in the research may be reduced to the minimum, applying alternative methods whenever possible.

Regarding the purpose of these experimental, scientific or teaching proceedings, in which animals are used, it highlights: prevention, diagnosis and treatment of diseases in humans, animals or plants; development and manufacture of pharmaceutical and food products or other substances or products, as well as carrying out tests to verify quality, effectiveness and safety; environmental protection to maintain biodiversity; assessment, detection, regulation or modification of the physiological conditions of man, in animals or in plants; medical-legal research and finally, education and training.

Secondly, we will refer to Law 8/2003, of 24 April, regarding animal health. This law intends to improve animal health and protect the maintenance and conservation of diversity of animal species. Therefore, it regulates the prevention, fight, control and eradication of animal diseases; animal exploitation; livestock health defence groups; livestock trade, transport and movement in national territory and finally, includes in several of its precepts, standards on zoosanitary products--veterinary drugs and other zoosanitary products. Said law was developed precisely by the previously quoted Royal Decree of 2005.

Finally, closely related to transgenic animals, Law 9/2003 of 25 April establishes the legal system of confined use, voluntary release and sale of genetically modified organisms. With said law, the Spanish legal system adjusts to the EU regulation, specifically, Directive 2001/18/CE and Directive 98/81/CE are incorporated into national law. Mainly, this law intends to take preventive and cautionary measures to prevent the potential adverse effects for human health and the environment resulting from these activities.

Aside from this regulation applicable to transgenic animals, there is no doubt that there is specific legislation regarding the minimum standards for the protection of animal species: protection of pigs (Royal Decree 11/35/2002 of 31 October); protection of calf (Royal Decree 1047/1994 of 20 May); protection of game species (Royal Decree 1095/1989 of 8 September), among others.

Finally, in view of the potential of transgenic animals in different scopes of application, it is necessary to develop a regulation for that purpose.

References

Emaldi Cirion, A. (2010). Panorama normativo del uso confinado de los animales trasngénicos, Transgénesis, clonación animal y xenotrasplante: aspectos científicos, éticos y jurídicos, ed. Chair in Law and the Human Genome-Editorial Comares, Bilbao-Granada, (In press).

Escajedo San Epifanio, L. (2010) Animal transgénico, Enciclopedia de Bioética y Bioderecho, ed. Chair in Law and the Human Genome-Editorial Comares, Bilbao-Granada.

Escajedo San Epifanio, L. (2003). Avances biotecnológicos y medio ambiente: implicaciones éticas y jurídicas de la Bioseguridad, Biotecnología y Derecho, Editorial Temis, Bogotá.

Hagen, K., M. Engelhard and M. Boysen (eds.) (2008). Genetic Engineering in Livestock: New Applications and Interdisciplinary Perspectives. Ethics of Science and Technology Assessment, Vol. 34. Springer, Germany.

Mellado Ruiz, L. (2004). Bioseguridad y Derecho, Comares.

Mir Puigpelat, O. (2005). Transgenicos y Derecho, Tirant lo Blanch.

Murray, J.D., G.B. Anderson, M.M. McGloughlin and A.M. Oberbauer (eds.) (1999). Transgenic Animals in Agriculture, CABI, UK.

O'Riordan, T., Cameron, J. and Jordan, A. (2001). Reinterpreting the Precautionary Principle, Cameron May, London.

Pinkert, C.A., (2002). Transgenic Animal Technology, 2nd edition, Academic Press, London.

Romeo Casabona, C.M. (2004). Principio de precaución, Biotecnología y Derecho. Comares, Chair in Law and the Human Genome, Bilbao.

Part 11. Animal research and animal disease control

Moral status of the African catfish

B. Bovenkerk and F. Meijboom
Utrecht University, Ethics Institute, Heidelberglaan 8, 3584 CS Utrecht, the Netherlands; b.bovenkerk@uu.nl

Abstract

The demand for fish is increasing world-wide, partly due to government directives to increase the intake of fatty acids by consuming more fish. However, we are experiencing global declines of wild fish populations and increasingly rely on farmed fish. Sustainable aquaculture, to comply with growing needs and biodiversity conservation, requires in-depth knowledge of biology, profitability, environmental aspects, consumer awareness and product appreciation. In aquaculture, function-based views of welfare (can fish cope with farming conditions?) prevail, disregarding feelings- and nature-based (allowing of natural behaviour) views. Our interdisciplinary research shows that these views are not necessarily mutually exclusive. The group consists of physiologists and ethicists who focus particularly on the African catfish, which due to its formidable growth and sturdiness is a favourite species in aquaculture. While the physiologists examine ways of reducing aggression and stress, the ethicists address societal and ethical questions surrounding sustainable aquaculture. The societal focus on improving welfare suggests that fish are part of our moral community and should receive moral consideration for their own sake. However, from a moral point of view we can ask what properties confer moral consideration and what it means to speak of moral status of animals. In this presentation we focus both on the question of why fish would have moral status as on the meta-ethical questions surrounding the concept of moral status. Ethicists with different theoretical backgrounds justify moral considerability of animals on different grounds; however, their focus on specific morally relevant properties pre-empts the question of what it means to be morally considerable and how exactly moral duties relate to the properties that animal ethicists deem morally relevant. We aim to show the relevance of explicating the conceptual assumptions that underlie the discussions on moral status and address the question whether we can justify moral status from a theory-independent perspective.

Keywords: animal ethics, fish, moral considerability

Introduction

As the world population is increasing, aquaculture is becoming the fastest growing animal production sector in the world. Government directives to increase the intake of fatty acids (omega 3) by consuming more fish also play a large role in the increasing demand for fish world-wide. At the same time we are experiencing global declines of wild fish populations, raising concerns about the sustainability of the fishing industry and causing us to rely more and more on farmed fish. Sustainable aquaculture, to comply with growing needs and biodiversity conservation, requires in-depth knowledge of biology, profitability, environmental aspects, consumer awareness and product appreciation (Leung *et al.*, 2007). In aquaculture we can distinguish three different groups of views on fish welfare. In the first group of function-based views on welfare the central question is whether fish can cope with farming conditions. The second group of feeling-based views works from the presumption that fish have feelings and that these are constitutive of their welfare. The third group of nature-based views presumes that welfare depends on the ability of animals to display their natural behaviour. While the first view seems to suggest that we should find out how best to adapt fish to their aquaculture surroundings, the latter two views suggest that we should adapt the surroundings to the feelings or natural needs of the fish. Function-based views currently prevail, disregarding feelings- and nature-based views. Nevertheless, our interdisciplinary research shows that these views are not necessarily mutually exclusive, depending on how one interprets them.

Our group consists of physiologists and ethicists who focus particularly on the African catfish (*Clarias gariepinus*), which due to its formidable growth and sturdiness is a favourite species in aquaculture. Dutch *Clarias* production in 2007 was 4,200 ton, realized in recirculation systems, and is increasing (Van Eijk, 2008). *Clarias* is piscivorous (fish eating) and aggressive (Martins, 2005; Van der Nieuwegiessen, 2009). Aggression is facilitated and evoked by farming stress. If aggression could be reduced this would offer a win-win situation for both welfare and productivity, and profitability. This was already successfully accomplished in Arctic charr, where reduced aggression improves growth performance through reallocation of energy flow (Jobling, 1995). The physiologists in our group examine ways of reducing aggression and stress in *Clarias*, by sorting for diminished aggression and improved production. As sorting itself is stressful and can evoke aggression, the physiologists will try to improve welfare through conditioning to stimulate a brain reward system. It is believed that reward will suppress stress, and reduce the risk and negative consequences of aggression. The hypothesis is that *Clarias* will react favourably to conditioning, as it is a pack hunter and social animals generally perform well in conditioning trials (Merron, 1993). The experiments will generate improved knowledge on adaptability, needs and emotional/cognitive capacities of *Clarias*. Therefore, the conditioning protocols will provide input to societal and ethical questions around sustainable aquaculture, which will be addressed by the ethicists in our group. Focus on welfare and sustainability by reduced aggression may help to address broader consumers' concerns on food safety and quality and issues of trust in aquaculture. The societal focus on improving welfare suggests that fish are part of our moral community and should receive moral consideration for their own sake. However, from a moral point of view we can ask what properties confer moral consideration and what it means to speak of moral status of animals in the first place. In this presentation we focus on these questions by firstly, discussing the question if and why the African catfish should be accorded moral status, and secondly, discussing meta-ethical questions surrounding the concept of moral status. As this research is still in its initial stages this discussion still has an explorative character.

Moral status

Ethicists with different theoretical backgrounds justify moral considerability of animals on different grounds. Within the analytical tradition, three general lines of justification can be discerned. First, in the utilitarian view of Peter Singer (1975), in short, entitities deserve moral status because, and in as far as, they have the ability to feel pain and pleasure. Even though Singer is a representative of preference utilitarianism and he, therefore, argues that we have an obligation to weigh the preferences of different entitities against one another and choose the course of action that maximizes the fulfillment of preferences, his basis for including entities in the moral community is that they have interests and they can only have interests if they are sensitive.

Second, in the deontological view of Tom Regan (1984), entitities deserve moral status if they are subjects-of-a-life, which could be taken to mean that they can experience their life subjectively. In order to experience ones' life as a subject of it, one needs certain characteristics, which include sensitivity, but also self-awareness. Regan, therefore, places stronger demands on entities' characteristics that confer moral status than Singer; for him it is not just about the capacity to feel pain and pleasure, but also about certain cognitive capacities. In his own words, those entities count as a subject-of-a-life who have:

> 'desires and beliefs, who perceive, remember, and can act intentionally, who have a sense of the future, including their own future (i.e. are self-aware or self-conscious), who have an emotional life, who have a psychosocial identity over time, who have a kind of automony (namely preference-autonomy), and who have an experiential welfare' (Regan, 2003: 18).

The research carried out by the physiologists in our group will help determine whether *Clarias* possesses the characteristics singled out by Singer and Regan. While the characteristic of sentience appears relatively straightforward and there appears consensus between physiologists that fish can indeed suffer, some still raise doubts about the question whether non-human animals actually consciously experience suffering. Bob Bermond (2003), for example, argues that the widely accepted view amongst animal ethicists that animals can suffer is based on research in animal welfare science that is founded on questionable assumptions and interpretations. For example, from the perception of behaviour of animals that is classified as pain behaviour we cannot automatically infer that animals subjectively experience pain. Bermond cites evidence of human patients with spinal cord lesions who exhibited pain behaviour but did not have the accompanying sensory experience. Whether these assumptions and interpretations are indeed questionable is a matter for philosophical reflection, particularly in the field of the philosophy of mind.

Bermond (2003: 79), furthermore, argues that to experience suffering, animals need a 'well developed prefrontal cortex and a right neocortical hemisphere', just like humans. However, recent research into intelligence in crows and jays suggests that these birds, like monkeys and apes, 'understand their physical and social worlds' and possess 'complex cognition', whilst having very different brain structures (Emery *et al.*, 2004); in fact, they lack a prefrontal cortex. The researchers conclude that convergent evolution has taken place whereby complex cognitive abilities have simply evolved in different areas of the brain in corvids. This research shows that we should be careful in attributing specific morally relevant properties to creatures simply on the basis of an analogy with human characteristics. Analogies with human characteristics are a common feature of the justification of moral status for animals, for example in the theories of Singer and Regan. The argument is usually that if specific animals possess a certain characteristic that humans possess and that we consider the basis of the moral status of humans, then we should also confer moral status to these animals. The question that the research into crow cognition raises, however, is what happens when the analogy does not hold. When we do not find this analogy in a certain species, such as fish, what does that mean? Do we then know that we should not confer moral status to them? Or does it mean that perhaps for that species other characteristics are morally relevant?

This resonates a more general criticism made of the theories of Singer and Regan. While these theories aim to include non-human animals in the moral community, they are still accused of anthropocentrism, because it is said that for them animals count in as far as they resemble humans. This criticism is too simplistic, because both philosophers offer justificatory reasons for why particularly these characteristics are morally relevant. However, the question remains as to why we should confer moral status on the basis of specific characteristics or capabilities of the entities in question. In other words, Singer and Regan's focus on specific morally relevant properties pre-empts the question of what it means to be morally considerable and how exactly moral duties relate to the properties that animal ethicists deem morally relevant. The fact that they take particularly these properties to be morally relevant appears to be dependent on their theoretical frameworks.

Meta-ethical perspective

Moral theories make assumptions about what morality is. We can distinguish between two ways of defining morality: substantive and formal ways. Singer and Regan both use substantive definitions; they define morality in terms of impartiality, meaning that if someone does not believe we should be impartial, in their eyes this person does not have a concept of morality at all. The third line of justification in the analytical tradition does not start from the assumption of impartiality and does not focus on intrinsic properties in the entity that confer moral status. Relational or care-ethical views in animal ethics suggest that we have obligations towards animals because we are in a specific relationship to them (Swart, 2006). This means, for instance, that we have different obligations to different types of animal. We have more responsibility towards a farm- or a laboratory animal than towards an animal in the wild, for example. While we have positive duties towards the former, we only have negative duties towards the latter.

The answer to the question of what entities possess moral status and why is, therefore, heavily influenced by one's theoretical ethical framework. Deontological theories could be defined as theories that take something specific as valuable in and of itself. For Kant this was autonomy; in his eyes there would always be a link between autonomy and morality. This starting point of morality prestructured his thinking and appears to also have prestructured Regan's thinking. While for Kant it meant that nonhumans could not possess moral status, for Regan it means that they can in as far as they have certain characteristics related to autonomy. It explains why he focuses more than Singer on cognitive capacities and even takes some sort of autonomy as a defining characteristic of being a subject-of-a-life. Utilitarian theories, on the other hand, assume that we have to promote a specific non-moral value, such as happiness. This explains why Singer focuses on the capability to suffer; the absence of suffering is a non-moral value that should be promoted. He does not want to take any moral starting point, like Kant does with autonomy. For this reason we want to problematise the current discussions in animal ethics about moral status. In these discussions it appears that one should side either with Singer or with Regan, but witness the more recent appearance of care ethics on the animal ethical stage, other theoretical frameworks may have other perspectives to offer. If we were to step outside of the analytical tradition we would have even more different perspectives on what entities deserve moral considerability.

This is not to argue that care ethics or other ethical framworks offer a more promising avenue for conferring moral status, but rather to take the discussion one step higher. A first step in our project is to identify what cognitive capacities can be discerned in *Clarias* on the basis of empirical research, and to determine to what extent these confer moral status. But, next, we want to ask how we should interpret moral status in the first place. Even if we would know that *Clarias* possesses certain moral status conferring properties, how should we interpret this? This interpretation is dependent on one's meta-ethical position. In other words, what determines towards whom we have moral obligations? Is moral status sufficient for this? Debates in animal ethics are strongly focussed on the question if and what animals have moral status, but the risk exists that we are begging the question. This debate is based on the normative assumption that if entities have moral status they should be treated in a certain way. However, this is an assumption that needs to be justified. We aim to show the relevance of explicating the conceptual assumptions that underlie the discussions on moral status and address the question whether we can justify moral status from a theory-independent perspective. Our project will focus on the interaction between the two levels that we have described; on the one hand the level of specific moral theories and the question whether *Clarias* possesses moral status according to these theories, and on the other hand the more meta-ethical question of what moral status means and on what normative assumptions the concept of moral status is based.

References

Bermond, B. (2003). The myth of animal suffering. In: Armstrong, S. and Botzler, R. (eds.) The animal ethics reader. Routledge. London, pp. 79-85.

Emery, N. and Clayton, N. S. (2004). The mentality of crows: convergent evolution of intelligence in corvids and apes. Science 306: 1903-1907.

Jobling, M. (1995). Feeding of Charr in relation to aquaculture. Nordic Journal of Freshwater Research 71: 102-112.

Leung, P, Lee C.S. and O' Bryen, P. (eds.) (2007). Species & system selection for sustainable aquaculture. Blackwell Publishing, Ames, Iowa, USA.

Martins, C.I.M., Aanyu, M., Schrama, J.W. and Verreth, J.A.J. (2005) Size distribution in African catfish (*Clarias gariepinus*) affects feeding behavior but not growth. Aquaculture 250 (1-2): 300-307.

Merron, G. (1993). Pack hunting in two species of *Clarias gariepinus* and *Clarias ngamensis*, in the Okavango Delta, Botswana. Journal of Fish Biology 28: 525-531.

Regan, T. (2003). The case for animal rights. In: Armstrong, S. and Botzler, R. (eds.) The animal ethics reader. Routledge. London, pp. 17-24.

Singer, P. (1975). Animal liberation. Avon Books, New York.
Swart, J.J.A. (2006). Ethiek uit het keurslijf. NVBE Nieuwsbrief 13 (3): 2-4.
Van der Nieuwegiessen, P.G. (2009). Welfare of African Catfish: effects of stocking density. Wageningen University, Wageningen, 137 p.
Van Eijk, W. (2008). Nederland op de bres voor duurzame viskweek. Visionair 8: 28-31.

Why don't they just pull themselves together and eat less? Public perceptions of animal use in obesity research

T.B. Lund, J. Lassen and P. Sandøe
Danish Centre for Bioethics, Department of Human Nutrition, Faculty of Life Sciences, University of Copenhagen, Rolighedsvej 25, 1958 Frederiksberg C, Denmark; tblu@life.ku.dk

Abstract

Opinion polls and previous research have demonstrated that, in Europe, public acceptance of laboratory animal use is depends heavily on the purpose for which the animals are being used. For instance, the use of animals in cancer research is widely endorsed, whereas their use in the testing of cosmetics is generally rejected, partly on the basis of the view that people should stop using cosmetics. A similar argument may form the basis of critical attitudes to the use of animals in research into obesity and other so-called lifestyle-related diseases. Previous qualitative research indeed suggests that animal research into obesity may be rejected by some because they believe that overweight people could just pull themselves together and eat less. Drawing on survey data from two representative Danish samples collected in 2010, this paper investigates the acceptability of animal experimentation relating to obesity. We also ask whether findings from qualitative research, where people stress personal responsibility as a reason for rejecting obesity research, can be replicated in a quantitative study. Further perceptions and ethical values lying behind an attitude of rejection are also examined. Our results show that obesity-related animal research occupies an intermediate position in public acceptability. It is clearly less acceptable than using laboratory animals in cancer and cardiovascular research but considerably more acceptable than cosmetics testing. Data also show that rejecters do not tend to have discriminatory views or less knowledge of the severe consequences of obesity. Their negative stance is based primarily on the view that obesity is self-inflicted and that people therefore must bear the consequences of their overeating.

Keywords: experiment, attitude, quantitative data, lifestyle disease

Introduction

Recent findings strongly suggest that a subgroup of the public does not accept the use of laboratory animals in obesity research. At the same time, this same subgroup does endorse use of animals in other research areas, such as cancer and cardiovascular diseases. However, the approximate size of this subgroup has not been investigated previously, so we lack a sense of the magnitude of public rejection specifically of this kind of research. Similarly, the reasons for rejection have not been investigated in a probability sample. Findings from focus groups suggest that a relatively prevalent argument for rejecting animal-based research into obesity is that obesity is self-inflicted (Lassen and Jamison, 2006; Lund *et al.*, 2009); but other reasons may also lie behind the rejection. Thus people might conceal their 'forbidden' feelings towards those who are obese in response to pressures, in focus groups, linked with social desirability. For instance, utterances of disgust and prejudice may not appear in the social context of a focus group. Similarly, lack of knowledge about the adverse and ultimately deadly consequences of obesity may also be underreported.

Thus, we consider three (in part competing) hypotheses as to why people reject obesity-related animal research. The rejection may occur as a result of (1) obesity prejudice/disgust, (2) lack of knowledge, and/or (3) the view that obesity is self-inflicted.

This article takes its point of departure in findings from the qualitative research referred to above and the resulting hypotheses advanced. We start off with a methods section, and then we continue with two modest investigations: first, we assess the relative acceptability of obesity-related animal experimentation and other research purposes. Secondly, we evaluate the three hypotheses advanced earlier by asking about the extent to which those in the obesity-rejecting subgroup (i.e. those rejecting animal-based research into obesity: I shall use the shorthand 'obesity-rejecting' henceforth) display obesity prejudice, ignorance as to the consequences of obesity, and the perception that obesity is self-inflicted.

Data and methods

Our data are drawn from two web-based questionnaire surveys of the Danish public carried out in February and March 2010. The samples were collected by a commercial survey company which holds a WEB panel consisting of approximately 120,000 Danish citizens. From this WEB panel two probability samples of Danish citizens between the ages of 18 and 70 were gathered. Analyses of the representativeness of the two samples show no difference from national census data in respect of gender and region. However, there is a clear tendency in both samples to over-represent more highly educated and older people. Despite this, we chose not to weigh the data. Education and age are unimportant theoretical factors in the analysis; neither contributes significantly to the variation in the dependent variable.

The first sample (N=1,246) is used to assess the relative acceptability of obesity-related animal research. Drawing on findings from the first sample, we designed a procedure for allocating respondents to subgroups in the second sample (N=1,019). The procedure is discussed in greater detail in the text. We report results from bivariate tests of the three hypotheses mentioned above with regard to the reasons those in the obesity-rejecting subgroup have for disapproving of obesity-related animal research.

The respondents allocated to the obesity-rejecting subgroup in the second sample were given additional questions in order to explore their reaction to the ethical dilemma between personal responsibility and the mortal consequences of obesity. All statistical analyses were conducted in SPSS (PASW Statistics 18).

Results

The relative acceptability of obesity-related animal research

Respondents in the first sample were presented with 18 cards with different examples of animal research. These examples varied in respect of the following five research aims: 'cosmetics safety', 'migraine medicine', 'cardiovascular diseases', 'cancer', and finally 'obesity'. At the same time we also varied the animal species ('mice', 'pigs', and 'dogs') and the degree of pain the animal suffers during experimentation ('no torment', 'moderate torment', and 'strong torment'), since this variable has been shown to affect the level of acceptability of animal research (Hagelin *et al.*, 2003; Aldhous *et al.*, 1999). In this paper we are focussing exclusively on the results pertaining to acceptability rates of the five research aim. Respondents were asked to report their attitude to every one of the 18 examples in one of the following 5 categories: 'Definitely should be allowed', 'Should be allowed', 'Should not be allowed', 'Should definitely not be allowed', and 'I am in doubt'.

For this study we have recoded the responses of all 45 examples (5 research aims × 3 animal species × 3 torment levels) into a dichotomous response variable indicating approval or non-approval of the particular examples. This simplifies the present analysis considerably, since we can readily compare variations in acceptability levels across the five research aims that were involved.

We calculated the approval rates of the five research aims as an average of the 9 approval rates, where the aims were evaluated by the respondents under different conditions (3 pain levels × 3 animal species). As such, the average is adjusted for any potential confounding of these two factors.

Results show a clear gradient of approval in which the most serious diseases (cancer and cardiovascular diseases) are rated most acceptable (with a 67.8% and 67.5% average approval rate, respectively). Cosmetic testing has, unmistakably, the lowest level of approval (36.2%). Migraine and obesity adopt an intermediate position (at 59.3% and 51.6%, respectively). This was also the basic gradient scale that appeared in a qualitative study presenting respondents with the same research aims. That study suggests that migraine research will on average be less approved of, because some people argue that migraine is not deadly. Hence, animals should not be sacrificed here (Lund *et al.* 2009). Obesity research is, on average, approved of approximately 16% less than cancer and cardio vascular diseases. Actually, this difference stretches between 10-25% percentage points depending on the animal species and level of pain involved. Thus, a crude assessment of the size of the obesity-rejecting subgroup would be that, as a maximum, it includes 25% of Danish citizens and, as a minimum, 10%. However, it is noticeable that the approval rate is considerably higher than the other lifestyle example, i.e. cosmetics testing (36.2% on average). This indicates that the severe consequences of obesity are acknowledged by many as a sufficient reason to accept animal-based research. However, a significant proportion of the population still does not accept animal-based obesity research.

Determinants of rejecting obesity-related animal experimentation

In this section we investigate the factors explaining why a subgroup of citizens holds the view that obesity-related animal experimentation should not be allowed.

We use as a dependent variable a classification of respondents into either an obesity-rejecting group or a non-rejecting group. Respondents who did not accept any animal research example are excluded from the analysis below. Thus, the dependent variable is binary. In order to obtain a relatively pure subgroup of rejecters we only place respondents in the obesity-rejecting subgroup if, on the one hand, they approve of the use of laboratory animals in cancer research at a non-critical level (that is, when the animal used in the experiment is a mouse and there is no torment associated with the experiment), and, on the other hand, they do not approve of animal-based obesity research at a similar non-critical level. This subgroup amounts to 11.7% of sample 1 and 10.3% of sample 2 (in which non-rejecters make up the remaining 89.7% (N=964)).

As outlined earlier, we are interested in investigating whether those in the obesity-rejecting subgroup display obesity prejudice, may not be acquainted with the severe consequences of obesity, and/or holds the view that obesity is self-inflicted. Below we present results of bivariate tests of the association between the dependent variable and the following three independent variables. As an indicator of obesity prejudice answers to the question: 'Fat people are a turn-off' is used. As an indicator of the belief that obesity is self-inflicted we use the answers to the question: 'Obesity is usually self-inflicted'. Both questions had the following answering categories: Definitely agree, agree somewhat, neither agree nor disagree, disagree somewhat, and definitely disagree. We also construct an index measuring knowledge of the consequences of obesity: the respondents were presented with 13 diseases of which 7 (cardiovascular diseases, coronary diseases, diabetes, cancers, pulmonary diseases, depression and knee/hip aches) are known consequences of obesity. They were told to mark any of the 13 diseases which they believed could develop as a consequence of obesity. We constructed a variable measuring knowledge of consequences of obesity by accumulating the correct responses the participants marked. This results in a variable stretc.hing from 0 to 7 correctly noted diseases.

In Table 1 below the percentages that agree with the indicators of prejudice and self-infliction are displayed in the two groups. In the final column the average correct response with regard to consequences of obesity is displayed.

As the results show, only the indicator of belief in self-infliction is significantly associated with the subgroup of rejecters. Thus, it can be concluded that those in the subgroup rejecting obesity-related animal research do not do this as a result of discriminatory perceptions of obese people or lack of knowledge. The only factor separating rejection and non-rejection is the propensity to believe that obesity is self-inflicted.

Taking a stance in the ethical dilemma: what do the rejecters do?

It could be argued that the findings just presented here were generated by an imperfect investigatory approach. For instance, it might be suggested that the rejecters did not think through the fact that obesity may be deadly at the time they answered the obesity research examples. Hence, they were not aware of the ethical dilemma between sacrificing animals and potentially saving human lives. To ensure that they were confronted with the dilemma, those in the obesity-rejecting subgroup were asked the additional question whether there are any special conditions under which obesity-related animal research should be admitted after all. In Table 2 the results are displayed. First of all it is noticeable that only 7% respond that life-threatening obesity research is an acceptable exception. Hence, even though the rejecters had the opportunity to (so to speak) bail out, they largely maintained their original opinion.

A majority in fact believes that there should be no exceptions; and roughly a third of respondents would only accept obesity research if it can shown that the individuals in question are not personally responsible for being overweight. In this respect it seems that the subgroup is characterized by strong ethical standards for personal responsibility, even at the cost of not saving human lives.

Table 1. Bivariate test of association between attitudes to obesity-related animal research and indicators of obesity prejudice, belief in self-infliction, and knowledge.

	'Fat people are a turn-off' (% agreement) *n.s.*	'Obesity is usually self-inflicted' (% agreement) **	Consequences of obesity (average correct responses) *n.s.*
Subgroup rejecting obesity-related animal research	20.5%	68.8%	3.9
Non rejection	21.6%	54.5%	4.6

n.s: not significant; **: $P<0.01$.

Table 2. Prevalence of views and arguments in the rejecting subgroup.

Are there special situations where obesity-related animal research is acceptable?	
Life-threatening obesity conditions	7%
If obesity is not self-inflicted	35.1%
Never use animals in obesity research	56.4%

Note: Frequencies do not sum to 100%, since respondents had multiple answering possibilities.

Discussion

According to the data presented in this article 10-25% of the Danish public reject animal-based research into obesity. We sought explanations for this, speculating that it was the result of obesity prejudice, ignorance of the consequences of obesity, and/or the belief that obesity is self-inflicted. Our results clearly show that it only is the perception that obesity is self-inflicted that influences the propensity to reject obesity-related animal research.

It is no doubt good news for those who are obese that disapproval of obesity-related initiatives involving animals is not deeply rooted in discriminatory thoughts and ignorance. On the other hand, the obesity-rejecting subgroup ascribed so much critical importance to personal responsibility that considerations about the severity of disease and its associated mortality rate were subordinated. The subgroup truly has the mind-set indicated in the title of this paper: *Why don't they just pull themselves together and eat less?*

In view of these results two recommendations seem appropriate. In the field of biomedical and bioethical research considerable attention should be given to discussion aimed at clarifying when and how obesity is self-inflicted. If it is self-inflicted, what arguments can be given for and against using animal for obesity-related research? These investigations might not only contribute to specialized discussion in academic circles, but also be of great interest to the general public.

In the field of social science it would be fruitful to investigate further examples of attitudes to obesity-related initiatives in order to fully understand the relationship between beliefs in self-infliction and discriminatory thoughts about, and actions towards, those who are obese.

References

Aldhous, P., Coghlan, A. and Copley, J. (1999). Let the people speak. New scientist, 22.

Hagelin, J., Carlsson, H.E. and Hau, J. (2003). An overview of surveys on how people view animal experimentation: some factors that may influence the outcome. Public Understanding of Science, 12, 67-81.

Lassen, J. and Amison, A. (2006). Genetic technologies meet the public. Science, Technology & Human Values, 31.

Lund, T.B., Lassen, J. and Sandøe, P. (2009). Bringing evidence into public deliberations on animal research: the case of research animals used in the development of food for obese people. In: Millar, K., West, P.H. and Nerlich, B. EURSAFE 2009. Ethical Futures: bioscience and food horizons. Wageningen Academic Publishers, the Netherlands, pp. 159-164.

Rite of passage to brute science with bias at the bench? Widening the ethical discussion of animal-based research

I.A.S. Olsson[1,2], O.E. Varga[1,2] and P. Sandøe[2]
[1]IBMC – Instituto de Biologia Molecular e Celular, Universidade do Porto, Rua do Campo Alegre 823, 4150-180 Porto, Portugal; [2]Danish Centre for Bioethics and Risk Assessment, Faculty of Life Sciences, University of Copenhagen, Rolighedsvej 25, 1958 Frederiksberg C, Denmark; olsson@ibmc.up.pt

Abstract

As biomedical researchers are well aware, animal experimentation is a highly controversial activity, with the main worries centring on the harm caused to non-human animals. Whether this harm is acceptable and can be balanced against benefits, as well as how in practice to minimize it, has been widely addressed in the ethics and laboratory animal science literature, and the scientific community has developed a number of mechanisms to handle these questions. However, scientists and ethics committees usually only look at specific experiments and do not raise more general questions about the value of animal experimentation. When we widen the discussion to consider these more general questions, research from other fields becomes relevant. The aim of this paper is to discuss examples of such research which may serve to broaden the ethical discussion of animal-based research. Within the philosophy of science there is a primarily theoretical epistemological analysis of the role of animal models in biomedical and biological research (e.g. LaFollette and Shanks' analysis of 'brute science'), which may provide a critical perspective on questions about the type of knowledge to be gained from research with animals. This is related to a more empirical analysis of actual animal research within at least some of the biological and biomedical research disciplines (e.g. the CAMARADES consortium analysis of 'bias at the bench'), challenging the idea that when animals are used, this is always done in the best way. The sociological and anthropological studies of the culture of the scientific research laboratory (e.g. Birke's descriptions of the 'rite of passage' students go through to become laboratory scientists) may help us to understand how the utility of animal experimentation comes to be widely accepted within the scientific community.

Keywords: animal experimentation, animal model, benefit, validity, relevance

Introduction

Of the public controversies over how animals are used and treated, that over animal experimentation is possibly the oldest and most well-known. For almost as long as animals have been used in experimental research, scientists have been confronted by disapproving members of the public. The 'father of modern experimental medicine', Claude Bernard was well aware of this issue and addressed it in his book on experimental methodology (1865), although he was dismissive of the point of engaging in a debate with laypersons and thought scientists needed only to listen to other scientists.

While some scientists may share Bernard's view, today society demands transparency and engagement in public dialogue over controversial issues. Understanding animal experimentation as the use of animals in research and testing with the objective of safeguarding human health, the main controversy is that over the harm caused to animals without any corresponding benefit for them. Whether this harm is acceptable and can be balanced against the *human* benefit as well as how to minimize harm in practice has been is widely addressed, and the scientific community has developed a number of mechanisms to handle these questions, including training and ethics committees.

However, scientists as well as ethics committees look at specific experiments from a predominantly technical viewpoint and do not raise more general questions about animal experimentation. The one aspect of such a broader discussion which is of more immediate relevance to scientists and ethics committees is the role of animal experimentation as a way of generating knowledge. This connects closely to the question of the benefit of animal experiments and thus is central to the ethical dilemma, but it is rarely addressed. As long as a research programme is expected to generate knowledge this is accepted as sufficient benefit. Only when that knowledge is considered of very peripheral or controversial benefit is it questioned. Little attention is given to the quality of the knowledge, to how it will be used, or to whether the proposed approach is the best way to generate the knowledge aimed at. It is assumed that science's own quality control looks after these issues.

To ask the right critical questions about the benefits achieved by animal-based research it is necessary to view that research at a distance – i.e. to come at the issues from fields outside the life sciences. This has in fact been done in a couple of publications, but although these papers made interesting observations, they were largely ignored by researchers using animals.

In this paper we will review these contributions and discuss how the approaches they represent could be re-vitalized in a renewed critical discussion of the benefits of animal based research.

The role of the animal model

The question of the role of the animal model is most visible in the rhetoric of the debate between critics and defenders of animal experimentation. Those opposing animal experimentation may of course do so on the ethical grounds that we have no *right* to use animals in this way. But often they query the usefulness of this research, arguing that animal use is unnecessary (as alternatives exist) or scientifically flawed (as critical differences between animals and humans preclude the extrapolation of data across species). Faced with these arguments, scientists typically respond by explaining in generic terms how alternatives cannot replace animals, and by pointing to results that have been successfully translated animal-to-human. This kind of discussion does not go beyond a rather simple confrontation of opinions relying on the same rhetorical tool. In view of the inquisitive, critical and sceptical nature of scientific research, there is scope for more serious discussion of the role of animal research.

Theoretical critique

A systematic critique of the *epistemology* of animal experimentation has been presented by the philosophers Hugh LaFollette and Niall Shanks (1996). Noting advocates both and opponents of animal use rely on rhetoric rather than critical analysis when they selectively pick examples that best fit their arguments, they set out to analyze whether it is theoretically and practically productive to use animals in studies where the aim is to generate knowledge that can be generalized to humans.

La Follette and Shanks describe animal use in biomedical research as a legacy of Claude Bernard, who in the nineteenth century proposed strengthening the role of experimental research in medicine so as to complement the hitherto dominant reliance on clinical observations. This perspective is important: troubled by risky and invasive experimental manipulations of humans, researchers needed to find a sufficiently similar but ethically less problematic organism as experimental subject. However, the topic remains humans, not animals, so the move to animals automatically raises questions about relevance: that is, is *this* species of animal a sufficiently similar and reliable model of *human* biological processes?

In attempting to answer this question, La Follette and Shanks start from Bernard's position. A major driver behind Bernard's 1865 book was his 'grand vision for physiology: he wanted to place it on as firm a scientific footing as chemistry or physics' (Bernard, 1865: 36). He meant that any living organism could be understood through the proper study of its parts and their interaction. This study was to take place under controlled conditions in the laboratory.

Possibly because it was new and controversial at the time, possibly because it was not based on laboratory observation, the theory of evolution was viewed sceptically by Bernard. This is a key issue in La Follette and Shanks critique, as they argue that the modern use of animals rests on Bernard's deterministic view of the functioning of the living organism. This is juxtaposed with the view of modern evolutionary theory, which tells us that '[biomedically significant] phenomena [in mammals] can best be understood as the dynamical characteristics of nonlinear, complex systems' (*ibid*.: 94). If regulatory genes with overarching functions have changed over evolution, the end result using the same building blocks may be vastly different in different species. Therefore, it is argued, even with significant knowledge of the biological building blocks making up different species, and with these blocks being highly similar across species, the end result may be greatly disparate. This, of course, is incompatible with Bernard's view.

Discussing animal experimentation, La Follette and Shanks distinguish between strong (causal analogy) and weak animal models. Of the conditions established for strong models, the one that proves particularly challenging to the authors is '(3), there must be no causally relevant disanalogies between the model and the thing modelled' (*ibid*.: 113). Both the 'modern evolutionary argument' referred to above and the empirical evidence presented by the authors favour the existence of relevant disanalogies. Consequently, the authors dispute the existence of strong causal analogy. In theory this does not exclude the validity and relevance of weak animal models – that is, those where analogy is only partial. However, the authors rule the latter out following a similar discussion: existing differences in combination with the complexity of the system of the mammalian organism mean that, with our present knowledge, we cannot predict what the outcome will be when we change or insert single variables. Put differently, small differences may become unpredictably enlarged, to the extent that the argument from analogy becomes invalid.

The animal model use discussed up to this point in the book involves direct extrapolation across species. By contrast, in basic research, about which the authors are more optimistic, '...research animals do not serve as [causal analogy models] – they are not causal test beds to confirm or falsify hypotheses about human biomedical conditions. ...instead, experiments on animals prompt the formation of hypotheses about biomedical phenomena in humans' (*ibid*.: 194). Labelling such models 'hypothetical analog models', the authors claim they might be useful as 'scientific devices that help spur fertile hypotheses about biomedical phenomena' as long as 'there are demonstrable functional similarities between the model and item modelled' (*ibid*.: 195). This holds in the presence not only of confirmed analogy, but also when disanalogy is exploited to understand underlying mechanisms. It is argued that in most cases where animal research successfully contributes to knowledge and understanding, it is through animals being used as this type of thinking device.

A few years later, psychologist Kenneth Shapiro built on the work of La Follette and Shanks in his critical analysis of animal models in psychology (1996). While describing a different scientific community, Shapiro's story of how the role of animal model research in psychology developed bears many parallels with the tale La Follette and Shanks tell about Claude Bernard's efforts to make medicine an experimental science. Psychology's aspiration to become a research field closer to the natural sciences than the humanities required, argues Shapiro, the social construction of the Laboratory and, in particular, the Laboratory Animal. The laboratory guaranteed that experiments could be run under controlled conditions; then '[a] ll that was needed to complete the laboratory ... was an object of study compatible with its physical and symbolic parameters. As none was available, one was constructed' (Shapiro, 1996: 63).

Shapiro observes that, once motivated to join biomedical researchers and adhere to their methodology, psychology researchers 'readily grafted the animal model strategy onto the construction of the laboratory animal. Psychologists selectively bred animals to model obesity, alcoholism, high blood pressure, depression' (*ibid*.: 98) among other things. But within the discipline of psychology this approach seems to be more controversial than it is in medicine: 'many psychologists argue that behavioral disorders ... are not illnesses in the biomedical sense' (*ibid*.: 99) because they involve a much more complex weave of underlying factors. Therefore, possibly even more than in the biomedical context, the only possible role Shapiro sees for animal models in psychology is as thinking devices – any analogy model would be irrealistic. In actual research, however, animal model use is not restricted to a heuristic role. Models are used as analogies, though in a way that rarely allows contact with the reality it claims to model, as '[a]nimal experimenters and clinicians are generally distinct groups who read and publish in different literatures and belong to different professional groups'. Within the professional group working with animals 'models and research procedures become established as areas of research within animal behavior psychology. A literature gets built up in which a model, say for human depression, is established through many variations on the basic model. Instead of checking the model against the original behavior for which it is a model, the 'validation' occurs within the paradigm of the model [by] testing the degree of consistency of the new findings with other research within the literature on the same model'.

In defence of this argument, Shapiro performed a more extensive analysis of a selected field of research: that using animal models of eating disorders (anorexia and bulimia). He criticizes the research for using a 'gross and simplistic model construction' (*ibid*.: 149) in which bulimia and obesity are equated with excessive food intake and anorexia with insufficient intake. This approach 'entails a deemphasis of social (cultural and familial), symbolic, and psychological levels of analysis in favor of simple behavioral variables (food intake or preference) and concomitant physiological processes' (*ibid*.: 152). This research bias is subject to criticism from within psychology, as is evident in several cited papers in the book. Critics point to the lack of evidence of physiological dysfunction as the root of eating disorder and the indications that sociocultural factors importantly affect this group of disease.

Empirical analysis

Shapiro supplements his theoretical critique of animal model research into eating disorders with a survey of clinicians and a citation analysis of selected papers. Altogether, the clinicians had little knowledge of the animal model and did not consider it a source of relevant information to guide their clinical interventions. The citation analysis supported the same conclusion – i.e. that animal model research has a very limited impact on clinical practice – as animal-based research was rarely cited in papers considered significant in clinical advances and never in the journals mentioned by clinicians as helpful. Later research, adopting Shapiro's approach and using citation analysis in an attempt to track how results from animal research are used, has also been highly critical of the apparently limited use biomedical research makes of animal model results in the development of therapeutic interventions (Knight, 2007; Digg, 2008).

The work discussed above was produced by scholars based chiefly in the humanities and social sciences or with an advocacy background. Although interest is growing, until very recently critical review of animal model research was rare within the field of biomedical research in which animals are actually used. One important exception is stroke research. In this field, researchers have highlighted the contrast between the many hundreds of substances with a therapeutic effect in animal models and the consistent failure to produce effective treatments for humans (e.g. Macleod *et al.*, 2009). This has lead to extensive discussion and a series of collaborative initiatives aiming to improve research strategies. One important observation based on empirical data analysis is that animal studies appear to be inappropriately designed in a number of ways. Systematic reviews of research approaches and tested drugs consistently report the same problems of experimental research. These do not mirror the clinical reality (potential therapeutics are delivered at a clinically unrealistic timepoint, and while the typical stroke patient is elderly, hyperglaecemic and/or hypertensive, the typical experimental animal is neither); they also have shortcomings in experimental design (sample sizes tend to be too small to generate useful results, and techniques such as blinding and randomization to avoid bias in the application of treatment and reading of results are underused) (e.g. Macleod *et al.,* 2009). These are not mere theoretical concerns; they affect research results, since publications not reporting randomization, concealment of treatment allocation, and blinded outcome assessment estimated treatment-effect as larger (Macleod *et al.,* 2008). Moreover, despite significant efforts in creating awareness and promoting better practice, the field has not freed itself of these quality problems (Philip *et al.,* 2009).

Discussion and conclusion

The publications reviewed in this paper present significant criticism of the current use of animal model research to make predictions about the effect of treatments in humans. Much of this criticism is external to biomedical research, and it has not been specifically considered within the areas of research where the animals are actually used. It is significant that of the 59 citations of La Follette and Shanks (1996) listed in Google Scholar, less than five are made in a biomedical science (as opposed to ethics or sociology) journal and context. From the outside, it appears that biomedical researchers are not too concerned about the criticism. One important explanation is probably that it comes from the outside. First, does not appear in publications read by biomedical researchers but rather in philosophy and social science journals and books. Second, researchers may well be less sensitive to criticism coming from persons they do not consider competent to criticize their activity – a tribute of sorts to Bernard (1865), who argued that 'a man of science should attend only to the opinion of men of science who understand him'.

The picture of researchers as individuals carrying on their activity as they have done before despite reasons to question current practice is nevertheless slightly puzzling given the critical and questioning nature of scientific activity itself. It becomes more understandable if viewed in the light of the 'socialization process' that is part of the training to become a life sciences researcher and through which researchers 'take on board the accepted ethics of animal use in science' and the 'consensus about what constitutes good research or what counts as medical advance' (Birke *et al.,* 2007: 11).

The present paper does not seek to criticize animal model research as such; it merely points out that there are critical questions to be asked about the use of animals in biomedical research which presently receive insufficient attention within this area of research. These include questions about the potential of different types of research with animals to generate biomedically relevant results, as well as the actual use that is made of these results to benefit human health.

References

Bernard, C. (1865). An introduction to the study of experimental medicine.

Birke, L., Arluke, A. and Michael M. (2007). The sacrifice: How scientific experiments transform animals and people. Purdue University Press.

Dagg, A.I. (2008). Blame and shame: How can we reduce unproductive animal experimentation. In: Castricano J (ed.) Animal subjects: An ethical reader in a posthuman world. Wilfried Laurier University Press.

Dirnagl, U. and Macleod, M. (2009). Stroke research at a road block: the streets from adversity should be paved with meta-analysis and good laboratory practice. British Journal of Pharmacoogy 157, 1154-1156.

Knight, A. (2007). The poor contribution of chimpanzee experiments to biomedical progress. Journal of Applied Animal Welfare Science 10, 281-308

La Follette, H. and Shanks, N. (1996). Brute science: dilemmas of animal experimentation. Routledge.

Macleod, M.R., van der Worp, H.B., Sena, E.S., Howells, D.W., Dirnagl, U. and Donnan, G.A. (2008). Evidence for the efficacy of NXY-059 in experimental focal cerebral ischaemia is confounded by study quality. Stroke 39, 2824-2829.

Macleod, M., Fisher, M., O'Collins, V., Sena, E.S., Dirnagl, U., Bath, P.M.W., Buchan, A., van der Worp, H.B., Traystman, R., Minematsu, K., Donnan, G.A. and Howells, D.W. (2009). Good Laboratory Practice: Preventing introduction of bias at the bench. Stroke 40, e50-e52.

Shapiro, K. (1998). Animal models of human psychology. Critique of science, ethics and policy. Hogrefe.

Philip, M., Benatar, M., Fisher, M. and Savitz, S.I. (2009). Methodological quality of animal studies of neuroprotective agents currently in Phase II/III acute ischemic stroke trials. Stroke 40, 577-581.

Assessing the benefit of animal experimentation

O.E. Varga[1,2], P. Sandøe[2] and I.A.S. Olsson[1,2]
[1]Laboratory Animal Science, IBMC, Instituto de Biologia Molecular e Celular, Universidade do Porto, Rua do Campo Alegre 823, 4150-180 Porto, Portugal; [2]Danish Centre for Bioethics and Risk Assessment, Faculty of Life Sciences, University of Copenhagen, Rolighedsvej 25, 1958 Frederiksberg C, Denmark

Abstract

Although animal experiments play an important role in biomedical research, their use is ethically challenging. The mainstream view is that such research is acceptable if the benefits (to humans) outweigh the harm (to the animals). In practice there is, of course, no simple way to weigh benefits against harm, but an attempt, at least, must be made to assess both harms and benefits. In this paper we will focus on the assessment of benefits. Benefits of animal research are evaluated at several points before (prospective) and after (retrospective) experiments. Prospectively, funding bodies evaluate the scientific merit of projects. Also ethics committees are supposed to assess benefits as part of their ethical review. Retrospective evaluations, which so far are rather rare, aim to understand the short- and long-term merits of animal research already undertaken. In this paper we will discuss the potential and limitations of prospective and retrospective benefit evaluation; we shall examine the current practice through real-life cases. Relevant regulation here lists a number of different purposes for which animal use in experiments can be approved, such as drug testing and prevention, and the diagnosis or treatment of disease. Within these defined purposes, project approval is usually decided on a case-by-case basis, leaving significant room for differences between ethics committees. Here we present and discuss examples of applications for ethical approval which were rejected on grounds of insufficient benefit. Since the animal research affects a wide range of human affairs, including science, health, education, commerce, its retrospective assessment is challenging and involves complex analysis. Currently, retrospective benefit assessment is rarely performed and there is little guidance on how to perform it. We present three types of retrospective analysis as presently available.

Keywords: ethics committees, harm-benefit evaluation, prospective benefit assessment, retrospective benefit assessment

Introduction

Although animal experiments play an important role in biomedical research, the fact that such experiments can cause harm to animals of a more or less serious kind makes this use ethically challenging. The mainstream view seems to be that such harm is acceptable if it is offset by sufficient benefit derived from the research. The weighing of harms and benefits is therefore considered an important aspect of the ethics review process. There is theoretical agreement that animal experiments must be beneficial, but the assessment of benefit remains, in practice, rather obscure.

Benefits of animal research can in principle be evaluated at several points, before (prospective) and after (retrospective) the experiments. In this paper, we use cases (published and made available to us by animal ethics committee members) to explore how the benefits of animal experimentation are regarded and can be evaluated. We argue that the assessment is complicated by the fact that there is no clear definition or shared understanding of which benefits are relevant, and of how benefits are to be assessed. Prospective evaluation is further complicated by uncertainty about the outcome of experiments. Retrospective evaluation is made more difficult by the considerable time-lag between the experiments and their effects; and in addition these effects are themselves complex (UK Evaluation Forum, 2006).

Prospective benefit assessment – background

Legislation regulating animal experiments generally lists the purposes for which research with animals may be approved. Within the biomedical field, relevant purposes typically include those related to the development, manufacture and testing of substances and products, as well as those related to research into the prevention, diagnosis and treatment of diseases. In some cases, this has been further narrowed down: e.g. the Cosmetics Directive (76/768/EEC) phases out cosmetics testing on animals in the EU (European Commission, 2009), and the UK Animals (Scientific Procedures) Act 1986 excludes the development of tobacco and alcohol products as well as offensive weapons from accepted uses of animals. With such limitations the legislature expresses its view that in these specific cases benefits are never sufficient to justify animal use.

In the majority of cases, however, no such categorical approach is taken; rather, decisions are expected to be taken on a case-by-case basis for individual projects. For this purpose, most countries in which animal research takes place have established some sort of review system, typically through institutional, regional or national ethics committees. In the majority of cases, these committees are required among other things to weigh benefits and harms in the decision-making process (Smith *et al.*, 2007). However, how this procedure actually takes place, and how reliable it is, have until now been investigated very little. It is known that the methods of cost-benefit analyses in different Member States vary, and there is inconsistency in the ethical conclusions of ethical committees (Plous and Herzog, 2001; Voipio *et al.*, 2004). In the Nordic-European Workshop on Ethical Evaluation of Animal Experiments participants (who were trained committee members) were asked to prepare ethical evaluations of two virtual cases. The results showed that there is a substantial variation in the way both costs and benefits are evaluated; however, the factors underlying this variation have not been explored.

Prospective benefit assessment – what is sufficient benefit?

One way to gain knowledge about prospective benefit assessment in practice would be to study proposals that have been rejected by ethics committees because they promise insufficient benefit. However, this information is not publicly available. Therefore, we designed an informal online survey for members of national animal ethics committees. An email message was sent out to various expert networks with which the authors have links encouraging colleagues with ethics committee experience to participate (and asking them also to invite other colleagues to take the survey). In the survey, the respondents were asked whether they were aware of applications rejected on grounds of inadequate benefit; they were asked to supply references to, or short descriptions of, these cases.

The 16 respondents were all present or previous ethics committee members: from France (4 respondents), Finland (3), Sweden (2), Italy (2), Norway, the Netherlands, Lithuania, UK and Denmark (1 respondent each). Of these, just under half (7) reported having no experience of committees rejecting a proposal on basis of insufficient benefit. Among the replies of all of the 16 respondents there was only one report of a case in which it seemed clear that the basis for rejection was insufficient benefit as such. In this particular case tests were required by the FDA to validate a new manufacturing process without further scientific aim. The remaining 14 reported examples suggest an unacceptable harm-benefit *balance* rather than a low benefit *per se*. It was not always clear whether the main problem was the large harm or the low benefit. Some of the examples are indeed suggestive of low benefit: e.g. several teaching situations, in which it is usually considered that alternatives exist, research into sexual preferences, and food safety testing. In other cases, it seems more likely that the concern was about harm: severe restraint, prolonged single-housing, organ necrosis, and death as an endpoint. Several of the examples involved wild animals. In one example a scientifically inappropriate approach was mentioned as a key reason for refusing approval.

No routine retrospective benefit assessment

Compared to prospective benefit assessment, retrospective assessment seems at first to be more solid, as it can be substantiated by actual facts rather than speculations about the future. In general, will not the benefits of a completed experiment be obvious and therefore easily evaluated? In reality, however, retrospective benefit assessment of animal experiments is complicated. It is not done routinely, nor, usually, is its purpose well defined.

The current European directive does not require national authorities regulating animal experimentation to undertake routine retrospective assessment incorporating consideration of costs and benefits. Only a couple of Member States (e.g. UK) regulate in this way, with ethical committees or expert groups performing 'cost-focused' retrospective evaluations. According to the 'Proposal for a Directive of the European Parliament and of the Council on the Protection of Animals used for Scientific Purposes' presented by the Commission in November 2008, once an animal research-based project has been completed, it should be assessed retrospectively by the competent authority. Apart from cost, retrospective assessment should evaluate whether or not the objectives of the project were achieved (Article 38) (European Commission, 2008). In the United States, the controlling role of post-approval monitoring, by which animal research activities are monitored for compliance against the protocols approved by the institutional animal care and use committees (Lowman, 2008), is more emphasized.

In funded research, scientists are usually required to provide a report on the outcome of their work to the funding agency. Although this could be considered a rudimentary retrospective evaluation, such reports are often not subject to any external scientific or other scrutiny, and sometimes the only evaluation is a self-prepared one made by the scientist when writing the report.

Although routinely performed retrospective benefit assessment is still rare, some examples of non-systematic assessments can be found. In the following section three types of assessment of this sort will be discussed: (1) non-proven statements on benefits of animal experiments, (2) reviews of the usefulness of particular animal models/strains, and (3) case studies involving retrospective assessment.

Non-systematic retrospective benefit assessment

1. A selective list of historical and recent biomedical discoveries that led to important knowledge for humans can be found on the pages of animal research advocate websites such as Animal Research Info (Animal Research Info, 2007). These lists give examples of questions for which animal research is argued to have been important: for example, mice and primate studies on HIV infection are described as contributing to the general understanding of immune-mechanisms. However, the detailed description of each example rather expands upon how that particular discovery was important than the extent to which animal experiments contributed to the final success. While these cherry-picking lists are popular among those seeking to prove the benefits of animal experimentation, they have mainly rhetorical value and are not based on real analysis (Matthews, 2008). In a somewhat indiscriminate manner, important biomedical results in the lists are treated as the direct results of animal experiments that were part of the particular research project. The extent to which the different approaches of each project actually contributed to the final outcome is not seriously considered.
2. The analysis of benefits which were achieved using a specific strain or species aims to assess, and sometimes demonstrate, the appropriateness of that particular animal model. Many reviews of animal models are published in scientific journals; they describe what we learnt from the model's use, or how the model contributed to a specific scientific outcome. More general reports analyzing the role of a particular type of animal use are also available in the scientific and technical literature. These kinds of publication can be very useful when their selected cases are carefully analyzed by

independent experts. One example is the Weatherall report, which addressed the role of non-human primates in biomedical research (Weatherall, 2006). This report was drafted after a hearing attended by experts from related scientific fields. While citing the concrete medical benefits of research using non-human primates, such as deep brain stimulation for Parkinson's disease, the experts argued that retrospective analyses of the origins of research leading to important scientific or medical discoveries were difficult because of the very limited number of studies that have attempted to define the relative roles of basic and applied science in major medical advances.
3. Retrospective analysis of animal experiments may be complex inasmuch it is performed on more than one occasion. If raw data and the conclusions of animal experiments are made available to those outside the research team, the results can be re-analyzed. This extended kind of retrospective analysis can bring important benefits, even if the analyzed animal experiment itself did not have medical benefits.

One of the most recent cases was that of the unfortunate series of events involving the drug TGN1412, which was intended for the treatment of chronic B-cell lymphocytic leukemia and rheumatoid arthritis(Expert Scientific Group (ESG) on TGN412, 2006). An independent Expert Scientific Group (ESG) was set up following the very serious adverse reactions observed in the first-in-man clinical trial of TGN412 in March 2006: here, six healthy male volunteers experienced severe systemic adverse reactions soon after receiving TGN1412 intravenously. This trial had been preceded by a number of safety and efficacy studies in non-human primates in which TGN1412 was found to be well tolerated. The expert report concluded that the adverse reaction was due to errors in the manufacture of TGN1412 or in its formulation, dilution or administration to trial participants. In their opinion, the most likely cause of the reactions in trial participants was an unpredicted biological action of the drug in humans.

The core question of the report and the following debate was whether the results of animal experiments can be extrapolated to humans in the pre-clinical development of a new medicine, and what limitations there might be when it comes to predicting the response, and the dose-response relationship, in humans. The availability of raw data on TGN 1412 reheated the old debate about the validity of monkey models by making it possible to calculate the predictive capacities. Cynomolgus and rhesus monkeys were considered the best species for safety and toxicology studies assessing any potential toxicity of TGN1412 administration to humans. Perhaps the most important conclusion is that first-in-man trials of any monoclonal antibody or biologic that targets the immune system via a novel mechanism, such as TGN1412, need to be considered carefully when *in vivo* tests in non-human primates are combined with *in vitro* screening using human cells (Stebbings *et al.*, 2009). The ESG suggested reducing the time pressure on this kind of trial. The ESG also pointed out the need to share preclinical safety data between different regulatory groups and stressed the need for early and open communication between product developers and regulators. Furthermore, making the raw data publicly available permitted anyone to analyze what caused the mistake and whether it would have been avoided by more careful interpretation of the results of the animal experiment. While the study *itself* may be seen as more harmful than beneficial, critical analysis *of it* produced additional benefits, including productive discussion of the statistical interpretation of data from animal experimentation (Senn, 2008) and the effect, on regulation, of the rules governing the enrolling of clients on trials(European Forum for Good Clinical Practice, 2008).

Discussion and conclusion

We present four approaches to the question of animal research benefits that are presently practiced in the research community, one prospective and three retrospective.

The prospective evaluation consists in the assessment of predicted benefit, and in the weighing of this against expected animal harm, in the ethics approval process. Our small survey suggests that reduced benefit is rarely a reason for refusing to approve an experiment. Only one example was given of a case in which reduced benefit was clearly the reason for withholding approval, and almost half of the respondents reported being unaware of any proposal being rejected due to insufficient predicted benefit. This is probably partly because ethics committees rarely reject applications *period* (Hagelin *et al.*, 2003). Benefit is not always within the mandate of the ethics committee, and even in the cases where it is members can sometimes be reluctant to consider it for a number of reasons: they may perceive their task as being to safeguard the animals' interests only, or they may consider that they are not in a position to influence benefits. There may also be an underlying assumption that, if the investigation has received funding, sufficient benefit should be assumed. In addition, evaluating benefit generally involves more unknown factors and is therefore more difficult than evaluating animal harm (Kalman *et al.*, 2010).

Of the three retrospective approaches, it is only the open analysis of cases with access to original data which really enables critical assessment of benefit. Thus, retrospective benefit assessment of animal experiments with the raw data available is not limited to a single occasion but is rather, potentially, an endless process. Re-analysis of raw data has scientific value and may have a controlling effect on scientists and the work of ethics committees. Although raw data availability is very limited, fortunately its provision is not connected exclusively with scandals. For example, members of a diabetes consortium have been uploading all raw data gained from their research for free analysis (Animal Models of Diabetic Complications Consortium, 2007).

Benefits of animal experiments can be assessed before and after the experiments. Both prospective and retrospective benefit assessments are important tools with which to prove the usefulness of animal experiments, but their practical application is highly problematic. In prospective harm-benefit analysis the potential benefits are weighed against predicted animal harm. Currently, there is no consensus as to what expectations in project proposals can be considered as benefits, and only a few public opinion surveys are reported on this issue. There is scope for further discussion of how best to talk about scientific promise and how to report achieved benefits. The retrospective benefit assessment is not simply a checking of benefits anticipated in the proposal, and it is not, primarily, designed to catch scientists violating the approved protocol, or to checking the work of ethical committees. Instead, routinely performed retrospective assessment will be important in making the benefits of animal experiments evident. Where the retrospective assessment of benefits is concerned, animal experiments might, on occasion, lead to benefits even when undertaken in the course of unsuccessful projects, provided that scientific results are made public and lessons are learned from them.

References

Animal Models of Diabetic Complications Consortium (2007). http://www.amdcc.org/. Accessed: 30. 11. 2009.

Animal Research Info (2007). Medical advances. http://www.animalresearch.info/en/medical. Accessed: 28.02.2010.

European Commission (2008). Proposal for a Directive of the European Parliament and of the Council on the protection of animals used for scientific purposes {SEC(2008) 2411}.

European Commission (2009). Council Directive of 27 July 1976 on the approximation of the laws of the Member States relating to cosmetic products. 76/768/EEC.

European Forum for Good Clinical Practice (2008). Animal experimentation and clinical studies: Ethical recommendations to ensure participants' safety in early drug development (Brussels).

Expert Scientific Group (ESG) on TGN412 (2006). Phase one clinical trials, final report (London).

Hagelin, J., Hau, J., and Carlsson, H.E. (2003). The refining influence of ethics committees on animal experimentation in Sweden. Lab Anim 37: 10-18.

Kalman, R., Olsson, I.A., Bernardi, van den Broek, F., Bronstad, A., Gyertyan, I., Lang, A., Marinou, K.A., and Zeller, W. (2010). Ethical evaluation of procedures for experiments involving animals: Recommendations for ethics committees. In The COST Manual of Laboratory Animal Care and Use: Refinement, Reduction, and Research B. Howard, T. Nevalainen, and G. Perretta, eds. (CRC Press).

Lowman, R.P. (2008). The institutional official and postapproval monitoring: the view from 10,000 feet. ILAR J 49: 379-387.

Matthews, R. (2008). Comment: Truth is the best defence. New Scientist, pp. 20.

Plous, S., and Herzog, H. (2001). Animal research. Reliability of protocol reviews for animal research. Science 293: 608-609.

Senn, S. (2008). Lessons from TGN1412 and TARGET: implications for observational studies and meta-analysis. Pharm Stat 7: 294-301.

Smith, J.A., van den Broek, F.A., Martorell, J.C., Hackbarth, H., Ruksenas, O., and Zeller, W. (2007). Principles and practice in ethical review of animal experiments across Europe: summary of the report of a FELASA working group on ethical evaluation of animal experiments. Lab Anim 41: 143-160.

Stebbings, R., Poole, S., and Thorpe, R. (2009). Safety of biologics, lessons learnt from TGN1412. Curr Opin Biotechnol 20: 673-677.

UK Evaluation Forum (2006). Medical research:assessing the benefits to society. A report by the UK Evaluation Forum, supported by the Academy of Medical Sciences, Medical Research Council and Wellcome Trust, London.

Voipio, H.-M., Kaliste, E., Hirsjärvi, P., Nevalainen, T., and Ritskes-Hoitinga, M. (2004). Nordic-European Workshop on Ethical Evaluation of Animal Experiments; Workshop Report on the Cost-Benefit Principle in Hanasaari, Helsinki, Finland 7-9 November 2003. Scand J Lab Anim Sci 31: 251-267.

Weatherall, D. (2006). The use of non-human primates in research. Wellcome Trust and Medical Research Council, London.

Attempt to typologize relations between different worlds (human, animal, living) based on legal and ethical questions surrounding the status of GM fish

L. Coutellec[1] *and I. Doussan*[2]
[1]*INSA de Lyon, Les Humanités-STOICA, rue des Humanités, 69100 Villeurbanne, France;* [2]*INRA, SAE2, CREDECO-Faculté de Droit, 7 av du Doyen Trotabas, 06050 Nice, France; leo.coutellec@insa-lyon.fr*

Abstract

This paper is the result of a contribution between ethics and law, which will be used as thought-process tools, to the complex issue of legal and ethical statuses of GM fish. To find answers, we propose to consider this issue from a wider angle, looking at relations between the human, animal and living worlds. In the face of the hurdles, which we will characterize, of dualism or monism to describe the relationship between Man and Animal or to account for the concept of living, we show that it is possible to construct other forms of intellectual logic that, without setting these worlds in opposition, do not lapse into relativism where boundaries are blurred. In this sense, we submit the hypothesis that ordered pluralism should help us to move past a mere Man-Animal relationship to reveal the entire complexity of relationships within mankind and between mankind and other (non-human) worlds alongside it and to which it belongs: the animal and living worlds. With this new logic, 'animal ethics' and 'animal law' are reembedded into a set of relationships. In practice, in the case of GM fish, we translate this by proposing a typology of legal and ethical considerations induced by multiple relations connected to this object. This work results in three types of relations: (a) between the human and animal worlds, (b) between the human and living worlds, (c) between human worlds. Some interesting results emerge: according to the type of relationship in question, the GM fish will not have the same status; it is a changing object that must be suitably apprehended; specificities regarding GM plants also arise. Hence, the law, the protector of values (ensuring animal well-being, protecting biodiversity, fair access to resources, etc.) will provide suitable answers according to the type of relationship. Tools exist in part but they have yet to be structured, made coherent, and prioritized. In this sense, ordered pluralism in law (ordering values, levels and systems of law) should be constructed closely with ordered pluralism in ethics.

Keywords: animal ethics, animal law, ordered *pluralism,* GMOs

Introduction

The question of GM fish is complex: there are a number of social, economic, ecological and food-related issues involved, and it is difficult to anticipate all possible trajectories, where multiple disciplines are involved, where the relations between cause and effect are not linear, where the consequences cannot be ascertained in the short term, and there is a storm of uncertainties at so many levels (Coutellec, 2009). In-depth reflection on the means necessary to shed light on the matter is therefore appropriate. And yet we can see that as far as ethics and law are concerned, the animal issue is sandwiched between two broad tendencies: a tendency to naturalize Man, which implies effacing the boundaries between Man and Animals, turning these two entities into the ingredients of ethical and legal analysis and the tendency to systematically think in terms of opposition (seeking separation criteria). In this paper, we propose another intellectual strategy, and put forward two hypotheses: it cannot be Man versus Animal but another necessary logic of interrelations between worlds; this new logic can be built up by adopting the epistemological principle of ordered pluralism (Delmas-Marty, 2006). The idea is to move past a mere Man-Animal relationship to reveal the entire complexity of relationships within mankind and between mankind and other (non-human) worlds alongside it and to which it belongs: the animal and living worlds.

Conceptions of law, conceptions of ethics, ordered pluralism

Law and ethics, as human and non-human fields of thought, unveil the inadequacy of two conceptions: dualism, where human and non-human are irreducible, and monism, which merges these two worlds.

In law, the animal issue is most often addressed in terms of legal subject/object categories. Yet, generally, legal experts tend to refuse to qualify animals, and anything non-human in general, especially components of the natural environment, as legal subjects. Nonetheless, placing animals under the category of legal objects does not seem satisfactory; hence, several authors propose that they be 'special things' or 'protected property', or even to create a new special category for animals between people and property (Suzanne, 2005). In any case, contemporary legal experts appear to admit that animals are no longer mere objects or movable property but sensitive living creatures, as we will see. But the question of animals may also be the opportunity to step outside the dichotomy of legal subjects/objects, because this leads to a dead-end (Farjat, 2002). This presupposes, on the one hand, heralding law as a system to acknowledge and protect values (in other words abandoning a conception of law that revolves exclusively around will) and, on the other hand, seeking to legally qualify not so much the animal itself but human-animal relationships. This position leads us to consider the pluralism of values and relationships to attempt to order the latter.

French and European positive law offers some clues. For example, the legal characteristic 'sensible creature' for animals indicates a moral value recognised by law: not to inflict unnecessary suffering (not justified by any interest or value deemed superior). The advantage of this classification is that it evokes and takes in hand the relation of power between man and animal. The law is based on a reality (the powers, including legal powers, of man over animal) that it takes in hand, for example by punishing actions resulting in the suffering of animals that are 'domestic, tamed or held in captivity' (e.g. Art. 521-1 of the French rural code). However, this relationship is itself inserted into a hierarchical order of values. Hence, the requirement not to inflict suffering upon animals is yielded when this suffering stems from research and scientific experiments, provided that they are 'necessary', that they cannot be 'effectively' replaced by alternative methods and that they pursue one of a limited number of goals (Art. R 214-87 of the French rural code). The conditions are even more restrictive for the patentability of genetically modified animals where 'the processes to modify the identity of animals that cause them to suffer without unsubstantial medical use for man or animal, and the animals resulting from such processes' are outside the realm of patentability (Art. 6 Directive No. 98/44/CE).

In addition to the relation of power of man over animal, other types of man/animal or man/non-human relationships can be identified in the field of law. Take, for example, the emotional relationship, where French law governing civil liability foresees reparation for the moral damages suffered by the loss of a pet. More embryonic, French law is opening up to acknowledge subjective losses suffered due to an offence against the natural environment (Tours Criminal Court, July 24, 2008, No. 1747), acknowledging the bonds between man and the natural environment.

In ethics, the question of animals is mostly addressed by considering Man and Animal as pertinent ingredients of ethical analysis. This conception often leads to the conclusion that there are no longer any boundaries between Man and Animal. This is a tendency to *renaturalize* Man according to the following movement: 'Man, like animals, is nothing more than a *natural*, in other words *biological creature*' (Sève, 2008: 166). The *all linguistic* (criticised by Hottois, 1979), the *all genetic* (challenged by Kupiecs, 2007) or the *all sensitivity* (advocated by utilitarian-inspired animal ethics) come under this same logic to *renaturalize* or *biologize* Man, whereby ethics are based on science (an aberration). Yet, if 'nothing 'peculiar to man' can explain so much newness in the human world, it is because all of this newness has not been produced by 'man', i.e. the human individual falsely taken as sufficient worth of its genus, but by this genus as a whole' (Sève, 2009: 300). So, what 'blatantly differentiates men from animals, which they are in fact themselves, is what is peculiar not to *man* but *mankind*, i.e. above all else, to the *human world* where it necessarily resides' (Sève, 2009: 300). This limitless and complex configuration, which is a distinctive feature not of man, but of *humanitas,* is irreducible and very difficult to define. Elisabeth de Fontenay expresses this by refusing to define humans: 'I object to the exalting of humans, rationality, freedom and all other characteristics generally listed to define man. I note that, without a doubt, and therefore without us being able to justify or elucidate it, humans are pre-eminent. But I refuse to define these humans, whose superiority I acknowledge – which has nothing to do with the right to total hegemony over everything not human. Indeed, as soon as we define, there are exclusions, humans excluded from rationality, from conscience, from freedom, from the representation of death' (De Fontenay, 2001: 12).

Inversely, there is another reductionist tendency to make Animals the cursor of ethical analysis, while we are dealing with an irreducible plurality of life forms: 'Neither one species, nor one individual, but an irreducible living multiplicity of mortals' (Derrida, 2003). By this, Derrida does not intend to efface the limits, but to express that 'among non-humans, and separate to non-humans, there is an immense multiplicity of other living things that are by no means becoming homogenised, barring violence and self-seeking ignorance, under the category of what is known as animals' (Derrida, 2003: 73).

Leaving behind these categories stemming from Machiavellian ethics we migrate to ethics of relationship (Glissant, 2009). We use the concept of (human, animal, living) 'world', as a non-homogeneous ontological element to translate the depth of a complex entity, its environment and all its relationships. It is therefore no longer a question of the depth of boundaries that bind rather than the narrowness of boundaries that exclude; ethics therefore become the study of human inter-links, this highly meaningful space, this collective part of humans (Berque, 1996). In this new plural logic, 'animal ethics' are reintroduced into a set of relationships to become *ethics of multiple relations and mutual interventions between human and non-human worlds*, and 'animal law' becomes the *legal apprehension of multiple relations and mutual interventions between human and non-human worlds.*

But leaving opposition behind to embark upon multiple relationships is very demanding, from an epistemological viewpoint. We believe that the principle of ordered pluralism is appropriate as an alternative, between separation pluralism (positivism, dualism) and merging pluralism (relativism, monism), for organising the relationships between worlds that these new conceptions of ethics and law lead us to ponder upon.

Attempt to typologize relations between different worlds revolving around GM fish

With this new logic, we propose to typologize the numerous relations linked to this object. The advent of the GM fish is therefore an opportunity to think about different types of relationships with which we associate a particular GM fish status (Table 1).

Human worlds versus human worlds

The relationship will be one of equity or iniquity between human worlds, linked to access to resources, to innovation; for example, in relations between North and South, between small family-based aquacultures

Table 1. Types of relationships associated with GM fish status.

	Type of relationships between human worlds	GM fish status
and human worlds	Relations of equity	Source of societal changes
	Relations of risk	Vector of health risk
and animal worlds	Emotional relations	Companion
	Symbolic relations	Cultural object
	Instrumental relations	Tool to serve Man
	Relations of power	Objects of duty
and living worlds	Relations of risk	Vector of environmental risk

and industrial structures. In these cases, GM fish will be a source of societal change, for example, through changes to the world's aquacultural development induced by monopolies, etc.

From this relationship, the law will have to order considerations connected to patent law and corporate responsibility. Here, ethics will analyse in the field of economical and social ethics (Arnsperger and Van Parijs, 2000).

Between human worlds, the relationship will also be a relation of risk (linked to food safety) between eaters and producers (or wholesalers and retailers). In this case, the GM fish is a risk vector (as is a new food substance). The law will utilize the consumer law corpus (food traceability) and corporate responsibility corpus; and ethics, the field of agricultural and food ethics.

Human worlds versus animal worlds

The GM fish reveals numerous relationships between human and animal worlds: emotional, symbolic, instrumental, or of exploitation, of power and of obligations (respect for well-being, intrinsic values, integrity). In this relationship context, the GM fish is an object of duty and responsibility (relation of power), a companion (emotional or symbolic relation) or a tool to serve man (instrumental relation). We notice that it does not differ from the condition of other animals. As we have already seen, the law will utilize rules, particularly revolving around the notion of the animal as a 'sensitive creature', which imposes an obligation not to inflict unnecessary suffering. Animal ethics will be there, at the core of the analysis, with work on how moral animal considerations are organised (Goffi, 1994; Sandoe, 2008, Marie, 2005). For example, utilitarian, deontological or contractualist frameworks can be utilized, but they cannot aspire to end the ethical analysis.

Human worlds versus living worlds (ecosystems, natural environments)

The GM fish induces between human worlds and living worlds a relation of risk in the sense that the GM fish is a vector of danger and threat in the event of dissemination in natural ecosystems (Cordis, 2009). As such, it can be compared to invasive species or hazardous substances, the production and dissemination of which are regulated by law. Other legal corpuses whose purpose is to protect biodiversity will also be involved. Ethics will utilize their corpus in the field of environmental ethics (Afeissa, 2007).

From this typology of relations, we can identify certain specificities of GM fish. First of all, we can say that the GM fish comes under these three types of relationship, just like any other breeding animal. However, within the first relationship between human worlds, the GM fish has its own specificity due to the patentability, which makes it a specific source of societal change. Next, what differentiates it from a plant GMO is the type of relationship with the animal worlds that it involves. Lastly, its generic particularity stems from the changing nature of its status according to the point of view taken.

Thanks to this typology, it is also possible to identify contradictions in the handling of the issue of GM fish. For example, in the relation between the human world and the animal world, considering the integrity of GM fish would imply refusing to take away any of their chief physiological characteristics such as reproduction. Yet, in the context of the relationship between the human world and the living world, we can see that the GM fish is a vector of risk in the event of dissemination and therefore that sterilisation is necessary to limit this risk. Hence, depending on which values are to be protected, sterilisation of GM fish will be more or less problematic.

The GM fish: a changing object for law and ethics

Thanks to this typology, we can see that understanding the GM fish object requires the use of a plurality of corpuses, already existing or to be invented, in law and ethics. The important thing is to order them. We can observe that, depending on the type of relationship at stake, the GM fish does not have the same status; in this sense, it can be considered as a *changing object*: an object of protection (as a sensitive creature), tools for Man (as a breeding animal), a vector of risk (as a living creature or as food), a source of societal changes (as an object of a patent or inducing a specific aquacultural structure).

The ever-changing nature of the GM fish object illustrates the current restructuring of the legal scene: the values that are now 'at the heart of the law' (Papaux, 2006) and which are concerned here (respect for animal life, biological diversity, equity between populations and in the access to resources) enable us to free ourselves of the subject/object categories and the paradigm of reciprocity to translate our relationship to non-human worlds. From this new perspective, the law unveils the plurality of this relationship's forms and kinds (instrumental, emotional, symbolic, risk, etc.). This still chaotic plurality nonetheless unearths responsibility ethics and law (Ost, 1995; Thibierge, 2004), emerging from and a corollary of the powers that the human world holds, powers concentrated in the GM fish, which is both a 'sensitive creature' and a vector of ecological risks and iniquity between populations. This concept of responsibility appears to be a sound candidate for giving a meaning to ordered pluralism. Also, in so far as the law both mirrors a society and is a means of affecting it, this ordered pluralism (of values, levels of involvement, scales and legal systems) must be built closely with ordered pluralism in ethics (food and agricultural ethics, animal ethics, environmental ethics, economic and social ethics, etc.). Like other authors have already suggested (Logar, 2004; Aerni, 2004), our work shows that in the case of GM fish, there is a need to rethink the strategies and the ties between ethics and the law to insert our analysis into a context more conducive to shedding light on complex objects.

Acknowledgements

This article is the result of research conducted for the purpose of the DOGMATIS project for 2007-2010, a project funded by the ANR (National Research Agency), coordinated by INRA (National Institute for Agricultural Research) and directed by Muriel Mambrini.

References

Aerni, P. (2004). Risk, regulation and innovation: The case of aquaculture and transgenic fish, Aquatic science 66 (3): 327-341.

Afeissa, H.-S. (2007). Ethique de l'environnement, Vrin, Paris, 394 p.

Arnsperger, C. and Van Parijs, P. (2000). L'éthique économique et sociale, La découverte, 123 p.

Berque, A. (1996). Etre humain sur la terre, Gallimard, Paris, 212 p.
Coutellec, L. (2009). New epistemologic perspectives in biotechnology ethics: the case of GM fish. In: Millar, K. *et al.* Ethical futures: bioscience and food horizons, Wageningen Academic Publishers, the Netherlands, 445 p.
Cordis (2009). EU-funded study urges caution on transgenic fish farming. European Commission, ERATS project (Ecological risk-assessment of transgenic salmon), European Union FP6.
De Fontenay, É. (2001). Question animale, question politique, Relations 670: 12-15.
Delmas-Marty, M. (2004). Le relatif et l'universel, Les forces imaginantes du droit (I), Seuil, Paris.
Delmas-Marty, M. (2006). Un pluralisme ordonné. Les forces imaginantes du droit (II) Seuil, Paris.
Derrida, J. (2003). De quoi demain ..., Flammarion, Paris, 320 p.
Descola, P. (2005). Par-delà nature et culture, Gallimard NRF, Paris.
Desmoulin, S. (2006). L'animal entre science et droit, Presse universitaires d'Aix Marseille, December 2006, No. 872.
Farjat, G. (2002). Entre les personnes et les choses, les centres d'intérêts. Prolégomènes pour une recherche, RTDCiv, p. 221.
Glissant, E. (2009). Philosophie de la relation, Gallimard, Paris, 176 p.
Goffi, J.-Y. (1994). Le philosophe et ses animaux, Jacqueline Chambon, Paris, 335 p.
Libchaber, R. Perspectives sur la situation juridique de l'animal, RTD civ. 2001, p. 239.
Logar (2005). Transgenic fish: Is a new policy framework for a new technology? Environmental Science & Policy 8: 17-27.
Marie, M., S. Edwards, G. Gandini, M. Reiss and E. von Borell (2005). Animal Bioethics, principles and teaching ethics. Wageningen Academic Publishers, the Netherlands.
Ost, F. (1995). La nature hors la loi, l'écologie à l'épreuve du droit, La Découverte.
Papaux, A. (2006). Introduction à la philosophie du 'droit en situation', LGDJ, coll. Quid juris?
Sandoe, P. and Christiansen, S.B. (2008). Ethics of animal use. Blackwell, Oxford.
Sève, L. (2008). L'homme, La dispute, Paris, 445 p.
Suzanne, A. (2005). Rapport sur le statut juridique de l'animal, Ministère de la justice.
Thibierge, C. (2004). Avenir de la responsabilité, responsabilité de l'avenir, Recueil Dalloz, p. 577.

Genomics and animal health: use of animals in biomedical research

B.M. Jugo[1], A. Larruskain[1], K. Garcia-Etxebarria[1], L.I. Mazón[1] and Miren Gorrotxategi Azurmendi[2]
[1]Genetics, Physical Anthropology and Animal Physiology Dpt. University of the Basque Country, Spain;
[2]Department of Constitutional Law. University of the Basque Country, Spain

Abstract

Although the use of animals in biomedical research leads to a strong debate regarding ethics, the study of animal models is extremely valuable to understand the genetic basis of infectious disease. Both the research animals and humans may benefit from the information obtained in the genetic association analyses, which can lead to the discovery of vaccines or treatments. Our group studies two retrovirus-induced infectious diseases affecting sheep, Ovine Pulmonary Adenocarcinoma (OPA) and Maedi-Visna (MV). Animals used in the research are protected by strict legislation which dictates that any distress or suffering caused to the animals must be minimized in housing, transport, scientific procedures and euthanasia. The amount of information obtained from the research also leads to a debate, regarding the ownership of the genetic data and the way it may be used.

Keywords: biomedical research, genetic association analysis, animal experimentation

Introduction

The use of animals in biomedical research is a highly controversial issue. Society is concerned with animal well-being, and there is an awareness that leads to demand animal welfare, where an animal's nutritional, environmental, health, behavioural and mental needs are met. However, animal experimentation has been and still is a basic activity in science. The impact of scientific research on animal welfare is as varied as the types of research, ranging from the little impact to serious consequences. It is therefore crucial to consider the pain, suffering or agitation that an animal experiment to evaluate the ethical implications of research to be carried out.

Some argue that animal experiments are used to understand the pathology of diseases and to develop new medicines, as well as testing the safety of other products. It cannot be denied that many vaccines, medicines and biomedical procedures would not be available had it not been for animal experimentation.

Advances in diseases such as cancer and AIDS (Acquired Immune Deficiency Syndrome) would hardly be possible without animal experimentation. In contrast, the critics oppose to the use of animals for research because it is cruel and subject test animals to suffering, and arguing that often experimentation is unnecessary and better alternatives to animal research are available.

Admittedly, there is an increasing tendency to reduce the number of animals used in research and to employ alternatives to animals are used whenever possible, such as cell cultures and computer modelling. In any case, researchers must comply with strict regulations when employing animals in scientific tests, to avoid animals being exploited or abused. As the adverse physiological and psychological states that animals may experience result from a number of factors, including transportation, accommodation, food and care, handling, identification, scientific procedures and euthanasia, accordingly, regulations have been established on the exploitation, transportation, testing and slaughter for animal care.

Being the use of animals in biomedical research such an important issue, our research group would like to contribute to the ongoing debate regarding ethics by sharing our experiences on the matter. We believe that animal models are an extremely valuable asset to shed light on the genetic basis of infectious disease, an opinion supported by the data obtained so far in our scientific research.

Genetic association analysis in domestic animals/mammals

Ovine pulmonary Adenocarcinoma (OPA) and Maedi-Visna (MV) are two chronic viral diseases that affect adult sheep. They are caused, respectively by Jaagsiekte Sheep Retrovirus (JSRV) and Maedi-Visna Virus (MVV). Neither treatment nor vaccination is currently available and therefore these illnesses result in heavy production losses and cause major welfare implications. Once an individual is infected and these two retroviruses cause a progressive deterioration of condition that ultimately leads to death. Furthermore, underlying aspects of disease pathogeny are not yet known or fully understood. In order to shed light on these aspects, and to characterize the retrovirus-host relationship, the association between genetic polymorphisms in the ovine genome and susceptibility/resistance to disease progression can be analysed by means of genetic association analyses. Moreover, OPA represents an animal model for a peculiar form of human cancer, Bronchioloalveolar Carcinoma (BAC), and MVV is a *Lentivirus*, the same genus as the HIV (Human Immunodeficiency Virus). Animal models have been extremely useful in the research of many genetic diseases, as the genomes are similar to the human genome.

Therefore, these tests can potentially benefit both humans and the animals themselves. Searching for susceptibility/resistance genes means that knowledge of the disease mechanism could be gained, leading to the development of vaccines, or to the selection of naturally resistant sheep for breeding, with the objective of controlling or eradication of OPA and MV.

Association analysis, step by step

The first step for an association analysis is the obtainment of samples, which includes both diseased (in our case, naturally infected) individuals and healthy individuals. Sources of the samples may be experimental sheep herds or the slaughterhouse, from animals destined to culling for various reasons such as illness of culling of breeding stock. For a successful association study, samples must be reliably classified as cases or controls, which due to the characteristics of OPA and MV involves a post-mortem histopathologic examination of the main target organ, the lungs, to search for injuries caused by JSRV and MVV. Euthanasia is carried out in authorized facilities following guidelines established by legislation in force. Euthanasia should not, if carried out properly, causing the animals cause any pain, suffering or distress. The method used for euthanasia must either kill the animal very quickly or instantly unconscious for slaughter before they regain consciousness, so that death occurs with the least fear, anxiety, pain and anguish. In the case of sheep from experimental herds, blood samples may be collected at regular intervals. This can sometimes cause discomfort or pain to the sheep, but under ideal circumstances, this procedure has only relatively minor effects on the welfare of animals.

There are other issues to be taken into account regarding the use of sheep for research. The handling and transportation of the animals may have adverse effects, as separation from familiar animals and surroundings, confinement in a shipping container, unknown sounds and smells, fluctuations in temperature and humidity and in the availability of food and water may potentially cause stress. The transport of animals used or intended to be used in procedures shall comply with the conditions laid down in existing rules on trade, health and animal welfare.

Regarding the personal participating in the research, they must have the appropriate education and training, both those carrying out the experiments and those who handle the animals. In particular, people caring for the sheep will perform their duties without resorting to violence or methods which could cause unnecessary fear, injury or unnecessary suffering.

The use of sheep is necessary in a genetic association analysis that aims to discover genetic polymorphisms that may confer susceptibility/resistance to OPA and/or MV. However, given the designs of this type of research the disturbance caused to the animals is essentially minimal. Samples are obtained from animals sent to slaughter from private exploitations or experimental herds, and therefore ethical issues mainly arise regarding transport and euthanasia, for procedures prior to the post-mortem analysis are minimised. However, as in any case where animals are employed in research it is necessary to critically examine the procedures used and, select the alternatives that cause less pain, discomfort or anxiety.

Once the part involving the handling of the animals or the laboratory work is over, more ethical issues, albeit of a different nature, arise. Genetic studies provide an enormous amount of information about the species under study. Being immersed in the era of genomics, the information available on different organisms is continually on the increase. In particular, model species of those which are economically important are being intensely studied, including both genomic and genetic analysis. However, this knowledge about the genomes, which can be considered the essence of an species, may result in ethical debates mainly for two reasons. On the one hand, there is the ownership of the acquired knowledge. If these genetic studies contribute to the improvement of the genetics of a species to make them resistant to disease or increase their productivity, should this knowledge be made public, or conversely be owned by the researchers, the institution these scientists work for or the owners of the sampled species? On the other hand, there is the impact on the biodiversity, another issue under debate. If genetic variants (such as specific domestic animal breeds) are described as less suitable for their use in agriculture, they will be in danger of disappearing, leading to a decline in biodiversity. Likewise, favouring the most interesting genetic variants, discarding the rest, will also lead to a reduction of the species' biodiversity.

References

Anonymous (1986). Ley 86/609/CEE. Disposiciones legales, reglamentarias y administrativas de los estados miembros respecto a la protección de los animales utilizados para experimentación y otros fines científicos.

Anonymous (2003). Ley 11/2003 de Protección de Animales. Legislación relativa a protección animal (bienestar animal). Animal experimentation ethics. Contemporary legal and ethical rules.

Anonymous (2003). Ley 8/2003 de sanidad animal.

Anonymous (2007). Ley 32/2007 para el cuidado de animales, en su explotación, transporte, experimentación y sacrificio.

Anonymous (undated). Declaración universal de los derechos animales

Nuffield Council on Bioethics (undated). The ethics of research involving animals.

Reilly, J.S. (2001) Euthanasia of animals used for scientific purposes.

Part 12. General questions on animal ethics

Do animals count for less?

T. Visak
Utrecht University, Ethics Institute, P.O. Box 80103, 3508 TC Utrecht, the Netherlands; t.visak@uu.nl

Abstract

It is broadly accepted that at least some animals disserve our direct moral consideration. However, it is controversial whether they should count as much as human beings do. David DeGrazia has explicitly defended that moral status can come in degrees and that even those animals that deserve our moral consideration might have less moral status than humans do. DeGrazia uses two models for arguing that animals have less moral status: the Unequal Interests Model and the Unequal Consideration Model of Degrees of Moral Status. I take issue with both ways of defending degrees of moral status. I argue that unequal interests should not be conceived as implying unequal moral status. In fact equal consideration of interests can lead to different treatment if interests are different, but then the considered beings do still have equal moral status. Furthermore, I argue that DeGrazia's defence of unequal consideration is unconvincing. I conclude that DeGrazia failed to establish degrees of moral status and less moral status for animals.

Keywords: moral status, DeGrazia, (un-)equal consideration of interests

Introduction

It is broadly accepted that the effects of our actions on animals cannot be completely ignored in moral decision-making. Animals deserve at least some moral consideration. For instance, there is much concern for animal welfare and it is broadly accepted that animal welfare matters, because it matters for the animals. However, it is unclear how much moral consideration animals deserve. Do animals have full moral status, or do they have less than full moral status? David DeGrazia has recently brought forward one of the scant explicit defences of unequal moral status. In this paper I will evaluate DeGrazia's arguments that moral status can come in degrees and that animals might have less moral status than human beings.

DeGrazia argues that unequal moral status might be justified, because morally relevant properties come in degrees and, correspondingly, moral status comes in degrees as well. I argue that the properties that come in degrees, such as self-consciousness or rationality, are not morally relevant. The property that determines whether a being should be granted equal consideration of interests is simply whether a being has any interests at all. This property does not come in degrees. Furthermore, contrary to what it claims, the Unequal Consideration Model does not live up to the alleged correspondence between what are taken to be relevant properties and moral status.

DeGrazia does also appeal to intuitions in defence of unequal consideration. Those appeals to intuitions are unconvincing. I conclude that there is as yet no convincing argument for moral status being a matter of degrees and thus for animals counting for less than human beings. Let me start with some preliminary clarifications.

Preliminaries

If a being has moral status that means that this being has to be taken into account in moral considerations. But what does *that* mean? Different moral theories give different answers to that question. Some focus on interests: taking someone into account means considering her interests. Some, instead, focus on welfare, rights, inherent value or dignity. DeGrazia understands 'equal moral status' in terms of 'equal consideration of interests'. The interpretation of moral status in terms of the obligation to take a being's interests into account is common. However one should be aware that it is not the only possible understanding of what granting moral consideration amounts to. Thus, there might be other ways for arguing for or against moral status coming in degrees and animals having less moral status.

Another thing to be clarified in advance is that this paper is about moral status, and the question whether animals might have less moral status than humans. It is not about how animals should be treated. Differential treatment might be justified, even if animals appear to have equal moral status. So, if I dismiss arguments for different moral status, I do not claim that differential treatment cannot be justified. I only claim that it cannot be justified by pointing to differences in moral status. There might be other justifications for unequal treatment, for instance the possession of different interests. In his *Unequal Interests Model* of Degrees of Moral Status DeGrazia brings forward unequal interests as an argument for unequal consideration. This model is conflating unequal treatment based on unequal interests with unequal consideration. The unequal interests model is not a model of unequal consideration, and therefore not a model of unequal moral status.[1]

A final preliminary remark is that unequal moral status is sometimes thought to be justified on the basis of species, simply because animals are no human beings. This *Speciesist Model* of unequal moral status is hardly defended anymore, although it might still be in the back of people's minds. Discrimination on the basis of species is as unjustified as discrimination on the basis of race, sex, or other morally irrelevant characteristics. Therefore, if animals have less moral status, it cannot simply be because they are not humans.

Can unequal consideration be justified?

DeGrazia's *Unequal Consideration Model of Degrees of Moral Status* captures what the idea of granting differing degrees of moral status is about: giving unequal consideration to equal interests.

One possible justification for unequal consideration is that characteristics that are deemed relevant for the ascription of moral status come in degrees. For instance, if one accepts that all sentient beings have at least some moral status, one can make further distinctions in the degree of moral status, based on the degree of sentience or based on additional characteristics, such as moral agency, autonomy, the capacity for intentional action, rationality or self-awareness. One might, for instance, hold that all sentient beings have moral status, but persons have more moral status, because they posses additional capacities that are deemed relevant. Persons are indeed defined by the possession of those capacities. Such a two-tier model distinguishes only between two tiers of moral status, namely persons and sentient non-persons. As an alternative one can accept the sliding-scale model of moral status. According to that model, there is a continuum of more or less moral status. The degree of moral status parallels the degree to which one or more of the relevant characteristics are possessed. It is assumed that moral status comes in degrees, because the possession of the relevant characteristics comes in degrees as well.

I think that such a model of degrees of moral status is logically and conceptually possible. It makes sense to accept that moral status comes in degrees, if the characteristics on which it is based come in degrees as well. Thus, it is necessary to explore whether the relevant characteristics, i.e. the characteristics that determine moral status, do indeed come in degrees. I will turn to this question in the next section.

---

[1] DeGrazia seems to agree with me on this point. In a footnote he admits: 'Now I am less confident that the language of unequal moral status will illuminate more than distort – if we are correct in asserting equal consideration.' It is surprising that DeGrazia still brings forward this model, though.

Do the morally relevant characteristics come in degrees?

So, the question is whether the relevant characteristics do indeed come in degrees. In order to explore this question, we first have to know what the relevant characteristics are. Assuming an interest model of moral status, having moral status means that one's interests have to be taken into account. The default position seems to be to consider having interests as the relevant criterion for moral status. Equal consideration means having one's equal interests considered equally. Equal interests should be considered equally, unless there are good reasons not to do so. So, having interests seems relevant for moral status, while other or additional characteristics seem not obviously relevant. A being's characteristics and capacities bear on what interests a being has. It is not obvious that a being's characteristics and capacities should bear on how much weight any given interest should get. Consider a normal human baby, which suffers a certain amount of pain. Consider a normal human adult, suffering the same amount of pain. Both have an interest in not suffering that pain. I do not see that the fact that the adult is also more rational, able to act morally, and more autonomous should count in favour of alleviating the adult's rather than the baby's pain. Maybe the adult has some additional interests, due to those characteristics. However, we are now concerned with their equally great interest in not suffering. Provided that this interest is equally strong, other characteristics are irrelevant for the weighing of this interest.

I think that much of the initial attraction of taking into account a broader range of characteristics comes from their role in determining a being's range or strength of interests. Possessors of those additional characteristics might have interests that merely sentient beings lack. However, as far as their shared interest in not suffering is concerned, this interest is, assumingly, equally present and equally strong. What is discussed here is the Unequal Consideration Model, not the Unequal Interest Model of Degrees of Moral Status. We are assuming the same interests. No convincing argument has been given for counting those same interests unequally. So, do the relevant characteristics come in degrees? I do not think so. After all, we assume an equal interest in, for instance, not suffering. And exactly this seems to be relevant.

DeGrazia (DeGrazia, 2008: 193) seems to have anticipated that sort of reply:

> 'If one responds that these personhood-constituting properties are largely irrelevant to moral status, sentience being the fundamental criterion, one might agree while insisting that even sentience comes in degrees'.

So, DeGrazia anticipates the possible reply that only sentience matters, but then, he claims, sentience as well is a matter of degrees. It might be true that sentience comes in degrees. Different animals might have different capacities of experiencing all kinds of mental states. We must remember, however, that sentience is deemed relevant, because of its connection with interests. We are assuming that the interests in question, for instance in not suffering are equal. Otherwise, we would talk about the Different Interests Model, rather than the Different Consideration Model of Degrees of Moral Status. If we assume that the amount of suffering of two beings is equally great, their different capacities of experiencing feelings is already taken into account. This capacity is relevant for determining a being's interests or the degree to which its interests are affected. However, if we already assume that they are equally affected, such differences have already been taken into account. They are no longer relevant for the weighting of the interests, unless a convincing reason for this can be presented. Until now, no such reason has been given. It seems that the characteristics that do come in degrees are not the relevant ones. What is morally relevant is the fact that the interests in question are equal and equally strong. Thus, having interests is a necessary and sufficient criterion for equal consideration of interests. Consideration of additional criteria does only make sense in as far as they affect the interests. Then, we are back in the Unequal Interests Model. In the Unequal Consideration Model, interests are assumed to be the same and equally affected. It has not convincingly shown that additional characteristics should bear on whether or not equal interests should get equal consideration.

Is there a correspondence between the degree of allegedly morally relevant properties and the degree of moral status?

Now, I want to bring forward an additional criticism of the Unequal Consideration Model. What makes unequal consideration initially look plausible is the alleged correspondence between the degree to which a being has a certain capacity and the degree of that being's moral status. The initially attractive idea is that the capacities come in degrees, and so comes moral status. I now want to argue that this correspondence does not really exist in the unequal consideration model. So, the models, as they are presented, do not exhibit the idea of correspondence. This is bad news for those models, as the alleged correspondence was meant to be their strong point.

Consider, first, the two-tier model of unequal consideration. This model accepts only two degrees of moral status. Either one is a person and has full moral status, or one is a sentient non-person and has a fixed, lesser degree of moral status. Let us assume, just for example, that sentient non-persons count for 50 percent. The relevant capacities that determine personhood are, for instance, rationality or self-awareness. If those capacities really come in a sliding scale, it seems that personhood must be a gradual matter as well. This, however, does not fit to a to-tier-model according to which one either has full moral status, or one has a fixed but lesser degree of moral status. The correspondence between the degree to which a being possesses the relevant capacities and its moral status is wanting. Fixing two tiers seems arbitrary, if one assumes that the relevant characteristics come in much finer degrees. The capacities can be possessed by different beings to a greater or lesser degree. So, the beings could be placed on a sliding scale, corresponding to the degree to which they possess the relevant characteristics. Their status, however, does not vary to the same degree, but has only two tiers.

This problem seems to be avoided in the sliding scale model of unequal consideration. However, even according to that model, persons are typically distinguished from non-persons. How does that person/non-person dichotomy fit to the alleged sliding scale of the possession of relevant characteristics? Typically, according to the sliding scale model, the finer grained sliding scale of moral status holds only for non-persons. Non-persons are indeed accorded moral status according to their capacities. Persons, however, in spite of gradual differences in terms of the allegedly relevant capacities, are all accorded equal moral status. So, above a certain threshold, the threshold of personhood, differences in the possession of the relevant capacities are not translated into differences of moral status. This model involves double arbitrariness. First, a strict person/non-person distinction has to be made somewhere on the sliding scale of allegedly relevant characteristics. This can only be made in an arbitrary way. Secondly, the differences of capacities among persons are not translated into different degrees of moral status, while the differences among non-persons are. This distinction is arbitrary as well. So, neither of those models of unequal consideration exhibits the correspondence between degree of allegedly relevant capacities and degree of moral status.

Now, assume that the idea of correspondence between degree of relevant characteristics and degree of moral status would be taken serious. This would result in a full-fledged sliding scale model. Individuals would be ascribed moral status on the basis of the degree to which they are rational, self-aware, and/or sensitive or whatever characteristics were deemed relevant. Depending on the amount of criteria, determining a being's degree of moral status would be a complex matter. The being would have to be scored on different scales, and then its moral status depends on the result of the calculation. In moral considerations that concern different individuals, all those individual scores would have to be taken into account. This might be a problem for the applicability of this theory. However, it is not a problem for a theory of moral status as such. A theory of moral status must only tell us what determines moral status. For practical purposes, simpler decision procedures based on that criterion could be offered. More importantly, such a theory of moral status would be morally implausible. Should, for instance, the equal suffering of two persons count differently, only because one of those persons is somewhat more rational than the other one? The differences in rationality seem irrelevant here. After all, they are assumed to have an equal interest in not suffering. So, while a full-fledged sliding-scale model would show correspondence between allegedly relevant characteristics and moral status, its plausibility still depends on the relevance of the alleged criteria. I have already argued that the relevant criterion is the degree to which the beings are affected in their interests (or, I would rather say, their welfare). That is taken to be the same. I have not seen a convincing argument that anything else should be relevant.

Appeals to intuitions in defence of unequal consideration

In the previous two sections, I have criticized the idea of moral status coming in degrees in a fundamental way. More commonly, unequal consideration is defended on more 'practical' or 'intuitive' grounds.[2] As an illustration, I will discuss DeGrazia's appeals to intuitions about cases as a justification for unequal moral status.

First, DeGrazia observers that, even among those who are willing to grant animals some moral status, there seems to be a consensus that persons count for more. According to DeGrazia 'the fact that there is consensus about persons' might show that 'persons are special in their status'. I do not see how the fact that most people consider themselves more important than animals can tell us anything about their relative moral status. Not long ago there was a consensus among men that they were very special and more important than women, and in some places this consensus does still persist. Slavery is another well-known example. In those cases, we cannot take the judgments of the dominant group to be 'considered' judgments, because after better consideration, the consensus tents to disappear. It is not clear, whether this will happen to the consensus among persons that they are more important than animals. The fact that the persons' interests in using animals are at stake should make us suspicious that their judgment might be biased. Maybe their reluctance to loose the benefits from their dominant status distorts their judgment. So, I do not think that this consensus allows us to draw conclusions about moral status. Furthermore, it is doubtful whether the consensus that persons have more moral status exists at all. I think that at least in Western countries, children, including young children and babies, are granted equal moral status than adults. There is, for instance, no consensus that a normal human baby should have less moral status than a grown up pig of dolphin. This shows that there is no consensus about persons having more moral status. There might be a consensus about humans having more moral status than animals, but that is another issue.

Conclusion

I have argued that DeGrazia's Unequal Interests Model might justify unequal treatment, but it is not about unequal moral status. It is compatible with the idea that equal interests get equal consideration. In order to establish degrees of moral status, it must be shown that unequal consideration can be justified. I criticised DeGrazia's correspondence model of degrees of moral status, which claims that morally relevant characteristics come in degrees, and so does moral status. I have argued that the morally relevant characteristics do not come in degrees, and the characteristics that come in degrees are not morally relevant. Furthermore, I have shown that the model does not really imply a correspondence between relevant characteristics and moral status, so this initially appealing feature is not given. I also showed

[2] Warren Moral Status.

that DeGrazia's appeals to intuitions in defence of unequal consideration are not convincing. I conclude that DeGrazia failed to establish degrees of moral status and less moral status for animals.

References

DeGrazia, D. (2008). 'Moral Status As a Matter of Degree?', in: Southern Journal of Philosophy, vol. XLVI, p. 181-198.

Duwell, M. (2009). 'Philosophical Presuppositions of Practical Ethics', in: Schaler, A. (ed.) Peter Singer Under Fire. The Moral Iconoclast Faces His Critics, Carus Publishing Company, USA.

Singer, P. (1995). Animal Liberation, Pimlico, London.

Warren, M.A. (2000). Moral Status. Obligations to Persons and Other Living Things. Oxford Univerisity Press, Oxford.

Williams, B. (2009). 'The Human Prejudice', in: Schaler, A. (ed.) Peter Singer Under Fire. The Moral Iconoclast Faces His Critics, Carus Publishing Company, USA.

Doing better: John Dewey's meliorism as a method for bringing animal ethics into practice

H. Grimm
Institute TTN, Marsstr. 19, 80335 Munich, Germany; herwig.grimm@elkb.de

Abstract

One of John Dewey's guiding ideas was to reconstruct philosophy in order to solve practical problems. He considered the methodology of empirical sciences as the most promising approach to guide this reconstruction. In his perspective, the pattern of *empirical inquiry* is not only an appropriate methodology in empirical sciences but also in the humanities and especially in ethics. However, he did not develop this particular methodology for ethical problem solving. The paper presents the results of a PhD project. Its aim was to translate the pattern of empirical inquiry into an ethical methodology and apply it to a subject of farm animal welfare. Against the background of John Dewey's work *Logic: The Theory of Inquiry* a melioristic methodology for ethical problem solving has been developed. In particular, the chapter *The Pattern of Inquiry* was used where Dewey describes a general structure of empirical sciences. It takes its starting point at the practical level of an indeterminate (problematic) situation, followed by four steps of effective problem solving: (1) *Indeterminate situation*; (2) *Institution of the problem*; (3) *Determination of the problem-solution*; (4) *Reasoning*; (5) *Testing the hypothesis by action*. In the following, the structure of ethical problem solving along Dewey's *Pattern of Inquiry* will be described and some of its potentials and shortcomings will be outlined. The resulting melioristic approach supports the transformation of abstract ethical principles into practical contexts of farm animal husbandry while taking the responsible actor's practical limitations into account.

Keywords: pragmatism, applied ethics, pattern of inquiry, farrowing crate

Introduction

Questions like 'Is the farrowing crate morally justified?' confront ethicists with a number of methodological problems. Due to the practical nature of such problems, the methods of applied ethics have to handle a number of specific characteristics. According to Bayertz (1999), problems of applied ethics can be characterized by four aspects: First, they originate in a loss of orientation in action and are therefore *normative* in nature. Second, they are *entangled with empirical facts* in such a way that they cannot be solved without taking empirical knowledge into account. Third, problems of applied ethics are *real life problems*. Neither are they problems of philosophical speculation, nor are they used to demonstrate ethical methodology or theoretical shortcomings only. Fourth, such problems have a rather *narrow thematic scope*. The aim of this paper is to present the main result of a PhD project. It dealt with the potentials of pragmatism to develop practical solutions for ethical problems with the mentioned characteristics and particularly as to whether Dewey's pragmatism can provide methodological support.

Pragmatism in applied ethics: diagnosis or cure?

Undoubtedly, the renaissance of pragmatism in philosophy (Sandbothe 2000; Hetzel/Kertscher/Rölli 2008) has had an impact on philosophical ethics as well. An increasing number of ethicists find valuable alternatives for abstract ethical theories in pragmatic approaches. We find a great number of authors referring to the classical theories of Peirce, James or Dewey and the pragmatistic tradition (e.g. McGee, 2003 [1999]; Light and Katz, 1996; McKenna and Light, 2004; Keulartz et al., 2002; LaFollette, 2000). Various reasons can be given for the revival of the pragmatistic tradition. One of them is the promise of pragmatism to solve practical problems of the outlined nature. This promise is especially tempting for ethicists working in the field of practice-oriented ethics. It is claimed that pragmatism can avoid deficiencies of ethical theories that Putnam pinpointed some years ago:

> ...when a philosopher 'solves' an ethical problem for one, one feels as if one had asked for a subway token and been given a passenger ticket valid for the first interplanetary passenger-carrying spaceship instead (Putnam, 1992: 179).

To design methods for solving ethical problems without losing sight of the practical circumstances could justly be called a *cure* for abstract and vague philosophical methodology. Consequently, more and more ethicists follow this line of thought and develop feasible tools, practice-oriented models or – and this was the focus of the PhD project – pragmatistic methods for solving problems in the field of applied ethics.

At second glance, we discover that the attempt to cure the illnesses of abstract ethical theory with pragmatic remedies sometimes falls short in theoretical foundations and methodological reflection (Beekman and Brom, 2007; Borchers, 2009; Grimm, 2008). This initiates the critique that pragmatism sometimes serves as a methodological 'anything goes'. One of the critics – being a pragmatist himself – sketc.hes the problem as follows:

> My hunch is that one of the best ways to get our colleagues in various fields of applied ethics to adopt more unpragmatic approaches to their work is to argue that they should consider becoming pragmatists (Light, 2002: 89).

The claim to work in a pragmatistic manner does not free authors from methodological demands and philosophical constraints. If authors are guided by the idea to solve ethical problems at whatever methodological cost and fall short in developing theoretically sound positions, pragmatism is not a cure but a *diagnosis* of a freehand undertaking that erodes the scientific basis of moral philosophy (Tollefsen and Cherry, 2003: 534). Following Light (2002), I am going to argue for a 'modest proposal' and that Dewey's pragmatism has – within limits – sufficient theoretical resources to found a valid methodology to solve practical problems in applied ethics.

John Dewey's *Pattern of Inquiry* as a model for applied ethics

Similar to Peirce, Dewey considers the methodology of empirical sciences as the most promising approach and the best model to guide the reconstruction of humanities and moral philosophy (Dewey 1920: 172-186). In general terms, he holds that inquiry is a human being's response to a break or interruption in a previously untroubled flow of action. The methodological response to such indeterminate situations is formulated in *The Pattern of Inquiry* (Dewey 1938: 105-122), a general structure of empirical inquiry:

1. *Indeterminate situation*: interruption of the previously untroubled behavior.
2. *Institution of a problem*: location and definition of the problem.
3. *Determination of a problem-solution*: suggestion of hypothetical solutions.
4. *Reasoning*: consideration of the bearings and subsequent adjustment of hypothetical solutions.
5. *Testing the hypothesis by action*: act according to the refined hypothetical solution.

Dewey himself never translated these steps into ethics, nor did his followers. Although there are several monographs on Dewey's *Ethics* (Dewey, 1908, 1932) – e.g. Welchman (1995), Fesmire (2003), Pappas (2008), Gouinlock (1972) – and a great number of articles – e.g. Caspary (1991), Gouinlock (1977,1978, 1993), Pamental (2000), Pappas (1997), etc. – *The Pattern of Inquiry* has not been translated into an ethical methodology. As a consequence, Dewey's approach is often stated and used without considering the methodological implications and limitations. This comes as a surprise, since ethical problems obviously show different characteristics than empirical ones. Not only are they normative in nature, entangled with empirical facts, narrow in thematic scope and real life problems, but also *ill-*

structured as opposed to *well-structured,* meaning that they have no definitive formulation and can only be defined during the process of problem solving. Moreover, ill-structured problems may have several alternative (good, satisfying, etc.) solutions, which are not easy to compare with each other (Dorst, 2006: 636). This is especially true in the field of applied ethics. It is not at all clear what e.g. the actual problem of the farrowing crate is. Is it the restricted behavior of the mother sow? Is it the squashed piglets? Or, is it the income of the farmer? Depending on how the problem is framed, the search for possible solutions will vary.

Deweys approach is likely to be suitable for overcoming such difficulties, since the *Pattern of Inquiry* has its origin in empirical sciences, where working hypotheses and inquiry into possible solutions play a major role. Therefore, the process of problem solving can be understood as a process guided by hypothetical solutions (working hypotheses) that are refined, adjusted and developed in reference to a certain problematic situation.

A pragmatistic methodology to solve problems in agricultural ethics

In the following, essential cornerstones of a pragmatistic methodology to solve problems in applied ethics are developed against the background of Dewey's *Pattern of Inquiry*. For this purpose, the core ideas of the five steps are presented and then translated into a methodology of ethics. After that, the method will be briefly applied to the farrowing crate. The resulting approach directs the transformation of abstract ethical principles into practical contexts of farm animal husbandry while taking the responsible actor's constraints and contextual circumstances into account. It is beyond the scope of this paper to elaborate the arguments and understanding of Dewey's thoughts in great depths, which has been done elsewhere (Grimm, 2010).

Problematic situations

According to the outlined *Pattern of Inquiry*, inquiry starts as soon as a situation is experienced to be confusing or ambiguous. As we face such problematic situations we experience an urgent need to find a way out (Murphey, 1988: xii). Doubt and perplexity indicate a problematic situation (Dewey, 1938: 109). The reason for this perplexity is that *habits* fail to direct behavior. 'Habits' can be understood as one's implicit routine way of doing things and the broad spectrum of tendencies and dispositions, or dominant ways of acting (Pappas, 2008: 122). Since pre-reflexive habits fail, they cannot guide us through such situations and reflective thinking and methodologically structured inquiry are called for (Dewey, 1938: 121).

The counterpart to the problematic situation in morals can be described as a pre-reflexive moral doubt. A vague intuition that there is something morally wrong with a certain situation:

> 'For in our moral life it is always a specific >felt< trouble, question, confusion that sets moral inquiry going and makes the moral qualities of different possible actions to concern of our focal awareness' (Pappas, 1997: 535).

Such situations *evoke* but also *control* inquiry since ends have to be formulated in reference to the problematic situation (Dewey, 1938: 109). In Dewey's own words:

> 'This intellectual search for ends is bound to arise when customs fail to give required guidance. And this failure happens when old institutions break down... And reflection upon what one shall do is identical with formation of ends. ...Thinking has to operate creatively to form new ends' (Dewey, 1932: 184).

Applied to the study case, the farrowing crate, it is easy to imagine that a farmer e.g. is confronted with critique and the question whether the fixation of the mother sow is morally justified. The following steps of ethical reflection provide a structured approach to give guidance in answering such practical questions.

Institution of a problem

In the second phase of inquiry the problem has to be formulated. Obviously, a problem is an intellectualization of a pre-reflexive indeterminate situation. In order to give this intellectualization a theoretical background and direction, Hayes' definition of *well-defined problems* can be used: 'move from I (initial state) to G (goal state) by O (operators) but without violating C (constraints)' (Hayes, 1981). In order to identify these dimensions in a morally relevant manner, the inquirer has to articulate the problem under an ethical perspective. Without going into depth, one can say that Dewey does not provide the methodology to elaborate such a perspective. However, starting at the moral intuitions in the problematic situation, an ethical perspective is gained if the intuitions can be reformulated in terms of a violated moral principle. In the case of the farrowing crate, the moral intuition that there is something morally wrong with the housing system can be linked with the moral principle to avoid suffering. Empirical facts, like behavioral constraints in the crate or risk of injuries can be plausibly related with suffering and the violation of the harm principle. Under this normative perspective, moral deficiencies of the problematic situation (initial state) and a vague goal state (less behavioral constraints) can be identified. In the study case, the restricted behavior of the sow was taken as the central moral deficiency to be dealt with. The missing dimensions of a well-defined problem are to be dealt with in the following steps.

Determination of a problem-solution

Dewey's third step is the determination of hypothetical solutions with reference to the problematic situation. The vague goal state has to be replaced by a *specific* goal state, that Dewey calls *ends-in-view*:

> 'Ends-in-view framed with a negative reference (i.e. to some trouble or problem) are means which inhibit the operation of conditions producing the obnoxious result; they enable positive conditions to operate as resources and thereby to effect a result which is, in the highest possible sense, positive in content' (LW13, 234).

At this stage, ends-in-view are hypothetical, because they are not proved to be realizable. Since the conditions of the situation frame and limit the range of possible solutions, the practical situation has to be taken into account. However, the nature of a hypothetical solution is to overcome the moral deficiencies of the initial situation without losing sight of empirical and normative constraints. Given that ends-in-view define a goal state that overcomes moral deficiencies that are identified through moral principles (phase II), they transform the normative content of a moral principle into an empirical solution. In the case of the farrowing crate one can imagine a pen without fixation and the possibility to turn around, group housing systems, free range, etc. where the behavioral constraints are reduced.

Reasoning: adjustment of hypothetical solutions

The fourth step in Dewey's *Pattern of Inquiry* is a thought experiment, that deals with the bearings of the hypothetical solution. In order to test whether it is morally and empirically reasonable, consequences of a realization are considered in thought. This phase is crucial because it determines whether an end-in-view can serve as an adequate solution or not. For instance, if significantly more piglets were squashed because of the sow's unrestricted movement in a group housing system (end-in-view), the hypothetical solution has significant moral shortcomings. If the alternative housing system is economically unbearable for the farmer, for instance if the alternative housing system implies a 20% loss in income without compensation, the alternative might be morally better, but unrealizable. This phase of inquiry is designed to elaborate moral improvements of the situation that are realizable. In order to structure such reflections, Dewey's methodology – the '*dramatic rehearsal*' (Dewey, 1932: 275; 1908: 93) – has been adapted. The procedure of a *dramatic rehearsal* aims at elaborating morally justified solutions in reach of the responsible actor. By using pragmatic criteria like 'practicability', 'feasibility', 'scientific soundness' and 'ethical soundness', hypothetical solutions can be adjusted and developed into credible solutions for the particular problem. In the study case, empirical data and economic calculations demonstrate that there are alternatives to the farrowing crate that do not cause more dead piglets and can be realized without major economical loss (Weber, 2000; Kühlberger and Jais, 2007).

Testing the hypothetical solutions by action

The last step in Dewey's scheme of inquiry is to test the hypothetical solution by action. The success of the process of inquiry depends on whether the hypothetical solution brings about the intended results.

> 'What empirical method exacts of philosophy is two things: First, that refined methods and products be traced back to their origin in primary experience, in all its heterogeneity and fullness; so that the needs and problems out of which they arise and which they have to satisfy be acknowledged. Secondly, that the secondary methods and conclusions be brought back to the things of ordinary experience, in all their coarseness and crudity, for verification' (Dewey, 1925: 39).

Conduct links the speculative process back to empirical perception. Such actions can only be carried out by responsible actors. Therefore, the actual solution of a problem is not in scope of an ethical reflection. Considering that this methodology is designed to guide actual changes in practice, the ethicist's responsibility in moral inquiry is to develop morally justified and feasible solutions. Hence, the responsible actor's constraints and possibilities have to be taken seriously into account. Otherwise inquiry will fail to formulate reasonable solutions.

Conclusions and methodological implications

The resulting pragmatistic methodology shows that Dewey's *Pattern of Inquiry* can (with adjustments) be successfully transferred to the field of ethics and be applied to practical ethical problems. It focuses on the individual actor's responsibility and helps to realize moral principles in practical terms. However, moral principles cannot be justified with this methodology. Further it cannot answer to fundamental theoretical problems in moral philosophy like the top-down/bottom-up debate. As a melioristic step by step approach, it aims to improve the actual situations in light of moral principles and provide feasible solutions that can in fact be realized by responsible actors. However, since the method focuses on individual responsibility it is not appropriate for normative problems on the societal scale. Accordingly, this pragmatistic methodology is not the universal remedy in ethics, but one stepping stone to bring ethics into practice.

References

Bayertz, K. (1999). Moral als Konstruktion. Zur Selbstaufklärung der angewandten Ethik. In: Kampits, P. and Weinberg, A. (eds.): Angewandte Ethik. Beiträge des 21. Internationalen Wittgenstein Symposiums. Wien, 73-89.

Beekman, V. and Brom, F. (2007). Ethical Tools to Support Systematic Public Deliberations About the Ethical Aspects of Agricultural Biotechnologies. Journal of Agricultural and Environmental Ethics 20: 3-12.

Borchers, D. (2009). Ethiktools für Güterabwägung oder: Wie pragmatisch dürfen Ethiker sein. In: Borchers, D. and Luy, J. (eds.): Der ethisch vertretbare Tierversuch. Kriterien und Grenzen. Paderborn, 15-52.

Caspary, W.R. (1991). Ethical Deliberation as Dramatic Rehearsal: John Dewey's Theory. In: Educational Theory 41/2, 175-88.

Dewey, J. (1908). Ethics (ed. J.A. Boydston, MW 5). Carbondale.
Dewey, J. (1920). Reconstruction in Philosophy and Essays (ed. J.A. Boydston, MW 12). Carbondale.
Dewey, J. (1925). Experience and Nature (ed. J.A. Boydston, LW 1). Carbondale.
Dewey, J. (1932). Ethics (ed. J.A. Boydston, LW 7). Carbondale.
Dewey, J. (1938). Logic: The Theory of Inquiry (ed. J.A. Boydston, LW 12). Carbondale.
Fesmire, S. (2003). John Dewey and Moral Imagination. Bloomington/Indianapolis.
Grimm, H. (2008). Pragmatische Leitkonzepte – Kriterien der Begründung angemessener Lösungsvorschläge in der angewandten Ethik. In: Grimm, H. and Zichy, M. (eds.): Praxis in der Ethik. Berlin/New York: 325-358.
Grimm, H. (2010). Das moralphilosophische Experiment. John Deweys Methode empirischer Untersuchungen als Modell der problem- und anwendungsorientierten Ethik am Beispiel einer tierethischen Fragestellung. in press.
Gouinlock, J. (1972). John Dewey's Philosophy of Value. New York: Humanities Press.
Gouinlock, J. (1977/78). Dewey's Theory of Moral Deliberation. Ethics 88, 218-228.
Gouinlock, J.(1993). Dewey and Contemporary Moral Philosophy. In: Stuhr, J. (ed.): Philosophy and the Reconstruction of Culture. Pragmatic Essays after Dewey. New York: 79-96.
Hayes, J.R. (1981). The Complete Problem Solver. Philadelphia.
Hetzel, A./Kertscher, J./Rölli, M. (ed.) (2008). Pragmatismus – Philosophie der Zukunft? Göttingen.
Keulartz, J. *et al.* (ed.) (2002). Pragmatist Ethics for a Technological Culture. Dordrecht/Boston/London.
Kühlberger, M. and Jais, Ch. (2007). Gruppenhaltung ferkelführender Sauen im Ökobetrieb – Vergleich zweier unterschiedlicher Buchten. In: Wiesinger, K. (ed.): Angewandte Forschung und Beratung für den ökologischen Landbau in Bayern. Tagungsband. Schriftenreihe der Bayerischen Landesanstalt für Landwirtschaft 3: 35-44.
LaFollette, H. (2000). Pragmatic Ethics. In: LaFollette, H. (ed.): The Blackwell Guide to Ethical Theory. Oxford: 400-419.
Light, A. (2002). A Modest Proposal: Methodological Pragmatism for Bioethics. In: Keulartz, J. *et al.* (ed.) Pragmatist Ethics for a Technological Culture. Dordrecht/Boston/London: 79-98.
Light, A. and Katz, E. (eds.) (1996): Environmental Pragmatism. London/New York.
McGee, G. (ed.) (2003 [EA 1999]): Pragmatic Bioethics. 2. ed. Cambridge, Mass./London.
McKenna, E. and Light, A. (eds.) (2004). Animal Pragmatism. Rethinking Human-Nonhuman Relationships. Bloomington/Indianapolis.
Murphey, M.G. (1988). Introduction. In: Human Nature and Conduct. John Dewey – The Middle Works: 1899-1924. Vol. 14. Hg. von Jo Ann Boydston. Carbondale, ix-xiii.
Pamental, M. (2000). The Structure of Dewey's Scientific Ethics. In: Philosophy of Education: 143-150.
Pappas, G.F. (1997). Dewey's Moral Theory: Experience as Method. In: Transactions of the Charles Sanders Peirce Society 23/3: 520-556.
Pappas, G.F. (2008). Dewey's Ethics – Democracy as Experience. Bloomington/Indianapolis.
Putnam, H. (1992). Realism with a Human Face. Cambridge *et al.*: 163-178.
Sandbothe, M. (ed.) (2000). Die Renaissance des Pragmatismus. Aktuelle Verflechtungen zwischen analytischer und kontinentaler Philosophie. Göttingen.
Tollefsen, Ch. and Cherry, M.J. (2003): Pragmatism and Bioethics: Diagnosis or Cure? In: Journal of Medicine and Philosophy 28/5-6: 533-544.
Weber, R. (2000). Gruppensäugen im Abferkelstall. Ein Vergleich zur Haltung in Einzelabferkelbuchten. FAT Bericht Nr. 549. Tänikon.
Welchman, J. (1995). Dewey's Ethical Thought. Ithaca/London.

The argument from marginal cases: a response

I. de Miguel Beriain
Inter-University Chair in Law and the Human Genome, University of Deusto, Avda. Universidades, 24, 48007 Bilbao, Spain; idemigue@deusto.es

Abstract

In the latest years, the Argument from Marginal Cases has been widely discussed by ethicists, as far as it is one of the most promising arguments against speceism. Most of those defending animal rights have used this tool in order to avocate for their recognitiionl. At the same time, those who try to hold the old paradimg have done their best to fight against it. However, most of the responses to the Argument from Marginal Cases lack consistence. This presentation tries to add a new response, based on the fact that we have not inherent moral obligations to human marginal cases, which obviously implies that we do not even have it to animals. Going one step further, it will be stated that the lack of intherent value makes it possible to treat human and animals in a different way without being unfair. If we have no previous moral obligations, talking about discrimination makes no sense.

Keywords: argument marginal cases, animal rights, moral status, specieism, marginal humans

The argument from marginal cases: an exposition

Stating that the Argument from Marginal Cases is crucial to the animal rights debate is not exaggerated. As a matter of fact, it is so important that one philosopher, Daniel A. Dombrowski, has written a whole book about it called Babies and Beasts: The Argument from Marginal Cases.

According to the Internet Encyclopedia of Philosophy, the Argument from Marginal Cases is an argument that attempts to demonstrate that if animals do not have direct moral status, then neither do such human beings as infants, the senile, the severely cognitively disabled, and other such 'marginal cases' of humanity. Since we believe that these sorts of human beings do have direct moral status, there must be something wrong with any theory that claims they do not. Evelyn Pluhar estructured the argument as follows:

1. Beings who are similar in all important morally relevant respects are equally morally significant.
2. Non-humans exist who are similar in all important morally relevant respects to marginal humans.
3. Therefore, those non-humans who are similar in all important morally relevant respects to marginal humans are maximally morally significant if and only if marginal humans are maximally morally significant.

It is very important to underline, even in these previous comments, that the definition of the Encyclopedia of Philosophy contains an affirmation which is not included in Pluhar's exposition. The idea that the infants, the senile, the severely cognitively disabled, and other such 'marginal cases' of humanity have direct moral status is a valorative sentence which is not included in a logical argument, as that described by Pluhar. In consequence, we may accept it or not.

The traditional responses to the argument: a critical analysis of their consistence

The most popular response to the Argument from Marginal Cases is the Argument from Species Normality, which holds the idea that if most of a species' members are moral agents then any member has the same rights and protections as the species. In brief, the moral status of an individual depends on what is normal for that individual's species. The problem of this response is that it goes directly against the logical rule of generality. Even if we consider that some human beings show some cognitive conditions we could never conclude from that part that all human beings have those qualities. Looking at the other side of the mirror, it is really hard to think that a monkey capable to develop most of the qualities of an adult human being should not have it because the specie it belongs to does not show that kind of skills (or, as James Rachels responded to Machan, if one adopts the idea that individuals of a species must be treated according to what is normal for that species, then it would imply a chimp that somehow acquired the ability the read and write should not enter a university since it is not 'normal' behavior for a chimpanzee). Those evidences make it really difficult to consider this objection as a real solid rock to build upon.

Less interesting are those other responses offered by contractualist philosophers such as Peter Carruther, who notes that if we do so extend Rawls's conception, animals will have no direct moral standing, or by those who merely try to reduce the number of human beings included in the category of 'marginal human beings', trying to demonstrate, for instance, that potential rationality is equally worth than actual rationality. In this respect, we must keep in mind that not everybody accepts Rawl's original position theory as the only valid contractualist approach, while it is quite obvious that lots of people do not consider 'potential' similar to 'actual'. Thus, we should conclude that the responses to the Argument from Marginal Cases have failed in their attempt to demonstrate its lack of consistence. That is why in the next pages another way of denying its conclusions will be explored.

The argument from marginal cases: a critical analysis

The Argument from Marginal Cases must face a serious problem which has to do with its starting point. As it was pointed out before, it presupposes a fact that is accepted as unquestionable: marginal human beings are endowed with moral worthiness. However, must we really think that a marginal human being necessarily posses that kind of value? In my opinion, the answer to this question depends on what we consider to be morally relevant. If we think that possessing rationality makes a being worth of respect, it is really hard to reject that not only marginal human beings but also some animals may be morally worthy. Moreover, if we consider that it is the capability to feel pain which makes beings morally relevant, it is hard to neglect that most of animals, and, of course, human beings, may be included in the category of morally relevant beings. However, if we limit the realm of relevant beings to those able to develop moral thoughts, it is quite obvious that only part of the human beings may be considered so. In other words, in those circumstances, neither animals nor marginal human beings should be endowed with moral worthiness. But, does it make any sense to sustain such a position?

The answer to this question is much more difficult to reach, but there is an important idea we should never forget: in the end, something is worthy if, and only if, someone capable to value considers it as such. Though, a being will be relevant from a moral point of view if and only if those capable to value consider it so. But what if this is not the case? Under those circumstances, we should conclude that the mentioned being would deserve no moral respect, that is, it would not be worthy. For adult humans, this means almost nothing. As most of them are moral agents, that is, a necessary condition for the existence of morals, they can consider themselves as worthy beings, they have inherent value (if they were not worthy, morals would never exist, so, only denying the value of morals we could deny the value of moral agency). However, in the case of marginal human beings it is quite clear that, as far as they are not moral agents, they cannot value themselves. Consequently, they do not have inherent worthiness. And consequently, we may conclude that they are not morally significant. But, if they are not so, it is quite obvious that neither do animals, keeping in mind that both, marginal human beings and animals have the same moral worthiness (none at all). So, finally, we could conclude that animals are not significant from a moral point of view, even if we accept the argument from marginal cases.

Going one step further

This is not the first time that someone states that marginal humans have not inherent value, which means that we do not have inherent moral obligations towards them. However, those who accept such statement usually accept as an inevitable conclusion that we must treat marginal human beings as if they were animals. My point is that we are not logically forced to arrive to such a cruel conclusion. As a matter of fact, such a thought would only be logically necessary if we had a previous moral obligation towards animals and marginal human beings, which is not the case (if they have no moral worthiness, we do not have direct duties with them).

In my opinion and, as far as it is not the case, we can treat marginal human beings as if they were moral agents and animals as animals without breaking any moral rule. My point relays on the acceptation of the idea that when some beings belonging to different categories have no moral relevance at all, we can discriminate them without being unfair. If the objects of our acts are not morally relevant, we have no obligations towards them. For instance, if someone usually prefers to eat bananas rather than strawberries, nobody will accuse him of doing something morally wrong, as far as neither bananas nor strawberries are considered morally relevant. Following the same logical pattern, it does not seem to be consistent to consider unfair to discriminate between animals and marginal human beings. As far as we have no previous moral obligations towards any of them, we may eat animals and give human rights to marginal human beings without breaking any moral rule.

Conclusion

The conclusion of this paper should be that even if we sustain the Argument from Marginal Cases, this does not mean that we should treat animals and marginal human cases in the same way. The absence of moral relevance in both cases makes it perfectly possible to discriminate between them without breaking any king of moral rule. Where no moral obligations exist, we can trace all the disctiontions we want without attempting against morals. That is why, in the end, we are free to give rights to animals, which does not mean that we ought not to do so. We could do it or not but, in any case, we would not act against the Argument from Marginal Cases: it is only necessarely aplicable to moral agents.

Acknowledgements

This work has been funded by a Grant to Research Groups by the Basque University System, ref. IT-360-07, Department of Education, Universities and Research of the Basque Government. It is directly connected with the aims of the EU-founded project XENOME: Engineering of the porcine genome for xenotransplantation studies in primates: a step towards clinical application, dedicated to xenotrasplantation.

References

De Miguel Beriain, I. (2009). '¿Derechos para los animales?', Dilemata. Revista de ética aplicada, n° 1, pp. 15-31.

Dombrowski, D.A. (1997). Babies and Beasts: The Argument from Marginal Cases. The University of Illinois Press. Champaign, United States of America.

Pluhar, Evelyn (1995). Beyond Prejudice: The Moral Significance of Human and Nonhuman Animals, Duke University Press, Durham.
Rachels, J. (1997). Created from Animals: The Moral Implications of Darwinism. Oxford University Press, USA.
Rawls, J. (2005). A Theory of Justice. President and Fellows of Harvard College. United States of America.
Regan, T. (2004). The case for animal rights. University of California Press. Los Angeles, United States of America.
Singer, P. (2002). Animal Liberation. Harper Collins Publishers, New York, United States of America.
Wilson, S. (2010). Animals and Ethics. The Internet Encyclopedia of Philosophy. Available at: http://www.iep.utm.edu/. Accessed 15 March 2010.

Part 13. Teaching ethics, animal welfare and law for the biosciences and agri-food sector

How *ethics* is taught by European veterinary faculties: a review of published literature and web resources

M. Magalhães-Sant'Ana[1,2], I.A.S. Olsson[1], P. Sandøe[3] and K. Millar[4]
[1]Laboratory Animal Science, IBMC – Instituto de Biologia Molecular e Celular, Universidade do Porto, Rua do Campo Alegre 823, 4150-180 Porto, Portugal; [2]Institute of Biomedical Sciences Abel Salazar, Universidade do Porto, Largo Prof. Abel Salazar2, 4099-003 Porto, Portugal; [3]Danish Centre for Bioethics and Risk Assessment, Faculty of Life Sciences, University of Copenhagen, Rolighedsvej 25, 1958 Frederiksberg C, Denmark; [4]Centre for Applied Bioethics, School of Biosciences, University of Nottingham, LE12 5RD, United Kingdom; mdsantana@gmail.com

Abstract

There is growing acknowledgement of the requirement to introduce future veterinarians to ethics. This subject, however, was not part of the original undergraduate veterinary programmes and there seems to be little consensus on how it should be included within the overall curriculum. In addition, *ethics* can be related to a number of different subjects covered in veterinary training, such as animal welfare science, professionalism, codes of professional conduct, and animal bioethics, to name but a few. Currently, there is no systematic approach that can be applied to identify what is understood by *ethics* in veterinary education, how it is taught, and how ethics teaching relates to the rest of the curricula. This paper presents an initial attempt to comprehensively map how undergraduate veterinary ethics is being taught across Europe. In terms of the data analysis, the data set derives from two information sources. Firstly, the web-based resources provided by 85 European Faculties out of the 99 available at the Federation of Veterinarians of Europe (FVE) website, are consulted (using a keyword search) to identify curricular units that relate to ethics. This data source is also used to retrieve information about descriptors, teaching methodologies, workload, European Credit Transfer System points (ECTS) and assessment methods. Secondly the relevant peer-reviewed literature on the subject was analysed. Our preliminary results suggest a wide range of ethics-related teaching activities underpinned by different pedagogical approaches and objectives. The diversity of combinations in which ethics appears to be conceptually framed could indicate that there is no explicit common aim in undergraduate veterinary ethics education across European Faculties. This finding may not be unexpected since within the regulation of veterinary training at a European level, in terms of ethics teaching, there is no clear description of the competencies that veterinary students are intended to acquire in this subject area by the end of their course.

Keywords: animal ethics, internet resources, ethics teaching, veterinary training, web search

Introduction

The importance of introducing veterinary students to ethics is increasingly acknowledged. Veterinarians face a number of conflicting expectations resulting in complex set of ethical dilemmas which they are expected to know how to deal with. Thornton *et al.* (2001) have described three objectives in the teaching of veterinary ethics: to promote an appropriate attitude to animals, clients and other parties; to equip students with the necessary skills in recognizing and dealing with ethical dilemmas; and to improve the public perception of the veterinary profession.

In a similar way to other life science courses, the place of ethics within the overall veterinary curriculum is far from undisputed (Downie and Clarkeburn, 2005). European Directive 2005/36/EC considers professional ethics as a core subject of veterinary education without, however, any clear description of the competencies that students are intended to acquire by the end of their course. In addition, there is very little understanding of how ethics is taught to veterinary students at the European level and, to our knowledge, only one European survey included questions relating to the teaching of ethics in undergraduate veterinary education. This survey was conducted between 2001 and 2002 and the results were presented in three different reports (Edwards, 2002; Gandini and Monaghé, 2002; Von Borrell, 2002).

The data presented in this paper is the first phase of an extensive European-wide analysis of how ethics teaching is integrated into university training (1st and 2nd cycle qualifications) of veterinary surgeons. On our previous work we focused on the methodological approaches which could be used to teach ethics (Magalhães-Sant'Ana *et al.*, 2009). The paper highlighted the possibility of addressing ethics using rule-based, value/virtue-based or skill-based approaches and drawing on three case studies, we explored how they might appear in combination with each other in some courses. Building on this previous knowledge, a web search was conducted that maps the inclusion of ethics within European veterinary curricula.

Methods

In order to identify ethics' related subjects within European veterinary curricula, a web search was conducted between 2 January and 17 February 2010. The aim of this analysis was to characterize the nature of the teaching of ethics, but it was not intended to assess the quality of the teaching process. The website of the Federation of Veterinarians of Europe (FVE) was used as a starting point to decide which faculties to consider and then used to find the website of each individual faculty. Taking into consideration that ethics is often part of another subject (since schools can understand the subject differently), a set of keywords was used to identify as much as possible of the content that could possibly relate to ethics. These keywords were: ethics/ethical, moral/morality, professional/professionalism, conduct/code of conduct, good practice/best practices, deontology/deontological, philosophy/ philosophical and animal rights.

When the computer-based search had identified a particular keyword, the related text was read carefully in order to confirm if the word was used within the context of ethics. Data sources were also used to retrieve information about descriptors, place in the curriculum (year), teaching methodologies, workload, ECTS and assessment methods. Only national veterinary curricula have been included, programs for foreign students (ERASMUS) have been excluded. This analysis refers only to compulsory disciplines as it was important to identify what were the ethics components that are included in the core curriculum and therefore common to the education of every student. Electives are considered separately in the general discussion.

The web search was performed by the first author of this paper, who is a veterinary surgeon with background in bioethics and experience in undergraduate veterinary education. Vernacular languages, including those known to the researcher, were translated to English using the electronic translation tool *Google Translate*. If the website was available in English as well as in vernacular language, information was sought using both languages to confirm consistency and extent of the information provided. If sometimes two conflicting findings arose, preference was given to the most recent information. Turkish Veterinary Faculties were not assessed at this stage. In total, 85 Veterinary Schools representing 32 countries are included in this study.

Results

The 85 schools can be divided into two groups on basis of the amount of ethics' teaching information available through the internet. Group A (n=55) refers to those faculties from which comprehensive information on the teaching of ethics was obtained or where the limited information retrieved enabled us to analyse the nature of ethics teaching. Group B (n=30) comprised those faculties where no website was available or where it was impossible to retrieve any reliable information on the teaching of ethics.

Within Group A, it was possible to identify at least one of the keywords within the title and/or descriptors of 98 compulsory units. This enabled identification of four underlying conceptual frameworks relating to ethics teaching: History of Veterinary Medicine (HVM), Animal Welfare (AW), Animal Law (AL) and Professionalism (P). Schools were distributed on the basis of this analysis (Table 1):

Following an analysis of course content, it was also possible to identify the teaching methodologies used in 66 of 98 ethics-related units. Lectures were documented as the basic pedagogical method used, whether alone (n=21), together with practical sessions (n=29) or with seminars (n=5). There is also a combination of lectures and practical sessions with seminars (n=6) and with field trips (n=2). Seminars are used alone only twice and a workshop once.

The analysis showed that ethics is taught in every curricular year (Figure 1) with the highest incidence in the fifth year followed closely by a notable peak within the first year. Ethics teaching is slightly more prevalent in the final (typically clinical) years (4th to 6th – 56%) when compared to the early (typically preclinical) years (1st to 3rd – 44%).

The nature of the workload was difficult to assess as on many occasions ethics is taught as part of a wider unit and the information available does not discriminate between the different components. The same difficulty occurs when trying to assess the ECTS credits allocated to the teaching of ethics. Credits

Table 1. Conceptual frameworks[1] found in the teaching of ethics at 55 European veterinary faculties.

Group A	HVM	AW	AL	P	HVM + AW + AL	AW + AL	AW + AL + P	AW + P	AL + P	HVM + AW + AL + P
55	3	4	25	3	1	5	2	1	8	3

[1] HVM = History of Veterinary Medicine; AW = Animal Welfare; AL = Animal Law; P = Professionalism.

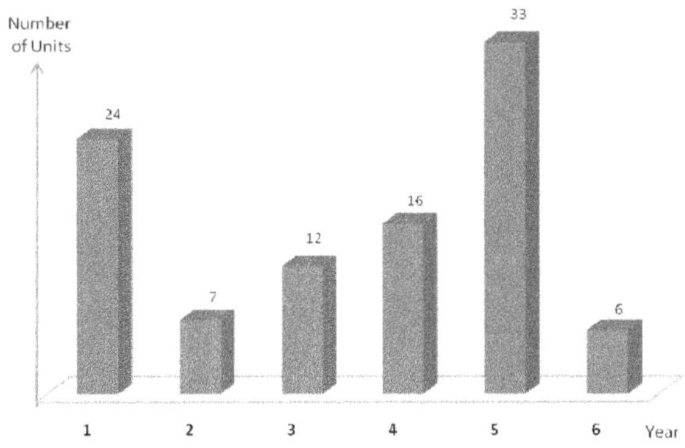

Figure 1. Curricular distribution of 98 compulsory ethics' related units.

granted to ethics' education ranged from 0.5 to 10 ECTS, although most of these figures refer to an entire syllabus and not to the teaching of ethics in particular. Additionally, some inconsistencies were seen between the number of ECTS and the disciplines' workload. One reason to explain this might be that sometimes information is referring only to the teaching sessions instead of the total workload, not including self directed study.

Of the 58 disciplines for which information on assessment methods was available, some type of formal assessment was always described. With only five exceptions, assessment involves at least one final exam (written, oral or both). Usually, assessment is mixed, involving combinations of exams, project/seminar work and involvement in classes.

Discussion

Study methodology

The website of FVE was chosen as the initial source of information for several reasons. First, the FVE is the European umbrella organisation that represents national veterinary boards. Second, the FVE, together with the European Association of Establishments of Veterinary Education (EAEVE), forms the European Committee on Veterinary Education (ECOVE) responsible for a trans-European evaluation system of veterinary faculties. And third, the FVE website offers in one resource (www.fve.org/education/index.html) contact information for 99 European veterinary schools.

The web search offers, however, some limitations. The way in which particular schools use and update their websites differs greatly and the quality and quantity of information provided is not always satisfactory, especially in terms of the unit descriptors. Although the majority of the information gathered makes reference to the present academic year (2009-2010), that was not always the case. The web search does not identify all of the relevant information that should be assessed and included in the data set (e.g. the academic background of the instructors and the exact amount of work dedicated to the teaching of ethics). In addition, as a result of using an electronic translation tool, some aspects of the ethics teaching in those languages not familiar to the researcher might have been missed.

General discussion

Although ethics was seen to be a part of the compulsory curriculum of 55 European veterinary faculties, attention given to this topic varies greatly from school to school. For example, the curriculum of the Faculty of Veterinary Science at the University of Thessaly (Greece) does not include a discipline of ethics but appears to provide some teaching regarding the ethics of laboratory animal use and the roles of the veterinary profession. On the other hand, the Latvian Veterinary Faculty and the University of Nottingham both include five different units with at least some ethical content. There is also great disparity in the contextual framing of the teaching of ethics, the time dedicated to it and its place within the curriculum.

The four units identified which include the teaching of ethics also seem to represent four different conceptual frameworks underlying the role of ethics in veterinary education. These frameworks are often used in combination with each other (Table 1). For some schools, ethics is addressed within the context of History of Veterinary Medicine while for others ethics is part of Animal Welfare Science and included within behavioural and/or welfare disciplines. A combination of ethics with Animal Law is frequently found. Here the normative and legal standards are given an ethical appraisal (codes of conduct and professional legislation, including welfare law, and forensic medicine). The word 'Deontology' is often found within this context with somewhat different meanings. It is not usually applied in a philosophical sense (e.g. referring to the approach of applying value principles over consequences), but to refer to the notion of the moral justifications underpinning normative regulations. Finally a group of units relating to 'Professionalism' were considered. Although there is currently no accepted definition of professionalism (Martimianakis *et al.*, 2009) in this context the term could be used to describe a number of topics connected with professional knowledge: sociology (including professional roles and professional behaviour); philosophy (including ethical theories); and management and organisation (including communication skills and conflicts resolution).

Some overlap was seen between a few of these concepts. Animal welfare legislation is found both within the context of Animal Welfare Science and of Animal Law. Codes of Professional Conduct were identified to be as much a part of the legal discourse as they were an important aspect of professional knowledge.

The study data indicates notable differences in the contextual approaches used in the teaching of ethics. There is, however, a considerable number of schools (n=25) addressing ethics purely in relation to Animal Law (Table 1). Although ethics is found in every curricular year it is especially taught at the beginning and at the end of the study programme (Figure 1). A combination of lectures with practical sessions is the teaching method most widely used (n=29), followed by the exclusive use of lectures (n=21). Mixed approaches combining lectures, practical sessions, seminars and field trips are less common. With only few exceptions, assessment involves at least one final exam.

The diversity of combinations in which ethics is seen to be conceptually framed (Table 1) could indicate that there is no explicit common aim in undergraduate veterinary ethics education across European Faculties. Formal teaching in ethics was not explicitly set out in the original undergraduate veterinary programme and there seems to be little consensus on how, when and where to include it within the overall curriculum. Our findings are in line with the results of the previous survey on the inclusion of animal bioethics courses in Europe (Edwards, 2002; Gandini and Monaghé, 2002; von Borrell, 2002). This previous study – which took place before the EU enlargement (accession of the ten new Member States in 2004 and 2007) and before the Bologna Process (2005) – also found inconsistencies in ethics education in terms of dedicated hours, availability and course content.

The inclusion of elective disciplines of bioethics is found in some veterinary faculties (e.g. Thessaloniki, Greece; Teramo, Italy; Torino, Italy). However, competition with more practical electives could reduce the impact of ethics with this educational strategy. The existence of some other interesting elective disciplines (e.g. 'Hunting Law and Ethics' offered by Budapest, Hungary and 'Animals, I care for eating them'[1] from Utrecht, the Netherlands) should also be noted.

Currently, there is no systematic approach that can be applied to identify what is understood by *ethics* in veterinary education. This web-based analysis is the first step in a broader study which aims to characterize ethics teaching to veterinary students, and the hypotheses generated from this work will be further explored using complementary qualitative approaches.

References

Bologna Process (2005). The framework of qualifications for the European Higher Education Area. In: Bergen Conference of European Ministers Responsible for Higher Education. May 19-20, 2005. Bergen, Norway. Available at: http://www.bologna-bergen2005.no/EN/BASIC/050520_Framework_qualifications.pdf. Accessed 22 March 2010.

[1] Free translation from the Dutch original: 'Dieren, ik vind ze om op te eten'.

Edwards, S.A. (2002). A synthesis of animal bioethics teaching in agricultural and veterinary courses in Northern Europe. In: Proceedings of the AFANet Workshop: Teaching Animal Bioethics in Agricultural and Veterinary Higher Education in Europe. May 23-24, 2002. INPL, Nancy, France, pp. 57-61.

Downie, J.R. and Clarkeburn, H.M. (2005). Approaches to the teaching of bioethics and professional ethics in undergraduate courses. Bioscience Education (BEE-j 5-2). Available at: http://www.bioscience.heacademy.ac.uk/journal/vol5/beej-5.2.pdf. Accessed 1 Dec. 2009.

Gandini, G. and Monaghé, A. (2002). Teaching animal bioethics in agriculture and veterinary higher education in Europe: a survey in Belgium, France, Italy, Portugal, Slovenia and Spain. In: Proceedings of the AFANet Workshop: Teaching Animal Bioethics in Agricultural and Veterinary Higher Education in Europe. May 23-24, 2002. INPL, Nancy, France, pp. 10-12.

Magalhães-Sant'Ana, M., Baptista, C.S., Olsson, I.A.S., Millar, K. and Sandøe, P. (2009). Teaching animal ethics to veterinary students in Europe: examining aims and methods. In: Millar K., Hobson-West P. and Nerlich B. (eds.) Ethical Futures: Bioscences and Food Horizons. Wageningen Academic Publishers, The Netherlands, pp.197-202.

Martimianakis, M.A., Maniate, J.M. and Hodges, B.D. (2009). Sociological interpretations of professionalism. Medical Education 43: 829-837.

Thornton, P.D., Morton, D.B., Main, D.C.J., Kirkwood, J.K. and Wright, B. (2001). Veterinary Ethics: filling a gap in undergraduate education. Veterinary Record 148: 214-221.

Von Borrell, E. (2002). A synthesis of animal bioethics teaching in agriculture and veterinary courses in West and Central Europe. In: Proceedings of the AFANet Workshop: Teaching Animal Bioethics in Agricultural and Veterinary Higher Education in Europe. May 23-24, 2002. INPL, Nancy, France, pp. 34-39.

Food ethics dilemma: an open learning resource for teaching ethics

A. Algers[1], M. Kaiser[2], C. Kallum[3], H. Loor[3], K. Wahlgreen[4] and S. Welin[5]
[1]Swedish University of Agricultural Sciences, Food Science, P.O.Box 234, 53223 Skara, Sweden; [2]University of Bergen, The Centre for the Study of the Sciences and the Humanities, P.O.Box 7805, 5020 Bergen, Norway; [3]Imcode Partner AB, Skeppsbron 24, 62157 Visby, Sweden; [4]The Danish Meat Trade College, P.O.Box 209, 4000 Roskilde, Denmark; [5]University of Linköping, Health and Society, 58183 Linköping, Sweden

Abstract

Ethical issues arise in any food chain and the depth and range of ethical aspects of the 'history' of foods are often hidden, but the potential for opening up this information is considerable. Climate change, animal welfare, fair trade, health and safety, novel foods and GM foods, sustainability and use of natural resources are all important dimensions within the complex concept of food ethics. It is also important to note the various levels of actor perspectives involved in food ethics. Stakeholders in the food chain prioritise and interpret ethical concerns differently. Consumers (micro-level) are increasingly worried about unethical aspects of the food chain. Restaurants, food chains and distributors (meso-level) will strongly influence market availability, and national and international policy (macro-level) can influence through agricultural production incentives/subsidies. For all these levels, ethical questions of good and bad decision-making can be asked. The different stakeholder perspectives will be the point of departure when developing an open learning resource on food ethics dilemmas. The web-based learning resource will be free of charge enabling open access to all members of society and with a multi-player function making it possible to choose an avatar (e.g. a farmer, a manager at a food industry, or a consumer) and play the game in real time against another player with a different stakeholder perspective. The avatar is placed in a situation with a dilemma and the player has to make choices between alternative actions, which bring the avatar to new and more intricate dilemmas. Having played the game from one avatar of stakeholders' view the user can choose to look at the same dilemma from another stakeholder's point of view. Through the learning resource on food ethics players will obtain a better understanding both of their own ethical views and those of others.

Keywords: teaching ethics, e-learning, open educational resources, decision-making

Teaching food ethics

Higher education is presently going through a period of rapid change including increasing globalization, the very rapid pace of knowledge creation, and the use of new information technologies which makes it much easier to share new knowledge globally. At the same time higher education institutions in general have to handle larger and more diverse groups of students which also call for new ways of organising learning (Biggs, 2003).

At the same time the need for education and training in ethical decision making is apparent and imperative (Crane, 2004; Giacalone, 2007). Individuals need to be assisted to become more aware of the complexities surrounding ethical decision making and more conscious of their own ethical orientation.

It is essential for all professionals involved in the food sector to be able to engage in dialogue about the ethical implications of food production, processing, policy, supply, consumption and health. The main target group for the learning resource is life science students in higher education but the target audience is unrestricted and the learning resource will help to inform societal debate about ethical issues associated with food consumption, food production and in general food cultures.

It is evident that most of the European consumers have concerns related to their choice of food (cf. e.g. Flash Eurobarometer 256, 2009; Special Eurobarometer 294, 2008). Passionate debates on biotechnology, stem cell research and cloning have shown that there is a need for open public dialogue on the ethical implications of scientific advances (EurActive Network, 2010).

The ethical debates over food are very diverse. There are many points of departure and this tool cannot cover them all. One part of the ethical debate is what values are most important and should be considered. Regardless if you are a Kantian or more utilitarian (or something else) you have to make up your mind about the importance of various values involved.

Climate change, animal welfare, fair trade, health and safety, novel foods and GM foods, sustainability and use of natural resources are all important dimensions within the complex concept of food ethics. The production, processing and trade of every food item often include value conflicts. One aim of the tool is to help the user to get a better awareness of his or her own value hierarchies. By 'playing' different stakeholders the tool will also give a deeper understanding of different value perspectives.

Open educational resources

A culture of sharing digital learning resources over the internet, openly and for free, is spreading internationally (OECD, 2007). The benefits of open educational resources (OER) are many including aspects such as democracy and preservation of public education. From a more individual point of view, open sharing increases publicity, reputation and the pleasure of sharing with peers. It also means broader and faster dissemination and enhances the motivation of students since they take part in problem solving and the intellectual development of learning resources.

An open learning resource that is accessible for free on the Internet not only addresses students in higher education but can also be used for capacity building of individual learners. This will give individuals an opportunity to reflect on their own ethical decision making and ultimately make individual reasoning more sophisticated and improved as the individual becomes more cognitively aware of moral and ethical principles. Increased knowledge combined with clarification of ethical issues may in the long run make consumers more cable of making informed choices of food items.

Video, audio and photo are media carrying a great amount of information which contextualise food ethics and therefore facilitate the understanding of food ethics dilemmas. Developing learning resources based on these media and making them accessible at the Internet can be of benefit not only to teachers and learners but may also have global impact on the way we produce food.

There are a variety of ways of motivating student learning that should be taken into consideration. Interactivity is one way to introduce reflexion and Reiss (2005) considers case studies to be highly motivating, because students perceive them to be relevant, especially when they are based on real life events. Through the interaction with other players the users will obtain a better understanding both of their own ethical views and those of others.

Food production is a global endeavour and every citizen is also a consumer of food. There is an increased discussion on food issues, among them also ethical questions. Higher education institutions and especially life scientists and philosophers play an important role in providing information and guidance on food ethics to students, citizens and consumers.

Multiplayer role game

In public debates on policy and ethics there are many possible standpoints. By setting up a multiplayer role game the user will not only have to make his or her own decision and argue for that, but will also meet other people's opinions and have to understand their way of thinking and be able to argue with them.

Arguing and discussing with other players also gives a stronger sense of reality, as the arguments given are not standard arguments created by a computer, but living arguments created by humans, thus being much more flexible and real.

Initially, the user will have to choose an avatar standing for a stakeholder in the food chain – e.g. a farmer, a manager at a food industry, a consumer or other people taking an active interest in the issue. Five different stakeholders are involved in each of the four cases and these stakeholders are described in a way that makes it possible for the user to identify him- or her self. Photographs and videos are used to strengthen the identification process. The learning resource also includes technical information about e.g. production methods, statistics on trade and consumption as well as knowledge on animal welfare and health issues.

A scenario based on an ethical dilemma involving every stakeholder is presented for the players. The situation is reinforced by figures showing statistics and engaging visual material. In order to learn argumentation technique, the whole process is transparent between the players. A player can always peek on another player's argumentation and thus learn.

The role game develops in three levels. On level 1 the users are asked to write a short analysis in a chat box on the consequences of the current situation on their organisation and argue for the position you wish your organisation to take in the matter. This analysis forces the player to go deeper into the material and reflect upon the consequences of the situation given. By asking the player to analyse in writing, two goals are achieved: the player will analyse the material using available sources and the players' views and argumentation can be evaluated by the other players. This evaluation takes place between every level. In this situation the players have to evaluate the strength in each stakeholder's argumentation, including own argumentation. The evaluations are transparent and every stakeholder will be able to see the evaluation they got from all other stakeholders.

On level two the scenario has developed into a more intricate dilemma. This new situation is presented automatically based on the evaluations in level one. After this evaluation the players meet in pairs for a real time interaction using chat boxes; making it possible to argue and defend your personal position. It is also possible to follow the argumentation between other pairs of stakeholders. Again, this is followed by an evaluation of each stakeholder's opinions about the strength of arguments.

At level three the player will receive the information that 'You have now been called to attend a discussion with all parties'. This real-time interaction will take place between all stakeholders. The discussion should lead participants to a decision or conclusion. On the third level the discussion is more complex and every stakeholder can bring in their emotions in the scientific argumentations.

The end of this multiplayer game depends entirely on the outcome of the discussion between the players. Once again the evaluation and argumentation done by the players will result in different endings of the case, and will follow the argumentation that had the strongest 'score' given by the users' evaluation.

Basically the game engine will be a discussion forum, with preset stories and open discussions. This engine will also contain modules for evaluation and for controlling the time spent by the users. It is essential that there is a mechanism for deadlines. Every level can close at a given time, thus forcing the players to act. If they do not act, their argumentation will be lost, thus receiving very low scores in the play. Having played the game from one avatar of stakeholders' view the user can choose to look at the same dilemma from another stakeholder's point of view.

From a pedagogical viewpoint, we expect that the combination of players that strongly identify themselves with their avatar and real discussions with other players will result in high motivation to go through the whole process. This in turn will create a high degree of understanding for the complexity that decision makers stand before.

Cases

A critical element is the creation of scenarios that present real-life ethical choices and which contain varying degrees of certain contextual variables like opportunity/risk, personal incentives and chance of getting revealed/caught (Perri *et al.*, 2009).

Three cases will be developed for the first version of the learning resource. The first case, the master case, will be developed by the steering group whereas the coming three cases will be developed by members of the Nordic Network of Agriculture and Food Ethics. The approach when writing cases is deliberately to use basic dramaturgical notions such as 'conflict', 'turning point', 'identification', and 'driving force' to enhance student motivation.

The master case will focus on ethics of aquaculture. Aquaculture is the fastest growing sector in global food supply, fish is an efficient feed converter and is healthy but several ethical problems are related to aquaculture according to Kaiser (2010):

- Natural resource from the oceans, that is already on the brink of overfishing, is used to feed aquaculture products.
- Pollution from aquaculture like discharge of nutrients and organic waste, excess fish-feed and medication.
- Aquaculture increase the outbreak and spread of diseases.
- Cross-breeding between wild and farmed stocks may lead to loss of genetic variation with negative effects on performance traits.
- Production sites for fish farms compete for the use of valuable areas of the coastal zone for which alternative uses may produce higher net-value.
- Increased food miles and energy use because products from aquaculture are mostly traded internationally.
- Animal welfare because fish most likely do feel pain and can suffer.
- Equity and fairness for the people working along the chain (corporate social responsibility) – fairly shared profit and good work conditions.

Aquaculture relates to fishery like agriculture relates to hunting and gathering. In other words, it is the controlled and planned production of aquatic bio-products for future harvesting, rather than going after what one finds in nature (Kaiser, 2010).

In recent years it has been found that fish most likely can feel pain and thus should be included in the moral circle (Lund *et al.*, 2007) and education about the nature and capacity of fish should be promoted. This is another argument for the development of a case on ethics of aquaculture. On the other hand, ethical evaluations are not that clear-cut or obvious and the case thus leaves room for various arguments.

Other cases that have been suggested are cases based on the production, processing and trade of tomatoes and broiler chickens. According to the idea of open educational resources users can in the final version of the role game develop their own scenarios.

Nordic Network of Agriculture and Food Ethics

The Nordic Network of Agriculture and Food Ethics is funded by NordForsk with the purpose to create a network, to develop cross disciplinary research education, to enhance knowledge, interest and participation in the ethical discussions and to develop an e-learning package tailored for the teaching activities addressed by the network.

Therefore, members of the network have been invited to collaborate on the development of case studies and approximately ten of the members have signed up for the task. During the coming year three cases will be developed by the network and added to the learning resource on these domains: www.foodethicsdilemma.net or www.fedilemma.net.

Every case has to be thoroughly worked through so that the activities and opinions reflected are realistic and clear. The cases also have to consider, and be created according to, basic dramaturgical rules. This requires a dynamic development group like people from the Nordic Network of Agriculture and Food Ethics, having diverse backgrounds and knowledge in philosophy/theology as well as natural and social sciences.

Acknowledgements

We acknowledge our gratitude to the Nordic Council of Ministers for funding the project activities (contract number HZ-2009_1-17559)

References

Biggs, J. (2003). Teaching for quality learning at university. What the student does. 2nd ed. The society for Research into Higher Education & Open University Press. Berkshire, UK. 309 pp.

Crane, F.G. (2004). The teaching of business ethics: an imperative at business schools. Journal of Education and Business. Vol. 79, pp 149-151.

EurActiv network (2010). Available at: http://www.euractiv.com/en/science/science-ethics/article-117537. Accessed 22 February 2010.

Flash Eurobarometer (2009). Flash Eurobarometer 256, 'European's attitudes towards the issue of sustainable consumption and production'.

Giuacalone, R.A. (2007). Taking a red pill to disempower ethical students: Creating ethical sentinels in business schools. Academy of Management Learning and Education. Vol 6, pp. 534-542.

Kaiser, M. (2010). The ethics and sustainability of aquaculture. Forthcoming in: David Kaplan (ed.), The Philosophy of Food, University of California Press.

Lund, V., Mejdell, C.M., Röcklinsberg, H., Anthony, R. and Håstein, T. (2007). 'Expanding the moral circle: farmed fish as objects of moral concern', Diseases of aquatic organisms, vol. 75: 109-118.

OECD (2007). Giving knowledge for Free – the Emergence of Open Educational Resources, no 3, pp. 153. Paris, France. Available at: www.sourceoecd.org/education/9789264031746. Accessed 22 February 2010.

Perri, D.F., Callanan, G.A., Rotenberry, P.F. and Oehlers, P.F. (2009). Education and training in ethical decision making: comparing context and orientation. Education + Training. Vol. 51, no 1, pp. 70-83. Available at: www.emeraldinsight.com/0040-0912.htm. Accessed 22 February 2010.

Reiss, M.J. (2005). Teaching animal bioethics: pedagogic objectives. In: Marie M, Edwards S, Gandini G, Reiss M and von Borell E (eds) Animal Bioethics: Principles and Teaching Methods, Wageningen Academic Publishers, the Netherlands. pp. 189-201.
Special Eurobarometer (2008). Special Eurobarometer 294, 'Europeans, Agriculture and the Common Agricultural Policy'.

New version of animal ethics dilemma allows students to make use of science, drama and images in ethics learning

A.J. Hanlon[1], A. Algers[2], T. Dich[3], T. Hansen[3], S.B. Christiansen[3], H. Loor[4] and P. Sandøe[3]
[1]School of Agriculture, Food Science & Veterinary Medicine, University College Dublin (UCD), Veterinary Sciences Centre, Belfield, Dublin 4, Ireland; [2]Swedish University of Agricultural Sciences (SLU), Sweden; [3]Faculty of Life Sciences, University of Copenhagen, Denmark; [4]imCode Partners AB, Visby, Gotland, Sweden; Alison.Hanlon@ucd.ie

Abstract

This paper presents and discusses a new feature in the use of the learning tool Animal Ethics Dilemma, which allows students to construct their own cases. The learning tool www.aedilemma.net aims to teach students to view animal issues from different perspectives of animal ethics by confronting them with ethical dilemmas where they have to choose between different options. They will be put into the role of a veterinarian or another animal science professional responsible for making choices, which typically involve potential harms to animals and potential benefits to humans. Through their choices the students will display and be confronted with utilitarian, animal rights or three other views on animal ethics. The tool also allows the students to reflect on their choices and the underlying ethical viewpoints. A new option in the use of the tool is to let the students construct their own cases. This option is made possible by a content management system. Thus teachers may set students the task to create their own case. This could serve as an exam task through which the students document their ability to understand and actively make use of different points of view in animal ethics and their ability to recognize and develop ethical dilemmas. This new feature gives students the opportunity to develop case studies on a topic relevant to their circumstances and from a pedagogic perspective may help to support deeper learning and understanding of animal ethics by requiring a kinesthetic or an active approach. It also gives the opportunity to work with animal ethics in a way, which involves competences in collecting relevant scientific information, and it enables students to be creative through the use of drama, images and graphical design. Thereby teaching of animal ethics may integrate scientific and philosophical disciplines and be focused on more than just abstract reasoning.

Keywords: teaching, open educational resource (OER), role play, case study

Introduction

Animal ethics is an obligatory component of veterinary training in Europe. As a philosophical discipline it differs to most of the other subjects within veterinary training, requiring new skillsets such as debating skills and thus is likely to require a different approach to teaching. Accordingly a new online teaching resource, www.aedilemma.net (AED) was launched in 2006. This programme has been described elsewhere (Hanlon *et al.*, 2007, 2009). Overall it has been designed to illustrate five ethical points of view (contractarian, utilitarian, animal rights, relational and respect for nature) towards five ethical issues involving animal use, presented in a case study format. The case studies were designed to challenge the student user, by demonstrating the conflicts and dilemmas arising from a variety of ethical issues. To further engage the students, a new online template has been developed to enable students to create their own case studies.

Pilot of the case study template

The template was developed by ImCode and made available on a development website in February 2010, to be piloted by 19 undergraduate students (Level 1, Stage 2) registered for an elective module entitled 'Animals in Society: Ethical Perspectives' (VET10020). The majority (16/19) of the students were registered for the Veterinary Medicine Programme at University College Dublin (UCD). The module was designed and sequenced to familiarize students with the five ethical points of view mentioned above and in addition, students were requested to complete two case studies in AED, as a specified learning activity (Figure 1). The case study was assigned 30% of the continual assessment marks for the module. The main recommended textbook, Ethics of Animal Use (Sandøe and Christiansen, 2008), provided a similar ethical framework to AED, thus further reinforcing the theory and application of animal ethics.

In January 2010, the authors started preparing a set of guidelines for using the AED case study template. The guidelines contained both text and screen shots from the programme, to help guide the students. They were further modified and published on Blackboard, the virtual learning environment used at UCD, on 23 February 2010. In addition an eLecture presenting the guidelines was prepared using Articulate Presenter (www.articulate.com), to further describe the functions of the template, and broadcast in Blackboard. The presentation contained audio, an attachment to the guidelines and a hyperlink to the development website.

To support the preparation of the case studies by students, two initiatives were taken:

1. Students were requested to submit a synopsis of their proposed topic for their case study into a 'wiki' created specifically for the case studies in Blackboard. This allowed students to start thinking about their case study well in advance of the submission deadline and view each other's topic, thus avoiding repetition.
2. A discussion 'thread' was created in Discussion Forum of Blackboard to give students an opportunity to discuss issues relating to the case study. However, students did not use this facility.

Template design

The template provides a systematic approach to ethical analysis. The first step is to establish the framework ('settings') of the case study. Students and teachers can decide on the complexity of the case study by selecting the number of levels (max. four levels). At each level students have to write a statement ending up with an ethical dilemma and four or five responses to the dilemma. Statements and responses have to be written from one of the ethical perspectives. There is also an option to deactivate one of the five ethical perspectives, if this is considered appropriate to the case. Once the framework has been established, the

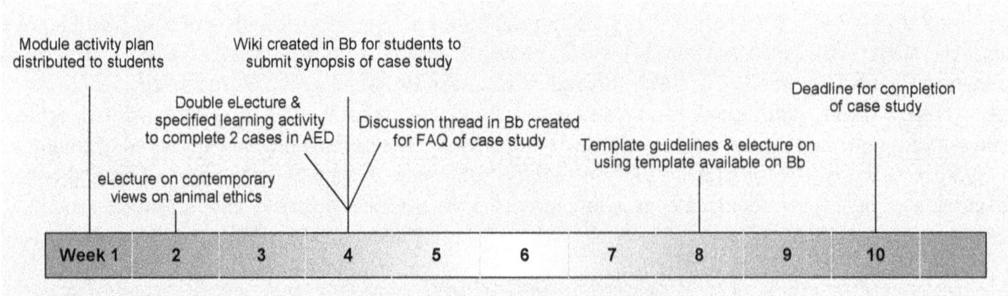

Figure 1. Sequence of the introduction of the online case study template of Animal Ethics Dilemma (AED) to first year students studying animal ethics (VET10020) (Bb=Blackboard).

students can start to add content in the 'Case Overview' page. The Case Overview page provides access to the Questions (ethical dilemmas) and Answers (ethical responses) for each level of the case study and to the 'Info-page' (provides background information about the case study).

An edit button provides access to the text editor and other tools such as creating hyperlinks to text and photographs. The editor tools have similar features to other word processors. Other features include 'Page Status' (i.e. indicates if the content is complete or incomplete) and 'View Page', which shows what the page will look like.

When the case has been completed, students should add a Case Code, which will be assigned by the teacher and then 'Activate' their case, both functions are on the Case Overview page. By assigning the same Case Code to the class, students will be able to view each others cases.

Pedagogic value of the case study template

One of the main objectives of the template is to support a greater understanding of ethical issues, ethical dilemmas and different points of view. The template enables an active learning or kinesthetic approach (Fleming, 2001) to engaging with the theory and application of animal ethics. It also has the advantage of allowing students to focus on a topic of interest or relevance to their life and thus may be more stimulating. Topics chosen by the 2010 students include zero-grazing of dairy cows, ethical issues involving animal shelters, fox-hunting, the price of a pedigree dog, Australia's red fox population, horse racing and animals in zoos.

From a teacher's perspective, the template provides a framework for ethical analysis and supports the expression of multiple ethical perspectives. By so doing, it should increase the ethical awareness and understanding by students. Special attention should be given to the guidelines, to provide simple instructions in both text and pictorial formats.

Conclusion

Overall, the case study template provides a structured framework for students to conduct an ethical analysis of an issue of their choice. It supports an active learning approach and may therefore help to improve the understanding and application of ethical theory relating to animal use.

References

Fleming, N.D. (2001). Teaching and Learning Styles: VARK Strategies.

Hanlon, A.J., Algers, A., Dich, T., Hansen, T., Loor, H. and Sandøe, P. (2007). 'Animal Ethics Dilemma': an interactive learning tool for university and professional training. Animal Welfare, Volume 16 (S): 155-158.

Hanlon, A.J., Algers, A., Dich, T., Hansen, T., Loor, H. and Sandøe, P. (2009). www.animalethicsdilemma.net: a learning resource for university lecturers in animal sciences and veterinary medicine. In: Ethical futures: bioscience and food horizons. Wageningen Academic Publishers, The Netherlands, pp.187-191.

Sandøe, P. and Christiansen, S.B. (2008). Ethics of Animal Use. Blackwell Publishing.

Shifting minds? The effect of training in laboratory animal science on researchers' attitudes to animal use in biomedical research

N.H. Franco[1,3], J.A. Nunes[2] and I.A.S. Olsson[1]
[1]IBMC-Instituto de Biologia Molecular e Celular, Universidade do Porto, Rua do Campo Alegre, 823, 4150-180 Porto, Portugal; [2]Center for Social Studies, University of Coimbra, Colégio de S. Jerónimo. Apartado 3087, 3001-401 Coimbra, Portugal; [3]Institute of Biomedical Sciences, University of Porto, Largo Prof. Abel Salazar, 2, 4099-003 Porto, Portugal; nfranco@ibmc.up.pt

Abstract

Animal use in biomedical research presents an ethical dilemma: between the potential benefits for human health and the costs borne by the animals. The 3Rs principle – *Replacement, Reduction, Refinement* – emphasizes the need not only to replace animal experiments with animal free alternatives, but also to reduce the number of animals used and promote their welfare as a compromise solution whenever animal experiments are inevitable, aiming overall for a more positive ethical balance. Even with existing regulation and supervision of animal use, the individual researcher's responsibility is still decisive in implementing the 3Rs. Training in laboratory animal science aims to raise researchers' awareness and increase their knowledge, but its effect on scientists' attitudes has not so far been systematically assessed. Sixty-one participants in three FELASA Cat-C courses in Portugal were surveyed in a self-administered questionnaire. Questions related mainly to the 3Rs and their application, attitudes to animal use and the ethical review of animal experiments. One year later, all respondents were asked to answer a similar questionnaire (46% response rate) with added self-evaluation questions on the impact of training. Overall, these preliminary results suggest that the course is effective in promoting awareness and increasing knowledge, in particular as regards Refinement. However, participation in the course did not change perceptions of the current and future need for animal use in research.

Keywords: 3Rs, ethics, education, survey

Introduction

> '...*the freedom of choice of the experimenter is often much wider than it at first appears. The full use of this freedom is the mark alike of humane and successful experimentation.*' (Russell and Burch, 1959, cited in Balls, 2009)

In their classic book, Russell and Burch acknowledged the important role that each researcher plays in applying the 3Rs of animal research. They urged researchers to express this *freedom* with responsibility – now often described as the '4th R of research' (Banks, 1995) – towards both science and animal welfare. Nowadays, this responsibility is distributed amongst various intervenients in the scientific process, including regulators and inspectors, animal caretakers and veterinarians, funding agencies, scientific journal editors and ethics committees (Smaje *et al.*, 1998). Nevertheless even with extensive regulation, scientists are ultimately those deciding how animals are treated; thus their attitude is crucial for the implementation of the 3Rs.

In order to work with animal models, researchers are required to attend specific training (EC, 1986); this training standardly consists in laboratory animal science courses following FELASA guidelines for categories B or C (Nevalainen *et al.*, 2002). These courses – which have the 3Rs as an overarching principle – aim to develop the skills considered necessary for the responsible use of animals in work maintaining high scientific standards (FELASA, 1995); they cover theoretical, practical and ethical aspects of animal research.

The present study aimed to assess the attitudes of FELASA course participants to the use of animals in the life sciences – as well as its regulation and supervision – and to evaluate the impact of this formal training on how the participants view and apply the 3Rs.

Material and methods

A self-administered survey was delivered to all participants of three FELASA Category C courses held in 2008 (N=61) at research institutes in the Portuguese universities of Porto (IBMC), Coimbra (CNC) and Braga (ICVS). The survey was divided in two parts, the first presented before the introductory lecture presenting the 3Rs and the other immediately after. In the first part, the participants were asked about their degree of awareness of the 3Rs. The second part covered general attitudes to animal use and the 3Rs, as well as the ethical review of animal experiments. One year later, the participants were asked to respond to the same questionnaire (46% response rate), which was presented online, and to answer a few additional questions concerning their evaluation of the course and its impact on their attitudes to the 3Rs.

Results and discussion

General attitude

At the start of the course, 62.3% of the researchers indicated that they were unaware of the 3Rs, 9.8% claimed to know of them but were unable to identify the 3Rs, and 27.9% were able to list or explain these principles. In the post-test, the percentage of respondents aware of the 3Rs increased to 96%, and researchers reporting that they had applied the 3Rs rose from 27% to 65%. Moreover, 39% reported that the course had made them more receptive to the 3Rs, 22% said that it helped them improve their previous efforts in implementing the 3Rs, and 22% stated that although they already had that concern before they lacked the appropriate knowledge to apply the 3Rs until attending the course.

Replacement, Reduction and Refinement

Attending the course does not appear to change perceptions of the current and future need for animal use in research. In their own research fields, the researchers classified animal research as either absolutely necessary (41% before, 39.3% after the course), very important (47.5% before, 50% after) or important (11.5% before, 10.7% after). Further, regarding possible replacement of animals in their research field over the next 50 years, before the course 29.5% (32% after) expected 'no changes', 62.3% (64.3% after) 'some replacement', 4.9% (3.6% after) 'replacement of most procedures' and 3.3% (none after) expected 'full replacement' of animal use. When faced with a hypothetical either/or Reduction/Refinement conflict (i.e. either 20 tests on one animal *or* one test on each of 20 animals), 59.5% (65.2% after) favoured the latter, i.e. refined methods. After the course, 74% of the respondents also declared that they preferred social housing over individual housing, even when that meant increasing the total number of animals used (question not asked before the course).

Ethics and review

When asked in the first survey to rank different species (*Dog*, *Octopus*, *Horse*, *Pig* and *Rat*) according to the severity of ethical dilemma presented by their use, 27.9% declared that they were not able to perform this task. Amongst those who established a hierarchy, the octopus was regarded as most or second-most ethically acceptable by 59.1% and 27.3%, respectively), while 27.9% and 54.6% respectively opted for the rat. The horse and the dog were ranked as 'Least ethically acceptable' by, respectively, 47.7% and 38.6%. The most consensual rank was that of the pig, positioned in third place (middle of the scale) by 70.5%. The great majority of the researchers (81.8%) cited their own 'emotional relationship' with the species as the main criterion used when estimating ethical acceptability. Interestingly, this human-centered justification was given much more importance than those referring to capacities of the animal species itself: 'Estimated degree of sentience' (27.3%), 'Phylogenetic proximity' (40.1%), 'Intelligence' (34.1%) and 'Self-Awareness' (36.7%).

On the regulation and supervision of animal use, prior to the course only 36.1% considered the 'National Veterinary Office' (DGV) as the body that should be responsible for the ethical evaluation of experimental protocols (despite this actually being the competent authority in Portugal). This rose to 52.2% after the course, in which, note, legislation and regulation was addressed. Nevertheless, 70.5% before, and 82.6% after, the course would prefer an institutional ethics committee to be responsible for reviewing protocols on animal experiments, and before the course 39% (43% after) considered that review by this committee should be mandatory, while 61% (57% after) favoured the committee making non-mandatory recommendations.

The great majority (91.3%) of respondents considered animal ethics to be a relevant subject for the course.

Conclusions

These preliminary data suggest that, prior to lab animal science training, a significant proportion of life sciences researchers are unaware of the 3Rs of animal research. This is a problem the courses seem to effectively overturn. In both surveys the relatively high percentage of respondents not expecting any changes in animal use in their field of research suggests a degree of reluctance to acknowledge alternatives to animal models in research.

When confronted with Reduction/Refinement dilemmas, the majority of respondents would rather refine, and thus use more animals, if that meant that each animal would suffer much less. Despite its subjectivity (or perhaps because of it), the affective relation with some species was seen as a more relevant *criterion* in ethical evaluation of their use than criteria based on the estimated level of cognition or sentience of the different species.

Overall, these preliminary results suggest that the FELASA-based course is effective in promoting awareness and improving knowledge, in particular as regards Refinement.

References

Balls, M. (2009). The Three Rs and the Humanity Criterion – an abridged version of 'The Principles of Humane Experimental Technique'. FRAME, Nottingham, UK, pp 89.

Banks, R.E. (1995). The 4th R of research. Contemporary Topics in Laboratory Animal Science 34(1):50-1.

European Commision (1986). Legislation. Oficial Journal of the European Council L358; Volume 29. 6/609/EEC; Article 14. Strasbourg.

FELASA (1995). FELASA recommendations on the education and training of persons working with laboratory animals: Categories A and C. Reports of the Federation of European Laboratory Animal Science Associations Working Group on Education accepted by the FELASA Board of Management. Laboratory Animals 29, 121-31.

Nevalainen, T., Blom, H.J., Guaitani, A., Hardy, P., Howard, B.R. and Vergara, P. (2002). FELASA recommendations for the accreditation of laboratory animal science education and training. Laboratory Animals 36(4):373-7.

Smaje, L.H., Smith, J.A., Combes, R.D., Ewbank, R., Gregory, J.A., Jennings, M., Moore, G.J. and Morton, D.B (1998). Advancing refinement of laboratory animal use. Laboratory Animals 32(2):137-42.

Policy, ethics and animals use; the need for training: a Dutch example of training civil servants in ethics

F.R. Stafleu[1], H.M.J. van Rij[2] and F.L.B. Meijboom[1]
[1]Ethics Institute, Utrecht University, Heidelberglaan 8, 3584 CS Utrecht, the Netherlands; [2]Ministery of Agriculture, Nature, and Food Quality, Direction Food, Animal and Consumer, Cluster Ethiek, P.O. Box 2040, 2500 EK Den Haag, the Netherlands; f.r.stafleu@uu.nl

Abstract

The Ministry of Agriculture, Nature and Food Quality in the Netherlands is more and more confronted with ethical issues concerning animal welfare, animal health and food production. The ministry wants to make the ethical and societal topics in its policy making more explicit and transparent and wants to communicate this to the outside world. For that reason it intends to incorporate ethical deliberation into its policy making process. The civil servants, working on animal welfare, animal health and food production, would like more ethical expertise to incorporate ethics in their work. For that reason the Ethics Institute was asked to develop an ethics course aimed at the practical implication of including ethics into policy making. The Ethics Institute developed a course focused on (1) the raising of consciousness concerning the ethical dimensions of policy making and the personal ethical orientation of the civil servants, (2) development of knowledge concerning ethical theories and how to incorporate these theories into the practice of policy making, (3) the development of skills on how to deal with the ethical dimension in the practice of policy making. The training was grounded in the idea of ethics as a reflection process represented in a model of a version of the reflective equilibrium. The participants of the course were asked to describe the matters of policy in which they were confronted with ethical issues. These descriptions were used as guidelines in the course. The course lasted three days: Two successive days of training and workshops. These two days were followed by training on the job concerning topical problems of law making. Although it appeared that civil servants work in an environment which is mainly focused on solutions and results rather than on underlying moral values, the result of the training was that they started to incorporate ethical reflection into their work.

Keywords: policy, training, reflective equilibrium

Introduction: moral plurality and government policy

In Europe there is a plurality of answers with respect to the question of how we ought to deal with animals and on best practices in food production. In daily practice this hinders (internationally) coordinated responses and decision-making with respect to a whole range of issues related to animal use. This situation is not so much a problem of knowledge and information, but has its origin in a fundamental plurality of moral values in European societies. This moral plurality becomes evident in the context of policy and law making, because decision-making asks for choices that cannot be value neutral. Especially for the policy makers this raises a fundamental problem. On the one hand, policy is often praised for making and preparing clear and effective solutions to societal problems. On the other hand, policy makers are not and should not be those who settle all moral diversity in society. As a consequence, policy makers need to deal with this tension between the imperative of the practical decision-making and the respect for societal plurality. This asks for an ethical competence of policy makers. Both the tension and need for ethical competence have been recognized by the Netherlands Ministry of Agriculture, Nature and Food Quality. This resulted in an open tender that invited a number of Dutch universities to offer a training program on ethics. At the end of this process, the Ethics Institute of Utrecht University together with Wageningen UR Livestock Research and Wing BV were asked to develop a training program in ethics with a special focus on issues of animal use and to train about 50 staff members of the Ministry.

Ethics as a process of reflection: awareness, knowledge and skills

Concrete questions and specific situations are often the basis for the need for ethical reflection. For that reason it is useful in practical ethics to start with these concrete questions. However to get a practical answer on these questions is not the only aim of the ethical process. Also the reflection on the quality of the underlying arguments is an integral part of it. This is also true for practical ethics aimed at policy making and therefore we define ethics as a reflection process. The outcome of policy making will not only be judged by the fact that a moral choice has been made, but mainly by the choices that have been made and the underlying considerations. So the focus should not be on the answers alone, but also on the process. For this reason we based our training on the concept of ethics as a reflection process consisting of three items: Awareness, knowledge and the development of skills

Awareness

Awareness is the first essential step. We use awareness in two senses: (a) awareness of the implicit moral dimension of a policy question and (b) awareness of the personal and societal moral convictions concerning the question.

a. When there are problems in moral decision making, it is important to make explicit the ethical elements that cause these problems. Sometimes problems that are at first sight non moral problems (e.g. seemingly communication problems) turn out to rest on underlying moral problems. It is important to recognize these and to make them explicit.
b. There is an enormous diversity in convictions concerning animal welfare, animal health and food production. These convictions stem from personal ideas, based on personal ethical beliefs and personal interests. Apart from the question to what extend these personal convictions may play a role in policy making, it is very important for a policy maker to know and make explicit his own convictions, as it is hard or even impossible to ignore them. In practice personal convictions are good or even inevitable starting points in the process of ethical reflection.

Knowledge

Ethics as a reflection process requires knowledge on at least three levels. The first level is knowledge of basic concepts like values, norms, principles, virtues etc. and definitions of moral problems and moral debate. This is necessary to develop a common 'language' and so to avoid communication problems. Secondly it is important to have insight in how these concepts get their meaning in ethical theories. These theories aim at finding, on a systematic and rational basis, answers to ethical questions and give rules how to judge the quality moral reasoning. Ethical theories show how ethical problems develop, why not everybody experiences the same problems and why different solutions are being sought after. Ethical theories may also function as tools to develop a rich array of possible arguments in specific cases. The third level of knowledge concerns the typical problems, arguments etc. which play a role in specific debates. To know these helps people to understand the debate and fruitfully participate in it. It is for example important to understand the relationship in the debate between animal welfare and animal integrity.

Skills

Ethical reflection requires skills, for example to recognize and analyze moral dimensions in concrete cases. Skills are also needed to reflect and discuss with others the moral dimensions of an ethical problem.

Special skills are necessary to integrate moral considerations into the complex reality of policy making in a political environment. Skills must be trained using realistic cases. Tools like discussion and decision models are useful for this training.

From reflection to a reflective equilibrium

As a background theory for 'ethics as a process of reflection' we used a version of the reflective equilibrium as a model for this process (e.g. Rawls, 1972; Van der Burg and Van Willigenburg, 1998). In this model the process of ethical reflection is represented as a process towards an equilibrium between three elements: Intuition/emotion, facts and ethical theories (principles/values). Intuitions are preexisting convictions stemming from culture, upbringing, emotions etc. The common characteristic of intuitions is that these convictions are not or only vaguely reflected upon but they come into mind looking at the moral problem at hand. In the reflection process these intuitions are challenged with ethical theories. Can the intuition be grounded in an ethical theory? Ethical theories are well reflected, connected systems which contain ideas about what is good, what are valid reasons, how one ought to act etc. Facts play a crucial role in this process. For example the facts and their interpretation direct intuition, the intuition tend to make a selection of the facts and in the theories certain facts are relevant and others not. So in this model of reflective equilibrium the ethical reflection is represented as a dynamic process alternating between the different elements and reaching an endpoint when these elements find a good mix: the equilibrium. When an individual uses this process she will probably (from a psychological point of view) start with her intuitions and search for fitting theories and facts. The use of the process is here that the person gets insight in her own moral convictions and how these can be grounded. This is a necessary condition to fruitfully participate in the discussion. But it is in the discussion with others, that the process of moral reflection has its added value: in these discussions the different equilibriums of the participants are mutually challenged. As a consequence, a new inter-subjective equilibrium may arise. The status of such an equilibrium is a subject of discussion. Some will deny the added value because of the contingent nature of the outcome, others may state the outcome of a ideally performed process is the moral truth per definition (Van der Burg, 2003) or that such a process comes as close as it may get to the moral truth. We believe that an inter-subjective outcome of a moral reflection process is, at least in a pluralistic society, the best defendable outcome. We ground this standpoint in a meta-ethical view on ethics of which a further analysis reaches outside the scope of this paper. In a practical sense the model has advantages when using it in the context of policy making. Because different intuitions and ethical theories can be brought into the process it is very suitable for a pluralistic discussion. As has been mentioned in the introduction, policy makers need to deal with a tension between the imperative of the practical decision-making and the respect for societal plurality. The model of reflective equilibrium is very suitable in dealing with this tension. It may be used to analyze the pluralistic social discussion by bringing the different intuitions, theories and facts into the process and to analyze which mixture results in which outcomes and how defendable these outcomes are. This can be done by the policy makers themselves, but ideally they have real discussions with the actual stakeholders. Subsequently the results of such an analysis serve as input into the process of developing the best defendable policy or to formulate defendable policy alternatives.

From theory to practice: training civil servants

Above we described the philosophical methodology behind the actual training of civil servants of the Netherlands Ministry of Agriculture, Nature and Food Quality. This still leaves the question: how to translate theory in practice?

Before the training started we asked the participants to choose one specific case from their work as input for the training and we asked them to make explicit what they saw as ethical problems in that case. These cases were used as a basis for discussion in the following two training days, thereby assuring a good fit between the training and the actual problems of the participants. Asking for such a case was also meant as a first step in awareness training.

The first day of the training started with the participants making a 'mental map' of the concept 'ethics': all intuitions, associations, thoughts, etc. were noted. This was followed by an introduction in ethics that explicitly referred to the mental maps in order to define concepts like ethics, values, norms etc. The reflective equilibrium was introduced as model for ethics as a process. This mental mapping and introduction was meant to develop one common conceptual framework. During the training this framework was discussed explicitly and implicitly several times and this resulted in a 'common language' in the group. After this introduction the participants split into groups and analyzed their own cases based on their personal dossier (see above). The aim was to become aware of the moral dimensions entailed in cases at stake and to train the ability to recognize moral problems. This was followed by an introduction in animal ethics. In this introduction three approaches to ethics of animal use were introduced: (a) An approach that takes respect for the individual animal as paramount (based on e.g. Regan). (b) The approach that consider welfare as the fundamental criterion for ethical reflection (based on e.g. Singer) and (c) an approach that stresses the importance of the human-animal relationship (based on ethics of care). These lines of reasoning were introduced in order to show the relevance of theoretical discussions for practical questions of animal use and to get a grip on the various arguments in the animal welfare discussion. After this introduction the participants split into subgroups again, now to analyze their cases with the help of the different lines of reasoning. The participants were asked to discover which line(s) corresponded with their own personal conviction. The background idea is here that 'ethics as a process' does not start in a vacuum or with a construction of idea's of others, but inevitably in one's owns ideas. These ideas should therefore be made explicit so they can be critically introduced in the equilibrium process. The first day ended with the presentation by one of the trainers of an actual case (Q-fever), to illustrate the different moral convictions that play in a topical case.

The second day of training focused on skills. The reflective equilibrium model is a conceptual model of ethics as a process and not a practical model for practical use. For that reason an introduction was given on several practical models of ethical reflection including their theoretical background and strong and weak points. Models used were the Ethical Matrix (Mepham *et al.*, 2006) and the Utrecht Roadmap for Ethical Reflection (Bolt *et al.*, 2003) and some variations on these models.

The follow up consisted of meetings of half a day on an actual topical policy problem. During the training the ministry was busy with shaping the new law on animal use. This was conceived as a relevant theme for ethical reflection. Therefore, five meetings were dedicated to an equal number of actual topics encountered in this law making process. These meetings were held with the civil servants who are accountable for one or more of these topics under supervision of one of the trainers of the Ethics Institute.

After two months all the participants and trainers met to discuss how the results of the training were incorporated in the actual work.

Results

In a ministry the work is aimed at solving problems. The traditional force of a ministry of agriculture, nature and food quality is in the technical approach of policy problems. In the practice of policy making the focus is therefore on facts. Intuitions/emotions are hardly spoken about. Principles are brought forward by different parties, but are not always recognized and thoroughly discussed. In the process of policy making it is not custom to make room for ethical reflection. It is often considered as time consuming and sometimes evaluated as no more than an exchange of subjective opinions, which is useless for civil servants who must ground policies on objective facts. We tried to deal with this context by grounding the training in the idea that policy making on subjects like animal welfare cannot be value free. Therefore implicit moral convictions must be made explicit. As a result the policy making will be of better quality, not only in an ethical theoretical sense, but also in a practical way: A policy in which pluralistic values have been taken in consideration (not necessary incorporated) will give a better fit with the social reality and can be better defended in e.g. the parliament. Against this background the results must be seen.

First of all the above described idea of ethics and policy making was recognized by many of the participants. The view on intuitions (about ethics and policy making) as relevant moral markers rather than as mere irrational emotions was evaluated positively by the trainees. Partly because they acknowledged the role their personal intuitions play in the policy process and partly because they are often confronted with the evident role intuitions play in the public debate. Furthermore, they acknowledged the added value of ethical reflection as it provided them with a better grip on and understanding of the (societal) discussion. Some participants started to bring in ethical reflection in their policy work and specifically asked for a follow up session dedicated to a specific question concerning lawmaking. We had several sessions on questions about rules for keeping animals, killing animals, weaning animals, etc.

Second, during these sessions the common moral language that we developed in the training was used and turned out to be useful. For example: during the session on weaning of production animals, the question arose whether the mother-child bond has an intrinsic value, apart from a welfare or respect value. Could such a bond be grounded in ethics of care reasoning? The civil servant working on this topic planned to incorporate this question into the preparations of law on this point.

Finally, an interesting discussion on the role and responsibility of a civil servant unfolded: should they give a clear unambiguous advise to the minister – as ministers often want – or should they give a range of possible and defendable arguments and let the minister make a (political) choice? This question touches upon a big problem concerning introducing ethical reflection into policy making: One can train a limited number of civil servants, but how can they make a difference in an environment still thinking along the traditional lines? But the Ministry of Agriculture, Nature and Food Quality in the Netherlands has established a group with the commission to develop the use of ethics in the ministry. The training course we described in this paper has been a successful start of this endeavor.

References

Bolt, L.L.E., Verweij, M.F. and Van Delden, J.J.M. (eds.) (2003). Ethiek in praktijk, Koninklijke Van Gorcum, Assen, Netherlands,

Mepham, B., Kaiser, M., Thorstensen, E., Tomkins, S. and Millar, K. (2006). Ethical Matrix Manual, LEI, The Hague.

Rawls, J., (1972). A Theory of Justice. Oxford University Press, Oxford.

Van der Burg, W. (2003). 'Dynamic ethics'. The Journal of Value Inquiry 37: 13-34.

Van der Burg, W. and Th. van Willigenburg (eds.) (1998). Reflective Equilibrium, essays in honour of Robert Heeger, Dordrecht: Kluwer Academic Press.

Transgenic foods: science, safety, law, and ethics

M. de Renobales[1], L. Escajedo[2] and M. Martínez de Pancorbo[3]
[1]Biochemistry and Molecular Biology, University of the Basque Country UPV/EHU; [3]Zoology and Animal Cellular Biology – Faculty of Pharmacy, Vitoria-Gasteiz, Spain; [2]Constitutional Law, Faculty of Labour Sciences, and Inter-University Chair in Law and Human Genome, Bilbao, Spain; mertxe.derenobales@ehu.es

Abstract

We present herein the objectives and organization of the course 'Transgenic Foods' (4.5 ECTS), an elective in the Master Program of 'Food Quality and Safety', taught at the Faculty of Pharmacy of the University of the Basque Country UPV/EHU since 2006-2007. The objective of this course is to give students enough scientific information about transgenic foods so that, as food specialists, they will be able to have their own opinions regarding many aspects of the current debate surrounding these crops and foods. The lecture part of the course (3.0 ECTS) comprises 3 modules. Module 1 includes a detailed description of the conventional plant-breeding techniques used to improve crops, followed by the description of the molecular biology and genetic engineering techniques to make a transgenic organism (plant, and animal). Module 2 describes de science behind the transgenic crops authorized in Europe and in the United States, as well as those in advanced states of development expected to be commercialized in the near future. These 2 modules give the students the necessary scientific background to understand the similarities and differences between conventional and transgenic foods. Given the lack of specific knowledge most food technologists have regarding transgenic foods, these 2 modules are a must for students to critically examine ethical issues regarding safety, environmental and legal aspects which are often brought up in the current social debate. Module 3 discusses how safety is evaluated in the European Union, environmental, and legal aspects regarding authorization and labelling of such foods, all from the perspective of Jonas' *Imperative of Responsibility*. In the laboratory part (1.5 ECTS), students get familiar with the cellular analytical techniques used in the food industry to determine the species of a food element and the identification of transgenic elements in foods.

Keywords: teaching, genetically modified foods, social issues, scientific issues

The master program in food quality and security

Offered at the Faculty of Pharmacia of the University of the Basque Country (http://www.mastercalidadseguridadalimentaria.ehu.es/p116-1000/es), it is designed to support students in their capacity to utilize different tools to control, design, develop or innovate in the field of food quality and security. The current social debate on transgenic foods, in addition to scientific aspects, covers legal issues, socio-economic questions, ethical, and even religious considerations. Food specialists are likely to have to give advice to the public regarding transgenic foods, or participate in regulatory local or state agencies. Thus, it was considered of great interest to include a course on 'Transgenic Foods' in the Module of 'Advanced knowledge related to quality and safety parameters in specific foods'.

The objective of this course is to give students enough information about all aspects of transgenic foods so that, as food specialists, they will be able to make knowledgeable decisions, or find the necessary information to do so. Just as it occurs in other important social issues with strong ethical and scientific components, such as research on embryonic stem cells, the social debate around transgenic foods is multidisciplinary. So, a multidisciplinary course was designed, considering that all students admitted to this Master program have a life science undergraduate degree, but few, if any, will have some formal education in law or ethics. The three professors who teach the course considered that in order to discuss ethical, legal or socio-economic questions, a solid knowledge of basic scientific aspects is absolutely necessary, as it occurs in other ethical and social questions with a strong scientific component, such as research on stem cells (for example). Appropriate responses to the use of any technology for the improvement of crops begin with understanding 'what is going on'. Due to time limitations, ethical issues are examined from the perspective of Jonas' *Imperative of Responsibility* (1984), as a reasonably adequate approach, because transgenic foods concern not only people, but animals, plants and the environment at large, as well.

Characteristics of the course 'Transgenic Foods'

The course has a lecture part (3.0 ECTS) and an experimental part (1.5 ECTS). The lecture part of the course comprises 3 Modules of approximately the same duration. Module 1 covers basic techniques of molecular biology and genetic engineering to create transgenic organisms. In addition, because conventionally improved plants are used as 'controls' for the risk assessment (safety and environmental) of transgenic plants, students must have a clear understanding of current, modern techniques used by seed producing companies, large or small. Module 2 describes the science behind the currently authorised transgenic crops in the world and those at advanced development stages. Module 3 deals with issues more closely related to the social debate: safety risk assessment, environmental impact, legislation (including labelling). Ethical issues related to the use of biotechnological crops consider our responsibility towards: (1) consumers (first, not to harm them, and then to provide nutritious food; second, the right to be informed and right to choose) and society (contribute to people's development in all countries); (2) all persons involved in transgenic food production and marketing (farmers, primary and secondary seed companies, distributing and marketing companies); (3) the environment. In the experimental part students learn the basic laboratory procedures to differentiate a transgenic from a conventional crop.

Teaching methodologies are varied. Formal lectures provide students with a structured overview of each lesson (or block of lessons) to facilitate understanding of key concepts. Exercises (some taken from resources freely available in the Internet) are designed to help students develop their critical thinking and analytical techniques. The objective of others is, for example, to reflect on how various issues regarding transgenic foods are presented to the public through the media: analyse the language used, how scientific concepts are dealt with, how social and environmental implications are explained. Is the article biased? Towards which position? How are the responsibilities of the various agents (including the writer of the article) dealt with? Then, they are asked to re-write the article backing their statements on testable, scientific data, and present it to the public without conveying their position (if they have any).

Because students do not have any background on legal questions, a special effort is made to get them familiar with the regulatory frameworks to conduct progress in Biotechnology. In addition to the risks, governing norms attend to a great many social interests which are often incompatible among one another. European regulations are amongst the most demanding and complete in the world. However, compliance levels with the rules has been very much uneven, namely because of the pressure of public opinion and the lack of general consensus on the part of the EU member countries. The question is, why?

The laboratory experiments are performed working in groups, with 1 report per group. They are expected to prepare a poster with their results, as if it were for a professional meeting.

Discussion

Students sign up for this course because it is a subject which appears quite often in the media and, as food specialists and, as they indicate in the initial survey at the beginning of the course, they realise that they cannot have an opinion because of their complete lack of basic scientific knowledge. In their course evaluation, over 95% show satisfaction with the level of knowledge gained and they would recommend it to others. In the last two years they did not suggest any modifications to the subjects included. Their only request is that more time be allotted to this course to cover other important aspects such as patents, and international trade. Unfortunately, this is not possible at the moment.

The main values that this course tries to promote are honesty, critical and independent thinking, and responsibility as future social agents through one or more of the following activities: (1) presentation and interpretation of data (preferably from more than one point of view); (2) critique of controversial cases; (3) interpretation of lab results (both their own, or from the literature) and legal texts; (4) analysis of current issues from several stakeholders' points of view. In all cases, scientific aspects are explained.

Lecture contents, laboratory activities and exercises are coordinated among the three faculty members who teach the course to avoid unnecessary repetition, or to emphasise certain difficult concepts. Because two of the three professors usually teach at the Bizkaia Campus, 80 km away, most of the teaching is done by the 3rd one. Contact through e-mail and/or telephone is necessary for the smooth running of the course. Students perceive an excellent coordination of the 3 persons involved.

For the 3 teachers involved, the experience has been highly enriching because each of us has been able to widen her perspective on the subject through discussions with the other 2 persons, or, when possible, attending each other's lectures or discussion of exercises.

References

British Medical Association, (2004). Genetically modified foods and health: a second interim statement.
Escajedo, L. (2007). Towards a sustainable and healthy development of Biotechnology.
Jonas, H. (1984). Imperative of Responsibility.
Muñoz, E. (2001). Biotecnología y sociedad: encuentros y desencuentros. Cambridge University Press.

European higher education area and teaching law and ethics in the biosciences

M. Gorrotxategi Azurmendi
Department of Constitutional Law and History of Political Thought, University of the Basque Country, Spain

Abstract

In the process of adapting Bioscience Degree Programs to the European Higher Education Area (EHEA), the Spanish government wrote a White Book defining learning outcomes for each of the three cycles. It established that among other competences, Bioscience students need to develop the ability to deal with ethical and legal issues regarding their professional areas. That means also being capable of reasoning about the environmental and social impact of advances in biosciences and some ability to arrive at reasoned decisions that take into account all perspectives. This paper presents the project being carried out by the Department of Constitutional Law and History of Political Thought in the University of the Basque Country (Spain) with a new subject called 'Law and Ethics in the Biosciences'. For three new bioscience degree programs (Biology, Biotechnology, Biochemistry and Molecular Biology), this subject is going to be compulsory from 2010/2011 on (6.00 ECTS). The goal of this subject is to give the students the necessary knowledge for the correct understanding of main social, legal and ethical issues regarding: ecology and environment, endangered species protection, environmental risks assessment, biomedical research, animal experimentation, biotechnology and agri-food technologies. This content will be developed theoretically and complemented by practical seminars in which students will have the possibility of being confronted by some of the legal and ethical issues. The application of ethical theory to specific topics will be done via case studies.

Keywords: teaching law and ethics, bioscience degree programs, constitutional law

Introduction

In the process of adapting Bioscience Degrees to the European Higher Education Area, the Spanish Ministry of Education wrote a White Book which defined the levels of competence that the graduates must achieve in Biochemistry, Biological Sciences and Biotechnology. In this White Book, it is stated that the professional performance of environmental biologists, clinical biologists and biotechnologists is being affected by the growing regulatory intervention of political power and it stresses the importance of developing in the graduates the capacity of confronting ethical-social and legal matters that this poses.

Therefore, in the university training of the bioscientists the objectives converge that have been used from the 1990s by the UNESCO (through the Ethics of Science and Technology Programme) and the European Union, by means of programmes such as ESLA (Ethical, Legal and Social Aspects of the Life Sciences and Technologies), which warn of the need for scientists of diverse disciplines and the society in general to participate in decision making in these areas.

This work presents the new course that the Department of Constitutional Law and History of Political Thought of the University of the Basque Country has planned for the purpose of responding to such need. The class of 'Law and Ethics in Biosciences' (6 ECTS), will be given progressively beginning with the 2010/2011 school year until it is taught in the three new bioscience degrees that are offered in the University of the Basque Country: 'biology', 'biotechnology' and 'biochemistry and molecular biology'.

Objectives of the course

In general, it intends for the students to become aware of the law as an element present in the implementation of their profession; it is hoped that the knowledge acquired will permit them to understand the social, ethical and legal impacts related to it and that they will be capable of recognising the legal element in their activity. Within the ethical debate, the values that have been formulated in the Law occupy a particular place; not because they should be considered, far from it, the best solution possible, but rather because of all the possible solutions to a conflict of values, the Law reflects that which is backed by the Parliamentary majority and, as such, can be required through the courts. One may propose beyond those minimum ethics, but in no case can one demand with the coercive instruments of the State.

The purpose is also to make them aware that criticism of the rule of law and of institutional functioning is perfectly possible and desirable. From this derives the importance of being alert, listening and even promoting participation in the debate of applied ethics, which will surely result in future legal reforms.

Specifically, it is intended for the students to have a basic knowledge of the questions related to their profession intervened by the law, as well as of the content of that intervention. This will be done by means of a detailed, in-depth study of the large subject groups. It also is aimed at the comprehension of the fundamental legal concepts involved and familiarisation with the activity of the regulatory bodies.

Focus of the course

The advance of science and technology makes it possible for human intervention in the natural and biological environments to acquire increasingly greater dimensions; this has aroused numerous reflections and debates on ethical, social and legal consideration of these advances. As a result of these debates and the cross between applied ethics and policy, the law, the instrument of public power for the organisation of human activity, intervenes more and more in scientific activities.

In this context it is important for the future researchers to know the obligation that this intervention involves; but it is also essential to empower them against the public intervention, making them aware of their rights. The scientific and technological advance is part of the common good, and the freedom of research, both of the scientific community in general and the individuals in particular, is a fundamental right sheltered by the Spanish Constitution and the Charter of Fundamental Rights of the European Union. The limits to this freedom, therefore, cannot in any case be rendered devoid of content. It is necessary to balance the good of science with the values that can be affected by the bioscientific advances.

Upon occasion science acts as an ally of the law; in these cases, scientific activities become a fundamental contribution within a legal process created to regulate a human activity. On many other occasions, the law conditions scientific activities; in these cases the ethical aspects of the intervention have a leading role, insofar as the law can be considered as the legally valid option among different ethical options from the weighting of the values involved.

The scientist is interested in understanding the impact of the legal intervention on his activity, which occurs from knowing:

1. The basis of the intervention (why and for what reason the legal intervention takes place). It deals with discovering the social, economic and ethical impact that certain activities have, underscoring the nature of the values committed to this activity and explaining the weighting and option that the law establishes, by which the biosciences are compromised.

2. The origin or the source of the intervention (who and how the intervention is carried out). This deals with identifying the bodies that have the capacity to create legal sources in the subjects that interest us, which is deduced from the distribution of competences, both vertical (involving the powers of different territorial entities: local, regional, national and supranational) and horizontal (in reference to the different organisations of the same territorial entity: Parliaments, Governments and Administrations, as well as Judges and Courts). In this section it is also interesting to recognise the legal form or the modality of the intervention (international treaties, EU legislation, laws, regulations, sentences).
3. The content of the intervention (what determines the intervention or what it requires). This deals with highlighting which aspect of the scientific activity is provided for, controlled or limited by the law.
4. The limits of the intervention. That is to say, up to what point an intervention of the public powers is legitimate in the scope of his professional field.

The viewpoint of Constitutional Law can be considered as a tool that allows taking on the contents and crossed competences of the described knowledge in a suitable manner. On the one hand, by providing the student with basic knowledge of the political organisation that dictates the legal rules that affect carrying out their professions. On the other hand, by assessing the weighting of the interests in conflict by virtue of the relative level of legal protection that the law provides them and that determines, in turn, the level of public intervention. Finally, with reference to the specific content of legal decisions.

All these elements for reflection will be a constant that will highlight the development of each of the subjects dealt with in themes that are going to be undertaken in this course: ecology and the environment, research in humans and in their biological samples, as well as in other organisms and microorganisms, biotechnology and human food.

Subjects that will be dealt with in the programme

1. Bioethics, Environmentalism, Biopolitics and Biolaw: an introduction for bioscientists.
2. Ecology and Environment: Conservation of nature and taking advantage of biodiversity: Protection of wild flora and fauna, continental hunting and fishing; Protection of the quality of the water, air and soil.
3. Research on human beings and on their biological samples: Rights of the subjects involved in the trials; treatment, storage and movement of biological samples; ethical control of biomedical research (the Research Ethics Committees).
4. Research on non-human microorganisms and organisms: Use of animals for experimental purposes; Biosafety in laboratories; biotechnological patents.
5. Ethical-social and legal questions that genetic advances pose.
6. Ethical-social and legal questions that biotechnology poses.
7. Ethics and Law of Agriculture and Food.

Teaching and evaluation method

In order to secure the acquisition of contents and promote the capacity of analysis and reflection of the students, teaching by lecture is planned along with classroom work and seminars. The teaching by lecture will be evaluated by means of mixed written tests (tests, with some short-answer questions) and 40% of the evaluation will be done according to the practice tasks and research carried out in the classroom work and seminars. In this classroom work, actual cases will be submitted to the students in which they will study the ethical and legal questions corresponding to each specific case.

Acknowledgements

This article has been made under the research project 'Modelos de reconocimiento de la diversidad lingüística entre coerción y libertad de lengua', granted by the Spanish Ministry of Education (SEJ2007-67178).

Part 14. Ethics, food policies and the case of GMO's

Widening the perimeter: economics, philosophy, ethics and the quest for vital questions

A. Tsioumanis
Democritus University of Thrace, Department of Agricultural Development, Lahana 31, 54638, Thessaloniki, Greece; steriost@auth.gr

Abstract

Scientific information always includes some degree of uncertainty and furthermore it is dynamic in the sense that via time, mainstream theories that were considered true for centuries were proved to be wrong. This inherent mortality of human knowledge means that emerging knowledge often presupposes the deconstruction of previous scientific 'truths'. Moving away from physical sciences towards social ones, considerations on the essence of scientific knowledge by no means decrease. It is very hard to see any parallels between 'economic laws' and 'laws of nature' and the centrality of any laws in economics should be questioned as they usually constitute mere causal mechanisms. Economics is often proclaimed to be a value-free science, presented as a technical issue of selecting among limited means of production in order to achieve given goals. However, this division among goals and means to achieve them is a serious, existing – as opposed to just methodological – mistake and lies in the heart of all existing debates on the predominance of science on decision making as well as on the inherent logic of economics seen as simply a technique of selection. Goals and means of a given society, of a specific economic system are highly homogenous. A given society will use its means in order to achieve its goals but, at the same time in a frantic circle, its goals will be nothing more than the steady produce of its means. It can not be otherwise as the goals are embedded in the very material existence, nature and organisation of the means. While it seems absurd to assess an economic investment via philosophical arguments, it is equally misleading to ignore that all tools and suppositions of economics are based on normative assumptions about the good human life and the proper societal structure. These influence directly all relating policy recommendations. It is not difficult to discern for example that 'efficiency' has precedence over 'equity' only in a non-egalitarian context and one does not need to turn to Aristotle and his conception of the good human life in order to question the very basis of the rat race that is apparent in modern human societies. Always leaving open the issue of alternatives in every single answer we have gathered so far, it may be time to ask a different set of questions altogether.

Keywords: means and ends, value-permeation

On the character of scientific knowledge

While 'sound science' can unquestionably provide objective facts and data as well as estimates on potential benefits and risks, it is a great conceptual leap to accept that hard certainties that science may produce could replace uncertainties generated from other domains of everyday life. The proponents of modern scientific elitism suggest that science, as an autonomous apparatus, will provide one absolute answer and this shall be accepted since it constitutes an objective outcome. However, this view seems to disregard defining characteristics of the essence of scientific knowledge.

It is not easy to disregard several examples throughout the history of modern science, which reveal that the prospect of a single scientific truth is at least deniable. In addition, scientific knowledge is not only uncertain, but also dynamic in the sense that via time, mainstream theories that were considered true for centuries were proved to be wrong. Sabbato (1951) approaches this dynamic character as inherent mortality of human knowledge pointing out that emerging knowledge often presupposes the deconstruction of previous scientific 'truths'.

Furthermore, the situation is aggravated by the fact that technological advances provide a picture of what is technically attainable, not necessarily socially desirable. Sabbato (1951) argues on the immoralist character in the modus operandi of scientific knowledge pointing out that *science on its own guarantees nothing as when it comes to the realization of a major achievement, its moral concerns are absent* while rejecting political neutrality of scientific knowledge as *in the development of science, humans act under a complicated set of ideas, emotions and prejudices that characterize their very nature.*

Economics and additional challenges

Moving away from physical sciences towards social ones, considerations on the essence of scientific knowledge by no means decrease. The very existence of universal laws that proved to be the golden arrow in the intellectual quiver of classical physics may be questioned in the case of economics. Although a body of literature advocates the opposite (e.g. Kincaid, 1996), the existence of numerous differences between natural and social phenomena makes it very difficult to see any parallels between 'economic laws' and 'laws of nature'. Thus, the centrality of any laws in economics should be addressed with skepticism as they usually constitute mere causal mechanisms.

While it is true that when the price of a given good falls its quantity demanded does not fall but it usually tends to rise, it is simultaneously either a conceptual leap or a mere word game to refer to a 'law of demand'. As Guerrien (2004) notes *...in fact, the use of the verb 'tend' fits what economists can at the very most hope to achieve: to detect some tendencies in the phenomena studied.*

The existence of regularities in the field of economics, usually fairly obvious and exception-laden, may not be addressed as 'laws', if the same term is used for the Law of Gravitation or that of planetary motion. It is much more modest and justifiable to seek causal mechanisms in economic reality rather than economic laws.

McKloskey (1985) in his influential work on the rhetoric of economics addresses the metaphorical character of terms used in economic theory: *To say that markets can be represented by supply and demand 'curves' is no less a metaphor than to say that the west wind is 'the breath of autumn's being'.* He used the example of the theory of production functions, underlying that the terms 'aggregate capital' or 'aggregate labour' entail an analogy with something that can be added meaningfully; in the case of labour that would be *hours of conscientious attentiveness. The very idea of a production function involves the astonishing analogy of the subject , the fabrication of things, about which it is appropriate to think in terms of ingenuity, discipline, and planning, with the modifier, a mathematical function, about which it is appropriate to think in terms of height, shape and single valuedness.*

While the 'metaphoric' character is apparent, economic analysis often opts to conveniently set the issue aside. However that does not change the fact that talking literally about production functions or opportunity costs is still engagement in *mere syntax* leading to C.S. Lewis' quotation:

The percentage of mere syntax masquerading as meaning may vary from something like 100 percent in political writers, journalists, psychologists and economists to something like forty percent in the writers of children's stories...the mathematician who seldom forgets that his symbols are symbolic may often rise for short stretches to ninety percent of meaning and ten of verbiage (Lewis, 1939).

Value-free or value-permeated?

Economics is often proclaimed to be a value-free science, presented as a technical issue of selecting among limited means in order to achieve given goals rather than a set of value or policy commitments. The debate over value-neutrality has already lasted several decades. Its proponents, which argue that values can be successfully separated from actual facts use, among other arguments, the distinction of economics into three separate entities; pre-scientific decisions, scientific analysis and post-scientific application (Blaug, 1980). It is thus possible to allow for value judgements before and after 'pure science' keeping the latter intact. The Humean guillotine, prerequisite for the conception of value-neutrality, is well entrenched distinguishing two different types of value judgements; characterizing and upraising ones. Nagel (1961) notes that a *characterizing value judgement* expresses an estimate of the degree to which some commonly recognized (and more or less clearly defined) type of action, object, or institution is embodied in a given instance while an *appraising value judgment* expresses approval or disapproval either of some moral (or social) ideal, or of some action (or institution) because of commitment to such an ideal. Using this distinction, some value judgments, necessary for the scientific process, are allowed as bearing no ethical significance.

It has also been argued that hard facts are out there for all to see, independent of any scientific theory. Brennan argues that *it is the teleological nature of explanation which adds evaluative judgements to empirical study as a facet of economics* (Brennan, 1979). However, this dream passage of objectivity to the empirical world is full of obstacles. Kuhn (1970), addressing natural sciences, points out that the empirical world can only be accessed through the filter of a theory, rejecting thus an important prerequisite of the value-neutrality position. Following this line of thinking in the field of economics, facts are theory-laden and a world view strongly influences the scientific paradigm including value judgements. Prioritising alleged problems, determining which question will be asked, forming research hypothesis, the whole scientific venture seems permeated by value judgements. As Wilber and Hoksbergen (1986) advocate *'what is' and 'what ought to be' are thus inextricably commingled in the data, the facts, the theories, the descriptions, the explanations, the prescriptions, and so on. All are permeated by the a priori world view.* To forget that economic theory contains a series of presuppositions about the good – individual and social – that directly influence policy recommendations seems deeply enrooted in a significant part of mainstream economic thought.

Means and ends and the neo-classical economic paradigm

Rational egoism is the standard behavioral postulate of neoclassical economics. Wilber *et al.* (1983) summarize the propositions of the world view of mainstream neo-classical economics. These include fundamental assumptions about human nature as humans appear to be rational and self-interested, choosing consciously among a variety of means in order to maximize their personal gain. Setting as the purpose of human life the pursuit of happiness, freely defined by individuals, satisfaction of self-interest becomes the motivating force in economic life. The economic environment, the general context in which each individual seeks its self-interest maximization, is characterized by scarcity leading thus to competition.

The rational egoism assumption has constituted a topic for discussion, as it is obvious that there are cases that people behave neither rationally nor primarily guided of self-interest. This seems like a never ending debate and certainly is out of the scope of this paper. However, it should be noted, as Boland (1981) reminds us that universal statements, such as 'all individuals are utility maximizers', can be refutable if false but not verifiable, even if true. The exact opposite is true for existential statements like 'there are some individuals that are utility maximizers'; the latter may verified if true but may not be refuted, even if false. The rational egoism theory is a universal assumption and thus may not be verified. The same conclusion stands, however, for every universal statement incorporated in different world views.

There have also been disputes over the necessary conditions in order for the consumer to actually be a utility maximizer. Shackle (1973), following Hayek, argues that in order to be able to determine which action maximizes utility, all related information need to be acquired, which is inconceivable, rendering thus deliberate maximization an impossible act.

There are also subsidiary value judgments incorporated in the neo-classical paradigm that need attention. Wilber and Hoksbergen (1986) point them out:

- Competitive market equilibrium is the ideal economic situation. Therefore, competitive market institutions should be established whenever and wherever possible and market prices should be used to determine value.
- Means and ends should be bifurcated into two mutually exclusive categories.
- Means and ends should be measured quantitatively.

While competitive market equilibrium constitutes both an ideal and a standard measuring value, the two value judgements concerning means and ends are of great importance.

The separation of means and ends is necessary in order to generate any meaningful result out of the paradigm. *If means and ends were not mutually exclusive, then neo-classical economics would be nothing more than a simple statement that humans do what they do because they wish to do it* (Wilber, 2003). Although an operational necessity, the division of human economic activities into means and ends entails a judgement of value. Consumption constitutes the end, good or bad inherent to the means becomes a secondary discomfort. Even more, *with means and ends separated, it becomes convenient to measure the satisfaction given by particular ends and the dissatisfaction (costs) resulting from employing various means. It becomes possible to measure how much better one situation is than another, by comparing numbers instead of concepts or ideas* (Wilber, 2003).

However, this division among ends and means is a serious, existing – as opposed to just methodological – mistake and lies in the heart of all existing debates on the predominance of science on decision making as well as on the inherent logic of economics seen as simply a technique of selection. The division is problematic because ends and means of a given society, of a specific economic system are highly homogenous. The ends will be nothing more than the product of the means in every case, as the ends are carved in the very materiality, the nature and organisation of the means. And the *telos* of each production system *...is neither production in general nor the production and reproduction of the material human life; it is the production and reproduction of the existing social system* (Kastoriadis, 1984).

Conclusive remarks

Embarking from the metaphoric character of terms used in economic science, question marks over the extent to which actual economic realities are depicted in economic theory are rising. McKloskey (1985) points out that *the world is said to be 'like' a complex model, and its measurements are said to be like the easily measured proxy variable to hand. The complex model is said to be like a simpler model for actual thinking, which is in turn like an even simpler model for calculation.*

While mathematical analysis may without doubt contribute to a better understanding of economic agents and processes, it is often dealt as the end itself. As long as the persistent mathematization of economics is not construed as a means towards a preselected end, it is not engaging in the quest for any kind of 'truth' and will only be good in order to increase the number of publishable articles. The latter would not contravene the theoretical background as it constitutes without doubt some kind of utility maximisation as far as the economist is concerned.

In addition to 'how' and 'why' we are counting, we should also ask 'what' the aggregate numbers do that involve costs and benefits refer to. Let us take as an example the term 'opportunity cost'. It may be defined as the most favourable price that can be commanded by a factor of production which thus tends to become the minimum cost at which that factor can be had by any entrepreneur. Trying to apply economic terms in real life situations reveal both its 'metaphoric' limitations and its failure to draw the whole picture.

If for example, a dozen productive humans, in their mid twenties decide to abandon their jobs and settle in a remote island surviving on basic agriculture and fishing, a 'social opportunity cost' related to their decision could be calculated. It would not be an easy task but it would still be feasible. What would the components of that calculation be?

Without getting into too much detail, I will just say that the sum of the production loss from their previous employment – or even better from the most effective employment using the most effective technique – will definitely enter the opportunity cost function. While at the very end of our calculations, we come up with a total of societal loss, we seem to forget that the result will solely depend upon the components that enter the calculation. Our result says absolutely nothing, not even provides tendencies, about the social benefit in its entire and real denotation. In order to do that considerations on the philosophy of having another dozen individuals sweating physically or mentally in jobs they least enjoy, in order to produce material objects that often no one needs and no one would probably buy if other individuals had not devoted their life and creative energy in order to persuade people to acquire them, may just not be omitted. That is if our results are to have any significance, other that of mathematical computation.

The third subsidiary value judgment incorporated in the neo-classical paradigm specifically states that means and ends should be measured quantitatively. Although conception and use of externalities is an honest effort, it still seems distant for the breeze of the autumn air in the forehead of an individual watching calmly the sunset in that remote island to fall into this category.

While it seems absurd to assess an economic investment via philosophical arguments, it is equally misleading to ignore that all tools and suppositions of economics are based on normative assumptions about the good human life and the proper societal structure. These influence directly all relating policy recommendations. It is not difficult to discern for example that 'efficiency' has precedence over 'equity' only in a non-egalitarian context. Little (2005) notes that *it seems fair to say that the ethical assumptions that neoclassical economics presupposes fall together into a family of normative ideals that privilege individualism, inequality, and the minimal exercise of public policy.*

Furthermore, one does not need to turn to Aristotle and his conception of the good human life in order to question the very basis of the rat race that is apparent in modern human societies. Always leaving open the issue of alternatives in every single answer we have gathered so far, it may be time to ask a different set of questions altogether.

References

Blaug, M. (1980). The Methodology of Economics: Or How Economists Explain, Cambridge: Cambridge University Press.

Boland, L. (1981). On the Futility of Criticizing the Neo-classical Maximization Hypothesis', American Economic Review, 71, 5 (December 1981), pp. 1031-1036.
Brennan, T.J. (1979). Explanation and value in economics, Journal of Economic Issues, vol. XIII, No. 4, December 1979.
Gurrien, B.A (2004). Science Too Human? Economics', post-autistic economics review, issue 24, article 4, available at: http://www.paecon.net/PAEReview/issue24/Guerrien24.htm. Accessed 20 March 2010.
Kastoriades, K. (1984). Crossroads in the Labyrinth, Cambridge, MA, MIT Press et Harvester Press.
Kincaid, H. (1996). Philosophical Foundations of the Social Sciences: Analyzing Controversies in Social Research. Cambridge; New York: Cambridge University Press, 1996.
Kuhn, T.S. (1970). The Structure of Scientific Revolutions, Chicago: University of Chicago Press.
Lewis, C.S. (1939). Bluspels and Flalansferes: A semantic nightmare in his Rehabilitations and other essays, reprinted in The importance of language, ed. M. Black, Englewood Cliffs, N.J. Prentice Hall, 1962.
Little, D. (2005). Philosophy of Economics in S Sarkar and J. Pfeifer, eds., The Philosophy of Science: An Encyclopedia, London: Routledge.
McCloskey, D.N. (1985), The Rhetoric of Economics, Madison: University of Wisconsin Press.
Nagel, E. (1961). The Structure of Science: Problems in the Logic of Scientific Explanation New York: Harcourt, Brace and World.
Sabbato, E. (1951). Hombres y Engranajes. Reflexiones sobre el dinero, la razón y el derrumbe de nuestro tiempo. Buenos Aires, Sur, 1951. Edición definitiva: Barcelona, Seix Barral, 1991.
Shackle G. (1973). Epistemics and Economics, Cambridge: Cambridge University Press.
Wilber, C.K. (2003). Ethics In Economic Theory, post-autistic economics review, issue no. 20, 3 June 2003, article 1, available at: http://www.paecon.net/PAEReview/issue20/Wilber20.htm. Accessed 17 March 2010.
Wilber, C. K. and Hoksbergen R. (1986). Ethical Values and Economic Theory: A Survey, Religious Studies Review, 12, 3/4, pp. 208-214.

Redefining the EU agricultural policy under market constitutionalism: from protectionism to competitiveness

A. Lasa López
Inter-University Chair in Law and the Human Genome, University of the Basque Country, University of Deusto, Bilbao, Spain; ainhoa.lasa@deusto.es

Abstract

This article attempts to analyse the role of the new paradigm represented by market constitutionalism and its fundamental principles, in the Common Agricultural Policy (CAP). Initially, the CAP was characterized by a kind of government intervention targeted at guaranteeing protected development of the agricultural sector against external competition, a more productive agriculture, while providing a reasonable standard of living for EU farmers. Nonetheless, the new basis of the Economic Constitution speculates on the community area and located in the autonomy of the market re-proposes its configuration in other terms. Moving away now from the parameters that have defined the institutionalization as well as the development of the CAP, some steps have just been taken according to the new paradigm, such as the consecutive CAP reforms or the political agreement reached by the agriculture ministers on 20 November 2008, known as 'Health Check.' All of these regulatory and political bases mentioned above, place the CAP in the way of promoting 'sustainable agriculture in a global environment.' Reinforcing a competitive and innovative agricultural sector that can respond to the challenges of the world market under the rules of a highly competitive social market economy, the new European policy paradigm seems to seek to balance competitive agriculture with social cohesion in a connection in which the second element is irreversibly bound to assume a subordinate position.

Keywords: state-assisted paradigm, market liberal paradigm, multifunctional paradigm

Preliminary approaches to market constitutionalism

Frequently, the study of reforms of the common agricultural policy (CAP), adopts the concept of policy change as a parameter. An idea that, in terms of public policy, is characterised by its dynamism compared to other traditional static categories of study. This peculiarity, which covers the aforementioned concept, implies approaching the reform process with its evolution in a relatively broad period of time, grasping all the possible policy changes in order to verify significant transformation in terms of a shift of paradigm.

However, this study perspective focuses on the concept of market constitutionalism, which transcends the scope of public policy to take on a greater reach centred on the analysis of the fundamental principles that provide the bases that guide regulatory action in the community legal system as a whole. Broadly speaking, market constitutionalism implies a confrontation with the material bases that make up the social State constitutionalised by the fundamental second post-World War texts. Specifically, opposite the recognition of private initiative in subordination to the political leadership of the economic process itself, originated in social constitutionalism around a system of economic relations based on the primacy of policy, in the market constitutionalism internalised by the European project, it is the market itself that legitimises government intervention in the economy to preserve the market and encourage the investment of capital.

The remercantilisation of the State places the market at the centre of the form of the State, sanctioning its autonomy and tying government activity to the protection and guarantees of the latter. In such a way that the integration of the markets of the member States into a single European market is done from the European level and its construction from the market in contrast with social constitutionalism.

The virtualities of this construction in the scope of the CAP unfold around a trio of considerations. In the first place, the CAP is presented as a foreign body to the constitutional community design of the origins. The peculiarities inherent to agrarian activity (market imbalances, unexpected climate conditions) shaped it into an exceptional sector in comparison to other sectors of the economy. Precisely this note of exceptionalness was the leitmotif of the state assisted paradigm for correcting the mentioned imbalances. To this we would have to add the priority given to maximising production resulting from the scarcity of supplies suffered during the Second World War. Maximisation that was articulated through a vast system of customs protections, export subsidies and minimum import prices.

In a parallel way, the decisive weight of the interests of the farmers present in the negotiations of the agricultural decision making process, embodied a certain primacy of the most favourable objectives for the latter, which in turn, contributed to homogenise the set of common agricultural policy objectives contemplated by article 39 of the Treaty of Rome. Specifically, 'to guarantee a fair standard of living to the agricultural population, especially, by increasing individual income.'

This market interventionist orientation contrasted, as we have just pointed out, with the model adopted by the constituent Treaties of the European Communities, which, far from surpassing the content inherent to the state-assisted, opted for adopting a model based on free competition and the market economy, in other words, a capitalism unconnected to any form of State programming or intervention in the economy.

In this way, the original CAP opposes the market constitutionalism model tied to the dynamic of negative integration and economic logic. Implicit in the coexistence of these contradictory logics were the resulting difficulties and disparities that began to manifest in the 90s, as we will see.

Second, the economic model designed by market constitutionalism that acts as a system of consciously designed global free market, and not the social commitment through a system of control checks, it takes shape as a decisive endogenous factor in the conditionality of the reform process of agrarian policy. Although from certain doctrinal sectors, it has been highlighted how the objectives of market constitutionalism (single market, monetary union), act as exponents of the change from the traditional CAP paradigm, exploring the consequences leads back to a mere list of the actions undertaken under the umbrella of the market-oriented paradigm, without going into depth on the disciplinary effects that this generates.

In the third place, there is a strong interconnection between the endogenous and exogenous factors that encouraged the CAP reforms. The effects of trade deregulation as a parameter of the new public policies spread through the process of community integration. In this regard, the legitimate basis of the reforms to the CAP that began in the 90s, places it in the international pressures (GATT), as well as in the existing consensus in the European Union itself that agriculture could not be excluded from the legislation regulating free competition. An orientation manifested in the secondary law regulation, such as the case of the Council Regulation (EC) No. 1184/2006 of 24 July 2006, applying certain rules of competition to the production of, and trade in, agricultural products (Codified version). Official Journal L 214, 04/08/2006 P. 0007-0009.

The close ties between international and community dynamics determine not only the imposition of a new 'competitive agricultural paradigm,' but also of the new organizational principles of the system that have been consolidated in the community area by way of its reception in the Single European Act, and fully with the Maastricht Treaty.

Moreover, the CAP reforms are not carried out autonomously, but rather taking into account whether the reform mechanisms are in tune with pronouncements in the international context. Which means that there is no separation between the international and the European areas, but rather these act as interconnected, defining new limits and new opportunities for common agrarian policy.

The winds of change of the MacSharry Reform

The 1992 reform is considered one of the most ambitious in the history of common agricultural policy. The change from price support to direct payments based on a regulation of supply through obligations to set aside land is the major issue of this reform.

At the same time, the introduction of more selective subsidies tied to the farmer's commitment to the environment and the rural landscape implies a significant change in the concept of aid. Economic justification together with social acceptance determines state intervention in the involvement of farmers in the fight to combat environmental imbalances.

Therefore the new guidelines for government intervention introduce the farmers' responsibility to protect the environment into the value scope of the productive system. The framework of reference that justifies attributing the aforementioned responsibility would be that of multi-functionality tied to the social functions of agriculture. A concept that shapes agriculture as an activity that is entrusted with fulfilling certain functions in society, and which we will touch on later.

In spite of the mentioned novelties, how reform 92 has kept the interventionist paradigm intact that has guided the CAP since its beginning, has been highlighted. Such that the introduction of new mechanisms has not been inherent to a change of the objectives to be achieved. Maximising production and safety of supplies are consolidated as fundamental guidelines of the agricultural project.

On the other hand, other reflections highlight the orientation of the MacSharry reform toward a 'more market-oriented CAP' which will lead to a gradual deregulation without reaching a situation of pure market. The loss of credibility in the illusion of the market by farmers, or the adoption of the small farmer's scheme, constitute some of the main arguments to justify this proposal.

Updating the CAP: toward the aim of a more competitive model

Emphasising the need to modernize the European model of agriculture in July 1997 the Commission presented some proposals that had as central challenge 'a leaner, greener European model of agriculture' focused on 'contented consumers, cleaner countryside, competitive farmers and stable spending.' Following these premises, the Europe's Agenda 2000, introduces new objectives that combine instruments already contemplated in the reform of 2000. This circumstance has been the object of important criticisms that accent the ambivalent relationship established between the ends and anticipated means to achieve them. In this regard, the delicate balance between maintaining the traditional paradigm and the demands of deregulation and greater competition demanded by the international market system and the internal European market are indicated.

Lower guaranteed prices, a continued commitment to stable farm incomes, decentralisation of responsibilities, flexibility of programming, a greener CAP, community enlargement, local decision-making and simplification of the rules, appear as the key elements which will help to develop 'a genuinely multi-functional, sustainable and competitive European model of agriculture.'

The road to reform continued in 2003 with the introduction of the compulsory aid modulation mechanism tied to the transfer of agricultural funds from the direct payment system to the Rural Development Fund, known as the second pillar of the CAP. This second pillar target to promote a sustainable development of rural communities by the way of improving the competitiveness of agriculture and forestry, supporting the environment and country side, promoting diversification of the rural economy and improving the quality of life.

Finally, the Health Check agreement of 2008, intended to 'allow a better response to the new challenges and opportunities faced by European agriculture,' only consolidated all the measures already contemplated in the reform process of 2003, such as cross compliance or shifting money from direct aid to Rural Development.

On the margin of this synthetic presentation of the main novelties contributed by the successive reform processes, what is really interesting to underscore is that in all of them, the already mentioned concept of multifunctionality takes on a special consideration and scope. In fact, in the policy definition documents, as well as in the reforms indicated in the previous paragraphs, multifunctionality appears as a precondition for reaching objectives of the renewed European agricultural model of the 21st century. To the point that it seems to acquire its own identity, which makes it the guiding principle of the reform activities of the late 90s.

Multifunctionality as a third way?

In early 2000, the Organisation for economic cooperation and development (OECD), began work on the concept of multifunctionality as a guideline for the structural analysis of the common European agricultural policy. The usefulness of using said concept was a given due to the inherent characteristics of agriculture as an industrial activity, especially emphasizing the high levels of protection provided to this sector. Specifically, 'the policy context for the work on multifunctionality is provided by Member countries' commitments to further progressive reductions in domestic agricultural support and border protection, and a shift away from policy measures that encourage higher levels of food production and input use, towards measures that are less distorting of markets and trade.'

A statement that approaches multifunctionality from the perspective of its meaning, scope of fulfilment and reach in the area of agriculture. In the first place, as it is taken from the work of the OECD, the concept of multifunctionality can be analyzed highlighting its positive connotation understood as the characteristic that makes it unique to an economic activity, or taking into consideration its role as social objective. This differentiation is interesting, given that in the first of the meanings, the connection is established with the economic value, while in the second, it leads back to social functions associated with agriculture. A duality of meanings that is parallel to the dialectic deregulation-intervention that was posed in the preparatory works for the CAP reforms. But, as it specifies, the accent on the positive approach does not exclude the multifunctionality approach as a regulatory concept. In such a way that multifunctionality seems to operate in practice as a wild card mid way between the competitiveness paradigm and the state intervention paradigm, in other words, as if a multiplicity of contradictory public policies were acceptable.

In the second place, multifunctionality is contextualised in the framework of the obligations and commitments acquired by the States in fulfilment of agriculture less distorted by competition. In this context the agreements reached in the Uruguay Round of 1995. Basically, because it entails the inclusion of agricultural products within the multilateral accords on merchandise trade, remaining therefore determined by the requirements anticipated in the aforementioned accords. In particular, according to the report from the WTO Secretary on trade policies and practices of the European Communities of 2009, to achieve a solid recovery and sustained growth, the Communities must continue to make key structural reforms, among others, continue to deregulate agricultural policy by reducing elevated customs tarries, incentives for production and exportation.

What we want to highlight with these considerations is that multifunctionality should be examined, taking into account its ties to free competition and the market economy supported by a more liberalised global market. Above all, because multifunctionality cannot be considered in a vacuum, but rather connected to the tie to free trade, and the guarantees that determine its true reach. Therefore, the social component of multifunctionality adopts a marginal position that translates to subordinate compatibility with the competitiveness paradigm.

References

Coleman, W.D. and Tangermann.S. (1999). The 1992 CAP reform, the Uruguay Round and the Commission: conceptualizing linked policy games. Journal of Common Market Studies 37: 385-405.

European Commission (1999). Europe's Agenda 2000. Strengthening and widening the European Union. Draft of Commission brochure for the general public on Agenda 2000. Available at: http://ec.europa.eu/agenda2000/public_en.pdf.

Organisation for Economic Co-operation and Development (2001). Multifunctionality: towards and analytical framework. Agriculture and Food. Available at: http://www.oecd.org/dataoecd/62/38/40782727.pdf.

Daugbjerg, C. (2003). Policy feedback and paradigm shift in EU agricultural policy: the effects of the MacSharry reform on future reform. Journal of European Policy 10: 421-437.

Rainaldi. F. (2006). Il policy change nella Politica Agricola comunitaria. Dal produttivismo del vecchio paradigma alla multifunzionalitá del modelo europeo di agricoltora. Rivista Italiana di Politiche Pubbliche 2: 5-31.

Daugbjerg, C. and Swinbank, A. (2009). Ideational change in the WTO and its impacts on EU agricultural policy institutions and the CAP. Journal of European Integration 31: 311-328.

Nielsen, H.O, Pedersen, A.B. and Christensen, T. (2009). Environmentally sustainable agriculture and future developments of the CAP. Journal of European Integration 31: 369-388.

The dignity of living beings with regard to plants: a brief commentary

I. de Miguel Beriain
Inter-University Chair in Law and the Human Genome, BBVA Foundation, Provincial Government of BiscayUniversity of Deusto, University of the Basque Country, Avda. Universidades, 24, 48007 Bilbao, Spain; inigo.demiguel@deusto.es

Abstract

Does it make any sense to talk about the dignity of plants? The article 120.2 of the Federal Constitution of the Swiss Confederation of 18 April 1999 seems to think so, as far as it mentions the *dignity of living beings*. The question which arises therefrom is whether plants should be included in those living beings and what the consequences of that inclusion should be. In order to clarify this issue, the Swiss Federal Ethics Committee on Non-Human Biotechnology (ECNH) published in April 2008 a Report entitled 'The dignity of living beings with regard to plants. Moral consideration of plants for their own sake'. That report stated that plants possess inherent moral value, that is to say, dignity. The aim of this paper is to explore the problems that such a conclusion implies, pointing out its deep incoherencies.

Keywords: dignity, deep ecology, biocentrism, plant sensation, moral status of plants, ECNH

Introduction

The article 120.2 of the Federal Constitution of the Swiss Confederation of 18 April 1999 states *that 'the Confederation shall legislate on the use of reproductive and genetic material from animals, plants and other organisms. In doing so, it shall take into account the dignity of living beings as well as the safety of human beings, animals and the environment, and shall protect the genetic diversity of animal and plant species'.* Since the first moment when this article was included above, it sparked great controversy regarding the interpretation it should be given. In April 2008, the Swiss Federal Ethics Committee on Non-Human Biotechnology (ECNH) published a report entitled 'The dignity of living beings with regard to plants. Moral consideration of plants for their own sake'. The purpose of this report was to make proposals forming an ethical perspective to concretise the constitutional term *dignity of living beings* with regard to plants. Its results are particularly interesting as they are (as far as the author of this paper knows) the most important document ever published regarding such an interesting issue as the inherent value – that is to say, dignity – of plants. In this paper, I will first try to present the key conclusions arrived at in this report. Then, I will try to explain the way such conclusions were attained, some of which may seem to be quite difficult to share. Finally, I will make some critical comments on the report, concentrating on its statement about the dignity of plants. In order to do so, I will first make a brief statement on the concept of dignity and its moral implications.

The 'dignity of living beings with regard to plants' report: a brief statement

The most important statement of the report is that plants have not only instrumental value (that is, a value related to the possibility of being useful for human beings), or relational value (that is, a value which rises from the fact that someone considers them to be worthy of protection), but also inherent value, which means, according to the report, that *'they should be protected for their own sake'.* The differences of these three concepts of value can be illustrated, as shown in the report, using the example of a rosebush. The rosebush has an instrumental value, because a rose hedge protects against undesired intrusions. The rosebush has a relational value, because its beautiful flowers remind you of your dead grandmother.

Finally, the rosebush has inherent worth, independently of whether it is useful or whether someone attributes a value to it. Plants, as has been said, pose, according to the report, inherent value, something which must be considered as a synonym of 'dignity' (in fact, the Report states that *'If something has inherent worth, this means it has something, which we also call «dignity»'*).

The consequences of this belief are not irrelevant. The report states that if plants have dignity, the complete instrumentalisation of plants – as a collective, as a species, or as individuals – requires moral justification. They should also be excluded for moral reasons from absolute ownership, which means that no one may handle plants entirely according to his/her own desires (as the majority of the Committee holds), and we may also think that the patenting of plants as such is morally impermissible and contradicts the dignity of living beings with regard to plants (as the minority of the Committee holds). If the plants have inherent value, arbitrariness against them should also be avoided. That is why the Committee members unanimously consider an arbitrary harm caused to plants to be morally impermissible. Finally, if plants have dignity, any action with or towards plants that serves the self-preservation of humans must be morally justified, as long as it is appropriate and follows the principle of precaution.

The Committee was aware that such conclusions may seem to be quite unrealistic for most people, and that a lot of people consider that the question of whether the treatment or handling of plants requires moral justification or not is a meaningless one. In fact, most of us think that some of these conclusions and, even, the main statement they are based on (the dignity of plants) go against our moral intuition. However, intuition is not a key factor in moral discussions. That is why in the following pages rational arguments will be presented which demonstrate why the report statement about the inherent value of plants and the conclusions that come from that fact should not be necessarily sustained.

Dignity: a brief description

The aim of this paper is to show why stating that plants have dignity makes no sense at all. However, in order to do so, it is unavoidable to take some time and space to refer to the concept of dignity. For the sake of the argument, it is necessary to keep in mind what dignity really means and what it implies, as it is, precisely, a defective knowledge of this notion that makes the report fall into confusion.

The concept of dignity has a long history. However, its meaning has changed a lot during the last centuries. In Rome, it referred to merit, integrity, the indifference to wealth, or loyalty. It was also linked to social rank, being associated with relevant personalities, such as the Roman magistrates. In the Middle Age and most of Modernity, this original sense of dignity was mixed with theological considerations, but, in any case, dignity was always considered a variable value, which could change according, for instance, to human deeds. Kant changed this entire framework proposing a new foundation of the idea of dignity. According to him, dignity finds its ground in the human autonomy, on the fact that the Man *'exists as an end in itself, not only as a means'*, a belief which was based on one of the formula of Kant's categorical imperative *('Act in such a way that you treat humanity, whether in your own person or in the person of any other, always at the same time as an end and never merely as a means')*. Since Kant's contribution, the meaning of dignity was directly associated to the idea of inherent value which human beings and other rational ones have due to their rationality, a quality that let them be autonomous.

Nowadays, we consider that there are different kinds of dignity that a being may have. Following Nordenfelt's classification, we could talk about dignity as a merit (the excellence and distinction we attribute to a person who has a rank or holds an office that entails a set of rights), as a moral stature (the dignity that depends on the thoughts and deeds of the subject), or as an identity (the one we attach to ourselves as integrated and autonomous persons with a history and a future, with all our relationships to other human beings, something which could be called self-respect). In all those cases, dignity exists in degrees and can come and go, according to different factors, such as the moral deeds of the subject, its fortune, the illnesses we suffer from, etc. There is, however, another kind of dignity, a dignity that pertains to all beings (usually human beings) to the same extent and cannot change as long as the being exists. That concept of dignity refers to a special dimension of value which is directly associated to the essence of a being, that is, to a quality or qualities which come from that essence.

The greatest discussions held nowadays refer to the quality or qualities that should justify the possession of dignity. The different views regarding this issue bring into existence different ontological paradigms, depending on their thoughts about which quality or entity is worthy in itself (that is, dingus). Following this pattern, we can distinguish between a lot of different paradigms: theocentrism, anthropocentrism, ratiocentrism, pathocentrism, or biocentrism. Someone holding theocentrism believes that God is the only source of value, that is, the only being who has inherent value, and that only those beings sharing His essence in some way share his dignity. Instead, for someone sustaining the theory of ratiocentrism, i.e. somebody who considers that the most important quality that could exist is rationality, the issue of whether beings count for their own sake depends on their (potential) capacity for reason and their capacity for abstract speech. An anthropocentrist, on the other hand, states that it is only human beings who hold dignity, for the mere fact of belonging to the species *Homo sapiens*. For someone who believes in pathocentrism, i.e. the paradigm based on the belief that the key factor in moral judgment should be the sentience of living organisms, all beings that feel pain should be considered to have inherent value. Finally, for those who hold biocentrism, i.e. those who consider life as the most important possession, all living beings must be considered to have inherent worth.

The dignity of plants

For the sake of this discussion, it is extremely important to keep in mind that all these paradigms will arrive at extremely different conclusions about the kind of beings that possess dignity. Most of the authors of the Swiss report took a biocentric position. This is an extremely important fact which should not be forgotten if we really want to understand the nature of its conclusions. Had they been theocentrists, for instance, the conclusions of the report would have been completely different. It is only from the biocentric point of view that the following paragraph of the report makes sense:

> 'Even if plants do not have their own interests, it is still possible to say that something is in their best interest. The precondition for this is that they have a right of their own. Organisms that have a right of their own may have good and bad things done to them. If plants do have a right of their own and something can be in their interest, then they must be given moral consideration. A plant can also be harmed even if this harm is not affected through the sentience of the plant and if it cannot express this directly'.

It is quite clear that if, and only if, someone believes mere life entails inherent moral worth, it can be stated that a plant which might not even suffer should still be considered as an inherently worthy being. So, the question which gives the title to this section, 'do plants have dignity?' can already be answered: yes, but only if we hold a biocentric point of view. However, will it be possible to face the consequences of such a statement? Does the report really do so? We will examine this issue in the next part of the paper.

The consequences of attributing inherent value to plants. A realistic framework?

The consequences of considering that a being possesses dignity are the following:

- It can not be considered as an object, but a subject of morals.

- It is equal to all inherent worthy beings (we can not make distinctions between these kinds of beings, as we can not make evaluative distinctions between beings which are all considered as inherently morally worthy).
- Its value will never change while the being is still alive.
- We should never treat this being as a mere means. Keeping all this in mind, it is quite clear that if we consider that plants have inherent value, we could never harvest lettuce while it is alive, as it would be contradictory to the categorical imperative we have described above, even if we could die of famine as a consequence of our abstinence (lettuce should never be used as a mere means to guarantee the survival of humans).

However, even the most convinced biocentrist could find it extremely difficult to sustain such types of conclusions. The report, in fact, does not do so. In order to avoid them, it states that *'Most ECNH members assume that the dignity of living beings is not an absolute value, but is achieved by the balancing of morally relevant interests: the good, or «interests», of plants should be weighed up against the interests or goods of other organisms'.* That statement is the real key factor to understanding the profound issues that the report had to face as a consequence of its attributing dignity to plants: the price they had to pay was accepting that dignity is not an absolute value, but rather achieved by the balancing of morally relevant interests: the well-being, or interests of a plant should be weighed up against the interests of humans, which is clearly contradictory with the idea of dignity as the inherent value that subjects of morals have. This few lines show clearly that the members of the Committee made a huge mistake about the real significance of the concept of dignity. Dignity, when defined as the incomparable worth that some beings have, is not a value such as justice, equality, etc. It is the inherent value that a being possesses, a value that cannot diminish while the being is alive.

We can, therefore, conclude that the acceptance of the dignity of plants may only be possible if we change the meaning of the term 'dignity'. In other words, only if we consider that dignity does not mean inherent value, or if we state that this inherent value is different from one being to another, something that in my opinion makes no real sense, we can consider that plants possess dignity. But, in that case, it is quite clear that we will have come back to the eternal discussion in ethics: what then should be the most important 'inherent value'? That of human beings, that of animals, that of plants, or what? These considerations should let us understand how meaningless it really is to try to use a concept such as human dignity in an attempt to solve all the problems that human beings have to face regarding our relationship with the rest of the inhabitants of the Earth.

Acknowledgements

This work has been funded by a Grant to Research Groups by the Basque University System, ref. IT-360-07, Department of Education, Universities and Research of the Basque Government. It is directly connected with the objectives of the EU-founded project XENOME: Engineering of the porcine genome for xenotransplantation studies in primates: a step towards clinical application, dedicated to xenotrasplantation and with the AUTBIO project: El concepto de autonomía en bioética e investigación biomédica: aspectos filosóficos y sociales.

References

Beyleveld, D. and Brownsword, R. (2002). Human Dignity in Bioethics and Biolaw. Oxford University Press, 1st edition, United States of America.

De Miguel Beriain, I. (2009). '¿Derechos para los animales?'. Dilemata: Revista de ética aplicada, no. 1, pp. 15-31.

Federal Ethics Committee on Non-Human Biotechnology – ECNH (2008). The dignity of living beings with regard to plants; Moral consideration of plants for their own sake. Publisher: Federal Ethics Committee on Non-Human Biotechnology ECNHA, Berne. Translation: Jackie Leach Scully. Available at: http://www.ekah.admin.ch/en/documentation/publications/index.html.

Kant, I. (1993). Critique of Practical Reason. Lewis White Beck (ed., 3rd. Macmilliam Publishing Co., New York, United States of America, 136 p.

Koechlin, F. (2009). The dignity of plants. Plant Signal Behav January, 4(1): 78-79.

Nordenfelt, L. (2004). The varieties of dignity. Health Care Analysis, Volume 12, Number 2, 69-81.

Pluhar, E. (1995). Beyond Prejudice: The Moral Significance of Human and Nonhuman Animals. Duke University Press, Durham.

Shifting schemes of naturalness

P.F. van Haperen[1], B. Gremmen[2] and J. Jacobs[3]
Wageningen University and Research centre, META, Hollandseweg 1, 6707 KN Wageningen, the Netherlands; paul.vanhaperen@wur.nl; bart.gremmen@wur.nl; josette.jacobs@wur.nl

Abstract

During the last thirty years, the public debate about modern biotechnology in plant breeding has mainly been concentrated around the possible societal & environmental risks of genetic modification. 'Naturalness' became a moral argument that was endlessly repeated in sterile discussions about acceptability. However, the latest developments in genomics and biotechnology are such that the traditional arguments on GM versus classical breeding no longer seem to be appropriate. New applications like cisgenetics, reverse breeding, and the use of oligonucleotides, cannot be categorized in these well-known schemes. A new analysis and understanding of the different societal images around naturalness seems necessary to enable a fruitful future debate. In this paper we discuss four methods of plant breeding in which the concept of naturalness plays an important role, but refers to different schemes that seem to shift with the context in which they are used: (1) Classical breeding. Many consumers consider products to be 'unnatural', not only because of their lack of taste but also because they were produced in a highly artificial and industrial way, which has a detrimental impact on the environment. (2) Genetic modification. Here, the cause of ethical concern is the very breeding process of the new cultivars. The GMO products were considered to be unnatural because species borders were crossed in making them. (3) Genomics assisted breeding. Compared to GMO's this method would be neutral with regard to the problem of species integrity. However, it is considered to be a violation of the 'natural speed of development'. (4) Organic plant breeding. Organic agriculture presents itself as a more 'natural' alternative to both high tech production and genetic modification. However, there are discussions about the boundaries of the organic concept of 'naturalness'. We conclude with proposing a dynamic model for analysis, based on three value spheres; ecology, culture and technology.

Keywords: biotechnology, genomics, plant breeding, public debate, value spheres

Introduction

Already thirty years, modern biotechnology, understood as the development and application of molecular biology knowledge and -technology in plant breeding, is a regular source of contention. The use of genetic modification has led to fierce debates in Europe, resulting in severe restrictions set on experiments and market introduction of genetically modified plants, legally defined in the European Union as organisms altered at the cell level in a way that would not be possible *naturally* through reproduction or recombination. In public debate about genetic modification, *naturalness* became, apart from being a legal and distinctive, also a moral argument, that was endlessly repeated in sterile discussions about acceptability which eventually landed in a complete stalemate. Nevertheless, in these discussions the legal and ethical distinctions were somehow clear: there were genetically modified plants and not-genetically modified plants and stakeholders were for or against GM.

Currently however, new developments have opened up the possibility that products appear to be clean from 'foreign' material but nevertheless were produced with the legally specified techniques, or that they do have altered DNA but were made with different methods than formerly specified as genetic modification. At the same time, new genome mapping techniques, like 'Next Generation Sequencing', soon will make it possible to reveal all direct interventions in the plant genome, possibly 'unmasking' all changes in plants caused by other methods than traditional breeding. Sooner or later these developments will make the old legislation – and the distinctions on which it is based – obsolete. Therefore, in the longer run, a new and contemporary classification seems unavoidable for setting new societal and legal boundaries. Images around naturalness seem to lie at the heart of the old legislation. Naturalness and the portrayal of nature carry in them the possibility of quickly becoming untouchable arguments that cause a stalemate in discussions on the impact of biotechnological innovation (Smits, 2002). In order to change successfully, and avoid a repetition of moves that led to a stalemate in the GM debates, a new but also dynamic analysis of how different groups see and use 'naturalness' in defence of their arguments, seems warranted to prepare for a process that negotiates these new societal and legal boundaries.

What does nature mean?

Much has already been written about the concepts of 'nature' and 'naturalness'. Williams, who remarks that it is perhaps the most complex word in our language, presents us with three areas of meaning for nature: (1) the essential quality or character of something; (2) the inherent force which directs either the world or human beings – or both; (3) the material world itself, taken as including or not including human beings (Williams, 1983). These concepts however, own more and different interpretations and foremost connotations that appear to be *shifting over time and according to the context in which they are used*. It seems hard to imagine now, but, in 1910, an American corn breeder referred to a method for hybrid crossing as 'doing violence to the nature of the plant' (Kloppenburg, 2004); a statement even organic agriculture would not underscribe nowadays. What is meant with nature ranges from romantic country side associations and ideas of the good, pure & pristine, to a resource for surplus use, a biological or ecological system that can be uncovered, a threat that must be mastered, a self healing wholeness, or a gradient scale of interrelations, to name just a few (Williams, 1983; Soper, 1995; Lovelock, 2000; Stephens, 2000; Verhoog *et al.*, 2003; Anderson, 2005; Gremmen, 2005; Hansen, 2006; Siipi, 2008). All these interpretations have one characteristic in common though and that is that they come to us as a binary, as standing in relation to an opposite, e.g. as nature versus culture, as the given versus the produced, or as the self regulating versus the managed. Nature *physically* cannot be seen but as an objectifiable instance (an entity, a force, a system), but at the same time will always come to us as mediated through our senses and as such can only be interpreted within the limitations of our (predetermined) cognition. Thus the above meanings of nature, separate or tied together, inevitably do not escape from (a cultural) interpretation or distortion and with that set the arena for a *moral* interpretation of what nature should be or do and how it is thought to be organized and understood; a moral interpretation of what the effect of nature is or should be on humans and society, and vice versa.

Let us take a closer look at the role of the concept of nature/naturalness as coming forward in four known discourses on plant breeding methods. In debates about these approaches, naturalness appears as an evaluation criterion, but as we will see, actually refers to different schemes, which on top of that seem to shift along with the context in which they are used.

Four discourses on breeding methods and agriculture

Classical breeding/conventional agriculture

In the late 1950's and throughout the 60's the concept of the green revolution was developed. Seed banks, developments in genetics and public – private partnership quickly led to greatly improved technologies for targeted breeding. Productivity in agriculture multiplied in a way that was unseen before, by concentrating breeding on yield characteristics -the so called bumper crops- and by compensating the lack of robustness of crops by applying chemical inputs for nutrition and disease control. This also marked the conversion from a rural to an industrial and capitalized production process. During the 70's and 80's the limitations of this approach became more and more apparent. The increased use of fertilizers, herbicides and insecticides led to environmental erosion and pollution and gave rise to the ecological movement, criticizing the chemical approach in agriculture. It was also found that the focus on production and quantity was leading to a loss of quality. During the 90's, consumers started to complain that products had lost taste and texture. Simultaneously, the spread of disease caused by mono-cropping started to cause concern as well. By that time the genetic base of popular cultivars had become so narrow, that breeding efforts to improve the quality aspects became difficult, giving rise to further environmental concerns, this time about biodiversity. (Goodman *et al.*, 1987; Juma, 1989; Kloppenburg, 2004; Murphy, 2007).

Genetic modification

In the 1990's, the advances in molecular biotechnology make it viable to bring genetically modified crops to the market, that offer highly attractive commercial qualities. The idea of modification at the cell level is a technical mimicking of natural phenomena as seen with viruses and the *Agrobacterium tumefaciens* (Van Wordragen and Dons, 1992). The development is a direct response to the lack of genetic diversity for breeding new cultivars. The introduction of these new crops quickly meets opposition from a variety of societal groups, ranging from environmentalists to religious groups, to consumers. Many believe genetic modification is tampering with nature, tantamount to playing for god. Crossing species borders is deemed irresponsible and asexual insertion of genetic material is seen as unnatural. There is the notion that in nature organisms are more than just the addition of molecular building blocks. (Cronk, 2002; Dupré, 2004; Sulmasy, 2006) Genetic modification would cause a further decline of biodiversity. As research into the long term effects of GM crops is still meager, the unknowns could be translated into environmental risks. There is fear for contamination of existing varieties or a disturbance of ecological equilibrium. GM food would cause harm to health.

The industry insists that GM crops are safe, but are distrusted. Critics of GM argue that it is simply the latest incremental intensification of agricultural production created by an exploitative industrial system that has steadily built up momentum over many decades. They question the integrity of commercial actors and their role in influencing the trajectory of the agricultural system. Here again the characteristic of the technology itself creates the opportunity for a further commodification and appropriation of plant material. This distrust is aggravated by (anti)globalization and IPR issues, where corporate control and protection measures are thought to lead to injustice (Rifkin, 1998; Kloppenburg, 2004; Campbell, 2006; Hansen, 2006; Murphy, 2007; De Vriend and Schenkelaars, 2008).

After the turn of the century, the industry starts to use naturalness as an argument too. Modification is presented as the timely solution to battle negative effects of agricultural production on the environment. GM crops do offer tangible benefits in reduction of resource use, as they promise less chemical use and less water and energy inputs, thus saving the environment. Now nature and sustainability have become conceptually linked as a moral argument to advocate GM, coming from the logic that being eco-friendly and sustainable *is* following the rules of nature (Karafyllis, 2003; Amman, 2008).

Marker assisted selection

As the knowledge of genomics and molecular tools progresses further, alternatives to –transgenic – GM have come up and are presented as 'more natural'. Marker Assisted selection, as compared to GM would be neutral in respect to the problem of species boundaries. It also is based on traditional cross breeding. It does not involve gene isolation, direct modification and asexual insertion of genetic material. Greenpeace says to support MAS, as the technique makes use of the 'natural systems' that were established and refined by the mechanisms of evolution (Vogel, 2009). However, other issues remain. To develop the markers themselves, genetic modification is still needed and the interest of breeders and researchers still seems to be aimed at isolated quality traits and not to the plant as a whole (Chagné, 2007; Collard, 2007).

Also, MAS does not escape from the suspicion that it is a technology which enables a system that treats genetic data, combined with biotechnology in- and outputs, as ordinary tradable factors under a global IPR system (McAfee, 2003).

Organic breeding

Organic agriculture presents itself as a more 'natural' alternative to both high tech production and genetic modification. Organic breeding/farming not only points to avoidance of chemical inputs, but acknowledges the integrity of life. The naturalness of the plant refers here to inherent nature, wholeness, completeness, species specific characteristics, and being in balance with the environment. The used technology should be in line with principles of fairness and care. The approach is holistic and ecological, trying to give weight to systems of internal self regulation through non-appropriation, agro-biodiversity, stability and resilience (Lammerts Van Bueren and Struik, 2004).

One problem in organic breeding is that the available varieties currently used as stock, are mostly derived from conventional agriculture and were developed for a production system where in crop protection and growth characteristics, chemical input has taken over from natural resilience. *Ipso facto*, they are also stemming from a breeding process that for ages effectuated chromosomal rearrangements in cultivars. This exact argument is used by Ammann (2008) to suspect an *ideological bias* in the values of organic agriculture.

Still, in the organic movement, there are now discussions about the boundaries of the organic concept of 'naturalness' as the growing need for improved cultivars for organic agriculture cannot be met by current breeding schemes with the limited and flawed varieties that are available. Under conditions, organic agriculture is now considering MAS as a possible technological opportunity to enhance cultivars for production. But, only in an organic breeding programme, only if 'clean markers' are developed, and only if marker screening is performed without enzymes originating from genetically-modified microorganisms and without radiation (Lammerts van Bueren, 2003; Verhoog *et al.*, 2003; Lammerts Van Bueren and Struik, 2004; Verhoog, 2004; Backes and Ostergard, 2008).

A dynamic model to analyse naturalness in debates

In the old debate naturalness was used as a stop argument, in a new debate this must change to avoid a new stalemate. To do this, we need to design a dynamic framework that permits an integrative approach to naturalness and that avoids binary oppositions. For this purpose it is also necessary to focus on the use of the concept of nature/naturalness as it appears in the four agricultural debates. In order to develop a dynamic framework, let us try to group together the references to naturalness in a couple of value spheres. There are references that point to ecological values. Notions like exploitation versus the nurturing of arable land, the use of chemical inputs balanced against resistant diseases and/or pollution, bumper crops related to biodiversity/resilience and a relation between crop quality and quantity seem directed towards perceived (or existing) natural balances or boundaries of an ecological complex. These references do come back in all four debates and are themselves hardly disputed, but discourses differ about the way technology and human activity are either correcting or disturbing these balances and boundaries. The main ethical question of the ecological sphere is related to the permissible degree of exploitation of nature and natural resources, and affecting *'our sense of a natural balance between exploitation and caretaking'* (Rifkin, 1998; McKibben, 2005).

Also, there are references to the very technologies self, that are used in breeding & agriculture. Ideas are brought forward that technology (and capital) has taken over from the rural, more natural, production systems and that the technology self is thought to interfere with or infringe on natural schemes. On the contrary, there is also the notion that technology helps to reinforce existing natural mechanisms. There is a distinction made between nature friendly and nature hostile technologies. Another notion refers to the idea that it follows from the internal logic of (bio)technology itself, the actual complex of technique, know-how and ownership -not just its use but its mere existence- that it now needs to appropriate life and/or its building blocks as a requirement for its self preservation. The main ethical question in the technological sphere relates to decision making, control and ownership of a new technology that can have a significant impact on our life and society – affecting *'our natural sense of fair ownership and distribution of wealth'* (e.g. Dyson and Harris, 1994; Murphy, 2007; Hardt and Negri, 2009).

Finally, there are references made that point to cultural values. Natural boundaries between species are to be respected. We need to respect the wholeness of life forms. Neither life, nor its building blocks can be appropriated as they are autonomous. Direct intervention at the cell level or asexual insertion of hereditary traits is overstepping given boundaries. The main ethical question in the cultural sphere is about whether humankind, given the means, should be allowed direct intervention in given biological processes – thereby affecting *'our sense of the natural order of things'* (Soper, 1995; Lovelock, 2000; Baird Callicot, 2001; Eden, 2001; Lammerts van Bueren *et al.*, 2002).

Most of the different meanings and associations of naturalness come up in similar forms in every debate. But they *do* appear in different contexts and timeframes and they *are* used to underpin different opinions.

Conclusion

What we propose is a model formed out of the interrelation between the three principle value spheres that we found in all 4 discourses: Ecology, technology and culture. In this model naturalness is then defined as a balance between them. The assumption at the base of the model is that when a breeding method and its agricultural application finds acceptance on a cultural, technological and ecological level, it will be perceived as natural. The model will allow a description of different approaches as more natural or less natural in relation to one or more of the spheres and the model can be used to clarify how arguments of naturalness can shift from one scheme to another. Partial arguments can be positioned under one of the three spheres and contribute to an overarching, unifying concept of naturalness, which can be dynamic, as the balance is changing along with the developments of conceptualization in each of the three spheres. Future research will be directed to expanding on the model and test its validity. The practical use of the model will be tested during workshops.

Acknowledgements

The research for this paper has been financed under the program of the Centre for Bio-Sytems and Genomics (Wageningen) and the Centre for Society and Genomics (Nijmegen).

References

Amman, C. (2008). Integrated farming: Why organic farmers should use transgenic crops. New biotechnology 00, 1-7.

Anderson, J. (2005). A Conceptual Framework for Evaluating and Quantifying Naturalness. Conservation Biology 5, 347-352.

Backes, G. and Ostergard, H. (2008). Molecular markers to exploit genotype-environment interactions of relevance in organic growing systems. Euphytica, 523-531.

Baird Callicot, J. (2001). Multicultural Environmental Ethics. Daedalus 130, 77-97.

Campbell, H. (2006). Consultation, Commerce and Contemporary Agri-Food Systems: Ethical Engagement of New Systems of Governance under Reflexive Modernity. Integrated Assessment Journa Vol. 6, 117.

Chagné, D., Batley, J., Edwards, D. and Forster, J. (eds.) (2007). Single nucleotide polymorphism genotyping in plants. Springer, New York.

Collard, B.C. and Mackill, D.J. (2007). Marker-assisted selection: An approach for precision plant breeding in the twenty-first century. Philosophical Transactions, 557-572.

Cronk, Q.C.B. (2002). Perspectives and paradigms in plant evo-devo. In: Cronk, Q.C.B., Bateman, R.M., Hawkins, J.A. (eds.), Developmental Genetics and Plant Evolution, Vol. 65. Taylor & Francis, London, pp. 1-15.

De Vriend, H. and Schenkelaars, P. (2008). Oogst uit het lab, biotechnologie en voedselproductie. Jan van Arkel, Utrecht.

Dupré, J. (2004). Understanding Contemporary Genomics. Perspectives on Science 12, 320-338.

Dyson, A. and Harris, J. (eds.) (1994). Biotechnology and Ethics. Routledge, London.

Eden, S. (2001). Environmental issues: nature versus the environment? Progress in Human Geography 25, 79-85.

Goodman, D., Sorj, B., Wilkinson, J. (1987). From Farming to Biotechnology. Basil Blackwell Ltd., Oxford.

Gremmen, B. (2005). Genomics and the Intrinsic Value of Plants. Society and Policy 1.

Hansen, A. (2006). Tampering with nature: 'nature' and the 'natural' in media coverage of genetics and biotechnology. Media Culture Society 28, 811-834.

Hardt, M. and Negri, A. (2009). Commonwealth. The Belknap Press of Harvard University Press, Cambridge, Massachusetts.

Juma, C. (1989). The Gene Hunters: Biotechnology and the Scramble for Genes, London.

Karafyllis, N.C. (2003). Renewable Resources and the Idea of Nature – What Has Biotechnology Got to Do With It? Journal for Agricultural and environmental Ethics vol:16 iss:1 pg:3.

Kloppenburg, J.R.J. (2004). First the Seed. The political economy of plant biotechnology, 1492-2000. The University of Winsconsin Press, Madison.

Lammerts van Bueren, E.T., Struik, P.C. and Jacobsen, E. (2002). Ecological concepts in organic farming and their consequences for an organic crop ideotype. NJAS – Wageningen Journal of Life Sciences 50, 1-26.

Lammerts Van Bueren, E.T. and Struik, P.C. (2004). The consequences of the concept of naturalness for organic plant breeding and propagation. NJAS – Wageningen Journal of Life Sciences 52, 85-95.

Lammerts van Bueren, E.T., Struik, P.C., Tiemens-Hulscher, M. and Jacobsen, E. (2003). Concepts of Intrinsic Value and Integrety of Plants in Organic Plant Breeding and Propagation. Crop Science 43, 1922-1929.

Lovelock, J. (2000). The ages of Gaia: A biography of our living earth. Oxford University press, Oxford.

McAfee, K. (2003). Neoliberalism on the molecular scale. Economic and genetic reductionism in biotechnology battles. Geoforum 34, 203-219.

McKibben, B. (2005). The emotional core of the end of nature. Organisation & Environment 18, 182-185.

Murphy, D. (2007). Plantbreeding and Biotechnology. Cambridge University Press, Cambridge.

Rifkin, J. (1998). The Biotech century, harnassing the gene and remaking the world. Tarcher, Los Angeles.

Siipi, H. (2008). Dimensions of naturalness. Ethics & The Environment, Indiana University Press 13(1), 71-103.

Smits, M. (2002). Monster bezwering: De culturele domesticatie van nieuwe technologie. Boom, Amsterdam.

Soper, K., 1995. What is Nature? Oxford, Blackwell Publishers Ltd.

Stephens, P.H.G. (2000). Nature, Purity, Ontology. Environmental Values 9, 267-294.

Sulmasy, D. (2006). The Logos of the Genome: Genomes as Parts of Organisms. Theoretical Medicine and Bioethics 27, 535-540.

Van Wordragen, M. and Dons, H. (1992). *Agrobacterium tumefaciens*-mediated transformation of recalcitrant crops. Plant Molecular Biology Reporter 10, 12-36.

Verhoog, H., Matze, M., van Bueren, E.L., Baars, T. (2003). The Role of the Concept of the Natural (Naturalness) in Organic Farming. Journal of Agricultural and Environmental Ethics 16, 29-49.

Verhoog, H. (2004). Organic agriculture versus genetic engineering. Louis Bolk Institute, Driebergen.

Vogel, B. (2009). Smart Breeding: Marker-Assisted Selection: A non-invasive biotechnology alternative to genetic engineering of plant varieties. In: Steve Erwood, N.T. (ed.). Greenpeace.

Williams, R. (1983). Keywords, A vocabulary of culture and society. Fontana Paperbacks, London.

Ignored bioethics? On the role ethical concerns in GM food policies

J. Lassen
Danish Centre for Bioethics, Department of Human Nutrition, Faculty of Life, University of Copenhagen, Rolighedsvej 30, 1958 Frederiksberg C, Denmark; jlas@life.ku.dk

Abstract

It is a well documented fact that European public concerns over agricultural biotechnology are based on arguments about health and environmental risks, usefulness, and what has been termed 'ethical', or, 'wider moral' concerns. These 'wider moral' concerns include worries about animal integrity, naturalness, justice, control, and fairness. The re-emergence of the GM food controversy in the mid 1990s can partly be explained by the failure to include public concerns over usefulness and wider moral concerns in the regulation of GM foods and agriculture. At both national and EU level, the political reaction to the re-emergent controversy involved, attempts to re-arrange existing regulation of GM foods and agriculture in response to this failure. The question, however, is this: To what extent have political discussions of GM foods and agriculture included issues beyond risk, and, thus, established a base for the development of regulation reflecting the width of public concerns? Drawing on a case study of debates in the Danish parliament, this paper examines and characterizes different storylines/ discourses about GM foods. Particular attention is paid to the inclusion and exclusion of arguments about ethics/'wider moral concerns' in what can be characterized as the cultural discourse. Drawing on the results of the case study, the paper will discuss the extent to which political discourses reflect public concerns, and, in view of this, the risk of a new re-emergence of social controversy over GM foods and agriculture.

Keywords: gene technology, discourses, debate

Introduction

Over recent decades many studies have analyzed the structure and content of public worries over GM foods. Some place the concerns in sub-categories, such as risks (pertaining to health and environment); utility and ethics (or 'other moral concerns') (Gaskell and Bauer, 2002; Lassen *et al.*, 2002); others pursue a more discursive approach and characterize concerns within different storylines or discourses of concern, over technological change (Lassen and Jamison, 2006). Three such storylines can be identified: a social story – in which the concerns address issues related to the environment and health, highlighting risks and uncertainty – an economic story addressing issues related to production and profitability as well as highlighting costs and benefits, responsibility and power – and a cultural storyline addressing moral and religious aspects such as ethics, rights and integrity.

Irrespective of the way concerns over GM foods are analyzed, many of the studies support the conclusion that the re-appearance of the European controversy over GM foods in the mid 1990s, following an attempt to introduce GM soy into the European food system, was the result of a mismatch between national/EU policies and the nature of the public concern. The political debates about GM foods tended to be restricted to risks and utility – and, thus, restricted to the economic and social storylines. The resulting public regulation within the EU and its member states was consequently restricted to regulating risks to health and the environment. By contrast, public concern, as noted above, also had a cultural dimension. Some studies even found that the moral concerns included in this storyline had a veto-like nature (Anonymous,. 1997). Others pointed to the fact that, within the social storyline, there was a strong public voice demanding evidence of the specific (societal) usefulness of GM foods before they could be accepted, and not just economic benefits to companies or individuals (Lassen *et al.*, 2006).

In many ways the second half of the 1990s mark a transition from political discussion of GM foods, which, at the programmatic level, now also included some of the concerns previously excluded. An example is the political discussion up to, and following, the so called *de facto* moratorium blocking approval of new GM foods in the EU in 1999, which resulted in a revision of the EU directives (Levidow and Carr, 2010). In this process new desiderata such as ethics, transparency and a focus democratic decision-making were included alongside the existing focus on environmental and health risks.

The aim of the paper is to study the extent to which this turn in political focus is reflected in the way politicians argue for and against GM food. Or, more precisely, to answer the question: Has the political discussion of the first decade of the twenty-first century mirrored the variety of public concerns identified in many empirical studies of public perceptions of GM foods?

To answer this question a case study is made of selected debates about GM foods in the Danish parliament. This is compared with an earlier study of the nature of public concerns in Denmark. (Lassen and Jamison, 2006). Danish parliamentary debate about GM food is a critical case (Flyvbjerg, 1998) and, is thus a particularly good case, since political processes at the end of the 1990s specifically focused on the inclusion of public concerns in the political process (for detailed discussion, see Lassen, 2004).

In spring 1997, shortly after the soya issue peaked, the birth of the cloned Dorset ewe, instantly famous as 'Dolly', was announced. Although strictly speaking Dolly was a medical application, the event fed into public as well as parliamentary debates about genetic technologies in Denmark. Whereas the Parliamentary debates following the introduction of GM soy were largely restricted to the economic and social storylines, Dolly invited the inclusion of the cultural storyline in debates about GM foods and other applications of genetic technology. Hence, issues concerning animal integrity and our right to make radical interventions in 'Creation' or 'Nature' were now taken up. In parliament 'ethics' was now a keyword in debates, and the result was a parliamentary decision instructing the Government '... *before the end of 1997 ... to find a method ensuring that intervention against developments offending ethical norms of society can take place in due time; and that a foundation for a renewed debate in the general public as well as in Parliament of ethical limitations to research, is created*' (quoted from (Lassen, 2004)). As a consequence the Ministry of Trade and Industry (*Erhvervsministeriet*) set up of the so-called BioTIK Committee, which was instructed to produce a paper that could provide a basis for balanced debate and increased understanding of public concerns. This paper (BioTIK Gruppen, 1997) is the first example in Danish biotechnology history juxtaposing benefits and worries within social, cultural and economic story lines in the area of GM foods. Apart from representing this discursive change, the paper also provided a vocabulary that could be used when talking about genetic technologies; this vocabulary included means to discuss ethical issues related to GM plants as well as GM animals and thus facilitated the discussion of important elements within the cultural discourse of concern.

Two years later the Minister for Trade and Industry produced a statement –largely based on the conclusions in the BioTIK paper – stating that it was the policy of the Danish government to work for the adoption of an international convention making the inclusion of ethical considerations in biotechnology regulations in the area of plants and food, mandatory. It furthermore stated that the government would work for the inclusion of ethical factors in the regulation of gene technology, specifically seeking to ensure that the pending revision of the EU directive on deliberate release included ethics; and finally that the government would develop guidelines for ethical assessments and support debate and information (The Ministry of Trade and Industry, 2000).

It can be seen, then, that by the turn of the century the Danish parliament had tools which, in principle, would enable them to debate GM food issues in a way that reflected public concerns. Below, an analysis of the 2002 parliamentary debates of a bill implementing the revised directive on deliberate release of genetically modified organisms is presented. Since one new aspect in the directive is a requirement to consider procedures regarding ethical problems raised by the application of genetic technologies, it can be safely assumed that this debate would provide an opportunity for the members of parliament to expand their discussion into the cultural story line.

Results

The following paragraphs briefly summarize the parliamentary discussion of the bill during its three readings in March-June 2002. The analysis of the parliamentary debate identifies the arguments used by the participating politicians. Toulmin's theory of the structure of arguments (Toulmin, 2003) is then brought into the picture.

The economic story line

Many arguments, both for and against, can be allocated to the economic story line, though it is never a key issue in the debates. The economic story line mostly appears in the shape of the hope that the passing of the act will provide a stable economic environment in future. Thus, in his opening remark, the Minister for the Environment argues about a simplified approval procedure included in the bill: '*When this is adopted, it will constitute a relief for industry, because it shall no more seek approval under different acts before a product can be approved*'. Others talk about the way the new technologies offer an opportunity to manipulate nature, opening up *'fantastic perspectives for agricultural production'*. Opposed to this optimism, one of the critics flags a concern over the economic impact on organic farmers of the uncontrolled spread of GMOs, and raises questions of liability.

Repeatedly, critics raise the issue of power, arguing that the proposed labelling is insufficient because there is no requirement to label meat products from animals that have been given GM fodder: '*...this opens up a barn door for the use of gene technology in products without the consumers knowing it*' Others welcome the way the bill facilitates transparency in the political process by enabling individual citizens to comment on applications, and to have better access to information.

Although the usefulness of GM technology is most often discussed in terms of the economic benefits for business, some participants, in accordance with the public call for societal usefulness, mention other kinds of usefulness. However, these points are never used actively as arguments for or against the bill – they seem instead to function as preliminaries, mentioned before the speaker engages in the (more important) arguments. An example of this is detectable the following remarks made a parliamentarian: '*In the view of the Socialist People's Party (the debates about) genetic technologies suffer from a lack of preceding public discussion of what societal needs (this technology) could meet. In stead we have witnessed a development where the technology has been applied in a way where commercial interests have been the driving force, not societal needs.*' Again it is interesting that this is a detached statement; it is not used in argumentation for or against the bill.

The societal story line

Compared with the economic story line, the social story line is strongly present in the debates. Not surprisingly, health and environmental risks are taken up by most debaters. These risk-related discussions most often revolve around the issue whether the bill will in fact provide sufficient protection from potential environmental risks. This concern is generally taken up by critics reflecting on worries as to whether long-term risks are adequately covered. A concern that by the proponents are met by the argument that the bill opens for approvals limited in time, and thus better will be able to take new knowledge into account.

Another prominent debate concerns the spread of GMOs in the environment, and the possible environmental consequences of this. Most agree that the escape of GMOs *per se* is a problem. Here a proponent of the bill questions a reply relocating the question of GMOs invading ecosystems to the economic story line by arguing: '*(The reply argues that GMOs polluting the fauna) is not an environmental problem, but solely a problem existing between different agricultural systems. I recognize that it is a problem for organic producers (...but) if we see a spread of GMOs it is a problem and an environmental problem.*'

The cultural story line

In light of the fact that the BioTIK report and Governmental statement, which specifically extended the GM food debate to include cultural issues, were only a few years old at the time of the debate, the cultural story line is strikingly absent. It must, however, be acknowledged that ethical issues are taken up at the very general level by several participants in the discussion. Thus many parliamentarians argue that they are happy that the bill requires decisions about the application of genetic technologies in the future to be based on a respect for ethical values, just as they acknowledge the instruction of the Danish Board of Ethics to review ethical aspects, when relevant. Ethics, however, remains a black box that none of the debaters seriously attempts to open; instead ethics is treated as a dimension of debate which most agree needs to be taken into account, but which nobody discuss in detail. An example is this contribution: '... *it is thus not strange that genetic technology sometimes stimulates very intense discussions. To some it creates fear that we are gradually losing control of the use of genetic technology. Others have concerns of a more moral or religious nature. But this is not, and should not be, a fundamental debate about genetic modification.*' This parliamentarian elegantly stresses that ethics is important, but, at the same time, she rules out the use of ethical arguments in the actual debate, since they are only relevant when debating the pros and cons of genetic technologies at another level. It is remarkable that nobody feels obliged to argue against this rather extreme position in the ensuing discussion.

Another spokesperson, however, calls for elaboration of the ethical values the bill mentions: '*Then there is this idea about including ethics in the deliberations. Ethics is by definition something good, but I believe that you need to elaborate the ethical principles one considers taking into account.*' Interestingly, this call for elaboration is not followed by any suggestion, by the proponent, as to what ethics could be about; nor does it prompt debate about the relevant ethical issues.

Conclusion

The 2002-bill implementing the EU directive on deliberate release in Danish GM-regulation is an extreme case for several reasons. First, the directive was one of the core pieces of regulation meant to account for the public concerns beyond the social and economical story line. Second, the bill specifically mentions ethics as a new issue to be included in the regulation. And third, the cultural story line about GM foods was opened few years earlier, and the Danish parliamentarians had even demanded, and received, means to talk about these issues. Despite this ethics, a core element within the cultural story line, never played any particular role in the debates – apart from rather programmatic statements arguing that ethics is/is not relevant it remained a black box.

The unwillingness to take up ethical issues cannot be explained by the parliamentarians lacking ability or will to discuss ethical issues – other debates in parliament about e.g. *in vitro* fertilisation, abortion or animal welfare clearly demonstrate that this is not the case. These debates, however, typically relate to issues involving humans or animals – areas, where there is a long tradition in Denmark as in other European countries for political debate and public regulation. Apart from the use of GM animals, this is not the case for crop plants and thus by far most of the GM foods covered by the directive and the bill.

A consequence of this failure to mirror the public concerns in the parliamentary debates may very well be that controversy over GM foods re-opens in the future. This may happen first, because there at present is no sign of dramatic changes in public perceptions of GM foods and second, because parliamentarians largely neglect public concerns within the cultural discourse and third, because the adopted bill – as the directive – does not include mandatory ethical assessments, but merely ethical reviews with no legal status in the approval procedure.

References

Anonymous (1997) Europe ambivalent on biotechnology. Nature 387: 845-47.

BioTIK Gruppen (1997). De genteknologiske valg. Erhvervsministeriet.

Flyvbjerg, B. (1998). Rationality and Power: Democracy in Practice. University of Chicago Press.

Gaskell, G and M. Bauer (2002). Biotechnology 1996-2000: the Years of Controversy. Science Museum Press.

Lassen, J. (2004). Changing Modes of Biotechnology Governance in Denmark. STAGE Discussion Paper no 3.

Lassen, J., K.H. Madsen, and P. Sandoe (2002). Ethics and genetic engineering – lessons to be learned from GM foods. Bioprocess and Biosystems Engineering 24(5):263-71.

Lassen, J., M. Gjerris, and P. Sandøe (2006). After dolly – ethical limits to the use of biotechnology on farm animals. Theriogenology. 65:992-1004.

Lassen, J. and A. Jamison (2006). Genetic Technologies Meet the Public: The Discourses of Concern. Science, Technology & Human Values 31(1): 8-28.

Levidow, L. and S. Carr (2010). GM Food on Trail. Testing European Democracy. Routledge

The Ministry of Trade and Industry (2000) The Danish government statement on ethics and genetic engineering. The Danish Ministry of Trade and Industry.

Toulmin, S. (2003). The Uses of Argument. Cambridge: Cambridge University Press.

Part 15. Ethical, legal and social issues concerning new systems, components and foodstuffs

The farm-to-fork approach of food security concerning chemicals (with special reference to REACH, a 'new' European Community Regulation)

E. Atienza-Macías
Inter-University Chair in Law and the Human Genome, University of Deusto, University of the Basque Country, Avda. Universidades N° 24, 48007 Bilbao, Spain; elena.atienza@deusto.es

Abstract

In the last few years, food security analysis has become a matter of great concern due to the existing controversies regarding to the way of production and interaction with the Environment. In the light of this, the European Union has given food security top priority within its policy agenda. Actually, this has turned out to be a horizontal objective to be taken into account in several areas falling under Community powers such as environmental awareness, amongst others. Furthermore, the publication of a White Paper on food security heralds the development of a legal framework covering the entire food chain – *'from farm to fork'* – using a global approach. On the basis of all this, a 'new' European Community Regulation concerning the Registration, Evaluation, Authorisation of CHemicals, better known as 'REACH', perfectly combines two prevalent challenges, namely, environment protection and food security. The main aims of this Regulation are, on the one hand to improve the protection of human health and of the Environment from the risks that chemical substances might pose for both, and on the other hand to promote the safe handling and use of substances across all industrial sectors, where food industry plays a key role. We should bear in mind that, in theory, the use of substances in this latter sector (for instance, food ingredients, which are already covered by the EU General Food Law Regulation 178/2002, do not have to be registered under the new REACH legislation) did not fall under the scope of that Regulation. Nevertheless, the impact that the EU's Chemicals Regulation has on food industry is subject to controversy. This is why EFSA produced a draft entitled *'Management Plan of the European Food Safety Authority for 2007'* where it announced that the impact of the new EU chemicals legislation on the food sector should be assessed and EFSA should liaise closely with the Commission and the new European Chemical Agency on these matters.

Keywords: chemicals, European Union, food safety, human health, the environment

Introduction and overview

REACH Regulation: a milestone in the development of the European Union´s environmental policy

After several years of development, debate and amendments, on 18th December 2006 the European Parliament and the European Council adopted the Registration, Evaluation, Authorisation (and the underlying principle of Restriction) of CHemicals (hereinafter referred to as 'REACH'), Regulation (Regulation No 1907/2006).

Though seven years were needed to write its 849 pages the final work produced has been described as the most comprehensive legislation in the European Union's history and the most important in the last 20 years. As a matter of fact, to date it is the strictest law regulating chemical substances and will no doubt have a great impact on industries all over the world. The European Parliament described REACH as *'the single most important dossier ever to be discussed within its walls'.*

The enactment of the Regulation, which entered into force on 1st June 2007, has become a significant milestone in the development of the EU's chemical regime. In practical terms, it is the starting point for a wide range of reports, assessment and management measures on chemicals which will be gradually transposed into the domestic legislations of the EU countries.

Moreover and importantly, the broad regulatory framework of REACH has replaced forty individual legislative instruments and has meant a fundamental shift in the way in which manufactured and imported chemical substances will be regulated within the EU.

In this respect, it could be argued that the new regulatory regime would not cover a large number of applications or preparations in their finished state since they are in fact regulated by a wide range of Regulations and Directives. Food or feedingstuff (including food additives, flavorings, animal nutrition) have a paramount position on this list of substances. As a matter of fact, this issue will be discussed in detail in this paper since both food and feedingstuff will actually fall within the scope of the new regime.

State of affairs before REACH: a patchwork of Regulations and Directives

The former legislative framework for chemical substances consisted of a network of Regulations and Directives described by some authors as 'notoriously dense and ineffective from an environmental and public health point of view'. These inherent weaknesses of the chemical regime were highlighted by the European Commission in its 2001 White Paper: *'Strategy for a Future Chemical Policy'*, along with the Commission's concern about the slow and reactive nature of the regime regarding substances linked to human and environmental damage, coupled with the obvious lack of innovation within the EU chemical industry.

As a result of this, the EC put forward the creation of a new, coherent chemicals regime to be applied on a uniform basis to both new and existing chemicals (a traditional distinction of substances superseded by REACH).

The scope of REACH: key objectives and challenges

REACH lays down a comprehensive framework to achieve three pivotal targets for European market regulation, namely, (1) offer of a high level of *health and environmental protection* from chemical risks, (2) facilitation of *market integration*, and (3) promotion of *innovation* in the chemicals sector. In this sense, REACH could be deemed to be an embodiment of three overarching programmes currently underlying policy developments within the Community Pillar of the EU.

Firstly, the purpose of the scheme is to ensure a high level of protection of human health and the Environment. In order to *reach* this overarching objective, the Regulation prohibits the manufacturer or sale of any substance in the EU that has not been previously registered with the European Chemical Agency 'ECHA', which will be dealt with later on in this paper. Pursuant to the Regulation this accounts for the saying: '*No data, no market*'.

Registration requirements of a substance will be dependent on the mass of the substance manufactured or imported per calendar year. The threshold level for the registration of a substance on its own in preparation or a substance in an article is one tonne per year per manufacturer, producer or importer.

A cornerstone of the EU's new regulatory regime is the obligation that it has placed on registrants to identify and manage substance risks, thereby shifting *the burden of proof* regarding safety away from the public sector and placing it on the *industry* itself. The issue of the burden of proof reversal is arguably better illustrated by the following words: *Traditional regulation has assumed that chemicals are innocent until proven guilty; REACH will take it for granted that chemicals, especially when used in large volume, are suspect until proven innocent by its supplier.*

Hence, as far as health and environmental protection expectations are concerned, the express reference to the precautionary principle is clearly the guiding principle underpinning this policy reversal.

Finally, the fact itself that the goal being pursued is the fostering of innovation and competitiveness is a clear indication of the key role of the 2000 Lisbon Strategy, which set as one of its main targets a scenario where by 2010 Europe should be 'the most competitive and dynamic knowledge-based economy in the world'.

REACH: a model for international risk governance

The impact of the EU 's new chemical regime goes far beyond its own borders and this is the reason why this section will briefly consider the broader impact of REACH.

The European expectations that REACH will actually become a world standard for chemicals management are well grounded and the passing of time has not really killed or even cooled down these initial expectations. In fact, countries such as Switzerland, Norway, Japan, Canada, Korea, New Zealand and China have clearly expressed their keen interest to 'learn from REACH', but even in those countries where their government has not expressly declared such intention, for instance the United States (USA, for short), REACH is being used as a benchmark to assess their domestic chemical risk regulations.

Extent to which the European chemical regime is expected to have an impact on the United States Regulations

The real issue at stake in the particular case of the USA is that chemicals exported to Europe are required to conform to the same environmental standards as chemicals produced within the European Union, and the amount of chemical exports to Europe that are subject to REACH is really high.

In practical terms, in the case of some USA states such as Massachusetts, Texas, California, Louisiana, West Virginia and Kentucky, the impact of REACH is far greater than that on the nation's chemical industry as a whole.

The initial question to be answered was whether REACH would impair American Chemical Industry due to the expensive regulatory burdens on its exports. At the beginning the chemical industry on both sides of the Atlantic was a detractor of REACH, but as the debate went on American firms have been slower to accept REACH than their European counterparts, and the reason accounting for this is obvious since as it has been clearly documented by a large number of studies, compliance with REACH does not really involve a substantial financial commitment, though maintenance of chemical sales is really paramount. Recent American experience has shown that the USA are likely to lose access to export markets, due to both their non-compliance with environmental standards and their failure to respond to the concerns of those countries buying American goods.

The impact of REACH on developing countries

On the basis of the Report produced and sent by the International Chemical Secretariat (Chemsec) to the European Parliament in March 2006, the scope of REACH for 'ACP countries' (i.e. the African, Caribbean, and Pacific Group of States, which is really an *ad hoc* group consisting of seventy nine developing countries having a special and long-standing relationship with Europe), should not be viewed as a rich country's luxury imposed on low income countries, since as a matter of fact some of the greatest beneficiaries of REACH could be businesses and workers in developing countries.

Thanks to REACH, businesses will gain access to key information about the potential effects of their products, and of the materials and substances they use, and this will, in turn, help them identify and implement safer alternatives, where appropriate, and also to be discharged from any future liability for damages. On the other hand, workers will also benefit as many chemicals pose greater hazards for those employees who handle them on a daily basis than for consumers of finished products.

Consequently, the European new regime could be particularly important for industries that are increasingly operating in developing countries.

A question subject to debate

Pursuant to Article 2(5) b (Application) of the REACH Regulation (EC) 1907/2006: '*...The provisions of Titles II, V, VI and VII shall not apply to the extent that a substance is used: in food or feeding stuff in accordance with Regulation (EC) No 178/2002...*'

Prima facie, it would appear that the core provisions of REACH, namely, the titles on Registration, Information in the supply chain, Evaluation and Authorisation would not apply to the food sector.

However, the use of other substances in food industry, for example in packaging, cleaning and sterilisation, and in some processes such as refrigeration, is regulated by REACH.

As a result of, and also due to the existence of a 'Memorandum of Understanding' which gives rise to a close collaboration between the forefront-Authority of EU food safety risk assessment (EFSA) and the keystone of Chemicals (ECHA), it might be possible that full exclusion were really inexistent, and this is an issue that will be dealt in this paper.

The role of chemicals and the REACH regulation in food security

Chemistry is a usual component of our meals as a large number of chemical substances are commonly added to food in order to increase their life span and/or improve their taste. Since ancient times, additives have been used to preserve food and make it more palatable. However, the increased processing of food in the 20th century created the need for new types of additives and eventually resulted in their increased use. Moreover, the use of chemicals in food production is subject to their full compliance with the relevant food security requirements.

Likewise, being aware of its importance, REACH focused on the principle of 'human health' as one of its main objectives. For that reason, the use of some products in the food industry such as in packaging, cleaning and sterilization, and in some processes such as refrigeration, is covered by REACH. Under EU legislation all materials coming into contact with food must comply with some health standards.

While chemicals used in packaging do fall under the scope of the REACH provisions, the inclusion of food and foodstuffs within this Regulation is subject to controversy. Thus, due to the lack of clear provisions in this regard, food industry has expressed its concern about the uncertainties relating to the scope of the legislation and, due to this has held a large number of workshops aimed at assessing the extent to which their sector may be affected by the law.

Despite the fact that pursuant to Article 2(5) b of REACH the use of substances in the food sector was exempted from compliance with it, the *'Management Plan of the European Food Safety Authority for 2007'* where announces: '*The impact of the new EU chemicals legislation on the food sector will be assessed and EFSA will liaise closely with the Commission and the new European Chemical Agency on these matters. The implementation of REACH will most likely require a reconsideration of risk assessment approaches at the European level*', and the *'Management Plan of the EFSA for 2010'* where this aim is confirmed, suggests that the role of food sector in REACH Regulation is a crucial one.

Joint Co-operation between ECHA and EFSA in the pursuit of synergy development

On the one hand, the European Chemicals Agency (ECHA for short), which was established on 1 June 2007 and currently is at the centre of the new regulatory system for chemicals in the EU, set out in the aforesaid Regulation (EC) No 1907/2006. Its goal is twofold since it is aimed at becoming the internationally recognized Agency on matters relating to the safety of industrial chemicals while it is also aimed at being a source for reliable, trustworthy and high quality information on chemicals. On the other hand, the European Food Safety Authority (EFSA), established by Regulation (EC) No 178/2002, is the keystone of EU risk assessment regarding food and feed safety.

Within the framework of ECHA's professional liaison policy, the Agency did not only intensify its liaison with the European Commission through its relationship with the Committee for Environment, Public Health and Food Safety but it also established one of its first contacts with EFSA regarding the sharing of best practices and procedures, which may result in an Inter-Agency Memorandum of Understanding (MOU) on the basis of specific areas of cooperation, in particular, the need to establish rules of procedure concerning substances.

'Risk assessment methodologies for food/feed and chemical safety' is one of the most relevant areas from those to which the MOU applies.

Conclusions and suggestions for further research

The adoption of REACH has become a landmark in the development of the EU's chemicals legislation, but its impact actually is far more reaching since REACH in actual fact its potential benefits are expected to have an international impact.

Despite the fact that at the initial stages food did not fall within the scope of REACH, our construction of the Regulation suggests that substances in the food sector do also fall within the scope of that regulatory regime. Nevertheless, there is still a great amount of uncertainty as to how REACH could be construed and also regarding the way in which it could be applied to food sector. As a matter of fact, such an uncertainty could have significant economic impacts on manufacturers, importers and producers of food.

In the light of this, it is our contention that institutions like ECHA and EFSA should not only have shed some light on this key issue but should have also set the production of guidance material for food industry as one of their top priorities in order to provide greater certainty for industry. This is actually the only means available to be able to assess the full impact and cost of REACH on the food sector.

Acknowledgements

This work has been funded by a Grant of the University of Deusto.

References

Ackerman, F. (2006). Implications of REACH for the Developing Countries. Report from the International Chemical Secretariat (Chemsec) to the European Parliament, March, 2006, available at: http://www.ase.tufts.edu/gdae/Pubs/rp/Implications_of_REACH.pdf.

Ackerman, F., Stanton, E., and Massey, R. (2006). European Chemical Policy and the United States: The Impacts of REACH. Tufts University. Global Development and Environment Institute Working Paper No. 06-06. September 2006.

Anonymous (2008). New European Union Law REACH regulating the use of chemical substances. European Food Information Council (EUFIC) Newsletter No. 63 – June/July 2008, available at: http://www.eufic.org/article/en/page/FTARCHIVE/artid/New-European-Union-law-REACH-regulating-use-of-chemical-substances/.

Bowman, D. and Van Calster, G. (2007). Reflecting on REACH: Global Implications of the European Union´s Chemicals Regulation. Nanotechnology, Law and Business 4 (3): 375-384.

ECHA (2008). General Report 2008. The year of pre-registration, available at: http://echa.europa.eu/publications_en.asp.

ECHA (2006). Regulation No 1907/2006 of the European Parliament and of the Council of 18 December 2006 concerning the Registration, Evaluation, Authorisation and Restriction of Chemicals (REACH), available at: http://eur-lex.europa.eu/lexuriserv/lexuriserv.do?uri=oj:l:2006:396:0001:0849:en:pdf.

EFSA (2009). Memorandum of Understanding between ECHA and EFSA. Available at: http://www.efsa.europa.eu/en/euagencies/document/MoU_ECHA-EFSA,0.pdf.

EFSA (2009): Management Plan of the EFSA for 2010. Adopted in Stockholm, Sweden on December 17, 2009, available at: http://www.efsa.europa.eu/en/mngtplan/publication/mngtplan10.pdf.

EFSA (2007): Management Plan of the EFSA for 2007 (Draft). Done at Parma, on 23 January 2007, available at: http://www.efsa.europa.eu/EFSA/DocumentSet/mb_managementplan_29thmeet_en_3,0.pdf.

European Commission (2001). Commission White Paper on Strategy for a Future Chemicals Policy, COM (2001) 88 final (Feb. 27, 2001), available at: http://eur-lex.europa.eu/LexUriServ/site/en/com/2001/com2001_0088en01.pdf.

Heyvaert, V. (2009). Globalizing regulation: reaching beyond the borders of chemical safety. Journal of Law and Society 36 (1): 110-128.

Heyvaert, V. (2007). No data, no market. The future of EU chemicals control under the REACH Regulation. Environmental Law Review 9 (3): 201-206.

In vitro meat: promises and responses: cooperation between science, social research and ethics

C. van der Weele
LEI, Wageningen UR, P.O. Box 29703, 2502 LS The Hague, the Netherlands; cor.vanderweele@wur.nl

Abstract

In vitro meat is muscle tissue grown from animal stem cells through tissue engineering. It does not yet exist but the idea offers great promise: it potentially helps to solve the growing problems of meat. Technological promises are typically met by moral responses in which scepticism and critical questions are prominent. But debates on *in vitro* meat so far deviate from that pattern. Society seems to call for its development, and moral evaluation tends to be very positive as well; clearly, antagonism between technology and morality is not inevitable. Since the development of *in vitro* meat is still in an early phase, it offers the opportunity of an early integration and co-development of biotechnological, social and ethical research. We hope to make full use of this opportunity in a joint and multidisciplinary research project of Utrecht University and Wageningen UR.

Keywords: moral debate, technology, yuck factor, tissue engineering

Introduction: the idea of *in vitro* meat

Meat has become a problem. Intensive husbandry has been the subject of moral criticism for decades, and meat consumption is now recognized as a big and growing problem in other respects as well. The FAO report *Livestock's Long Shadow* (Steinfeld *et al.*, 2006) has been a milestone in raising wide awareness about the impact of livestock on the environment. The report presented shocking facts and figures, for example that livestock are responsible 18% of greenhouse gas emissions and 70% of agricultural land use, and contribute hugely to pollution, deforestation and water depletion. Adding to the alarm is the expectation that global meat consumption will double in the coming decades.

Within the affluent world's (over-)consumption patterns, meat is probably the most urgent and problematic element, and efforts directed at dietary change are being launched. The Dutch government, for example, is looking for ways to accomplish a transition towards the consumption of more sustainable sources of protein (Ministry of LNV, 2009). Such a transition will be far from simple, since most people enjoy meat very much. While meat consumption in developed countries is now more or less stable, it is greatly increasing in China and other upcoming countries. Big global increases in meat consumption appear to be inevitable.

In this situation, there is room for new ideas. One such new idea is to produce meat with the help of tissue engineering: *in vitro* meat. The basic idea is simple: you take stem cells, either embryonic stem cells or adult (muscle) stem cells, and once you have established the right conditions, you let them specialize to muscle cells and grow into edible muscle tissue. Potential benefits are conspicuous: no animals are involved, there will be gains in sustainability, and much land will become available for other uses. *In vitro* meat is an idea under research, not yet a reality. In principle, the idea is feasible, and progress is being made, though many obstacles are still to be overcome: finding the right cell lines, suitable media for culturing the cells and ways to obtain a three dimensional structure are still challenges (Haagsman *et al.*, 2009; Langelaan *et al.*, 2010).

Though an edible product that can be produced reliably and in sufficient quantity may be years away, the idea of *in vitro* meat has met with much societal response right from the start. This paper will take a look at those responses, taking the usual pattern of moral debates about new technologies as a background expectation. It is becoming clear that *in vitro* meat discussions deviate from this pattern. Is this an indication of shifts in technology-morality relations?

New technologies and moral debate: roles and patterns

After several decades of moral debates on new technologies, patterns can be discerned in how new (bio-) technology is received and discussed from moral points of view. The first thing to notice is that science and technology tend to set the agenda for such debate, by developing new technologies at the frontiers of their efforts. Ethics has been following these frontiers through programs such as ELSI, which stands for Ethical Legal and Social Implications (of new technologies). ELSI-genomics is the best known example, but ELSI stands for a more general phenomenon: whenever a new technology emerges, ethics comes in and looks at the ethical implications, aspects or issues. The debates are often characterized, certainly in the early phases, by a division of roles. Technologists sketc.h the promises and potential benefits of the new technology, while ethicists subsequently argue that this framing is far to narrow and fails to take several problems into account. Ethics has thus become associated with a critical role and a focus on unknown risks. The reasons for this critical atmosphere in ethics are diagnosed in different ways. Philosopher of technology Verbeek (in press) emphasizes the humanistic background assumption that the human sphere and the technological sphere are and should be separate, an assumption which turns the moral role into a guarding one: the life sphere should be protected from technology. Others focus on social dynamics, pointing out that technologists, in their optimism as well as their need for funding, create waves of hype-like expectations, which are one-sided and need to be complemented. Swierstra and Rip (2007) analyzed patterns of moral argumentation about New and Emerging Science and Technology (NEST) along these lines. The role division is an element of a habitual pattern of argumentation, in which ethics has got the name to put a brake on technology, though in fact many of the arguments in favour of technologies are just as ethical.

Swierstra and Rip detect a sequence of moves in the early phases of debates. First, a new technology full of options and promises is announced. In response, people pleading for prudence and precaution warn that novelty comes with uncertainty and ignorance about effects. The promoters then often react by downplaying the novelty. For instance, they argue that genetic engineering is really nothing new; nature itself has been doing it through the millennia and we are just doing it faster. Next, technology critics refocus on problems in the existing situation. So the moves continue. The steps in this pattern have become expected as moves in a game, the authors write, and the pattern creates the positions of proponent and opponent, in which every move will be interpreted as a pro- or counter-move. The innocent bystander who inquires into side effects thus receives the label of an opponent (Swierstra and Rip, 2007: 9).

Thus, there is a clear pattern of expectation for debates on a new biotechnology, and the first steps of the *in vitro* meat debate seem to fit: big promises have been announced. But let us take a better look.

Promises

Calls for *in vitro* meat go back to Winston Churchill, who in the 1930s called it an absurdity to raise a whole chicken in order to eat just a breast or a wing. Some five decades later, the Dutch business man and researcher Willem van Eelen also came up with the idea of growing muscle tissue *in vitro*, and together with two allies acquired a patent on a not-yet-realizable route to *in vitro* meat in 1999 (In Vitro Meat Foundation, 2010). In 2003, the Australian artists Oron Catts and Ionat Zurr succeeded in keeping muscle tissue from a frog alive for three months and let it grow somewhat. The projected, called 'Disembodied Cuisine' yielded very tiny steaks that were marinated in calvados, fried in honey and garlic and served as pieces of 'victimless meat' during a dinner performance in a museum in Nantes, France (Tissue Culture and Art Project, 2003; Van der Weele, 2007). In 2004, Jason Matheny, an American student of public health, visited some farms and studied the statistics of meat. Alarmed by what he learned, and convinced that there was not enough time to wait for a massive dietary transition, he put his hope on technology and founded *New Harvest*, an organisation to promote research into meat substitutes, specifically *in vitro* meat (Schonwald, 2009). Matheny's influence helped to get a four years research project off the ground in the Netherlands, which was the first serious research effort into the development of *in vitro* meat. The project ran from 2005 to 2009 and made progress on the way to *in vitro* meat, but at its completion, a product was still a promise for the future.

Brief though this overview is, if suffices to see that the early promoters of *in vitro* meat are not the usual suspects. These were not scientists who believed in their research and needed more money, but people form other fields: a journalist-politician, a business man, bio- artists, a public health student. Voices from society, rather than responding to new technology with scepticism, have been pointing out the promise, calling for science to step in.

Responses

After atypical origins of the promise, it need not surprise that moral responses too are deviating form the expected pattern, according to which they should be critical. In fact, many moral and morally inspired responses can be seen as a continuation of the societal call for action. Animal protection organizations, for instance, have reacted with interest. In 2008, the American organization People for the Ethical Treatment of Animals (PETA, 2008) announced a $1 million prize to whoever is able to make *in vitro* chicken meat that is indistinguishable from real chicken, and sell it commercially by June 30, 2012. Though it is not very likely that anyone is going to win this money, the initiative illustrates that from animal welfare quarters, at least some of them, the prospect of *in vitro* meat is actively welcomed. Ethicists have joined the call. Patrick Hopkins and Austin Dacey (2008) were the first to review moral arguments for and against *in vitro* meat. They framed it as a potential solution for people who want to eat meat yet do not want to contribute to animal suffering. The arguments in favour, they write, are clear and straightforward and essentially show the hopeful outlook of a technological fix: technology may allow us to change the physical constraints of the world so that we can better avoid the bad and pursue the good. So what about the counter-arguments? The paper discusses no less than 13 real and potential objections. Let me pick out the two that seem to be most real (as opposed to potential) and describe briefly how the authors deal with them: the 'yuck factor', which is certainly present in some public responses to *in vitro* meat, and moral objections to technological fixes.

With regard to the yuck factor, Hopkins and Dacey acknowledge that feelings of disgust should be taken seriously, but they also argue that these feelings do not present the ultimate wisdom. First, feelings should be supplemented and modified through rational analysis to yield moral arguments. Second, first reactions of disgust are bound to change through familiarization. With regard to technological fix, they discuss the idea that a technological fix amounts to moral cowardice. A social problem, according to this criticism, should be solved through genuine moral work; it calls for virtue, not for technology. The authors' response is consequentialist and pragmatic. If moral arguments or moral virtue could do the work, fine. But if this turns out not to work, why not choose a more effective way to reduce animal suffering? Reducing this suffering should be the main goal, they argue, not cleansing our souls or boost our moral self-regard (Hopkins and Dacey, 2008: 589)

Hopkins and Dacey are strongly in favour of the development of *in vitro* meat. In their view, this development is not just interesting, but something that we may be morally required to support. They add that this also involves a recognition that 'morality is not something that must simply respond to new technologies as they arrive, throwing us into confusion, but rather that morality may champion and assist in the development of new technologies' as a step toward a better world (Hopkins and Dacey, 2008: 595).

A need for integrative knowledge

Hopkins and Dacey's paper not just illustrates a morally positive response; the authors give this response a wider meaning by pointing to a constructive role for ethics in the development of technology for a better world. This remark implies, rightly I think, that the unusual pattern of moral debate on this new technology is more than an interesting deviation. It can be understood as a symptom that cooperative interactions between morality and technology may be, or have become, somewhat marginal, but are now being (re)discovered. The type of debate in which science frontiers optimistically set the agenda and society and ethics respond with critical scrutiny is increasingly diagnosed as being unsatisfactory. For example, Alfred Nordmann (2007) has criticized the 'speculative ethics' associated with this pattern. Such ethics suffers from what he calls an 'if-and-then' syndrome, in which potential future moral dilemmas are discussed as if they were real. In a paper on moral agendas for genomics I made a similar argument (Van der Weele, 2008), starting from the perspective that attention is a scarce thing, in ethics as much as anywhere. Why spend so much moral attention on potential future dilemmas, while the world is so full of urgent actual problems? When the priority is to make the world a better place, an important role for morality could be to wonder and study how and where science and technology might be helpful. Such an inspirational role is associated with the need for integrative rather than fragmented knowledge. It seems to me precisely what the *in vitro* meat process exemplifies.

This is not to say that the future of *in vitro* meat is unproblematically bright. First, it is not yet there. The great expectations make it vulnerable to what Nik Brown (2003) has called the 'cycle of hype and disappointment, awe and loathing.' Even though Brown described the 'hype dilemma' as a dilemma for researchers who needed to attract money, varieties of this process of hype and disappointment may just as well take place when the hope originates in other citizens than scientists. The great hopes may even give rise to varieties of the speculative ethics described by Nordmann. Second, even if further research results in an edible product which is positively evaluated from a moral point of view, this is not enough for *in vitro* meat to succeed. The question is not just whether we approve of the idea, but whether enough people will really want to eat it. In other words, *in vitro* meat should not just be attractive as a contribution to a better world, but it should also look well, smell well, taste well, feel well. There is a potential gap, therefore, between moral considerations and eating considerations, a gap that resonates with the well known distinction between citizens and consumers. Neither the yuck factor, nor worries about the technological character of *in vitro* meat may amount to a decisive moral argument. But the realisation that there are no good arguments for dislike is not enough to make it go away.

When it comes to attractiveness of *in vitro* meat, there are mainly open questions at present. Much will depend on what a final product looks like. But these questions can also be turned around: what should a product (and the production process) be like in order to be appreciated? This is where the need for an integrative development of *in vitro* meat shows.

An interdisciplinary research project

Since *in vitro* meat does not exist yet, it affords an opportunity to accomplish an early integration of social, moral and technological research. In a multi- and inter-disciplinary research project, we, researchers from Wageningen UR and Utrecht University, aim to grasp this opportunity and contribute to further development of *in vitro* meat as 'a co-con-construction of science, ethics and social science', as the research proposal puts it. The project includes, first, biological research on how to grow muscle cells from stem cells in an efficient way. Second, we will conduct social research on how people respond to (stories and images of) *in vitro* meat in various forms, and on their subsequent appreciation. Third, we will study patterns and processes in moral evaluation of *in vitro* meat. Most importantly, we will have regular interchange and feedback between these lines of research. Results of consumer studies and moral evaluation will be integrated, and will become an input for the technological design of products as well as production processes. In this process we want to develop and strengthen an atmosphere of cooperation and integration. This includes the identification not only of opportunities but also of obstacles – to *in vitro* meat as well as to cooperation – and the encouragement of dialogue between various perspectives – again, both on *in vitro* meat and on inter-disciplinary relations.

Results are not yet there. But a brief conversation with my hairdresser, which was very similar to some other conversations I had, may serve as an example of the kind of preliminary findings and experiences that help to become sensitive to the issues. After I explained the idea of *in vitro* meat and asked what he thought, my hairdresser said: 'Yuck'. Can you explain, I asked. Well..., he said, what are they going to do to our food this time? I wondered what his associations were. 'It makes me think of GM food', he said, 'and also of hot dogs. I saw a documentary about hot dogs recently, you never can be certain what they are putting into them. But I must say that it makes a lot of differences how you talk about it. When you think of what it could mean for animals, it looks a lot more positive.'

The conversation is suggestive in various ways. The first thing that strikes me is how quickly a shift from yuck to wider considerations can take place. Further, GM food and minced meat products are mentioned in the same breath and both laden with distrust. The discomfort with what 'they' are doing perhaps suggests that a lack of transparency of technology is a problem, and an accessible, transparent means of production may be important for *in vitro* meat.

In conclusion: we intend this integrative project as a contribution to the question whether *in vitro* meat can be developed as a product and process that is attractive from many perspectives. In doing so, we will also wonder how an atmosphere of cooperation in relations between science, ethics and social research can be strengthened and developed.

References

Brown, N. (2003). Hope against hype – Accountability in biopasts, presents and futures. Science Studies 16(2): 3-21.

Haagsman, H.P., Hellingwerf, K.J. and Roelen, B.A.J. (2009). Production of animal proteins by stem cells. Faculty of Veterinary Medicine, Utrecht. Available at http://www.invitromeatfoundation.eu/images/Studie%20Haagsman.pdf.

Hopkins, P.D. and Dacey, A. (2008). Vegetarian meat: could technology save animals and satisfy meat eaters? Journal of Agricultural and Environmental Ethics 21: 579-596.

In Vitro Meat Foundation (2010). What is cultured meat. Available at http://www.invitromeatfoundation.eu/info_en.html.

Langelaan, M.P.L., Boonen, K.J.M., Polak, R.B., Baaijens, F.P.T., Post, M.J. and Van der Schaft, D.W.J. (2010). Meet the new meat: tissue engineered skeletal muscle. Trends in Food Science and Technology 21(2): 59-66.

Ministry of Agriculture, Nature and Food Quality (2009). Policy Document on Sustainable Food. Available at http://www.minlnv.nl/portal/page?_pageid=116,1640321&_dad=portal&_schema=PORTAL&p_file_id=40703.

Nordmann, A. (2007). If and then: a critique of speculative nanoethics. Nanoethics 1(1): 31-46.

People for the Ethical Treatment of Animals (2008): PETA offer $ 1 million reward to first to make *In vitro* meat. Available at: http://www.peta.org/feat_in_vitro_contest.asp.

Schonwald, J. (2009). Future Fillet. University of Chicago Magazine, May-June. Available at http://magazine.uchicago.edu/0906/features/future_fillet.shtml.

Steinfeld, H., Gerber, P., Wassenaar, T., Castel, V., Rosales, M. and De Haan, C., (2006): Livestock's long shadow, environmental issues and options. Food an Agricultural Organization of the United Nations. Rome. Available at: http://www.virtualcentre.org/en/library/key_pub/longshad/A0701E00.pdf.
Swierstra, T. and Rip, A. (2007). Nano-ethics as NEST-ethics: Patterns of moral argumentation about New and Emerging Science and Technology. Nanoethics 1: 3-20.
Tissue Culture and Art Project (2003): Disembodied Cuisine. Information available at http://www.tca.uwa.edu.au/.
Van der Weele, C.N. (2007). A taboo on moral solutions. Simulacrum 15 (3/4): 28-39.
Van der Weele, C.N. (2008). Moral agendas for genomics: how to find the blind spots? Configurations 16: 117-135.
Verbeek, P. (in press). Moralizing technology: understanding and designing the morality of things. Chicago, University of Chicago Press, USA.

Legal and ethical challenges regarding edible *in vitro* meat production

E.J. Armaza-Armaza[1] *and J. Armaza-Galdos*[2]
[1]*Inter-University Chair in Law and the Human Genome, University of the Basque Country, Sarriko Campus (Zubiria Etxea), Lehendakari Aguirre N° 83, 48015 Bilbao, Spain;* [2]*Catholic University of Santa Maria, School of Law – Department of Criminal Law, P.O. Box 2235, Arequipa, Peru; julioarmaza@gmail.com; emilio.armaza@deusto.es*

Abstract

In view of the high probability that in less than a decade the first generation of edible animal muscle (meat) *in vitro* will be ready to be launched on the market, it is essential to initiate a multidisciplinary discussion that addresses a series of issues derived from this reality. Thus, throughout the paper, the authors deal with some of the most important legal and ethical issues arising from the possible large-scale production of *in vitro* meat. As we know, at least in Europe, there are no rules regulating the processes of production, trade and consumption of this new product. Therefore, one research objective has been to establish some basic points to be taken into account during the process of development and structuring of the rules that regulate this material. In this manner, the precautionary principle, (bio) safety in the production of this food, the problems that eventually may result from the protection of copyright and industrial property, the applicable laws (e.g. GMO's Laws, in view of the absence of a specific regulation), among others, are some of the issues discussed throughout the article. Finally, from a utilitarian perspective, the authors analyze and suggest some answers to the ethical issues arising from the eventual marketing of this new food product.

Keywords: tissue engineering, animals, food, legal regulation, anthropocentrism

Preliminary words

The first vampire film released in Spain this year (*Daybreakers*, starring Ethan Hawke – the same actor who, oddly enough, also starred in the now classic *Gattaca* which, as we know, set a precedent in that which regards audiovisual and cinematographic references to Biolaw) brings with it a background that, without a doubt, leaves many questions for those of us who specialise in the study of Bioethics and Biolaw, principally in relation to the principle of equal consideration of interests, experimentation with human and non-human animals, and finally, the exploitation of sentient beings for food. Indeed, the film is set in a time (2019) in which the most powerful civilisation on the planet is composed of vampires (that were once human) who struggle with the scarcity of the only nourishment they can consume: human blood. There are almost no living human beings left; a few live in hiding, while the majority has been imprisoned in farms specialising in the cultivation and extraction of the prized tissue. In this setting, a vampire scientist (Hawke) works tenaciously on the search for a human blood substitute that will cover the uncontrollable demand that exists for it with the purpose of halting the unlawful massacre and exploitation to which the few living human beings remaining are submitted. The ending of the film, however, occurs when Hawke discovers the form to turn the vampires back into humans, thereby bringing about the resurgence of the human race.

However, if we may be permitted, we could let our imaginations soar and add a couple of scenes to the aforementioned film:

a. Once the vampires become extinct and human beings once more take control of the planet, the farms into which human beings were packed, formerly used for the 'cultivation' of their blood, are replaced by new ones in which other sentient beings are exploited (cows, pigs, chickens, rabbits, etc.) for the satisfaction of a human caprice (not a necessity), that is to say: the consumption of meat from non-human animals.

b. Hawke, after turning into a human again, and becoming vegan, has continued his research in relation to alimentary biotechnologies, although this time he has dedicated his efforts to the search for a substitute for animal meat that does not require the massacre of billions of sentient and conscious individuals for the purpose of satisfying the whims of the other members of his species, and after much work has discovered the form to artificially produce human meat through the cultivation of stem cells in a capsule especially designed for that purpose (we may add that he has also won the prize of one million dollars that PETA has offered since 2008 to anyone who is capable of producing *in vitro* chicken meat).

In this sense, it is important to realise the fact that neither of the two scenes we have just imagined could be labelled as fictitious situations or situations of impossible implementation in contemporary societies. In fact, while on one hand it is calculated that the number of non-human animals massacred by the meat industry amounts to the impressive figure of five billion individuals per year; on the other hand, recent advances in biotechnology have made it possible for us to consider the possibility of launching on the market 'real' meat that has not come from the killing of any individual, given that it is cultivated in bioreactors especially designed to favour its growth from an initial biological sample. In this way (and given that consumers, in recent periods, have shown a growing interest in the consequences of production and consumption of animal meat, such as sicknesses caused by its consumption, the contamination of the environment, and, most importantly, the attacks on the interests of the nonhuman animals exploited in factory farms) in the present article we will analyse the ethical and legal implications that the imminent production of *in vitro* meat would bring to the societies in which this new product would be commercialised. Therefore, while in Epigraph 2 we will briefly describe the technical viability of the production and commercialisation of this product (the state of technology, economic costs); in Epigraph 3 we will see to the challenges that the effective commercialisation of this new product would raise for the field of Law; finally, in Epigraphs 4 and 5, as far as they are concerned, we will attempt to synthesise the implications of an ethical nature that arise from an analysis of the matter. Let us begin, therefore, as indicated.

Situation of scientific research concerning cultured *in vitro* meat

The literature has shown that Biotechnology has already the development sufficient to produce, through tissue engineering, edible animal muscle (meat). In this sense, there are two techniques that, presently, are available for such a purpose:

a. scaffold-based techniques;
b. self-organising techniques.

The scaffold-based technique takes place by means of the extraction of embryonic myoblasts or adult skeletal muscle satellite cells (for which the killing of the source subject is not required). The next step consists in attaching the obtained biological sample (cells) to a biocompatible and reassimilable collagen scaffold and inducing its proliferation by means of the artificial recreation of a development medium located in a rotating bioreactor, this is the device which deals with the recreation of environmental cues, such as providing the necessary nutrients for the development of myotubes, and later, of myofibers. By means of this technique, meat in the form of processed grains (the product would be similar to boneless and ground meat) may be obtained, perfectly suitable for the production of hamburgers and sausages. On the other hand, it is advisable to note that some studies have indicated that only one cell, cultivated by means of this technique, could be multiplied to the point of producing the necessary quantity of meat to meet the annual demand of this product in the entire world. Finally, we should indicate that the procedure has been efficiently attempted with sample cells from chickens, turkeys, pigs, lambs and cows.

In turn, the self-organising technique, unlike the previous one, would make no longer the production of grains of meat, but of compact pieces similar (in size and shape) to any *in vivo* structure (e.g. steak) possible. The process is based on the placement of an explant of animal muscle in an appropriate growth medium (during the procedure tests it has been observed that, in certain conditions, the initial sample could experience an increase in surface of up to 79%). However, the literature points out that this technique has not yet reached its maximum level of development, principally due to the impossibility of recreating the mechanisms of blood circulation that make a homogeneous and constant distribution of nutrients possible; henceforth, the only technique that could successfully face up to the demand for this product would be the scaffold-based technique – it must be remembered that tissue engineering for therapeutic purposes has already exceeded the technical requirements for producing edible muscle with nutritional and aesthetic properties equal to, or even better than many traditional meat products.

In short, in spite of the fact that there are still some technical deficiencies to be overcome in both processes (nothing special in relation to the scaffold-based technique, and some of greater magnitude with regards to the self-organising technique), everything seems to indicate that in a few years we will have the surprise of seeing this product in the corner supermarket, for the price of approximately € 3.3/kilo (cf. The *In Vitro* Meat Consortium. Preliminary Economics Study, 2008).

Basis for legal regulation of the production and commercialisation of *in vitro* meat

In Europe, the lack of specific regulation and/or norms that can be applied by analogy to this *novel food* (in the terms established by Regulation [EC] No. 258/1997) is evident. In this sense, in spite of the fact that the legislation that appears to boast the greatest proximity to our subject is that which refers to genetically modified food (e.g. Regulations [EC] numbers 1829/2003 and 1830/2003), at least generally speaking, the analogical application of said norms to the case of *in vitro* meat does not appear to be advisable, given that they present substantial differences in that which refers to the circumstances and motivations that make up the base of their implementation (it should be remembered that the primary motivation of these norms is to guarantee biosecurity in the production and commercialisation of these products).

Thus, there is an obvious need for the local and national legislator to develop a series of specific sets of laws, which have as their primary function the regulation of this subject, principally influencing the following matters:

a. The particularities that should be had with regards to the question of patents, especially in relation to the limitation of the capacity of corporations (e.g. Monsanto, which already controls various patents related to genetically modified food) for the use and exploitation of this novel and useful technology, as we should not forget that the *principle of justice* can be violated by means of the establishment of highly elevated prices for the use of this technology, which could make it impossible for people with fewer resources to have access to this food.
b. The regulation of the process, authorisation and accreditation requirements of the companies that wish to make an incursion into this area of the market, whether through production or by means of the commercialisation of this food.
c. The matter of establishing the necessary means to guarantee that this new food is safe or beneficial for human health (*precautionary principle*) by means of the evaluation of the nutritional effects of the product, the food's direct effects on health and, finally, the effects of the possible introduction of new components (nutritional or not) in the product.

d. The question of protecting the consumer's right to receive adequate and sufficient information regarding this new product (adjustment of the product's particularities to the demands established in relation to its labelling and traceability).

Finally, we should keep in mind that legislators, aptly tend to incorporate in such norms an article that contains the definitions that are adopted for the purposes of legal interpretation. In this sense, in case that it is decided to devise a norm which specifically regulates the process of production and commercialisation of the so-called *in vitro* meat, we believe convenient the inclusion, apart from the traditional definitions (consumer, operator, traceability, etc.), of the definition of the term *cultured meat*, which, in our opinion, would have to be the following: 'edible skeletal animal muscle produced by means of tissue engineering techniques.'

Ethical issues concerning traditional and *in vitro* meat: should we have preference for any?

The commercialisation of *in vitro* meat does not appear to be, in itself, ethically questionable. What's more, if the guidelines and principles described in the preceding epigraph were at least incorporated into the set of laws (especially in relation to the question of patents, the principle of justice and the protection of the consumer's right to receive adequate and sufficient information regarding this product) we could affirm that the commercialisation and consumption of this new product, more than being ethically acceptable, would be a moral duty of the omnivorous citizen of the 21st century.

In fact, this affirmation is based on the fact that there exist solid reasons to certify that the consumption of traditional meat (that is to say, that which comes from slaughterhouses) constitutes a seriously questionable activity given that, throughout the process of its production, a great quantity of legitimate ethical and legal interests are harmed or put in danger of harm. Below we will describe, briefly, the most important ones:

The violation of the principle of equal consideration of interests: the unequal consideration of the interest in not suffering

The traditional meat industry massacres, as we indicated in the beginning of this article, no less than 5 billion nonhuman animals per year; this fact reveals that the breeding of animals for food causes more suffering to more animals than any other activity in the world. As we know, the principal attacks on the health, physical integrity and life of the animals submitted to this kind of exploitation are materialized:

a. *Through the conditions in which they are raised:* These animals are artificially reproduced to be later castrated, mutilated, and branded *without anaesthesia*; they are packed into small spaces or cages – which causes them to have grave disorders of both a psychological (e.g. social isolation, stereotypy, self-injury) and a physical nature (e.g. deformity, muscle atrophy and hypertrophy) –, where they receive no natural light, and not even the necessary precautionary health measures to successfully avoid the sicknesses derived from overcrowding; as the majority eventually fall sick, they are given various doses of antibiotics with the purpose of avoiding their death before its due time, that is, before they reach the adequate size for their commercialisation.
b. *During the slaughtering process:* The individuals who have reached the size desired by the farm owner are transported to the slaughterhouse in trucks adapted for such a purpose; the journey may last several days (on no few occasions the animals are taken to slaughterhouses in other countries to make sure that the meat is fresh in the moment of its commercialisation) during which, more often than not, they are not offered food or water; given the composition of the floors of the trucks and the treatment they receive from the workers, an important amount of animals suffer serious physical traumas (fissures, fractures, cuts, etc.); once the stressful journey is finished, the animals are

> led (that is to say, those which can still walk, given that those which have been rendered disabled or handicapped during the journey will be dragged with ropes or chains to the place of execution) to an environment in which workers try to stun them to facilitate the process of bleeding and cutting them to pieces; many times the stunning process is not effective, for which, in these cases, the bleeding or dismembering is carried out when the animal is still conscious and crying in pain; the case of fish is no less horrifying, as they are captured by the nets at great depths and then dragged to the surface at high speeds, which causes their eyes and internal organs to burst due to the decompression they suffer; hundreds of millions are crushed to death by other fish, while many others are cut by the nets themselves; those that survive the process of capture writhe in agony on the surface for several minutes and the majority are cut up and processed while they are still alive.

It is advisable to note that modern sets of laws (European or national) regarding the welfare of animals raised for food have legitimised each and every one of the practices described in the preceding paragraph under the label of 'humanising' the methods of breeding and slaughtering (it must be recognised that, in the past, the situation of these animals was even more deplorable).

The environmental degradation

In a time in which human beings have started to become conscious of the importance of preserving the environment (and, specifically, that of the increasing scarcity of water) it turns out to be somewhat paradoxical its not being considered in the fact that the reproduction and intensive breeding of animals for food constitutes one of the primary sources of waste and contamination of our hydrobiological resources. In effect, without counting the fact that the meat industry consumes the bulk of the water used by man (for the watering of the cereals which serve as food for the animals – 70% of farmland is destined for such purposes – as well as for the feeding of the animals and other activities related to the production process), it is evident that the dumping of manure, which annually contain the waste of five billion animals used by this industry, is responsible for a large part of the contamination of water and subsoil. It is estimated, also, that this activity is the cause of 18% of the gas emissions which produce the greenhouse effect and 64% of the ammonia that contributes to acid rain and, as a result, the acidification of the ecosystems (cf. FAO report *Livestock's Long Shadow*, 2006).

The violation of the principle of justice: world hunger

The estimations of the FAO indicate that presently (based on data compiled in 2009), there are approximately 1.02 billion people on the planet who suffer from chronic hunger (that is to say, that their habitual food consumption is insufficient to cover their minimal energy needs), principally due to the scarcity of food as well as the lack of financial resources necessary to solve this problem.

Nevertheless, as previously indicated, the fact that 70% of the planet's farmland is destined to the production of fodder used in the meat industry is ethically criticisable. In this sense, if we keep in mind, as indicated by Singer, that 9 kilos of vegetable protein (an amount that would feed 70 people for a day) are needed to produce just half a kilo of meat (a quantity that 2 or 3 persons would consume in a day) it is clear that Frances Moore Lappé hit the nail on the head when she affirmed that the meat industry, in reality, is 'a protein factory in reverse' (Singer, 2002: 265). With these assessments as a base, we can affirm that the principle of justice (that, generally speaking, involves the rational use of available resources with the purpose of permitting all human beings – especially those in greatest need – an equal access to them) is being seriously infringed, due to, obviously, the stubborn human attachment to the satisfaction of this caprice which does nothing more than waste valuable resources that could be used to alleviate hunger in the less fortunate countries.

Ethical implications of the commercialisation and consumption of *in vitro* meat

As indicated at the beginning of the previous epigraph, the commercialisation and consumption of *in vitro* meat, in itself, does not appear to be ethically objectionable. In fact, unlike that which occurs with the exploitation of nonhuman animals in the meat industry, during the process of production and commercialisation of *in vitro* meat there seem to be no principles, rights or moral obligations being violated, for which, in principle, there would not exist a conflict of interests that could give cause to a discussion regarding the possible affectation of some or other ethically relevant interest (from which, for now, we cannot qualify it as anti-ethical).

However, we had also indicated that the use of this technology and the consumption of the products obtained from it would rather constitute a moral duty of all omnivorous citizens than otherwise. Indeed, in view of the fact that, at least for the time being, the majority of human beings have showed little or no interest in changing their nutritional habits with regards to meat consumption (in spite of the fact that, as we have seen, said activity is ethically illegitimate due to the consequences which can be derived from it), it is important to keep in mind that once this novel product is launched on the market the classic excuses that are frequently invoked to justify the active participation in the exploitation of other animals will cease to be valid.

It must be remembered, therefore, that for some time nutritionists have insisted on the idea that traditional meat, that is, meat obtained from slaughterhouses, is not an indispensable source of nutrition for the bio-psychological development of the human being (regarding this matter in more detail: Singer, 2002: 180-183), for which we could affirm that its consumption corresponds to a clearly unnecessary caprice or whim that the average human being finds himself incapable of giving up.

In this way, it is clear that the consumption of *in vitro* meat would bring a series of advantages that we should not forget:

a. Given that the ratio of saturation of polyunsaturated fatty acids can be controlled, the risk of suffering from a cardiovascular illness is considerably reduced.
b. The incidence of food poisoning can be significantly reduced given that substances (organic and inorganic) detrimental to the human organism can be reduced or even eliminated.
c. Replacing the consumption of traditional meat with that of artificial meat, we would cease to worsen the damage we are causing to the environment since, as we know, the process of *in vitro* meat production does not need the resources which are wasted in the traditional meat industry, nor does it produce the same waste material or contaminating agents derived from the aforementioned.
d. As it does not require the presence of a sentient individual to serve as a meat factory, the massacre of the billions of animals that are used for this purpose would be considerably reduced. Precisely for this reason, we could today affirm the full validity of the words of Winston Churchill with regards to this matter: 'We shall escape the absurdity of growing a whole chicken in order to eat the breast or wing, by growing these parts separately under a suitable medium.'
e. If, finally, this product were to beat the market of the traditional meat industry, millions of hectares of fertile land would be free to be used to alleviate the hunger of those in most need as well as to reforest our emaciated planet.

Nevertheless, there still exists a series of questions that must be dealt with from the perspective of education and social communication. In effect, the literature has indicated that objections to the production of cultured meat could be raised for a series of sociological reasons, among which (Hopkins and Dacey, 2008: 585-595) stand out:

a. *That this technique could bring about cannibalism by means of the cultivation of human meat*: In our opinion this fact, in itself, does not constitute an ethical dilemma given that, if human meat were in fact to be produced from a biological sample donated for such a purpose, as there would exist no conflict of interests to serve as a base to initiate a debate of an ethical nature, the autonomy of the consumer would have to be respected above all. However, it would be advisable to initiate a debate with regards to the matter, so frequently discussed in the literature, of the commercialisation of human biological material.
b. *That the in vitro meat is not real, but artificial, and that its consumption would be an unnatural activity:* These last two objections lack a solid means of support given that *in vitro* meat is neither the first nor the last artificial food that humans have consumed (among many others, for example, GMO's) and that it would neither be the first nor the last unnatural activity in which man has engaged (in fact, the consumption of milk coming from other animals is a habit which has only accompanied man for a little more than 10,000 years, which seems little if we compare it with the time that *homo sapiens* has inhabited the planet).
c. *That the mere fact of thinking about 'in vitro' meat causes a sensation of repugnance:* To this criticism, which is known by the name of *yuck factor*, it would be necessary to object that, once *in vitro* meat is launched on the market, no one will be forced by law to consume it. Many persons are disgusted by the idea of *civet coffee*, but this fact does not appear to be sufficient to prohibit its commercialisation and/or consumption.
d. *That the technological fix is moral cowardice:* This argument, sustained by a sector of animal rights activists, makes reference to the fact that reducing the exploitation of nonhuman animals by the introduction of *in vitro* meat to the market does not solve the problem of speciesism, as animals will stop being exploited not because of a change in the attitude of human beings, but due to the resource of new technologies. It would be necessary to insist that, although it is certain that the elimination of speciesism is a desirable objective, it would seem that another important thing, on a short-term basis, is the implementation of a series of mechanisms which would cause the *de facto* stoppage of the massacre to which nonhuman animals are being submitted, independently of the fact that, for the moment, it may be impossible to obtain that change of attitude so desired by many of us (we could explain it in another way with an example: the important thing is to achieve the elimination of any racist actions, not that racists cease *to think* the way they do, as *cogitationis poenam nemo patitur).*

In short, in the near future, the work of the agents involved (producers, sellers, governments, associations that defend animal rights, etc.) would constitute a fundamental factor to keep in mind so as to obtain the maximum benefits that this technology is capable of offering, for the development of our societies as well as to confront the ethical dilemmas that surround the consumption of meat which comes from traditional intensive breeding farms.

Acknowledgements

This work has been funded by a Grant to Research Groups by the Basque University System, ref. IT-360-07, Department of Education, Universities and Research of the Basque Government. We would also like to thank José Valle from the organisation *Animal Equality* (www.igualdadanimal.org) for helping us to clear up several doubts with regards to the *de facto* situation in which animals exploited in European, and principally Spanish industrial farms, live.

References

Armaza-Armaza, E.J. (2007). Los animales y la biotecnología, aportes al debate de carácter ético-jurídico. In: Palacios, M. (ed.) Book of Lectures and Oral Presentations of the V World Conference on Bioethics. International Society of Bioethics (SIBI), Gijón, Spain, pp. 183-185.

Churchill, W. (1952). 'Fifty years hence', *Popular Mechanics Magazine*, January, 1952, p. 326.
Edelman, P.D., McFarland, D.C., Mironov, V.A. and Matheny, J.G. (2005). *In Vitro*-Cultured Meat Production. Tissue Engineering, Volume 11, Number 5/6. Mary Ann Liebert, Inc.: 659-662.
Hopkins, P.D. and Dacey, A. (2008). Vegetarian meat: could technology save animals and satisfy meat eaters? Journal of Agricultural and Environmental Ethics, Number 21. Springer: 579-596.
Romeo-Casabona, C.M. (ed.) (2004). Principio de Precaución, Biotecnología y Derecho. Inter-University Chair in Law and the Human Genome – Comares, Granada/Bilbao, Spain, 501 pp.
Singer, P. (2002). Animal Liberation. Harper Collins Publishers, New York, United States of America, 324 pp.
Singer, P. (2003). Ética práctica. Cambridge University Press, Madrid, Spain, 320 pp.
The *In Vitro* Meat Consortium. Preliminary Economics Study, Project 29071, V5, Study commissioned by Prof. Stig William Omholt. eXmoor Pharma Concepts, March 2008.

Safe and socially robust development of nanofood through ISO standards?

E.M. Forsberg
Work Research Institute, P.O. Box 6954 St. Olavs plass, 0130 Oslo, Norway; ellen-marie.forsberg@afi-wri.no

Abstract

The use of nanotechnology in the area of food and agriculture is progressing rapidly. New products and applications appear, while regulation lags behind. There is vast scientific uncertainty related to the effects of nanotechnology applications on human and animal health, as well as on the environment. In 2005, the International Organization for Standardization (ISO) set up a technical committee for developing nanotechnology standards. These standards may form the basis for national and EU regulation of nanotechnology and nanofood, and may therefore play an important role in the governance of nanofood. There are, however, challenges related to both scientific quality control and the consideration of societal dimensions in the ISO nanotechnology work. In order for ISO standards to function as a tool for socially robust and sustainable development of nanofood a greater variety of researchers, NGOs and other stakeholders should get involved in the ISO work. As this presupposes funding, greater attention should be paid to the ISO work by relevant authorities. In general, the role of ISO in the development of nanotechnology should be subject to broader public discussion.

Keywords: nanotechnology, food, voluntary standards, governance

Introduction: nanofood

A short and simple definition of nanotechnology is 'the study of the controlling of matter on an atomic and molecular scale' (Wikipedia). A more elaborate definition is that '[n]anotechnology is the manipulation or self-assembly of individual atoms, molecules, or molecular clusters into structures to create materials and devices with new or vastly different properties' (Nanoforum Report, 2006: 2). Nanotechnology involves manipulation or study of objects with at least one dimension at the nano-scale. Researchers and other stakeholders disagree as to the exact dimensions of this scale and as to whether one should define such a scale at all. Often the nano-scale is defined from 1 nm to 100 nm, but could also be expanded to up to 300 nm or even 1000 nm (Friends of the Earth Australia, 2008). Some claim that a narrow size based definition may exclude larger particles that behave like nanoparticles, in particular, by showing novel properties.

Nanotechnology is used for many purposes. It is used in making materials with novel characteristics, like being disinfectant, dirt repellent, ultra strong, etc. It can also be used in ultrasmall sensors that can monitor their environment, in fuel cells and solar cells, and in various other applications. Nanotechnology is used in areas like health, information technology, cosmetics, defence and energy. Food and agriculture is also thought to be a promising area for the use of nanotechnology. In a 2006 Report Nanoforum defines nanofood as food where 'nanotechnology techniques or tools are used during cultivation, production, processing, or packaging of the food' (Nanoforum Report, 2006: 7). Nanotechnology can apparently enhance the flavour and nutritional quality of the food and can also contribute to new 'novel foods' where medicine and nutrients are delivered through ordinary foodstuffs (for more applications see ETC, 2004). Nano agriculture presently involves fertilisers, herbicides and soil enhancers, sensors for monitoring crops and livestock, etc.

Although proponents of nanofood claim great benefits of nanotechnologies in the food and agriculture sector, nanofood is an area of nanotechnology where one can expect – based on the experience from GM foods – a high level of consumer and public concerns. Packaging is perhaps the area of nanofoods that provokes the least amount of concern among consumers, and this is an area in fast progress. Nano packaging may for instance block UV-radiation or indicate if the food is becoming contaminated. In spite of a certain cautiousness about marketing nanofoods, nanotechnology is thought to revolutionise the food and agriculture sector and the development of nanofood is accelerating. Nano-werk (2006) claims that big food industry locomotives like Nestle, Altria, Heinz and Unilever have started investing in largescale R&D on nanofood.

The main concern about nanotechnology is about the lack of scientific knowledge of processes and objects at the nano level, and thus about potential health effects to consumers and to workers, and also consequences to the environment. The great promise of nanotechnology lies in the fact that objects may obtain radically different characteristics at the nano scale, – and this is also where the threat lies. Several materials become more toxic at the nano scale and as they may penetrate the skin and enter the human blood stream they may cause unintended, undesired and unknown effects. If deliberately released into the environment (for instance as part of so-called precision agriculture) they may similarly have uncertain adverse effects to living beings and ecosystems in general. Thus, with regard to nanotechnology in general, and nano food and agriculture specifically, the precautionary principle may be held to apply (see e.g. Throne-Holst, 2008, and for a definition of the precautionary principle see COMEST, 2005). This raises issues of responsibility (see e.g. Ewald, 2000) and governance (see e.g. Stirling *et al.,* 2006). Several parties may take this responsibility. Public authorities have of course a responsibility, and some initiatives have been taken to regulate nanotechnology and nanofood. The EC Novel Foods Regulation (No 258/97) applies for nanotechnology in novel foods and processes. However, there are here no specific criteria to consider particle size (see for instance Foss Hansen, 2009). The Codex Alimentarius will also eventually include standards on nano food. The European Food Safety Authority has started work on nano food and has published a report on the potential risks arising from nanoscience and nanotechnologies in food and feed. The main conclusions from the opinion are that the current risk assessment paradigm is appropriate for nanomaterials, but that 'the risk assessment of engineered nanomaterials in the food and feed area should consider the specific properties of the ENMs in addition to those common to the equivalent non-nanoforms' (EFSA, 2009: 2). It is also acknowledged that there are limited data on oral exposure to nanomaterials and any consequent toxicity and limited methods to characterise, detect, and measure nanomaterials in food/feed. Moreover, it is noted that one cannot extrapolate toxicity profiles of nanomaterials from data on their equivalent non-nano forms. One may claim that it is paradoxical that the current risk assessment paradigm is deemed to be appropriate when at the same time it is acknowledged that there is such limited data and methods available.

It is clear, then, that current regulation is not sufficient to ensure the safe development of nanotechnology and nanofood and that a more comprehensive scientific knowledge base needs to be developed. The International Organization for Standardization (ISO) is among those organisations that have taken on responsibility for developing such a knowledge base, and we shall now look at the initiative taken by ISO. The main focus in this paper is to discuss the role of ISO in the governance of nanotechnologies and nanofoods, and to identify challenges with way the work is currently undertaken. I will focus on ISO's work on nanotechnology (TC 229). ISO is also developing a standard on social responsibility (ISO 26 000) that will be applicable also in the food and agriculture area. However, I will not discuss the potential role of this standard for nanofood governance in this paper. The European Committee on Standardization (CEN) also has a technical committee on nanotechnology. However, as this is closely integrated with ISO's work I do not consider it necessary to comment upon CEN specifically in this paper.

The role of ISO in the governance of nanotechnologies

The aim of the ISO technical committee on nanotechnology (TC229) is to develop a framework of standards that will 'help create a smooth transition from the laboratory to the marketplace, promote progress along the nanotechnology value chain ... and facilitate global trade.' (from the ISO Nanotechnology Business plan). The tasks of this committee are divided into working groups. There are working groups for developing adequate standardised terminology and nomenclature (WG1), as well as measurement and characterisation standards (WG2). On this basis, standards on health, safety and the environment are being developed (WG3). Finally, a fourth working group has started developing material specifications for specific raw materials in order to secure standardised products.

The work in ISO is important because no common framework has yet been established for scientists to discuss and compare their work, and there is thus no common framework for risk assessment. Research performed in different countries or in different institutes cannot accumulate into a common knowledge pool unless it can be compared in a reliable way. In order for such comparison to take place one must agree on a common terminology and metrology. This work is usually done in traditional scientific *fora* like international scientific conferences and journals, but ISO seems to take a prominent role in this work in the nanotechnology area. Even if industry is more present in ISO than in other scientific *fora*, it may initially seem like this is acceptable, at least when it comes to such apparently purely technical issues as agreeing on some basic concepts. Friends of the Earth Australia has shown, though, that even this terminological stage is an important arena for conflicts of interests. They say: 'The size range within which the ISO has defined nanoparticles will have significant implications for health and safety regulation at a national level' (Friends of the Earth Australia, 2008: 1). They go on: 'the use of an arbitrarily defined size range to act as an index of novel properties is problematic. Particles that fall outside the size range deemed to encompass nanoparticles – even if they are not much bigger and also exhibit novel, nano-specific behaviour – will not be assessed as new chemicals. These particles will not trigger new health and safety assessments where substances have previously been approved for use in larger particle form.' (*ibid*.: 2). Also metrology is important: 'It is hoped that exposure and commercial use will be measured using nanoparticle-appropriate metrics (e.g. number of particles or particle surface area) rather than mass as per conventional materials' (*ibid*.: 1). Value questions thus arise also in these seemingly purely technical areas and interests may play out in the discussions.

It seems superfluous to argue that the health, environment and safety aspects (HES, WG3) are important in the governance of nanotechnology. These are exactly those aspects that are the main concerns. The ISO seems to take a leading role in coordinating research on HES into a common pool of knowledge to be used for risk assessment and management. When ISO does this, more industry influence can be expected than if it was done by a public actor or the research society itself. This pool and the resulting guidelines and standards will likely form an important basis for national and international regulation in the nanotechnology area. An important reason for why ISO's work likely will influence regulation is that the WTO acknowledges ISO standards and will not regard legislation built on these standards as protectionist technical barriers to trade (Murphy and Yates, 2009: 44-45). Accepting ISO standards may therefore be perceived by some nations as a simpler way to regulate in this area than operating with their own national standards. Balanced participation in the process in order to secure an unbiased output is therefore crucial.

The societal legitimacy of the ISO process

Because the ISO is likely to have such influence and because it is close to industrial interests, it is important from a legitimacy point of view that other stakeholders engage in the work critically. However, the work in the TC 229 consists mostly of researchers from government and industry, as well as bureaucrats and industry representatives. There are also liaison organisations, like the OECD Working Party on Nanotechnology. There are some representatives from other interested parties, in particular consumers. The ISO has organised a special committee for consumer affairs, the Copolco, and has set up a system for enhancing the inclusion of consumer representatives in ISO processes in general. In addition, there are other interested organisations, like the European Environmental Citizens Organisation for Standardisation (ECOS). However, ECOS has so far not been much involved in the TC229 work. The main reason for the lack of broader participation is that all the work in ISO is based on self funding. Combined with the fact that the meetings are moved around the world, this makes it difficult for some parties to attend. Those who invest in ISO processes are often the ones who have most to gain financially. NGOs seldom have the funding to participate in such meetings. Admittedly, in some countries there is some funding available, in particular for consumer representatives. Labour representatives may also be funded by their organisations, and there is some institutional funding for ECOS. However, it is not only the travelling that requires funding. In order to engage in depth in the issues quite a substantial amount of time is also required. All this results in quite limited stakeholder participation.

The ISO work goes on in semiannual TC meetings, and by internet correspondence in between the meetings. When the work is about to result in a standard, guideline, technical specification or technical report, it is sent out on hearing to all the member countries in the TC. All countries that are actively engaged in the work (so-called P-members) have one vote, and they can also submit comments during several stages of the work process. The countries each have their own expert committee (mirror committee) that is supposed to function as a quality control and peer review of what takes place in the working groups. However, in many cases, the mirror committees lack competence on many of the areas discussed in the working groups. The actual quality control is therefore not necessarily present. Still, countries may vote for a proposal even if they do not have adequate expertise in their committees. Different mechanisms may explain this, for instance that the national standardisation body is financed based partly on how many standards it contributes to making. The real quality control of the process may therefore be questioned, which again may reduce the legitimacy of the output.

With regard to the role of ISO in the societal governance of nanotechnology, it should be mentioned that two task groups have been set up to ensure that sustainability issues, as well as consumer and societal issues, are taken into account in the TC 229 work. The mandate of the task group on consumer and societal issues (TGC&S) is e.g. to advise the TC on priorities for standards development and ensure that TC 299 has mechanisms for receiving input from relevant consumer and other societal communities. The TG will also identify relevant issues that the TC should address. Moreover, the TG intends, among other things, to explore:

- The need to improve upstream engagement and public participation in the decision making process on directions of technology development, gaining and maintaining public trust and confidence – and then validate that these tasks are within the TC scope.
- The need for a TC229 committee that examines the ethics of new work item proposals.
- How to highlight the existing transparency of standards development process.

Although the mandate and work plan of this TG seems to position the TC work in a context of societal governance, the experience of this author (being a participant in the TG since 2009) is that the work in the TG progresses so slowly that it raises the question of the real commitment of the TC to this kind of work.

The task group on sustainability (TGSus) uses the Brundtland Commission's definition of sustainability, stresses the ecological perspective (i.e. that it is not only impact on humans that is important), and include Life Cycle Assessment and Social Responsibility in their scope. They do not want to develop separate guidelines for sustainable nanotechnology, but want to supply input to TC229 about sustainability issues associated with the TC's work program and how these might best be resolved and to other technical committees so that nanotechnology is considered during the drafting of their standards. The group will also advise the TC229 on for instance how to include sustainability as a strategic priority for TC229 and how standards might contribute to the successful implementation of sustainability solutions. The TGSus commits to supporting the TC in making it a priority to involve developing nations in the standards process, and to assist the national bodies in involving stakeholders. The current discussion in the task group is, however, whether there is indeed a need for a separate task group on sustainability, or whether this is covered by TGS&C and WG 3 (health, environment and safety). In the latter case the task group on sustainability can be dissolved.

With regard to the two task groups, it seems that they are not well integrated with the technical work in the working groups. It seems the interaction takes place at the top level and the plenary, and not as engaging cross disciplinary and reflexive discussion in the working groups. All in all, there seems to be challenges concerning participation and broad peer review of the technical working groups, and the task groups on consumer & societal issues and sustainability are not well functioning. One may still claim that the ISO takes an important societal role in building a common framework for technology development and risk assessment. However, critics object that nanotechnology – including nanofood – is embedded in such vast scientific uncertainty that the establishment of standards in itself is premature. Moreover, they claim that this work will mislead the public, consumers and decision makers into believing that this is an area eligible to risk assessment when it in reality is too uncertain. To be fair, however, it should be said that most of the documents produced in TC 229 stress the scientific uncertainty and lack of data. In any case, disagreements abound, but what is certain is that the quality of the outcome of the work will depend upon the quality of the working processes within ISO.

Concluding remarks on the safe development of nanofoods

ISO's work on standardising terminology, measurement and characterisation, as well as its work on developing standards on health, environment and safety and material specifications, will most likely form an important basis for how risk assessment on nanofoods will be carried out in practice. It is therefore crucial for the safe development of nanofoods that researchers, NGOs and other interested parties engage in the standardising processes. The different member countries' standardisation bodies nominate experts and it is possible to become involved by contacting these. ISO is in principle an open organisation and it explicitly states its intention to ensure that stakeholders are involved (see for instance its strategic plan 'Standards for a Sustainable World').

All stakeholders must take responsibility for the development of safe and socially robust application of nanotechnology in food and agriculture. This includes governments, who should not only take a proactive regulatory stance on nanotechnology nationally, but should also fund a wider selection of expertise to participate in the ISO processes. ISO itself – and participating national standardisation bodies – are responsible for following up on its own policies by ensuring adequate stakeholder involvement and making sure the commitments to sustainability and consumer and societal issues are properly integrated in the TC work. When this succeeds ISO may play a proactive role in the development of socially robust and sustainable nano food and agriculture. Finally, there is a need to increase public awareness of both nanotechnology issues and the importance of ISO. As individual citizens and as a society we should be more attentive to the places where decisions are taken that may impact on both our own health, the health of our children and grand-children, and the environment. Only in this way can it be ensured that work in global institutions like ISO is indeed contributing to a socially robust and sustainable technology development that also benefits those that need it the most.

References

COMEST (2005). The Precautionary Principle. World Commission on the Ethics of Scientific Knowledge and Technology. http://unesdoc.unesco.org/images/0013/001395/139578e.pdf [Accessed Jan 2010].

ETC. Group. (2004). Down on the farm. The Impact of Nano-scale Technologies on Food and Agriculture. http://www.etc.group.org/upload/publication/80/02/etc._dotfarm2004.pdf [Accessed March 2010].

European Food Safety Authority (2009). Scientific Opinion. The Potential Risks Arising from Nanoscience and Nanotechnologies on Food and Feed Safety. The EFSA Journal 958, pp. 1-39. http://www.efsa.europa.eu/en/scdocs/doc/958.pdf [Accessed March 2010].

Ewald, F. (2000). 'The precautionary principle: between responsibility and policy'. Bulletin de l'Académie nationale de médicine, 2000;184(5):881-94.

Foss Hansen, S. (2009). Regulation and Risk Assessment of Nanomaterials. – Too little, too late? Ph.D thesis. Technical University of Denmark, Depart. of Env. Engineering.

Friends of the Earth Australia (2008). Discussion paper on nanotechnology standardisation and nomenclature issues. http://www.ecostandard.org/downloads_a/2008-10-06_foea_nanotechnology.pdf [Accessed March 2010].

ISO. (2004). ISO Strategic Plan 2005-2010. Standards for a Sustainable World. http://www.iso.org/iso/isostrategies_2004-en.pdf [Accessed March 2010].

Joseph, T. and Morrison, M. (2006). Nanotechnology in Agriculture and Food. A Nanoforum report. http://www.nanet.nu/upload/centre/nanet/links_ikoner/rapporter%20under%20links/landbrug%20og%20f%C3%B8devarer/nanotechnology_in_agriculture_and_food_nanoforum.pdf [Accessed March 2010].

Murphy, C.N. and Yates, J. (2009). The International Organization for Standardization (ISO). Routledge, London and New York.

Stirling, A., Renn, O. and Van Zwanenberg, P. (2006). 'A framework for the precautionary governance of food safety: integrating science and participation in the social appraisal of risk'. In Fisher, E., Jones, J and von Schomberg, R. (eds.) Implementing the Precautionary Principle. Edward Elgar Publishing Company.

Authors index

Abilleira, E. 129
Ahrens, W. 184
Albisu, M. 129
Algers, A. 447, 453
Almaas, V. 215, 269
Amores, G. 129
Anthony, R. 55
Armaza-Armaza, E.J. 513
Armaza-Galdos, J. 513
Arranz, J. 129, 138
Atienza-Macías, E. 501
Aursand, M. 239
Ausín, T. 202
Barrón, L.J.R. 129
Battaglini, L.M. 132, 144
Beekman, V. 91
Beltrán de Heredia, I. 129, 138
Benard, M. 287
Blázquez Ruiz, F.J. 189
Börnhorst, C. 184
Bovenkerk, B. 390
Bracke, M.B.M. 323, 360
Brochado Teixeira, A.C. 194
Bursztyn, M. 307
Cano, W. 141
Christiansen, S.B. 453
Clotet Ballús, R. 25
Colomer Xena, Y. 25
Copier, M. 323
Cornale, P. 132, 144
Coutellec, L. 412
Dagevos, H. 275
Damsgård, B. 116
De Bakker, E. 275
De Cock Buning, T. 287
De Henauw, S. 184
De Lattre-Gasquet, M. 307
De Lauwere, C. 91
De Miguel Beriain, I. 435, 483
De Renobales, M. 104, 129, 228, 464
De Tavernier, J. 251
Díaz-Méndez, C. 207
Dich, T. 453
Díez-Unquera, B. 138
Domínguez-Trejo, L. 141
Döring, M. 234
Doussan, I. 412
Driessen, C.P.G. 323
Duée, P.-H. 307
Dürnberger, C. 313
Emaldi Cirión, A. 384
Engelhard, M. 149
Escajedo, L. 104, 228, 464
Essien, E. 263
Formowitz, B. 313
Forsberg, E.M. 521
Fortina, R. 132
Franco, N.H. 456
Freire de Sá, M.F. 194
Frohlich, X. 221
Gamborg, C. 295, 372
Garcia-Etxebarria, K. 418
Gjerris, M 372
Glitz, F. 245
Gorrotxategi Azurmendi, M. 418, 467
Grafström, L. 184
Gremmen, B. 488
Grimm, H. 313, 429
Guiral, E.F. 165
Hallazi Méndez, L.A. 121
Hanlon, A.J. 453
Hansen, T. 453
Hardiman, J.W. 97
Haugen, A.S. 215
Heerwagen, L. 337
Henseleit, M. 257
Huber, R. 159
Ibañez, M.E. 198
Jacobs, J. 488
Jelenkovic, A. 198
Jensen, K.K. 171
Jokinen, P. 342
Jorqui Azofra, M. 378
Jugo, B.M. 418
Kadri, S. 116
Kaiser, M. 31, 215, 269, 447
Kallum, C. 447
Karlsson, F. 348
Korthals, M.J.J.A. 153
Kupsala, S. 342
Landera-Luri, M. 354
Landeweerd, L. 301
Larruskain, A. 418
Lasa López, A. 478
Lassen, J. 395, 494
Létourneau, L. 281

Libano Beristain, A. 65
Loor, H. 447, 453
Lord, F. 281
Lund, T.B. 395
Lussiana, C. 144
Magalhães-Sant'Ana, M. 441
Mandaluniz, N. 129, 138
Marild, S. 184
Martínez de Pancorbo, M. 464
Martinez, I. 239
Mauleón, J.R. 129
Mayor Zaragoza, F. 25
Mazón, L.I. 418
Meijboom, F.L.B. 87, 360, 390, 459
Mejdell, C.M. 116
Millar, K. 295, 441
Mimosi, A. 144
Molnar, D. 184
Morales-Garza, M.R. 141
Moreno Aznar, L.A. 184
Muñoz, E. 46
Myskja, B.K. 110
Myszczuk, A.P. 245
Nájera, A.I. 129
Nari Lloveras, M. 79
Neeteson-van Nieuwenhoven, A.-M. 97
Nicholls, S. 184
Nicolás, P. 176
Nunes, J.A. 456
Olsson, I.A.S. 159, 400, 406, 441, 456
Osseweijer, P. 301
Pérez Elortondo, F.J. 129
Pérez-Martín, R.I. 239
Pigeot, I. 184
Poveda, A. 198
Rebato, E. 198
Renna, M. 132, 144
Rocandio, A.M. 228
Röcklinsberg, H. 366
Romeo, S. 176
Romeo Casabona, C.M. 37
Ruiz de Gordoa, J.C. 129
Ruiz, R. 129, 138
Sandøe, P. 295, 395, 400, 406, 441, 453
Siani, A. 184
Sodano, V. 73
Sotelo, C.G. 239
Stafleu, F.R. 459
Stirn, S. 234
Ten Pierick, E. 91
Thompson, P.B. 329
Timmermann, C.A. 153
Tornaritis, M. 184
Tsioumanis, A. 472
Ugarte, E. 138
Uhl, A. 313
Urruela Mora, A. 65
Van Asseldonk, M. 91
Van den Belt, H. 153
Van der Staay, F.J. 360
Van der Veen, M. 301
Van der Weele, C. 507
Van Haperen, P.F. 488
Van Rij, H.M.J. 459
Varga, O.E. 400, 406
Veidebaum, T. 184
Vermersch, D. 307
Vinnari, M. 342
Virto, M. 129
Visak, T. 423
Wahlgreen, K. 447
Welin, S. 447
Whitelaw, C.B.A. 159
Wickins-Drazilova, D. 184
Williams, G. 184
Yamashita, M. 239
Yamashita, Y. 239
Zichy, M. 313

Keyword index

A
agricultural research 307
agrofuels 121
alliances 275
Alpine
- areas 144
- valley 132
alternative methods 378
animal 384, 513
- breeding 97
- cognition 323
- ethics 348, 390, 412, 441
- experimentation 366, 400, 418
- husbandry 159
- law 412
- model 400
- protection 384
- rights 435
- welfare 37, 87, 97, 263, 287, 348, 360, 378
anorexia 194
anthropocentrism 513
applied ethics 429
aquaculture 141, 239
aquaponic 141
argument marginal cases 435
attitude 269, 395
authenticity 239

B
balanced choices 97
behavioural change 91
benefit 400
biocentrism 483
biodiversity 132
bio-economic modelling 116
bioenergy 295
biofuels 121, 295, 301, 307
biology 372
biomedical research 418
bioscience degree programs 467
biotechnology 37, 46, 153, 159, 165, 488
boulimia 194
bounded rationality 31

C
capabilities approach 301, 323
capacity 194
carbon sources 149
case study 453
certification 215
chemicals 501
children 184
civil society 263
climate 79
- change adaptation 153
- ethics 55
code of conduct 215
competition policy 73
constitutional law 467
consumer 116, 245, 269
- information 228
- survey 257
- understandings 342
corporate social responsibility 263
corporation 73
criminal
- law 37
- liability 176

D
dairy products 228
debate 494
decision-making 447
- democratic 366
decision models 31
deep ecology 483
DeGrazia 423
deontology 275
developing countries 159
development ethics 153
diet 207
dignity 483
discourses 494
dual-purpose 144

E
ECNH 483
education 456
efficiency 171
e-learning 447
embryonic stem cells 37
empirical ethics 55
empowered consumers 221
energy from biomass 313

equality 189
ethical
- consumption 251, 337
- index 215
- matrix 215, 313
- responsibility 165
- standard 215
ethics 79, 281, 295, 307, 456
- committees 406
- of the culture of peace 25
- teaching 441
ethos 307
European
- law 37
- legislation 354
- patent office 37
European Union (EU) 337, 501
- legislation 65
- regulation of food 234
experiment 395

F
fair competition 354
fairness 171
fair trade 215
famine 202
farm animal 354
- welfare 342
farming 138
farrowing crate 429
fatty acids 144
fish 390
- welfare 116
food 73, 184, 269, 295, 513, 521
- citizenship 251
- classification systems 228
- consumption 251
- ethics 215, 251
- labelling 221, 337
- obesity policies 207
- safety 87, 207, 360, 501
- security 55, 65, 149
- sustainability 263
- system 79, 129
free
- movement of goods 354
- trade agreement 121
functional foods 360

G
gene technology 494
genetically modified foods 464
genetic association analysis 418
genomics 488
geographic origin 239
global
- food security 31
- warming 73
GMO 245, 412
- regulation 281
governance 269, 521
guideline 215

H
harm-benefit evaluation 406
health
- and nutrition claims 234
- mental 194
- policy 198, 207
healthier food product 129
hermeneutics 366
heuristics 31
human
- development 25
- dignity 189
- feeding 25
- health 501
- rights 189
hunger 79, 189
hydroponic 141

I
informational environment 221
informed consumers 337
intellectual property 121
interests, (un-)equal consideration 423
internet resources 441
interventions 184

L
landscape 87
legal
- guardianship 176
- regulation 513
legitimacies 281
lifestyle disease 395
livestock production 329

local
- breed 132, 138
- cows 144

M
marginal humans 435
market liberal paradigm 478
means and ends 472
medical hermeneutics 348
milk 144
minors 176
moral
- considerability 390
- debate 507
- standing of animals 342
- status 423, 435, 483
- vocabularies 301
multifunctional paradigm 478

N
nanotechnology 521
nanovaccines 110
neglect 176
nutrient profiling 228
nutrition 198

O
obesity 176, 184
open educational resource (OER) 447, 453
ordered pluralism 412
organic agriculture 104
Organismus syntheticus 149
outdoor feeding 129
overweight 176

P
pattern of inquiry 429
perfectionism 329
philosophy 372
pig
- housing enrichment 323
- husbandry 287
plant
- breeding 488
- sensation 483
- varieties 37
play design 323
policy 459
- making 281
politics of consumption 342
polyculture 141
post-humanist ethics 323
poverty 202
pragmatism 429
prevention of disease 234
price premium 257
primates 378
product quality 257
property rights 37
prospective benefit assessment 406
public
- debate 488
- engagement 55
- perception 46
- policies on drinking water 202
- policies on food 202
pure ethical motives 257

Q
quantitative data 395

R
Rawls 301
recovery program 132
reflective equilibrium 459
regulation 329
relevance 400
respect 360
responsibility 110, 171
retrospective benefit assessment 406
risk management 65
role play 453
Roma people 198
3R's principle 378, 456
rural traditions 132

S
salmon farming 110
scepticism 372
science journalism 165
science-society dialogue 287
scientific issues 464
seafood safety 239
second generation 295
SES 198
social
- controversy 46
- issues 464

– sustainability 129
Spain 384
specieism 435
stakeholders 215
standard setting 329
state-assisted paradigm 478
study of food and society 234
survey 269, 456
– questionnaire 184
sustainability 87, 91, 97, 138, 215, 275, 301, 313
sustainable
– agriculture 104
– breeding 287

T
teaching 453, 464
– law and ethics 447, 467
technology 507
the environment 501
theory of planned behaviour 91
tissue engineering 507, 513
training 459
transgenic 384
– organisms 121
– seeds 104
transition 275
transparency 110
trustworthiness 110

U
unnatural 372
unnecessary suffering 354
utilitarianism 275

V
validity 400
value 269
– frames 287
– permeation 472
– spheres 488
veterinary training 441
virtues 275
voluntary standards 521
vulnerability 245

W
web search 441
welfare 372
– actions 116
welfarism 329
Wemelsfelder 348

X
xenotransplantation 378

Y
yuck factor 507

www.ingramcontent.com/pod-product-compliance
Lightning Source LLC
Chambersburg PA
CBHW081123180926
51611CB00004B/41
* 9 7 8 9 0 8 6 8 6 1 5 4 5 *